Systems Science

Methodological Approaches

Advances in Systems Science and Engineering

Series Editor: Lida Xu

PUBLISHED

Systems Science: Methodological Approaches
by Yi Lin, Xiaojun Duan, Chengli Zhao, and Lida Xu
ISBN: 978-1-4398-9551-1

FORTHCOMING
Enterprise Integration and Information Architecture: A Systems Perspective on Industrial Information Integration
by Lida Xu
ISBN: 978-1-4398-5024-4

Systems Science

Methodological Approaches

Yi Lin
Xiaojun Duan
Chengli Zhao
Lida Xu

CRC Press
Taylor & Francis Group
Boca Raton London New York

CRC Press is an imprint of the
Taylor & Francis Group, an **informa** business

CRC Press
Taylor & Francis Group
6000 Broken Sound Parkway NW, Suite 300
Boca Raton, FL 33487-2742

Printed in the United States of America on acid-free paper
Version Date: 20121030

International Standard Book Number: 978-1-4398-9551-1 (Hardback)

Library of Congress Cataloging-in-Publication Data

Lin, Yi, 1959-
Systems science : methodological approaches / Yi Lin, Xiaojun Duan, Chengli Zhao, Li Da Xu.
pages cm. -- (Advances in systems science and engineering (asse))
Includes bibliographical references and index.
ISBN 978-1-4398-9551-1 (hardback)
1. System theory. I. Title.

Q295.L525 2013
003--dc23 2012030745

Visit the Taylor & Francis Web site at
http://www.taylorandfrancis.com

and the CRC Press Web site at
http://www.crcpress.com

Contents

Preface

It was in the year 1978 when Chi Xu published his head-turning reportage, entitled *The Goldbach Conjecture*, that the Chinese people with great national enthusiasm learned about Jingrun Chen, a mathematician, and his life-long attempt to prove the problem of 1 + 1 = 2, a shining star on the mathematical crown. However, as of the present day, the public is still not adequately acquainted with systems science and the fundamental idea behind 1 + 1 > 2.

Within the landscape of modern science, at the same time when disciplines are further and further refined and narrowed, interdisciplinary studies appear in abundance. As science further develops and human understanding of nature deepens, it is discovered that many systems interact nonlinearly with each other and do not satisfy the property of additivity. Their emergent irreversibility and sensitivity cannot be analyzed and understood by using the methodology of traditional reductionism. Facing this challenge, systems science appeared in response. The most fundamental characteristic of this science is the concept of "emergence": The whole that consists of a large number of individuals that interact with each other according to some elementary rules possesses some complicated properties. That is, the whole is greater than the sum of its parts (1 + 1 > 2). The basic tasks of systems science are the exploration of complexity and the discovery of elementary laws that govern complex systems of different kinds so that by making use of the principles of systems science, one can explain many complicated and numerous matters and events of the kaleidoscopic world and provide different control mechanisms.

Since 2004, the second author of this book has been teaching first-year systems science graduate students the course named Systems Science. Throughout these years of first-hand teaching, she has referenced the relevant books available in the marketplace and employed mainly the qualitative theory of differential equations to investigate the behaviors of static and dynamic systems. However, as she attempts to cover more materials, she realizes that because the amount of contents covered is huge, it is very easy for the instructor, including all the references used, to face the problem of mentioning a collection of unrelated pieces instead of an organic whole of knowledge. Feeding students with dead fish is surely not the same as showing the students how to fish. As teachers, we not only profess knowledge but should also focus more on motivating students to comprehend on their own, to learn how to learn, and to think how to think independently. When the first author of this book visited the National University of Defense Technology in 2009, they discussed how such a course as Systems Science could be taught cohesively as a whole. Considering the successes of the first author's monograph and textbook *General Systems Theory: A Mathematical Approach* (Kluwer Academic and Plenum Publishers, New York, 1999), they agreed that the rigor provided by set theory is the right framework for presenting systems science. As a result of the discussion, they decided to supplement the existing lecture notes, which were based on the qualitative theory of differential equations, from a new angle with the mathematical framework and foundation of set theory so that the basic concepts and characteristics of systems could be systematically reformulated and analyzed. By doing so, they expected to be able to present systems science as an organic whole of knowledge instead of a collection of unrelated materials.

Starting in 2009, the second author of this book introduced the thinking logic of set theory in her teaching by reformulating the basic concepts of systems in the new chosen language and by providing different versions of proofs for systems' characteristics. She encouraged students to describe and to comprehend systems and their properties in terms of set theory. This first-hand classroom teaching indicates that processes of resolving either real-life or real-life-like problems can lead students to deeper thinking and much closer touch with what they are learning. At the same time, when their interest in knowledge exploration is greatly stimulated, students immensely enjoyed their sense of achievement through their own discoveries. After many rounds of improvements, the lecture notes were reorganized into this manuscript.

Historically speaking, the approach we take in this book is very important. In particular, since the time when the concept of general systems was initially hinted at by Bertalanffy in the 1920s, the systems movement has experienced over 80 years of ups and downs. After the highs of the movement in the 1960s and 1970s, the heat wave started to cool gradually over the last 20 or 30 years. In the past few years, one has witnessed a fast disappearance of major systems science programs from the United States (in the rest of the world, the opposite has been true). By critically studying the recent and drastic cooling of the systems movement in the United States, one can see the following problems facing systems research:

1. With over 80 years of development, systems research has stayed mainly at the abstract level of philosophy without developing its own clearly visible and tangible methodology and consequent convincingly successful applications in the traditional disciplines. This end is extremely important because systems scientists are humans, too; they also need a means to make a living. Only when some of the unsolvable problems in the traditional science, together with the capability of solving new problems that arise along with the appearance of systems science, can be resolved using systems methods will systems research be embraced by the world of learning as a legitimate branch of knowledge.
2. Those specific theories, claimed to be parts of systems science by systems enthusiasts, can be and have been righteously seen as marbles developed within their individual conventional areas of knowledge without any need to mention systems science. This fact partially explains why programs in systems science from the United States have been disappearing in recent years.
3. Systems engineering considers various practical systems that do not really have much in common even at the level of abstract thinking. This end constitutes a real challenge to the systems movement. Specifically, when a systems project or idea needs public support, such as locating reputable reviewers for a research grant application, the organizer in general has a hard time locating a base of supporters other than a few possible personal contacts. Due to this reason, it generally takes a long time, if it ever happens, for a new rising star in systems research to be recognized. Without a steady supply of new blood, the systems research effort will surely stay on the sideline and secondary to the traditional sciences.

Comparing this state of systems science to that of, for instance, calculus, one can see the clear contrast. The former does not have a tightly developed system of theory, which newcomers can first feel excited about and consequently strongly identify themselves with, and scientific practitioners can simply follow procedures to produce their needed results, while the latter gives one the feeling of a holistic body of thoughts where each concept is developed on the previous ones in a well-accepted playground, the Cartesian coordinate system. Beyond this, calculus possesses a high level of theoretical beauty and contains a large reservoir of procedures scientific practitioners can follow to obtain their desired consequences. In other words, by working on further developing calculus and related theories or using these theories, thousands of people from around the world in the generations both before us and after us in the foreseeable future have made and will continue to make a satisfactory living.

It is on the basis of this understanding of the history of the systems movement that the yoyo model is introduced in this book to play several crucial roles. For more details about this model, please consult the work of Lin (2008).

1. It will be the intuition and playground for all systemic thinking in a similar fashion as that of Cartesian coordinate systems in modern science. To this end, of course, each of the characteristics of the model needs to be specified in each individual scenario of study.
2. It will provide a means and ground of logical thinking for one to see how to establish models in traditional fields, the first dimension of science, in an unconventional sense in order to resolve the problem in hand.

3. It will help to produce understanding of nature from an angle that is not achievable from traditional science alone. Compared to items 1 and 2, it is this that will provide the needed strength for systems science to survive the current slowdown of development in the United States and to truly become an established second dimension of science, as claimed by George Klir.

The final version of this book is jointly written by us on the basis of the lecture notes of the second and third authors. In particular, Dr. Duan's notes were embedded into Chapters 2–7 and 11, while lecturer Zhao's notes, in Chapters 8–11. Dr. Lin was responsible for the drafting of Chapter 1 and expanding and polishing in great detail each of the other chapters using the commonly accepted notations of set theory. At the final stage, Dr. Xu provided critiques on the overall quality and composition of the entire manuscript. As a reader of our book, if you have any comment or suggestion, please contact us by dropping a line or two to any of us at Jeffrey.forrest@sru.edu or Jeffrey.forrest@yahoo.com (for Lin), xj_duan@163.com (for Duan), chenglizhao@gmail.com (for Zhao), and lxu@odu.edu (for Xu).

Yi Lin (also known as Jeffrey Yi-Lin Forrest)
Xiaojun Duan
Chengli Zhao
Lida Xu

[illegible]

[illegible]

[illegible]

MATLAB® is a registered trademark of The MathWorks, Inc. For product information, please contact:

The MathWorks, Inc.
3 Apple Hill Drive
Natick, MA 01760-2098 USA
Tel: 508-647-7000
Fax: 508-647-7001
E-mail: info@mathworks.com
Web: www.mathworks.com

Acknowledgments

This book contains many research results previously published in various sources, and we are grateful to the copyright owners for permitting us to use the material. They include the International Association for Cybernetics (Namur, Belgium), Gordon and Breach Science Publishers (Yverdon, Switzerland and New York), Hemisphere (New York), International Federation for Systems Research (Vienna, Austria), International Institute for General Systems Studies, Inc. (Grove City, Pennsylvania), Kluwer Academic and Plenum Publishers (Dordrecht, Netherlands and New York), MCB University Press (Bradford, U.K.), Pergamon Journals, Ltd. (Oxford), Springer-Verlag (London), Taylor and Francis, Ltd. (London), World Scientific Press (Singapore and New Jersey), and Wroclaw Technical University Press (Wroclaw, Poland).

Yi Lin expresses his sincere appreciation to many individuals who have helped to shape his life, career, and profession. Because there are so many of these wonderful people from all over the world, he will mention just a few. Even though Dr. Ben Fitzpatrick, his PhD degree supervisor, has left this material world, he will forever live in Dr. Yi Lin's works. Dr. Fitzpatrick's teaching and academic influence will continue to guide Dr. Yi Lin for the rest of his professional life. Dr. Yi Lin's heartfelt thanks go to Shutang Wang, his MS degree supervisor. Because of him, Yi Lin always feels obligated to push himself further and work harder to climb high up the mountain of knowledge and to swim far into the ocean of learning. Yi Lin would also like to thank George Klir—from him, Yi Lin acquired his initial sense of academic inspiration and found the direction in his career. He would also like to thank Mihajlo D. Mesarovic and Yasuhiko Takaraha—from them, Yi Lin was affirmed his chosen endeavor in his academic career. Yi Lin would also like to send his sincere appreciation to Lotfi A. Zadeh—with personal encouragement and words of praise, Yi Lin was further inspired to achieve scholastically at a high level. His heartfelt thanks also go to Shoucheng OuYang and colleagues in their research group, named Blown-Up Studies, based on their joint works; Yong Wu and Yi Lin came up with the systemic yoyo model, which eventually led to the completion of the earlier book *Systemic Yoyos: Some Impacts of the Second Dimension* (published by Auerbach Publications, an imprint of Taylor & Francis in 2008). Yi Lin also thank Zhenqiu Ren—with him, Yi Lin established the law of conservation of informational infrastructure. Also, his heartfelt thanks go to Gary Becker, a Nobel laureate in economics—his rotten kid theorem has brought Yi Lin deeply into economics, finance, and corporate governance.

Xiaojun Duan would like to use this opportunity to send her sincere appreciation to those individuals who have been helpful in shaping her career and profession. In particular, her heartfelt thanks go to Dr. Zhengming Wang, her master's and PhD degree program supervisor, who has guided her throughout her professional life; to Dr. Dongyun Yi, with whom Xiaojun Duan discussed constructively how to organize the various course materials and related topics of systems science; and to Haiyin Zhou, for providing Xiaojun Duan with his materials of instruction for the systems science course. Dr. Jubo Zhu has specifically instructed her and pushed her to work harder in different professional aspects. Dr. Yi Wu has been so kind that he always spared moments of his busy professional life to consult with Duan with various thought-provoking insights. In addition, her thanks go to many of her students, including Lijun Peng, Xu Liu, Bin Ju, Shuxing Li, Kai Jiang, Haibo Liu, Xueyang Zhang, Junhong Liu, and others, who provided interesting materials, modified versions of which eventually helped to enrich the presentation of this book.

Acknowledgments

Authors

Yi Lin, also known as Jeffrey Yi-Lin Forrest, holds all his educational degrees (BS, MS, and PhD) in pure mathematics from Northwestern University (China) and Auburn University (U.S.A.) and had 1 year of postdoctoral experience in statistics at Carnegie Mellon University (U.S.A.). Currently, he is a guest or specially appointed professor in economics, finance, systems science, and mathematics at several major universities in China, including Huazhong University of Science and Technology, National University of Defense Technology, and Nanjing University of Aeronautics and Astronautics, and a tenured professor of mathematics at the Pennsylvania State System of Higher Education (Slippery Rock campus). Since 1993, he has served as the president of the International Institute for General Systems Studies, Inc. Along with various professional endeavors he organized, Dr. Lin has had the honor of mobilizing scholars from over 80 countries representing more than 50 different scientific disciplines.

Over the years, he has served on the editorial boards of 11 professional journals, including *Kybernetes: The International Journal of Systems, Cybernetics and Management Science, Journal of Systems Science and Complexity, International Journal of General Systems, and Advances in Systems Science and Applications*. Also, he is the editor of the book series entitled *Systems Evaluation, Prediction and Decision-Making*, published by Taylor & Francis since 2008.

Some of Dr. Lin's research was funded by the United Nations, the State of Pennsylvania, the National Science Foundation of China, and the German National Research Center for Information Architecture and Software Technology.

Professor Yi Lin's professional career started in 1984, when his first paper was published. His research interests are mainly in the areas of systems research and applications in a wide-ranging number of disciplines of the traditional sciences, such as mathematical modeling, foundations of mathematics, data analysis, theory and methods of predictions of disastrous natural events, economics and finance, management science, philosophy of science, etc. As of the end of 2011, he had published over 300 research papers and over 40 monographs and edited special-topic volumes by such prestigious publishers as Springer, Wiley, World Scientific, Kluwer Academic (now part of Springer), Academic Press (now part of Springer), and others. Throughout his career, Dr. Yi Lin's scientific achievements have been recognized by various professional organizations and academic publishers. In 2001, he was inducted into the honorary fellowship of the World Organization of Systems and Cybernetics.

Xiaojun Duan received her BS and MS degrees in applied mathematics and her PhD degree in systems engineering in 1997, 2000, and 2003, respectively, from the National University of Defense Technology, Changsha, China. She also had one year of visiting scholar experience in the School of Earth Sciences at The Ohio State University during 2007 and 2008. Currently, she is an associate professor at the Department of Mathematics and System Sciences, College of Science, National University of Defense Technology, Changsha, China. She has taught systems science, linear algebra, and probability and statistics courses for over 8 years. Additionally, she also teaches mathematical modeling and, as a faculty advisor, trains undergraduates to participate in the Mathematical Contest in Modeling, which is held by the Society for Industry and Applied Mathematics in the United States.

Some of Dr. Duan's research was funded by the National Science Foundation of China, the Program for New Century Excellent Talents in University, and The Project—sponsored by the Scientific Research Foundation for Returned Overseas Chinese Scholars, State Education Ministry, in China. She has also been a principal investigator of more than 10 research projects at the national, provincial, and ministerial levels. By the end of 2011, she had published one monograph and over 30 research papers in professional journals, such as *Kybernatics, Circuits, Systems, and Signal Processing, International Journal of Infrared and Millimeter Waves, Journal of Geodesy, Geophysics Journal International, Defence Science Journal, Science in China* (*Series E*), etc. Her research interests cover areas such as general systems theory and applications, data analysis, and complex system test and evaluation.

Chengli Zhao received his bachelor's degree in applied mathematics and his MS in systems analysis and integration from the National University of Defense Technology, Changsha, China, in 2002 and 2004, respectively. He has been a visiting scholar at the Department of Statistics of Oakland University, Michigan. At present, he is serving as a lecturer in the Department of Mathematics and Systems Science at the National University of Defense Technology.

Zhao's scientific interests and publications cover studies of complex systems, massive data computing, and web data mining. He has also directed and/or participated in more than 10 research projects at the national, provincial, and ministerial levels. Over the years, he has published over 10 research papers.

1 A Brief History of Systems Science

The basic concepts of modern systems science were first introduced by Bertalanffy (1901–1972). In the 1920s, Bertalanffy (1924) started to realize that there are similar properties existing in different disciplines and scientific areas, so he began to explore the possibility of describing systems of various objects by using a unified language, which are the thoughts of modern systems. In 1932, Bertalanffy (1934) developed the theory of open systems, constituting one pillar of the general systems theory. In 1937, he established the key concept of general systems (Bertalanffy 1937). By the end of 1940, the International Association of General Systems was established. Bertalanffy's (1968) *General System Theory: Foundations, Development, Applications*, published by George Braziller, has been treated as the first monograph of systems science. His theory of general systems has constituted the theoretical foundation of systems science and is a qualitative research of systemic thoughts.

1.1 WHAT IS SYSTEMS SCIENCE?

According to *Ci Hai* (*Ocean of Words*, if translated word for word from Chinese), published by Shanghai Press of Dictionaries in 1989 (p. 1291), systems science is the study of classes, characteristics, mechanisms, and laws of motion of systems. It appeared initially at the end of the 1940s and contains fives main areas of materials (see Figure 1.1):

1. The concepts of systems, that is, the general thoughts and theory of systems.
2. The theory of general systems, that is, the pure mathematical theory of systems structures and behaviors using the formal language of mathematics.
3. Theoretical analyses of systems. They are particular theories developed to investigate specific systems' structures and behaviors, such as graph theory, game theory, queuing theory, cybernetics, information theory, etc.
4. Systems methodology. It consists of methods and procedures developed to analyze, plan, and design specific systems by using the theory and technology of systems. It mainly means systems analysis and systems engineering.
5. Applications of systems methods, that is, to apply the thinking logic and methods of systems science to various scientific disciplines.

Since the 1960s, the appearance and development of the theory of nonequilibrium self-organizing systems have helped in enriching the contents of general systems theory. The epistemological significance of systems science is that it has promoted and accelerated the development tendency of modern science from a spectrum of relatively isolated disciplines into an organic whole.

Systems science, according to Qian Xuesen (2007), stands for such an area of science and technology that investigates systems and various applications. Like natural science, social science, mathematical science, etc., it is a new branch of the system of modern science and technology.

Qian Xuesen, a well-known Chinese scientist, has played an important role in the initiation and development of systems science. He explored the general laws of evolution of various matters and social events by using systems logic of thinking and systems methods. On the basis of the existing achievements of systems research, he proposed, at the end of 1970s, the leveled structures of

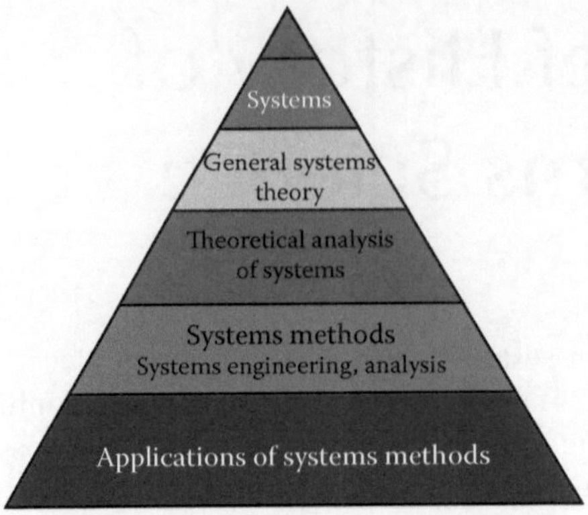

FIGURE 1.1 Triangle of systems science.

systems science. He recognized that systems science is composed of three layers, and multiple branches and different technologies. The following is paraphrased from Automatic Control and Systems Engineering, *Encyclopedia of China* (Press of Encyclopedia of China, Beijing, 1991).

1. The layer of engineering technology: Systems engineering, techniques of automation, and technology of communication are the knowledge that directly help to reshape the objective world, where systems engineering contains techniques useful for organizing and managing systems. Corresponding to different types of systems, there are different systems engineering, such as agricultural systems engineering, economic systems engineering, industrial systems engineering, social systems engineering, etc.
2. The layer of technological science: Operations research, information theory, control theory and cybernetics, etc., are theories that help guide engineering techniques.
3. The layer of basic science: Systems research stands for such a science that investigates the fundamental attributes and general laws of systems. It is the theoretical basis of all systems works. Systems science is still in the process of being established. The bridge connecting systems science and philosophy is the so-called systems theory or known as systems points of view. It belongs to the category of philosophy. The establishment and development of systems science will definitely widen the coverage of human knowledge, strengthen the human capability to fight against nature for its very own survival, and enhance the development of science, technology, and economies.

1.2 DEVELOPMENT HISTORY OF SYSTEMS SCIENCE

As an area of knowledge, systems science has experienced its very own process of appearance, growth, and formation. And because it is a cross-disciplinary endeavor, it has touched upon a wide range of traditional disciplines, such as mathematics, physics, chemistry, and other branches of natural science, many areas of engineering technology, and quite a few disciplines of social science. Hence, the history of systems science is closely related to the entire development history of the human race. Its ideas can be traced back to the primitive societies of antiquity. The ancient people started their knowing of nature from its wholeness. In other words, the thinking logic of systems science is the very first theory that has guided the man in his exploration of nature. As for the establishment of systems science, one has to mention the modern achievement of science and technology. Almost every theory of modern science has been absorbed by systems science and modified to become a part of systems theory. When looking at the history comprehensively, the development of systems science can be roughly divided into three stages: appearance and formation of systems

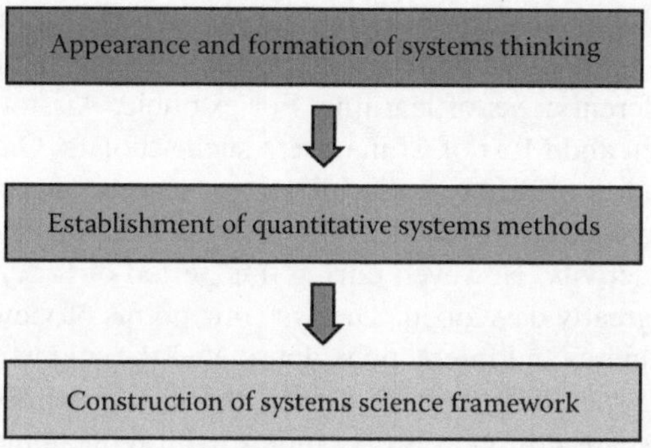

FIGURE 1.2 Development stages of systems science.

thinking, establishment of quantitative systems scientific methods, and construction of the comprehensive framework of systems science (see Figure 1.2).

1.2.1 Formation of Systemic Thoughts—The Elementary Stage

In ancient times, the level of productivity was low. Man had great difficulties fighting against nature. So, human understanding of nature was staggering at the level of "systems thinking." People intuitively employed the concept of wholeness to look at the physical world. They related the phenomena of birth, age, sickness, and death with other relevant natural occurrences, forming the world view of unity of man and heaven. Within such a world view, one can find the logic thinking of systems. Philosophies of Lao Tzu and Zhuang Tzu have reflected such a way of reasoning. For instance, both the *Book of Changes* (Wilhalm and Baynes 1967) and *Tao De Ching* (English and Feng 1972; Lao Tzu, unknown) have described the integration and transformation of things and events by pointing out that "all things under heaven are born out of 'have', and 'have' is born out of 'don't have'," "having no name is the start of heaven and earth, and having name is the mother of all things," and "Tao breeds one, one breeds two, two breeds three, and three begets all things of the world." Later, Wang Anshi (1021–1086) studied the evolutionary order of the world as "tian nian sheng shui," which in Chinese is

天年生水

and pointed outin his work "Wu Xing (Five Elements): Heaven Names Everything" that the heaven first gives birth to water, the earth then gives birth to fire, the heaven afterward gives birth to wood, the earth follows with its birth to gold, and then the heaven produces earth. For more about Wang Anshi, please see the work of Mote (1999). Wang believed that the heaven and the earth first give birth to the five elements: water, fire, wood, gold, and earth, which then led to the formation of everything else of the world. So, ancient Chinese employed such concepts as yin and yang, five elements, and eight trigrams to unify the explanation of various natural phenomena and to form the unity of man and nature. That we can see as realizations of the perspective of wholeness, a viewpoint of movement and change, holistic point of view of the world, and other concepts of systems thinking. Ancient Greek philosopher Democritus (Guthrie 1979) saw the universe as a unified whole and studied it holistically. At the same time, he considered that the world was composed of individual atoms and that it was the movements and interactions of these atoms that constituted the motion and change of the entire universe. He published a volume on the large system of the universe, which can be seen as the earliest work that used the word "system."

No matter whether it was ancient Chinese thinkers or Western philosophers, they addressed worldly problems from the angle of wholeness. These scholars achieved high in several disciplines and were specialists in different areas of learning. For example, Aristotle and Archimedes of the ancient Greece, and Lao Tzu and Mozi of China, were such scholars. One characteristic of the science of that time is that studies of different disciplines are closely intertwined together. Studies of science and philosophy are written together. The holistic development of science is a product of the then low levels of productivity. However, during this period of time, in terms of science itself, the systemic thinking was greatly developed. The systemic points of view of wholeness, movement and change, interconnectedness, and interactions dominated the mainstream of the then scientific research. What needs to be specifically pointed out is that the systemic thinking of the time was forced upon the scholars, because the causes of complex natural phenomena could not be fathomed so that they had no choice but to look at the phenomena from the angle of wholeness, macrocosm, and dialectical thinking. The appearance and development of these early systemic thoughts are closely related to the low levels of productivity and the primitive stages of scientific knowledge and technology.

Although early man was forced to understand nature by using systemic thoughts and reasoning, these thoughts and logics of reasoning have indeed helped mankind to achieve magnificent successes in science, technology, and commercial goods production. For instance, in the area of civil engineering, Li Bing and his son constructed during 256–251 BC the well-known Dujiangyan Water Project (Zhang and Hu 2006), which is an irrigation system located in Dujiangyan City, Sichuan. It has not only been a magnificent achievement in the world history of water conservancy construction but also a successful application of systemic thoughts. The entire irrigation system consists of three main constructions:

1. *The fish mouse levee* that divides the water of Min River into inner and outer streams, where the inner stream carries approximately 40%, rising to 60% during flood, of the river's flow into the irrigation system while that outer stream drains away the rest, flushing out much of the silt and sediment.
2. *The flying sand weir* that has a 200-m-wide opening that connects the inner and outer streams. This opening ensures against flooding by allowing the natural swirling flow of the water to drain out excess water from the inner to the outer stream.
3. *The bottle-neck channel*, which Li Bing and his son gouged through the mountain, is the final part of the system. This channel distributes the water to the farmlands to the west, while the narrow entrance works as a check gate, creating the whirlpool flow that carries away the excess water over the flying sand weir, to ensure against flooding.

That is, Dujiangyan is a large engineering construction that has the capabilities of flood control, sand drainage, water diversion, and others. Along with over 120 ancillary constructions, the system has also played the role of irrigation. It can be recognized that without the fish mouse levee, a large amount of sand and gravels would not be drained into the outer stream; without the bottle-neck channel, no swirling circular flows would be formed so that the sediment could not be drained by being rushed over the flying sand weir. And without the flying sand weir to take away the sediment, the bottle-neck channel would be filled with the sediment so that it could not play the role of irrigation and the water could not be transported into the Chengdu plain. Because this project was designed and constructed holistically, it has played the expected multiple roles. It is also because of its holistic design that, since its initial construction of more than 2000 years ago, Dujiangyan has been in effective use in the agricultural production and producing economic benefits for Sichuan plain.

In the area of medicine, the theory of Chinese medicine has sufficiently embodied the thinking logic of systems science. Yellow Emperor's Inner Canon of the ancient Chinese medical theory (2696–2598 BCE) emphasizes on the connection of various human organs, relation between psychological and physiological symptoms, and interaction between human bodily states and the natural environment (Liu 1994). Each human body structure is seen as a component of nature with all the organs working together as an organic whole. The concepts of yin and yang and the theory of the five elements are employed to illustrate the interactions and mutual constraints of the five internal organs. The cause of a disease is jointly analyzed from bodily symptoms, physiological phenomena, and spiritual states of the patient. When treating a disease, the therapeutic principle of correspondence between man and universe is developed by jointly considering the laws of health and nutrition and those of nature. This end in fact represents an emphasis on the interactive effects of subsystems within the human body system and of the system and its environment.

Feeling the pulse, that is, treating the human body as a whole, is one of the methods used in Chinese medicine when a patient is diagnosed. Based on the fact that when the body is ill, its blood circulation is somehow affected, the doctor employs the speed, strength, and other characteristics of the wrist arterial pulse to determine where the pathological change is and its severity. The method of acupuncture is developed on the basis that all organs work together closely as a whole; and problems with different organs can be treated through needling some of the acupuncture points on the ears. That is, no matter whether it is diagnosing or treating a sickness, Chinese medicine sees the human body as a whole, recognizing that there are close connections between different body parts. What is interesting is that the material supports of these connections include not only the nerves, blood vessels, and various joints between organs, as observed in human anatomy, but also the so-called meridian channels. According to Chinese medicine, the meridian system connects all parts of the human body into a unified whole and constitutes the important channels of communication between the parts; only when the body is metabolically alive and forms a whole the meridian system appears and functions. Otherwise, the meridian system would disappear. To this end, it is a known fact that as long as a body is opened for the purpose of observing the organization of different parts, the meridian system becomes invisible. The traditional Chinese medicine, which is developed on the viewpoint of wholeness of systems science, has been widely used in China and forms an important part of the world medicine.

Along with the development of productivity, human understanding of nature has been deepened further and further. However, at the same time, there has been a tendency of ignoring the research of the holistic characteristics of systems. For instance, let us look at the situation of the human body again. Through anatomy, not only have various body parts been well recognized, along with a deep understanding of their individual structures, but also blood circulations, nerve systems, and lymphoid tissues that connect the parts are found. While our comprehension of each body part is deepened and each specific connection of the parts is discussed, the world of learning seems to have forgotten the need to analyze the whole body and ignored the roles of various parts in the operation of the whole system. Along with the constantly refining knowledge of anatomy, human understanding of muscles has reached a delicate level. At the same time when various physical and chemical experiments on muscles are designed and conducted and when various characteristics of muscle cells have been noted, some special attributes of the muscles that are shown along with the existence of the whole body are rarely considered. The situation has been so extreme that some extraordinary capabilities of the whole body, which cannot be observed in laboratory experiments of local muscle studies, are considered not trustworthy. When some professional researchers saw the performance of a qigong master lying down his naked upper body on a bed of standing nails with extra weights placed on top of his body, they could not believe the fact that the master did not get hurt from the sharp nails. Because no matter whether it is in their theoretical computations or laboratory experiments, they could not make a small piece of muscle stand extra pressure without being pierced through; they deny what they saw in one way or another. However, when this situation is seen from the angle of systems science, it is not difficult to see that experiments on regional muscles cannot reflect the characteristics of the whole system; it is quite normal to observe different attributes of the whole body from those of a piece of local muscle. Ignoring essential differences between the whole and parts has been a negative effect of deepened and refined scientific researches on the development of systems science.

With the deepening understanding of nature, scientific divisions have been finer and finer over time. Scholars in one narrow area no longer know much or anything from another refined area. They can only be known as specialists in their small field. Leonardo da Vinci, an Italian scientist of the middle ages, was a physicist, architect, and an expert in many other scientific areas. After that historical period of time, it became impossible for anybody to be proficient, like da Vinci, in so many different scientific fields as astronomy, physics, and others. Even within the field of physics, modern experts of electromagnetics no longer know much about atomic physics, and theoretical researchers in general are not good at experiments, while laboratory technicians and experts do not have a strong background in theoretical analysis. Speaking macroscopically, the reason for the occurrence of this situation is that the more the in-depth knowledge is established, the more the contents there are. So, even if a person does nothing in life but learns different areas of knowledge, it is still impossible for him to become proficient in many different disciplines. At the same time, because scientific disciplines are so finely defined, it is no longer necessary for any scientific worker to acquire knowledge beyond what he or she is doing to make a living within his or her narrow field of specialized knowledge. The abundant existence of narrowly trained scholars is one of the reasons why the development of systems science has been slow.

When seeing from the angle of economic development, no matter whether it is a long-lasting slavery system, a feudal society, or the beginning of capitalism, the scale of production is generally small, and the equipment of manufacturing is simple. So, it is relatively easy to establish harmonic development in economic productions. As long as the intensity of labor is increased, the level of production can be consequently raised. It seems that natural resources are plentiful and could last forever. Any environmental pollution caused by manufacturing commercial goods can be corrected through the effects of nature. So, during this period of time, people focus more on improving individual machines, while ignoring the overall benefits, and on increasing the levels of production, while not considering integrated utilization and resource allocation. That explains why no objective demand exists for the research of systems science that investigates wholes of systems. As a matter of fact, after the magnificent development during the initial period of time, other than some isolated progresses in some regional, industrial, and particular aspects, the overall research of systems science has been in a state of stagnation. As a form of thinking and reasoning, it has been gradually replaced by metaphysics. The situation is like what Friedrich Engels pointed out (1878) that to the Greeks, it is exactly because they were not advanced enough to dissect nature and analyze the resultant parts that nature was still treated as a whole and observed as a whole; holistic connections of natural phenomena were still not confirmed with comprehensive details so that these connections to the Greeks were only consequences of intuition; that is where there were deficiencies in Greek philosophy, and because of these deficiencies, it later had to submit to another point of view.

1.2.2 Quantitative Systems Science Developed for Applications

Just like any scientific theory, its appearance and development of systems science are also originated in practical applications and motivated by calls of solving real-life problems. During the time periods both before and after World War II (WWII), systems science developed quickly at the height of applications. In particular, cybernetics, operations research, and information theory, branches of systems science at the height of applications, gradually appeared after WWII. Before the war, the research of possible applications of systems science had started with some major progress. However, these works were isolated, scattered, and regional. Two particular cases stood out the most. One is the telephone traffic model (Erlang 1909) proposed by Agner Krarup Erlang (1878–1929). In the twentieth century, the enterprise of telephones developed and grew quickly. When telephone lines were installed, one needed to consider the service efficiency of telephones. If there were too few lines, many telephones would have to share one line, creating jams of conversation; if there were too many lines installed, many of the lines would be idling so that the limited resources would be wasted. By using the method of comparison, Erlang resolved this problem by establishing a clever model. He compared a telephone communication system with a vapor–liquid equilibrium system of water by identifying each telephone that is being picked up to a water molecule that changes from a vapor state to a liquid state. When the telephone is returned to its idling state, it is compared to the molecule that returns back to the vapor state. Assume that the probability for a telephone conversation to finish within a unit time interval is λ. That is, the length of conversation is $1/\lambda$. And further assume that the probability for a telephone to be used within the unit time interval is μ. Then, according to the fact that at the vapor–liquid equilibrium of water, the number of molecules that move from the liquid state to the vapor state is the same as that of molecules that migrate from the vapor state to the liquid state within a unit time interval, Erlang obtained that

$$P_{i-1}\,\mu = p_i i\lambda$$

where p_i stands for the probability that the system contains i molecules in the liquid state. By applying this general equilibrium relation repeatedly, one obtains

$$P_n = \frac{\mu}{\lambda n} P_{n-1}$$

$$= \left(\frac{\mu}{\lambda n}\right)\frac{\mu}{\lambda(n-1)} P_{n-2} = \ldots = \left(\frac{\mu}{\lambda}\right)^n \frac{1}{n!} P_0.$$

Because probability distributions satisfy the condition of summing up to 1, that is, $\sum_{i=0}^{n} P_i = 1$, one has

$$P_0\left(1+\frac{1}{1}\left(\frac{\mu}{\lambda}\right)+\frac{1}{2!}\left(\frac{\mu}{\lambda}\right)^2+\ldots+\frac{1}{n!}\left(\frac{\mu}{\lambda}\right)^n\right)=1$$

and

$$P_0 = 1\left|\left(1+\frac{1}{1}\left(\frac{\mu}{\lambda}\right)+\frac{1}{2!}\left(\frac{\mu}{\lambda}\right)^2+\ldots+\frac{1}{n!}\left(\frac{\mu}{\lambda}\right)^n\right)\right..$$

Letting $\rho = \mu/\lambda$ leads to the Erlang formula:

$$P_i = \left(\frac{\rho^i}{i^i}\right)\Big/\left(1+\rho+\frac{\rho^2}{2!}+\ldots+\frac{\rho^n}{n!}\right).$$

This Erlang formula, which is derived by using statistical computations, indicates that if the probability for a telephone to be used is μ and the average length of telephone conversation is $1/\lambda$, then when the probability for a line is open when a telephone is picked up is given, such as $\rho = 0.9$, i telephones can be served by n telephone lines. Generally, $\mu > \lambda$ holds true. If the total number of telephones is fixed, increasing the number n of telephone lines in the system will make the probability of telephone connections greater; if the number of telephones is increased while all other parameters stay invariant, the probability of telephone connection will drop.

This example not only serves as a representative case of applications of systems science but also possesses practical significance. The previous Erlang formula is still employed today as a basic reference on which telecommunication networks are designed. The later studies of such problems become the main content of queuing theory. In terms of the methodology of resolving practical problems, the process of model building also provides a useful reference for establishing systems models when systems science is applied in real-life situations.

Another example is the input–output model (Raa 2005) proposed by Wassily Leontief (1905–1999). The relationship between various departments of a national economy is very complicated. Their individual products serve mutually as raw materials, energies, etc. The available amount of one product more or less affects the production of many other products; at the same time, it is constrained by the availability of many other products. However, for most products, the amount of inputs corresponding to the unit production output, that is, the amount of consumption of the various supplies for the production of one unit of products, is relatively stable within a short period of time. That is because the consumption of various raw materials, supplements, energies, etc., is determined by the level of production and technology, quality of management, relevant natural conditions, and other factors. Within a short period of time, these factors cannot be substantially changed. Even if the consumption coefficients of some products change relatively significantly, the changes tend to follow a pattern that can be determined quite easily.

Leontief's input–output model is established by using mathematics and employs computers to investigate the quantitative relationship between the input and output of various economic activities. In particular, this model plays an important role in studying and analyzing the quantitative dependency relationship between the products and consumptions of various departments (or various products) of a national economy. It has greatly helped to scientifically arrange, predict, and analyze economic activities. This method was initially published in Leontief's paper of 1936, entitled "Quantitative Input and Output Relations in the Economic System of the United States," in *Review of Economics and Statistics* (vol. 18, pp. 105–125). Later in 1941, he published *The Structure of American Economy 1919–1929* (Oxford University Press, New York), and in 1953, he published *Studies in the Structure of the American Economy* (Oxford University Press, New York). In these and relevant papers and books, Leontief not only developed the input–output method but also constructed the input–output tables for American economy for the years of 1919, 1929, and 1939 based on the published economic statistics. Considering the important roles the input–output method played in economic analysis, Leontief was awarded the 1973 Nobel Prize of economics. And his method has also been widely applied by various nations.

During WWII, various nations demanded the resolution of how to bring about the optimal overall effects with reasonable local results. Studies of these relevant problems greatly enhanced the development of operations research, various control methods, and game theory. Due to the needs of the war, various real-life problems were posed, such as: How can a defense system be arranged in order to better deter and destroy the enemy's airstrikes? How can targets be searched so that submarines can be found? How much advanced time is needed to fire artilleries in order to deal with high-speed flying objects, such as planes? These and related problems not only represent extreme practicality but also possess high levels of theoretical values. To meet the needs of the war, a large number of scientists shifted their focus of work to serve the national defense and military operations. Because of the surge in scientific manpower, some of the problems that were related to military operations were resolved quickly along with many new scientific concepts introduced.

When the war ended, these scholars continued their works at the theoretical heights. They established various applied branches of systems science, such as operations research, management science, cybernetics, information theory, etc. In particular, Norbert Wiener, who during the war investigated radar and air defense fire control systems, established cybernetics by publishing *Cybernetics: Or Control and Communication in the Animal and the Machine* (Hermann & Cie, Paris; MIT Press, Cambridge, MA) in 1948. Claude Elwood Shannon laid down the foundation for information theory by publishing the paper "A Mathematical Theory of Communication" (*Bell System Technical Journal*, vol. 27, pp. 379–423, 623–656, 1948). In the 1950s, various papers and monographs on the methods of operations research appeared, among which Harry H. Goode and Robert Machol (1957) published the first monograph in systems engineering, *System Engineering: An Introduction to the Design of Large-Scale Systems* (McGraw-Hill, New York).

With efforts that followed the initial publications, these and related works have grown into technical disciplines with their respective mathematical foundations. And the establishment of these disciplines made systems science evolve at the height of epistemology into such a quantitative field of knowledge that is constructed on the basis of mathematical science.

The development of a science must find its motivation in economies and technology. After World War I, the economic and scientific developments once again placed systems science in the front of the world of learning from two different angles. One is the expansion in the scale of production, which could no longer be improved by merely increasing the intensity of labor. Due to the increasing ratio of automated production lines and more diversified manpower involved in various projects, organizational modernization was needed. In particular, the completion of a specific production task might need the collaboration of many different departments. American Apollo space program involved over 1 million people of different specialties and more than 200 universities. It involved over 100 commercial firms and cost over US$24 billion. In this huge project, the number of personal achievements accomplished by the creative works of individual scientists was small. What accounted the most were the comprehensive goal-oriented researches of large groups of specialists from many disciplines. For the success of these researches, the key was the organization and coordination of people and machines, the smooth connection of various production procedures, and how to optimize the whole. That is, the key was about how to employ the thinking logic and methods of systems science to coordinate research and productions.

On the other hand, due to the limitations of natural resources and working spaces, any improvement of production corresponding to the economic development is not merely for the purpose of economies. Problems related to the production also need to consider the optimization of the whole, allocation of resources, sustainable development of the society, etc. These problems need to be addressed from multiple angles, such as production and consumption, capital and labor, and resource utilization and environmental pollution, by using combined approaches of natural and social sciences. In terms of the economic development itself, the free competitions of early capitalism have gradually evolved into strengthened macroscopic national controls. Redistributions of production wealth, especially the international trades, have arranged the economic developments of different nations into an organic whole.

All these scenarios indicate that the relationships between the whole and parts, global optimizations, and other related concepts, as advocated by systems science, have become the significant theoretical support for the development of scientific production. During this period of time, the development of systems science can be characterized as follows:

1. Large amounts of mathematical tools are employed so that systems science has been changed from an epistemological discipline into a rigorous science that is practically applicable.
2. The fundamental and applied researches of technology have developed quickly. Aiming at resolving various kinds of particular problems, different methods are established. Although these methods are purposefully designed to resolve practical problems with their individual advantages and constitute a general methodology, no generally applicable theory is formulated and no theoretical system is constructed.
3. In terms of the guiding principle, scholars have either consciously or unconsciously done their thinking and reasoning with the whole in mind. They start with an initial understanding of development of things and events and follow by applying effective controls of the system until reaching the eventual optimization of the whole. The process can be summarized as from understanding to application, from that the whole is greater than the sum of parts to that imperfect parts can be arranged to form an optimal whole.

1.2.3 Construction of Comprehensive Systems Science

Dissipative structures and synergetics of physics offer a theory and a method to investigate natural complex systems and a possible way to unify natural and social systems. They provide materials for systems science. Physics is used to discuss reversible, degenerative systems. Both Newton's second law and the second law of thermodynamics describe the evolutionary direction and characteristics of systems. However, the evolutionary directions of these natural systems contradict the phenomena of developments and evolutions widely existing in the biological world and studied in social science. That is, no unified methodology existed for scholars to investigate natural and social systems jointly. To this end, Eugene Paul Wigner (1902–1995), a physicist and mathematician, a cowinner of 1963 Nobel Prize in physics, expressed his opinion (Wigner 1969):

> What is the most important gap in modern science? It is obviously the separation between physical science and human science. As a matter of fact, there is nothing in common between physicists and psychologists. Perhaps an exception is some of the tools physicists provided for the superficial psychological attentions, while psychologists warn the physicist about how not to get his or her thinking and discovery affected by his or her hidden desires.

If one places such a comment in the perspective of the scientific history, he or she could see that there used to be a barrier that separated the heavenly bodies from earth. Then with the coming of modern science, the celestial bodies and earth are now seen together in a unified whole, known as the universe, without any separation in between. However, the mainstream literature indicates that modern science is largely a science of quantities; it describes a wonderful geometric world in which no man lives. For more details, please consult with Lin and OuYang (2010). That is, there is a barrier between the natural scientific world and the physical world with people that is studied in social science.

IIya Prigogine proposed the theory of dissipative structures. When an open system is far away from an equilibrium state, due to exchanges of matter, energy, information, etc., with the environment, it can possess some kind of order structure. In naturally existing physical, chemical systems, phenomena of evolution, similar to those seen in biological systems, can also be observed. These phenomena of evolution can be investigated using the unified language of dissipative structures.

Hermann Haken proposed the theory of synergetics, believing that phase transitions of complex systems are originated in the interactions and coordinated effects of subsystems. The concept of order parameters and the enslaving principle of synergetics are proved to be an effective method to address evolutionary directions of systems.

When discussing the problem of why natural systems evolve in the direction of strengthened orderliness, catastrophe theory, stability theory of differential equations, hypercycle theory, etc., are employed. When investigating the evolution of nonlinear systems, such concepts as chaos, bifurcation, fractal, etc., are introduced with relevant theories developed. One characteristic common to all these new concepts, new methods, and new theories of natural science is the new phenomena of evolution emerging out of the nonlinear interactions between various subsystems of complex natural systems. Relevant studies show that these new concepts, methods, and theories are equally applicable to the investigation of natural phenomena, from which of course they are initially proposed, and some social events. So to a degree, they can serve as a unified theory for the study of nature and society, two totally different objective phenomena. Although these contents are not themselves parts of systems science, they can be seen as some of the main materials to be included in the foundations of systems science.

As a matter of fact, the development of systems science over the past century has indicated that only when the study of natural science has deepened to a certain level, and only after some of the theories as mentioned above are successfully established, a corresponding theory of systems science can be developed. At the same time, when such new theories are proposed, the world of learning also likes to enrich them so that they can become systems theories that are able to deal with the widely existing general complex systems. In the rest of this book, we attempt to enrich and improve these new materials of natural science in order to make them parts of systems theory. Of course, our works along this line represent only the start that needs to be followed by many colleagues in the years to come.

Additionally, along with the appearance of complicated production tasks and the new problems arising from the studies of large systems, giant systems, and distribution parameter systems, etc., the development of systems science at the level of application has been accelerated. The classical control theory, which focuses mainly on single machine automation by using the method of transfer functions, has been evolved into such a modern theory of control, which consists of the concept of state spaces, dynamic programming, Kalman filter, maximum principle, etc., useful for addressing the problem of automation of multiple machines and of the whole factory.

Nonlinear programming, integer programming, game theory, stochastic processes, and other great many new methods of nonlinear operations research, along with modern information theory and technology, make the branches of systems scienceat the level of techniques, such as control theory, operations research, and information theory, more mature. That of course helps provide a large amount of materials for the foundation of systems science. In the area of applications, major attention has been given to the study of social systems, economic systems, and other similar systems that involve people. For such systems, theoretical analysis has been combined with computer works so that the method of metasynthesis can be developed by joining both quantitative and qualitative methodologies. By combining men and machines through the use of scientific discussion halls, the problems of evolution of giant social systems can be possibly resolved both quantitatively and qualitatively.

From the discussions above, it can be seen that in the three levels—fundamental theory, technical basis, and application technology—systems science has grown tremendously in recent years. And in the 1980s, Qian Xuesen (1981) put forward a framework for the system of systems science and analyzed the contents of the science at different levels and their interactions. This work has surely brought systems science into a new era of integrated development. The theoretical system of systems science is currently growing in a rapid growth spurt. The characteristics of this period of development are as follows:

1. Various theories that are closely related to systems science are developing tremendously in their respective fields. Such nonlinear studies as chaos, bifurcation, fractals, etc., have become part of the scientific frontier. Stochastic control and hierarchical control of complex systems of multiple levels and multiple targets have also become hot research topics in the area of automation.
2. The work of constructing the theoretical framework for systems science by synthesizing advanced theories of different disciplines has been started. Differences between these theories are analyzed, and commonalities are sought. On the basis of the thinking logic of evolution of complex systems, theories and methods from different fields are summarized into a unified general theory, leading to the creation of new concepts and new methods.
3. Due to the need of production development, the research of various real-life systems has been deepened under the guidance of the thinking logic of systems science. Although such systems as artificial intelligence, economic operation, human brain, etc., have their respective characteristics, they can surely be analyzed using new logic of reasoning. When one addresses problems from the angle of systems science, he or she can stand high and go deeper under the surface. When studying complex systems, other than practically applying systems theories, all the new methods developed and new conclusions obtained in the study can also help to enrich the systems theories themselves, making systems science truly able to address the evolution of complex systems.

1.3 SYSTEM OF MODERN SCIENCE AND TECHNOLOGY

According to Qian Xuesen (1979, 1983a), the system of modern science and technology can be depicted as in Table 1.1, where the basic reason and criteria of classification are the following.

Since the start of the twentieth century, the rapid development and the tendency of refiner division and synthesis of modern science and technology have brought forward many new problems and new scenarios, creating difficulties and complexity for any attempt to classify the spectrum of science. When seen from the angle of the objects investigated in various traditional disciplines, physics has currently been studying the structures and attributes of large molecules and low-dimensional molecules, which were once considered within the realm of chemistry. It has also been studying the structure of the human brain and laws of economic development. The research of psychology has also expanded its scope from attributes of the human body and mind to various computational schemes of computer science, the psychological characteristics of intelligent machines with interacting humans, etc. In terms of the methods of research employed in various disciplines, almost every new method established in one discipline is quickly applied to other disciplines so that there is no longer any obvious difference between the methods used in any particular discipline. Inconsistencies existing in the traditional division of disciplines have become evident and have caused some inconveniences for the further development of science. So, the problem of classifying the spectrum of science needs to be analyzed from a new angle in order to satisfy the new need. That is, the world of learning needs to establish a new hierarchical structure for the system of science and technology in order to adapt to the current trend of development. The particular criteria of establishing the new system of scientific disciplines are outlined as follows.

TABLE 1.1
Classification of Modern Science and Technology

Bridge	Branch	Foundation	Technical Basis	Applied Technology
Mathematical philosophy	Mathematics	Geometry, algebra, mathematical analysis	Computational and applied mathematics	Coordinated planning speed computing
Natural dialectics	Natural science	Physics, mechanics, biology, chemistry	Principles of chemical and electrical engineering, principles of mechanics	Sulfuric acid production process, gear technology
Historical materialism	Social science	Economics, sociology, ethology	Capital economics, socialist theory	Business management, social works
Systems research	Systems science	Systemology	Cybernetics, operations research	Systems engineering
Views on heaven and man	Human body science	Physiology, psychology, neurology	Pathology, pharmacology, immunology	Counseling techniques, internal medicine
Epistemology	Cognitive science	Studies of thoughts, informatics	Intelligence studies pattern recognition	Cryptography, artificial intelligence
Social theory	Behavioral science	Ethics		
Study of behavior	Moral theory socialism	Public relations, human relations		
Aesthetic standards	Literature/arts	Aesthetics	Theories of music, art	Literary techniques, methods of drawing
Military philosophy	Military science	Study of strategies	Art of commanding	Tactical training, military engineering

1. The research objects of each discipline are parts of the objective reality. The difference between disciplines is not about how varied their research objects are. Instead, it is the different angles and emphases from which the investigations are carried out. The classification of disciplines according to the differences in the objects studied indicates an early stage of scientific exploration of the whys regarding various mysteries of nature. No matter what is considered, what is faced directly in the inquiry is almost always the object of interest. Only after a period of time of thinking and theoretical abstraction, will it be discovered how this specific object is related to other objects, matter, and/or events, and how methodological clues are obtained so that the particular method can now be applied to the research of other objects. As soon as connections between different things are established and as soon as a similarity between the respectively developed methods is materialized, it is necessary to analyze and reconsider the classification of disciplines. It is because in today's science, the disciplines of which have been well intertwined and synthesized, it is no longer appropriate to distinguish disciplines according to the difference of the objects they study. That is, by using the difference in angles and emphases instead of the difference in objects of research to classify the spectrum of science, it will help one to understand the separation of disciplines.
2. There are longitudinal differences between different pieces of knowledge within one discipline. Generally, each discipline needs to be classified into three layers such as fundamental theory, technical basis, and practical applications. By fundamental theory, it stands for the most basic knowledge of the discipline and reflects the essential contents of the objective world. For example, the basic knowledge in physics that reflects the essence of electromagnetic phenomena is electrodynamics. It describes the attributes, characteristics, and laws of electromagnetic fields. The concept of fields represents the most essential aspect of electromagnetic phenomena. The often-talked-about electric circuits in fact are only an approximate theory of the knowledge related to fields. By technical basis, it means the analysis and discussion of particular attributes and characteristics that appear when the fundamental theory is specified to a set of given conditions. For instance, if the previous example is continued, then around the parts with conducting wires connected, an electromagnetic field exists mainly in the neighborhood of the wires; electrons move no longer within the entire field but within the wires. As far as the properties of this electromagnetic field are concerned, the discussion will take place within electrics or electrical engineering. Electrics is not only a specified and approximate electromagnetic theory under a particular set of conditions but also the theoretical foundation for the large amount of phenomena of electricity-carrying wires.

   By a practical applied field, it most likely means a particular technique or technology that can be applied in real life. For example, the principle of electric engines, transformers, etc., focuses more on addressing technical problems on the basis of some basic technical discipline. In terms of the teaching of disciplinary knowledge, universities of science mainly profess the knowledge at the level of fundamental theories, while engineering schools focus more on the courses of technical bases. Some of the knowledge on the level of practical applications is taught as major courses in engineering schools, while some as courses in vocational schools.

   Classifying scientific knowledge into three layers longitudinally can help one understand the development of disciplines. Generally speaking, the knowledge on the level of practical applications grows the fastest, contains the most detailed contents, and is closely related to manufacturing productions. And it is often the case that the formation of a discipline initially starts with the development on the level of practical applications. The knowledge at the level of fundamental theory evolves the slowest. And as soon as the theory at this level is established, the discipline is considered to have a theoretical framework and is mature. On the other hand, the establishment of the fundamental theory generally motivates more mature development of the discipline at the levels of technical basis and practical applications.

3. The classification system of disciplines should not be fixed forever. Along with the development of science, human knowledge is constantly evolving and gradually forming various structures. So, any classification of disciplines should not be a prefixed framework into which all known knowledge is shelved. All the disciplines in the spectrum of science should be constantly synthesized, enriched with the development of the society. Disciplines that have relatively short history or are still immature should be more finely classified, while those relatively more mature disciplines should be more synthetic, and cover wider areas. Based on what is currently available, it can be seen that researches on man himself represent an ill-defined problem, although it has been a hot problem. Therefore, in terms of the classification of knowledge, the classified disciplines regarding researches on man himself would bear the following characteristics:
   a. Relatively fine divisions. There are many areas of study that focus on man himself. This fact reflects how hot such research topics have been.
   b. Major differences exist from one classification to another. Different organizations and different experts have proposed their respectively different classifications. This fact implies that further and in-depth considerations are needed.
   c. Each classification should be based on the objects—man—of study. It is because such investigations are still in their primitive stage. As the research deepens, the classification of the related works will change. Some will dissolve into multiple pieces, while others will combine, and new disciplines might even appear. In short, the classification of disciplines is also a process that evolves constantly and forever, just like any discipline itself.

Based on the previous criteria, Qian Xuesen classified modern science and technology into nine major departments of disciplines. He placed the Marxist philosophy at the most abstract, most fundamental level, which is connected to the nine departments through the so-called "bridges." Each department is stratified into three layers: fundamental theories, technical bases, and applied technologies. Historically, the formulation of these nine departments of disciplines went through a process of constant improvement. In 1986, Qian Xuesen suggested six major departments, includingmathematics, natural science, social science, systems science, human body science, and cognitive science. With further enrichment and perfection, by 1990, he supplemented the original list with three additional departments: behavioral science, literature/art, and military science, along with the specific contents in each of the nine departments.

According to the criteria of the classification, as listed earlier, these nine major departments, respectively, reflect the researches of the objective world from different angles. Mathematics investigates the objective world from the angle of quantitative relationships of numbers and shapes without considering the qualitative difference of the objects. So, it can be applied to study either the lifeless world or the lively world. It can be said that mathematics appears everywhere. Natural science looks at the objective world from the angles of movements of things and of energy transfers and changes. Current science indicates that the interactions between objective matters can only be one of the following four kinds: gravitational, electromagnetic, strong, and weak. They cause various kinds of movements in matters. And natural science studies the characteristics and attributes of these interactions.

Social science considers the objective reality from the angle of human societal development and movement by studying the social behaviors of people. Although the development of human society can change unexpectedly in millions of different ways like the unpredictable winds and clouds in the sky, there must be some inherent laws underneath all the changes. Economic foundations determine the infrastructure of the entire society; productive forces promote and confine reforms in the relations of production. These and other basic laws determine the tendency and speed of societal development. Systems science investigates the objective reality by using the connection between wholes and parts. It addresses the overall optimization of systems, systems' structures, the relationships between systems' functionalities, stabilities of systems, etc., without paying much attention to such problems as how energies are transferred or conserved.

Human body science looks at the realistic world from the angle of people. It discusses the characteristics, attributes, behavioral patterns of man himself, as well as the impact of the objective world on human bodies and their mutual influences. Cognitive science researches the objective reality from the angle of how man understands the world. Cognition represents particularly a human activity; it stands for the flash contents of the human brain. Although cognitive science is closely related to the study of human brain, it does not consider the material processes that underlie human brain activities. Instead of looking at the phenomena of brain energy transfers, it aims at uncovering the characteristics of thoughts and the laws of thinking. Currently, there are also a good number of cognitive science research problems in the area of computer science.

Behavioral science studies societies and cultures from the angle of human sociality by addressing group behaviors and relationships between individual behaviors and group performance. In the biological world, many different animals live in groups, so there are inevitably various contacts and communications between the individuals. That is, to different extents, group behaviors must emerge in the populations of these animals. Because of this understanding, behavioral science considers not only humans but also the organizational structures of insects, the life patterns of mammals, and the laws underneath other phenomena of life. Literature and art study the objective world from the angle of beauty and ugliness. On the surface, both beauty and ugliness seem to be only human sensations. However, the reality shows that it cannot be any further from the truth. Many expressions in animal courtship contain aesthetic phenomena. In particular, harmonic beautiful music can facilitate cows and sheep to produce additional amounts of milk. Speaking aesthetically, it can be concluded that if the objective reality is looked at from the angle of beauty and ugliness, many phenomena of life will be touched upon.

Military science addresses the objective reality from the angle of confrontations between organized masses. The phenomena of conflicts, invasions, and wars do not just appear within human societies. It seems to be more prevalent in the animal world. Furthermore, even the struggle for survival among plants can be part of the research of military science.

The previous departmental classification of disciplines has covered mostly all known knowledge. However, what we need to point out is that due to the complexity of science and technology, it is possible that a certain discipline cannot be easily placed in any defined class. Such a discipline generally represents a cross-disciplinary field. For example, for architecture, its contents on building structures, materials, etc., belong to the realm of mechanics of the natural science and can be considered as an important direction of application of static mechanics. The building designs and floor plans seem more likely to belong to the department of literature and art, where aesthetics represents the main concern of architecture. At the same time, the sizes and arrangements of buildings and arrangements of rooms are also related to human body science and behavioral science. Therefore, architecture stands for a comprehensive field that connects with many sciences. At the same time, it should also be noted that according to the new method of classification, the traditional disciplines, which are defined by using the objects of concern, need to be transformed so that they can reflect the characteristics of their respective research angles and methods.

1.4 DISCIPLINARY CHARACTERISTICS OF SYSTEMS SCIENCE

Systems science not only deals with the fields of science and technology but also touches on such problems as sociality, globality, and even universality. Included in the main problems of concern are

1. Social, geographical, and ecological problems that involve interdependences
2. Social problems with labyrinths of relationships
3. Various particular systems, such as aerospace systems, human body systems, human brain systems, etc.

1.4.1 Schools of Systemic Thought

The contents of systems science cover a very wide range of topics, including human behavioral systems, social systems, mechanical systems, computer and network systems, intelligent systems, simulation systems, biological systems, aerospace systems, the earth system, commercial systems, administrative systems, etc. Currently, the research of complexity is one of the foci of systems science.

As for the schools of thought in systems science, they can be roughly grouped into three clusters:

1. The American school. It consists of Santa Fe Institute, Center for Complex Systems and-Brain Sciences, Center for Complex Systems Research (University of Illinois at Urbana-Champaign), and International Institute for General Systems Studies (IIGSS), Pennsylvania. The first two centers emphasize on the complexity of particular natural systems, while the IIGSS establishes the general theory of multirelational systems and applications in the evolution of systems.
2. The European school. It consists of Brussels' school and Haken's school. The former was headed by Ilya Prigogine until his death. It pushes for complexity and dissipative structures and advocates to give up the belief of a simple world. The latter promotes synergetics and maintains that the complexity of the objective world is derived from gradual evolution out of the initial simplicity through self-organizations.
3. Other schools of thought include, but are not limited to, (i) ARC Centre for Complex Systems, Australia; (ii) Complex Systems Research Centre, Sydney, Australia; (iii) Centre for Computational Intelligence, Singapore; and (iv) Institute for Systems Science (ISS), Singapore.

For a quick grasp of the landscape of systems research, one is advised to visit the Website of the International Federation for Systems Research and that of the International Society for the Systems Sciences.

As a matter of fact, the main stream research institutions of systems science consist of nonprofit international organizations, social groups, and universities. Their common goal is to conduct research and to provide educational programs at the same time. By excavating commonalities out of different disciplines, these institutions promote the cross-fertilization of various scientific fields and coordinate the activities of scholars who otherwise would be scattered in their respective disciplines by providing them with a basic framework.

Due to the characteristics of systems science that cuts across traditional scientific fields, university research centers and known research institutions have very different starting points and directions, covering a wide spectrum of topics. Even so, their works can be grouped into two main classes: one is on the foundations, including such topics as set-theoretic systems theory, systems dynamics, multiscale analysis, systems modeling, feedbacks, and controls; the other belongs to applied research, such as systems theory applied in finance, military, commerce, disease control, pharmacology, biological genes, sociology, industrial decision making, etc.

In short, the characteristics of systems science require scholars of different backgrounds to talk and to conduct research together so that they can potentially discover implicit connections underlying the artificially separated disciplines. Through integrations of multiple fields, commonalities of systems can be discovered so that practical problems can be resolved.

1.4.2 Suggestions from World of Learning

In terms of philosophy and applications, establishing systems science meets the historical need of dialectical unity of modern science. Developing systems science is one of the consequences of altering the traditional logic of thinking and reasoning. Systems science represents an intelligent tool for developing an information-based society. So, speaking of science, the importance of developing systems science is clear without expressing it in words.

Scholars in systems science have defined its goals and methods from different directions and in different fields. The most important include the following [this subsection is based on Umbach (time unknown)]:

1. Employ the concepts of systems and models within a discipline and between disciplines.
2. Look for cross-disciplinary structural similarities and integrated synthetic concepts, such as those of open systems, hierarchies of systems, homeostasis, controls, feedback, equilibrium, self-organization, emergence, networks, etc.
3. Construct the general language of science that can go across disciplines.
4. Emphasize on the thought of wholeness.
5. Apply quantitative tools and mathematical models.
6. Try to resolve the problems the world is most interested in.
7. Supplement systems science with practically functional concepts from catastrophe and chaos theories.
8. Synthesize the scientific views of one world and apply them in as many areas as possible, such as in biological spheres, evolution theory, human social systems, state of human survivals, etc.

As of this writing, the development of systems science has greatly contributed to the world of learning and brought forward enormous economic benefits. Even so, as any human endeavor, scholars still have criticisms toward, and constructive suggestions for, systems science, including the following:

1. The concepts of systems and models are too general, mathematical, and formal. Different definitions of systems might even be inconsistent.
2. It is difficult to sufficiently express the concepts of structural similarity and metasynthesis based on empirical studies.
3. It is impossible to construct any general commonly usable language of science based on systems science.
4. The innovativeness of wholeness thinking cannot be shown by using empirical philosophy.
5. Not all systems can be treated quantitatively.
6. Expecting to resolve the problems the world is most interested in is not realistic because the collected data might not be sufficient, the methods employed might not be adequate, mathematical models and computers are given too much credit, economic principles and laws are not sufficiently considered, the damages of disasters are exaggerated, etc.
7. The general adaptability of chaos theory is questionable.
8. The integrated scientific world outlook is not clear enough to be meaningful.

No matter whether or not these criticisms have any ground, they still indicate that in all areas of fundamental research, methodology, and applications, systems science needs to go further with visible and convincing results.

1.5 SOME PHILOSOPHICAL THOUGHTS OF SYSTEMS SCIENCE

Systems science studies general systems as its objects for the purpose of uncovering isomorphic structures of systems. That is, it attempts to unearth the unified laws that govern the forms of existence and forms of motion of systems of various characteristics, and the laws all systems comply with. According to the difference between thinghood and systemhood, systems science focuses on the properties of systemhood of systems instead of properties of thinghood.

Systems science, especially complex systems science, understands the world using its unique logic of thinking, view of nature, view of time and space, and meaning of life. It explores the territories that traditional science has never touched before. It attempts to reveal a much greater and much more truthful world than the one that traditional science has shown.

The essence of science to a great extent is to reveal the uncertainty of the world. To this end, the cognitive approach of systems science provides a brand new way to comprehend the world.

1.6 PHILOSOPHICAL SIGNIFICANCE OF SYSTEMIC EPISTEMOLOGY

Other than the motivation provided by the societal development and background preparation of prior scientific advances, the establishment and development of systems science also need to have philosophical supports in order to break out the constraints of the traditional ways of thinking and accepted beliefs and to formulate a new scientific norm and methodology. There have been various kinds of philosophical thoughts covering almost all mainstreams of philosophy that systems experts have been making use of. The dialectical principles of universal connections and movement and development are the premises of systems science that are not clearly stated.

1.6.1 Ludwig von Bertalanffy

For systems thinking to become a historical must and a trend of thoughts, Bertalanffy (1968) has made great contributions. However, as a matter of fact, this trend of thoughts has been explored and developed either before or after Bertalanffy independently by various scholars in many different fields or research topics. Some of these works were even more specific, more systematic, and more in-depth than what Bertalanffy proposed. This trend of thoughts appeared widely in linguistics, psychology, mathematics, control theory, philosophy, sociology, etc. The only thing missing was to use a set of unified language of systems to describe all the relevant concepts and results. For instance, dialectics has long ago emphasized on the viewpoints of connection, wholeness, and development. If the forthcoming systems theory only restated these dialectical views of point using different terms, then it would not represent any intrinsic progress. And if systems theory simply reuses the high-level popularized works of mathematics or operations research, then it would not count as a true development, although promoting and making a scientific work known is important in terms of social advancement. The opening up of a new research field is always related to the establishment of new concepts, new visions of some accustomed senses, discovery of previously unknown structures, and new abstraction. In some aspects of opinions, conceptualization, principles, methods, schemes, theorems, techniques, etc., the new field has to have some substantial amount of innovations instead of entirely terminology switches. For meaningful applications to appear, each scientific field and technology has its conditions and ranges, within which the knowledge can be utilized freely, and beyond which the validity of the knowledge becomes questionable. For example, a theory that is developed on a certain level for some particular physical or biological systems is generalized into different levels and different fields; special attention has to be given to the basic requirements, which generally go way beyond comparisons and analogies, under which the generalized theory would hold true. Comparisons and analogies can only be used as a motivation for the generalization to take place instead of the judgment and verification of scientificality. This warning of course is a conversational rut; however, it should be well referenced to when the achievements of systems science are evaluated, analyzed, and popularized across disciplines.

Bertalanffy's general system theory attempts to investigate the universal aspects, uniformities, and isomorphic structures of various systems for the purpose of illustrating and deriving the models, principles, and laws that are applicable to general systems and their subsystems. This goal is great and represents the purpose of many other scientific works. However, the key is to investigate systems innovatively while taking care of these three purposes. As limited by historical conditions, Bertalanffy only looked at organic systems, open systems, and dynamic systems. In his theory of organic systems, he criticized the mechanistic and vitalistic points of view, believing that each biological being is a stable open system with its wholeness, dynamic structure, originality, and organizational hierarchy. Although such excellent views on biological science can be seen as a declaration and introduction of a series of new explorations, the truly systematic works were still waiting to be carried out. Looking back, what has been achieved in biology in the past decades is still not quite the same general systems theory as what Bertalanffy imagined, although the future and potential importance are very optimistic. Bertalanffy's theory of open systems considered systems with inputs, outputs, and states; it illustrated systems' stability, increasing orderliness, etc. To this end, the classical and modern theories of control, automata, mathematics of automatic control, etc., have done much better, and they are no longer based on natural or seminatural language comparisons and inductions. Instead they have developed many practically applicable operating systems. Bertalanffy's theory of dynamic systems attempts to sentimentally illustrate some of the typical properties, such as the wholeness, additivity, competitiveness, mechanization, centralization, etc., of systems through using a few simple ordinary differential equations. What is considered are still some particular systems. Although they can be treated as typical in order to provide clues for future research, in terms of the general systems as what Bertalanffy imagined, they are still quite a distance away. One advantage of Bertalanffy's work over what the general mathematician would do is that he paid great amount of attention to the backgrounds to make sure what he did could potentially be generalized to general systems and to new levels of abstraction of concepts instead of being trapped in engraving computational details.

One of the main theoretical contributions of Bertalanffy is the introduction of emergence into systems science. He borrowed Aristotle's proposition that the whole is greater than the sum of parts to describe this concept. This concept has created an overall and long-lasting influence to the development of systems science. Its philosophical foundation is dialectics. Reductionism and holism, and analytical thinking and synthetic thinking, have always been pairs of opposites in the development of science. In the past several hundred years, science has advocated reductionism and denounced holism and emphasized on analytic thinking and despised synthetic thinking. To establish systems science and to open up the research of complexity, there is a need to revisit these two pairs of opposites and to go beyond reductionism and analytic thinking in order to reestablish the dominance of holism and synthetic thinking on the basis of modern science.

1.6.2 Ilya Prigogine

Prigogine's (1961) theory of dissipative structures, studies a kind of nonequilibrium thermodynamics by employing the local equilibrium assumption, which was earlier used by continuum mechanics and equilibrium statistical mechanics, the description of continuum mechanics, the theory of Lyapunov's stability, the theory of bifurcation, and the fluctuation theory. Its focus is the investigation of the characteristics and conditions under which the so-called dissipative structures appear. In particular, through analyzing the diffusion processes of mechanics, physics, and chemistry and chemical reaction processes, he discovered the typical, mostly physical connections between structures, functionalities, fluctuations, systems' openness, being far away from equilibrium, etc. He then applied such connections to biological, social, economic, and other nonphysical systems.

Prigogine did not use new advances in physics to deny dialectical materialism. Instead, he used these new developments to help the dialectical materialism's critics of mechanism to reach a deeper level. The main philosophical pairs of opposites that helped guide the establishment of dissipative structures include existence and evolution, order and disorder, equilibrium and nonequilibrium, reversible and irreversible, determinism and indeterminism, etc. Having full grasp of these opposites of course requires him to comply with the thinking logic of dialectics. From his works, it can be seen that existence and evolution are the most central opposites. Studying closed systems belongs to the science of existence, so only researching open systems can possibly be part of the science of evolution; studying equilibrium states is the science of existence, so only researching nonequilibrium states can possibly be seen as the science of evolution; studying linear systems is the science of existence, so only researching nonlinear systems can possibly be part of the science of evolution; studying reversible processes belongs to the science of existence, so only researching irreversible processes can possibly be in the science of evolution; studying deterministic systems is the science of existence, so only researching nondeterministic systems can possibly belong to the science of evolution; and so on. For the transformation of physics into a science of evolution, Prigogine's work has made the determining stride. His philosophical basis is exactly the dialectical materialistic thought of historical development of nature and Whitehead's process philosophy.

1.6.3 Hermann Haken

The original purpose of Haken's (1983) synergetics is to study the synergetic processes of the subsystems of a general system. However, what is actually addressed stands for the system's synergetic process or self-organization process that can be expressed by using equation(s) satisfied by the time-dependent probability distribution of the system. Through using mathematical modeling, synergetics derives a principle of dominance, which in terms of general systems is simply a principle of general symmetry, where slow-change variables dominate the motion of fast-change variable. The argument of this conclusion is based on a particular way to simplify the system of equations. Just as what Haken (1983) himself said, "In physics or chemistry, the determination of the order parameters and the determination of the dominated variables that decays fast are relatively simply. However, when we deal with a complex system, such as the production process of a factory, or the dynamics of a mass, or an economic system, one of the important tasks, maybe it will be a future task, is how to distinguish slow and fast variables."

Different from Prigogine, in Haken's works, he did not clearly evaluate dialectical materialism. However, he firmly believes in scientific materialism. As a German scholar, he is deeply influenced by the classical philosophies of Germany. That makes his descriptions of systems' evolution contain rich content of dialectical thinking and extremely inspiring. Haken emphasizes that synergetics covers different layers and aspects. In the area of philosophy, synergetics relates to how to explain and understand nature. He admits that synergetics is based on two major philosophical bases: one is the law of unity of opposites, and the other is the law that quantitative changes lead to qualitative changes.

Haken has been interested in many categories of unity of opposites: the relationship between parts and the whole, whether to use the analytic or the synthetic method to deal with complex systems, the relationship and transformation between quantities and qualities, the relationship between organization or controlled self-organization and self-adjustment, how order occurs out of disorder, whether the process of order development is determined by unidirectional causation or by circular causations, etc.

Among all the pairs of opposites, what is most noticeable and unique is Hakan's analysis of self-organization and hetero-organization. His thought contributions can be summarized into four points:

1. In systems research, Haken is the first person who unearthed the opposites of self-organization and hetero-organization.
2. On the one hand, he emphasizes on the opposition between self-organization and hetero-organization; on the other hand, he also emphasizes on the collaboration of the two.
3. He points out that the separation of self-organization and hetero-organizationis relative without any absolute boundary.
4. He also points out that self-organization and other organization can be transferred into each other with mathematical models to describe such transfers.

1.6.4 Manfred Eigen

Eigen's (1971) hypercycle theory studies the mechanism of the self-organization of biological large molecules. Its emphasis is on the exploration of the role hypercycles play in the evolution from nonlife molecules to an entity of life, where a hypercycle stands for a new level of organization in which self-replicative units are connected in a cyclic, autocatalytic manner. The hypercycle model is developed on a solid mathematical–physics foundation. It could be seen as an evolution model in the areas of molecular biology and molecular biophysics and a theory of particular systems.

1.6.5 James Grier Miller

Miller's (1978) general living systems theory investigates the differences and commonalities of all particular living systems. From small to large, he defines eight levels: cells, organs, organisms, groups, organizations, communities, societies, and supernational systems. Although such classification of systems of different scales is not new, what is most important is that he reveals some of the commonalities of the particular systems of these levels. These systems have to deal with the input, circulation, and output of materials, energy, and information. Even though this realization is still not novel, Miller's contribution is that he specifically lists 20 functional subsystems for these known commonalities. Some of these subsystems possess the capabilities of observing and controlling materials, energies, and/or information, while others play the roles or serve the roles of collecting or spreading out the materials, energies, and/or information in order to actualize specified functionalities of observation and control.

1.6.6 Qian Xuesen

Qian's (1979, 1981, 1983a, 1983b, 2007) contribution in systems science has been multifaceted. First of all, he traced the history of systems thinking and concepts on the basis of Marxism philosophy by summarizing the experience of developing systems science, and how systems thinking evolves from empirical grounds to philosophical heights, from sentimental thinking to qualitative reasoning and then to quantitative analysis. Second, he applied Marxism philosophy to guide his research of systems science. The main idea is that when dealing with open complex giant systems, one should never forget about dialectical materialism, while he or she should be warned against mechanical materialism and idealism. In all his works related to systems, he has clearly shown the underlying philosophical thoughts. Third, he maintained that systems scientific dialectics should be a part of systems theory, qualitativeness and quantitativeness should stand for a dialectical process, and reductionism and holism should be united dialectically.

1.6.7 Sunny Y. Auyang

Auyang (1999) proposed the method of synthetic microanalysis and clearly distinguished isolationism, microreductionism, and synthetic microanalysis. When speaking ontologically, each composite system is made up of components; the components can be comprehended through the concepts and theories that represent them.

1.6.8 Others

Along with the systems movement that started over a half-century ago, there have appeared a good number of well-established scholars of systems science, such as Mario Bunge, George Klir, Mihajlo D. Mesarovic, Yasuhiko Takahara, A. Wayne Wymore, etc., to name just a few. In order not to make this chapter too long, all the relevant details are omitted here. For those readers who are interested, please consult with Klir (1985) and the references therein.

1.7 SOME UNSETTLED QUESTIONS

1.7.1 Are Systems Thoughts Commonly Usable Scientific Views?

Is it possible to construct a language of science on the basis of systems science? Many scholars think that it is impossible. Because the concepts of systems and models in systems science are very general, mathematical, and formal, the meaning of systems becomes abstract so that it will be difficult to produce multiplicity and particularity. Because of this reason, the practicality of systems science will have to be discounted. On the other hand, the view of a synthetically integrated world is not well defined either.

1.7.2 Can Systems Research Help with the Understanding of Complex Systems?

Because many systems cannot be quantitatively dealt with, it is not realistic to expect the research of systems science to be able to directly address the problems of complex systems the world is interested in. The available data might not be enough; each of the available methods could be limited; too much depending on mathematical models and computers could be misleading; relevant laws of sociology, patterns of economic phenomena, and others might have not been considered. There are just too many areas where things could go wrong.

1.8 ORGANIZATION OF THIS BOOK

This book contains a total of 11 chapters. In particular, Chapter 2 presents the basic concepts, characteristics, properties, and classifications of general systems. Chapter 3 is devoted to the study of nonlinear systems dynamics and the theory of catastrophe. Both dissipative structures and synergetics are introduced in Chapter 4. Chapter 5 focuses on the studies of chaos, including such details as logistic mapping, phase space reconstruction, Lyapunov exponents, and chaos of general single relation systems. In Chapter 6, different aspects and concepts of fractals are developed, including a presentation of L systems analysis and design. Complex systems and complexity are the topics addressed in Chapter 7. Other than introducing the basic concepts, the reader will also learn how the phenomena of "three" and complexity are related, and how various cellular automata can be constructed to generate useful simulations and figurative patterns. Chapter 8 deals with complex adaptive systems. Chapter 9 presents topics of open complex giant systems. As a proposed method for dealing with open complex giant systems, the yoyo model is introduced, followed by some practical applications. Complex networks and related concepts and methods are introduced in Chapter 10. In Chapter 11, several case studies are presented to show the reader how various concepts and the logic of thinking of systems can be practically applied to resolve real-life problems.

1.9 OPEN-ENDED PROBLEMS

1. In your own language, state the relationship between systems science and other scientific disciplines.
2. What is your enlightenment gained from learning the development of systems scientific methodology?
3. What do you feel about the future development direction of systems science?
4. What do you expect to gain in terms of your way of thinking from studying systems science?
5. What is the motivation for systems science to appear?
6. Try to locate such a phenomenon that appears in physical or chemical systems, biological systems, and social systems. What are the difference and connection when this phenomenon appears in different systems?
7. A person is often blinded by details so that he/she could no longer see the overall situation. Please apply a personal experience to analyze this phenomenon by using the viewpoint of systems.
8. Find out how systems and their concepts were stated in your national history and then analyze the background for such studies to ever appear.
9. What is the difference between systems science and philosophy, and between systems science and systems thinking?

2 Concepts, Characteristics, and Classifications of Systems

2.1 FUNDAMENTAL CONCEPTS

2.1.1 Systems and Nonsystems

The English word "system" is originally adopted from the ancient Greek "system α." It means a whole that is made up of parts. *The Great World System*, by the ancient Greek philosopher Democritus, is the earliest book in which this word is employed.

Due to specific and varied focuses of study, the concept of systems is often defined differently from one discipline to another. Systems science investigates those concepts, properties, and characteristics of systems that are not discipline specific. As for the concept of systems well studied in systems science, it has two widely accepted forms.

Definition 2.1

Each system stands for a whole that is made up of mutually restraining parts and that possesses certain functionality that none of the parts possess (Qian 1983b).

Systems introduced in such a way are technology-based and emphasize on the attributes, properties, and functionalities of the systems not shared by their parts. Speaking in terms of technology, the purpose of various designs and organizations of management systems is to materialize and to actualize some predetermined functionalities of the specific systems. Therefore, the essential characteristics that separate one system from another are the systems' specified functionalities. ■

Definition 2.2

A system is an organically combined entity of some elements along with their connections and mutual interactions. In particular, if in the set of elements, there are at least two distinguishable objects and all these elements are associated together through a recognizable expression, then this set is seen as a system.

The concept of systems defined in Definition 2.2 is introduced on the level of fundamental science, emphasizing on the mutual interactions of the elements and the overall effect of the system on each of the elements.

Based on what is presented above, a more comprehensive definition for the concept of systems can be established as follows: ■

Definition 2.3

Each system is an organic whole, which is made up of some elements and their mutual interactions, with certain structure and functionality. ■

In order to develop both mathematically and scientifically sound results for the concept of systems of systems science, let us see how systems can be defined and studied using set theory.

A set is defined as a collection of objects, also known as elements. Let X be a given set. This set is ordered if a binary relation, called an order relation and denoted $\leq$, exists with X as its field such that

1. For all $x \in X$, $x = x$. That is, the relation $\leq$ is reflexive.
2. For $x, y \in X$, if $x \leq y$ and $y \leq x$, then $x = y$; that is, the relation $\leq$ is antisymmetric.
3. For $x, y, x \in X$, if $x \leq y$ and $y \leq z$, then $x \leq z$. That is, the relation $\leq$ is transitive.

The ordered set X will be denoted as the ordered pair $(X, \leq)$. The notation $x > y$ means the same as $y < x$ and is read either "y precedes x" or "x succeeds y." The field of the order relation $\leq$ is often said to be ordered without explicitly mentioning $\leq$. It should be noted that a set may be ordered by many different order relations. An order relation $\leq$ defined on the set X is linear if for any $x, y \in X$, one of the following conditions holds true: $x \leq y$, $x = y$, or $y \leq x$. If X is not linearly ordered, it is known as partially ordered.

Two ordered sets $(X, \leq_X)$ and $(Y, \leq_Y)$ are similar, if there is a bijection $h: X \to Y$ satisfying for any $x, y \in X$, $x \leq_X y$ if and only if $h(x) \leq_Y h(y)$. This similarity relation divides the class of all ordered sets into pairwise disjoint subclasses, each of which contains all sets that are similar to each other. Two ordered sets $(X, \leq_X)$ and $(Y, \leq_Y)$ are said to be the same order type if they are similar. Each order type will be denoted by a lower-case Greek letter.

An ordered set $(X, \leq_X)$ is well ordered if it is linearly ordered and every nonempty subset of X contains a first element with respect to the relation $\leq_X$. The order type of this well-ordered set X is known as an ordinal number. This concept of ordinal numbers is a generalization to that of any (either finite or infinite) sets of the finite ordinal numbers of 1st, 2nd, 3rd,....

With the necessary background of set theory in place, let us see how the concept of systems has been carefully studied in the language of set theory since the late 1980s.

Definition 2.4

A system S is an ordered pair of sets, $S = (M, R)$, such that M is the set of all objects of S, and R a set of some relations defined on M. The sets M and R, respectively, are called the object set and the relation set of the system S. Here, for any relation $r \in R$, r is defined as follows: There exists an ordinal number $n = n(r)$, which is a function of r, called the length of the relation r, such that $r \subseteq M^n$ where

$$\begin{aligned} M^n &= \underbrace{M \times M \times \ldots \times M}_{n \text{ times}} \\ &= \{f : n \to M \text{ is a mapping}\} \end{aligned} \tag{2.1}$$

is the Cartesian product of n copies of M (Lin 1987).

Remarks

Based on the conventions of set theory, what this definition says includes the following:

1. When a specific study is given, the set of all the elements involved in the study is definitely determined. Although a specific element itself can be a system at a deeper level, in the said study, it is simply an element. For instance, in a research of a social organization, the focus

is the people involved in the organization so that these particular people will be considered as the elements of the system investigated, although each of these human beings can be further researched as a system of cells.

2. The relation set R contains all the existing descriptions regarding the connections and mutual interactions between the elements in M. In particular, the relations in R together as a totality describe the internal structure of the system S. In the physical world, it is the internal structure that represents the visible attributes and characteristics of the system S.
3. From the discussions that will follow in the rest of this chapter, it can be seen that each system as defined in this definition is static. It represents the state of affairs of the system at a specific moment of time. That is, if the evolution of the system is concerned, one has to connect a sequence of such defined systems together along the time axis so that the concept of time systems will be needed; see below for more details.

Given two systems $S_1 = (M_1, R_1)$ and $S_2 = (M_2, R_2)$, S_2 is said to be a subsystem of S_1 if $M_2 \subseteq M_1$ and for each relation $r_2 \in R_2$, there exists a relation $r_1 \in R_1$ such that $r_2 \quad r_1|M_2 = r_1 \cap M_2^{n(r_1)}$. That is, system S_2 is a part of S_1. It is neither a general part of S_1 nor a simple component of the overall system. Instead, it itself also forms a system. At the same time, as expected, the overall system S_1 also contains other subsystems. When compared to the overall system S_1, each particular subsystem focuses more on a specific region or part of the overall system. Different subsystems represent their distinct spaces, structures, and internal connections.

An entity N is considered a nonsystem or a trivial system, if it satisfies one of the following conditions:

1. The objects involved with N are not clearly distinguishable.
2. The objects of N are not associated with each other in any way.

Using the language of set theory, a nonsystem N can still be written as an ordered pair of sets $N = (M,R)$ so that either $M = R = \varnothing$ (the empty set) or, even though $M \neq \varnothing$, either $R = \varnothing$ or $R = \{\varnothing\}$ holds true. Similar to the concept of zero in mathematics, in this case, when N satisfies the former condition, N is referred to as a trivial system; when N satisfies the latter condition, N is referred to as a discrete system.

Theorem 2.1

There does not exist such a system that its object set contains all systems (Lin and Ma 1993).

This is a restatement of the Zermelo–Russell paradox: By definition of sets, consider the set X of all those sets x that do not contain themselves as elements, that is, $X = \{x: x \notin x\}$. Now, a natural question is whether $X \notin X$ or $X \in X$. If the former, $X \notin X$, holds true, then the definition of the set X implies that $X \in X$, a contradiction. If the latter, $X \in X$, holds true, then the definition of X implies that $X \notin X$, is also a contradiction. These contradictions constitute the well-known Zermelo–Russell paradox.

Proof

Consider the class V of all sets. For each $x \in V$, define a system by $(x, \varnothing)$. Then a 1–1 correspondence h is defined such that $h(x) = (x, \varnothing)$. If there is a system $S = (M, R)$ whose object set contains all systems, then M is a set. According to the Zermelo–Russell paradox, this end is impossible. ■

2.1.2 Components, Elements, and Factors

When one places himself or herself inside a system in order to carefully comprehend the internal makeup of the system, he or she has to deal with various concepts, such as components, parts, elements, factors, etc. Although one needs to investigate the system holistically and intuitively, merely doing so is not sufficient. He or she also needs to "enter" into the inside of the system in order to determine what components and parts the system contains.

Components are different of parts, while parts may not be the same as components. Any entity that belongs to a system and that is smaller than the system is a part of the system. However, a part might not have any structure so that it does not represent any structural unit of the system. On the other hand, a component stands for a structural unit of the system and must possess its own structure. When a system is randomly cut up into small pieces, what are obtained are generally parts instead of components of the systems. When people say that it is difficult for a broken mirror to be repaired back to its original quality, it means that the broken pieces of the mirror are not components of the mirror; they do not possess the characteristics of the mirror's structural units. Only the parts singled out according to the structural characteristics of the system constitute components of the system. For instance, the bones, muscles, blood, and the like of the human body are components of the human body system.

The smallest parts, which cannot be or no longer need to be further divided in the study of concern, are the elements of the system. The fundamental feature of elements is their indivisibility; they represent the most fundamental level of the system. For instance, the elements of a human body are the cells; the elements of a social system are the people involved; the elements of a chemical system are the atoms. In general, elements are parts, while parts may not be elements.

Elements are the basic concept in the "exact" sciences, such as physics, chemistry, etc., because all the traditional scientific disciplines are developed on specific classes of things. Due to its focus on systemhood, systems science focuses more on relationships and the concept of factors, while placing great energy and attention on analyzing factors instead of elements. To this end, systems science considers the concept of factors in two different fashions. One is that of factors in terms of elements, where important elements are considered factors, while other elements can be ignored. However, there are many systems for which it is difficult either to single out their realistically meaningful components or to establish effective descriptions for the system using the components specified according to the requirements of the study. Introducing imaginary, nonrealistic factors that affect the systems' behaviors often makes it possible to develop meaningful descriptions of the systems. For instance, for a specific research paper, it might be hard to single out one word that is most important. However, such factors as the topics covered, bases of arguments, and the resultant conclusions can be realistically employed to analyze the work presented in the paper. Hence, another way of considering factors is to look at the most important factors, while insignificant factors can be ignored. It might be difficult or impossible to determine accurately the realistic components of complex systems, especially those involving human factors. It might also be difficult or impossible to acquire the meaningful and holistic understanding of such systems by analyzing only realistic components. For such scenarios, it is often necessary to introduce and analyze some imaginary factors. As a matter of fact, systems science often treats certain variables of systems as factors. Through observing the characteristics, changes, and mutual relationships of these variables, the researcher gains the needed holistic understanding of the systems.

In general, there are two kinds of components or factors of systems: constructive components and connective components. For illustrative purposes, let us look at the example of building a house. The constructive components include bricks, concrete blocks, etc., while mortars, plasters, nails, and others are connective components. When elementary systems are analyzed structurally, all the constructive components are the focus of attention, with the connective components ignored. However, when investigating complex systems, the connective components have to be considered. In some situations, the systems' connective components are even more important than the constructive components. For instance, the transportation network of an urban system plays the role of connective components.

For systems involving humanistic social problems, it is very necessary to separate the soft and hard factors. If a military conflict is treated as a system, then the armed forces, equipment, and so forth are the hard components, while the forces' morale, educational backgrounds, etc., are the soft factors. Generally, it is relatively easier for people to pay attention to hard factors than soft factors. However, in a great many scenarios, the soft factors can play a more important role than the hard factors.

2.1.3 Structures and Subsystems

Components (and parts) and structures stand for two closely related but very different concepts. The former represents systems' basic or important building blocks (or hard factors) or constructive factors (soft factors). Parts do not involve relationships. However, structures and only structures deal with the relationships between parts.

The concept of structure represents the totality of all connections and interactions between parts. Its focus is how parts are combined together to form the whole system. All physically existing associations and relative movements of the parts belong to the structure of the underlying system. However, in practical investigations, it is both impossible and unnecessary to uncover all possible associations and relative movements of the parts. In such a case, the word "structure" stands for the totality of only those associations and movements that play the dominating role in the development of the system, with most of the other minor relationships ignored.

The concepts of structures and parts, as just described, are related but different. Structures cannot exist independently from parts, and their appearances can be carried only by parts. When parts are considered without the context of structures, they are no longer parts. Without a comprehension of structures, there will be no way to determine parts; without an understanding of parts, there will be no base to talk about structures. That explains why in the concept of systems $S = (M,R)$, as defined using the language of set theory, the object set M and relation set R cannot be talked about separately without considering the other.

As for the structure of a system, it generally contains intensions of many different aspects. Thus, it needs to be studied from various angles. For instance, let us look at a nation as a system. Then this system needs to be analyzed from such diverse angles as the structure of people, that of classes, that of administrative organizations, etc.

When studying systems' structures, one needs to pay attention to generative and nongenerative relations. In a school, seen as a system, the relationships between the teachers and students, those between students, those between teachers, etc., are generative relations, while such relations as those of coming from the same hometown are nongenerative. For a biological body, when seen as a system, generative relations might be the biological associations and functions between cells and between organs, while all connections and functionalities in the sense of physics are nongenerative. Speaking generally, each decisive connection stands for a generative relation. Of course, some nongenerative relations may also produce nonignorable effects on the system of concern. Hence, when analyzing systems, one should also pay attention to nongenerative relations. For instance, within a large social organization, there are often unorganized groups, which are created by some nongenerative relations of the social organization. Theories of organizations often treat these unorganized groups as an important research topic.

Generally, systems are structurally classified according to the following two aspects.

1. Frameworks and movements
   - Frameworks: the fixed ways of connection between components. They stand for the fundamental ways of association between various components when the system is situated in its inertial state and not affected by any external force.
   - Movements: the relationships of components when the system is in motion due to the effects of external forces. They represent the mutual reliance, mutual supports, and mutual constraints of the components when the system is in motion.

For example, in an automobile, when seen as a system, the relative locations, connections, spatial distributions, etc., of its body, engine, steering wheel, and other parts are the framework (structure) of the system. The way all the components coordinate with each other when the automobile is moving stands for a movement (structure).

2. Spatial and time structures

Each physical system and its components exist and evolve within a certain space and time period. Any association of the components can be shown only in the form of space and/or time. Hence, there are the concepts of spatial and time structures.

- Spatial structures: the arrangement and distribution of components in space. They stand for the way components are spatially distributed and the consequent relations of mutual supports and mutual constraints between the components.
- Time structures: the way of association of the components in the flow of time. They represent the development of the system, treated as a process, in the time dimension. They can be associations, connections, or transitions of the components from one time period to another.

For instance, the structure of a house is a spatial one. The transitions between childhood, youth, middle age, and old age are time structure. There are, of course, situations where time and spatial structures are mixed. For instance, the growth rings of trees are mixed time and spatial structures.

Although there are many different kinds of associations between a system's components, the associations can be roughly classified into two classes. One is explicit; it can be easily felt, described, and controlled directly. This class is referred to as hard associations. The other kind of association is implicit, difficult to feel, describe, and control. Each such association is known as soft. The totality of the former associations is known as the hard structure of the system, while the latter, the soft structure. In principle, each system has both a hard and a soft structure. The framework of the system is a hard structure, while the movement reflects a degree of its soft structure. For example, the physical connections of the hardware of a computer represent the system's hard structure, while the associations of the software of the computer stand for the soft structure. For mechanical systems in general, there is no need to consider their soft structures. However, for organic systems, their soft structures generally have to be considered. In particular, the soft structures of humanistic social systems are extremely important. For instance, the success or failure of any management system is very often determined by the system's soft structure.

Typical structures of systems include chains, rings, nests, pyramids, trees, networks, etc.

If one uses mathematical symbols, the concept of structures has been well studied in universal algebra (Gratzer 1978). At this junction, let us take a quick look at this concept.

Let A be a set and n a nonnegative integer. An n-ary operation on A is a mapping f from A^n into A. An n-ary relation r on the set A is a subset of A^n. A type τ of structures is an ordered pair

$$\left((n_0,\ldots,n_\nu,\ldots)_{\nu<O_0(\tau)},(m_0,\ldots,m_\nu,\ldots)_{\nu<O_0(\tau)}\right) \tag{2.2}$$

where $O_0(\tau)$ and $O_1(\tau)$ are fixed ordinal numbers, and n_ν and m_ν are nonnegative integers. For every $\nu < O_0(\tau)$, there exists a symbol f_ν of an n_ν-ary operation, and for every $\nu < O_1(\tau)$, we realize r_ν as an m_ν-ary relation.

A structure U is a triplet (A, F, R), where A is a nonempty set. For every $\nu < O_0(\tau)$, we realize f_ν as an n_ν-ary operation $(f_\nu)_U$ on A; for every $\nu < O_1(\tau)$, we realize r_ν as an m_ν-ary relation $(r_\nu)_U$ on A, and

$$F = \{(f_0)_U, \ldots, (f_\nu)_U, \ldots\}, \nu < O_0(\tau) \tag{2.3}$$

$$R = \{(r_0)_U, \ldots, (r_\nu)_U, \ldots\}, \nu < O_1(\tau). \tag{2.4}$$

If $O_1(\tau) = 0$, U is called an algebra, and if $O_0(\tau) = 0$, U is called a relational system.

For the rich variety of results on mathematical structures, as defined above, please consult Gratzer (1978).

2.1.4 Levels

Comparing to the concept of subsystems, the concept of levels is more important in comprehending systems. This concept constitutes an important content of the part–whole relationship of systems. At the theoretical level, there are several ways to study this concept. For example, each system has at least two levels: one at the overall system level and the other at the parts' level. Physically existing systems, almost without exception, possess multilevels. Here, the classification of levels can be, but does not have to be, closely related to that of subsystems.

Because parts themselves have different senses of levels, it makes the concept of levels extremely difficult to define. As of this writing, there is still no complete level theory of systems. To this end, in this subsection, we will look at one specific concept of level systems, where each system contains only two levels, one at the overall system level and the other at the element level, where some elements are also systems.

Symbolically, a system $S = (M,R)$ is said to have n levels, where n is a fixed natural number, if

1. Each object $S_1 = (M_1,R_1)$ in M is a system, called the first-level object system.
2. If $S_{n-1} = (M_{n-1}, R_{n-1})$ is an $(n-1)$th-level object system, then each object $S_n = (M_n, R_n) \in M_{n-1}$ is also a system, called the nth-level object system of S.

Definition 2.5

A system $S = (M,R)$ is centralized if each object in S is a system and there exists a nontrivial system $C = (M_C,R_C)$ such that for any distinct elements x and $y \in M$, say $x = (M_x,R_x)$ and $y = (M_y,R_y)$, then $M_C = M_x \cap M_y$ and $R_C \subseteq R_x|M_C \cap R_y|M_C$. The system C is called a center of S (Lin 1999, p. 99).

The concept of centralized systems captures the realistic phenomena that in a physical system, it is very likely that the system has a central part such that when this part changes slightly, major effects are felt throughout the entire system. ■

Theorem 2.2

Let κ be a natural number and $\theta > \kappa$ the cardinality of the set of all real numbers. Assume that $S = (M,R)$ is a system satisfying [ZFC (Lin and Ma 1993)]

1. $|M| \geq \theta$.
2. Each object $m \in M$ is a system with $m = (M_m,R_m)$ and $|M_m| < \kappa$.

If there exists an object contained in at least θ objects in M, there then exists a subsystem system $S' = (M',R')$ of S such that S' forms a centralized system and $|M'| \geq \theta$.

The abbreviation ZFC in this theorem and some of the following theorems means that this result holds true in the Zermelo–Fraenkel axiomatic set theory with the axiom of choice.

The proof of this theorem is quite technical and is omitted. For interested readers, please consult Lin (1999).

This theorem is employed in the study of civilizations by Lin and Forrest (2010). These authors provide an insightful theoretical explanation for why any closed society has to go through periodic turmoil in order to adjust itself in its evolutionary development. ■

Theorem 2.3

Let $S_0 = (M_0, R_0)$ be a system and $S_n = (M_n, R_n)$ be an nth-level object system of S_0. Then S_0 cannot be a subsystem of S_n for each natural number n [ZFC (Lin 1999, p. 192)].

Proof

The theorem will be proved by contradiction. Suppose that for a certain natural number n, the system S_0 is a subsystem of the nth-level object system S_n.

Let $S_i = (M_i, R_i)$ be the systems, for $i = 1, 2, \ldots, n-1$, such that $S_i \in M_{i-1}$, $i = 1, 2, \ldots, n$. Define a set X by

$$X = \{M_i : 0 \le i \le n\} \cup \{S_i : 0 \le i \le n\} \cup \{\{M_i\} : 0 \le i \le n\}. \tag{2.5}$$

From the axiom of regularity of axiomatic set theory, which states that every nonempty set A contains an element B that is disjoint from A, it follows that there exists a set $Y \in X$ such that $Y \cap X = \varnothing$. There now exist three possibilities:

1. $Y = M_i$, for some i
2. $Y = S_i$, for some i
3. $Y = \{M_i\}$, for some i

If possibility 1 holds, $S_{i+1} \in Y \cap X$ for $i \le n-1$, and $S_i \in Y \cap X$ if $Y = M_n$. Therefore, $Y \cap X \ne \varnothing$, is a contradiction. If possibility 2 holds, $Y = S_i = (M_i, R_i) = \{\{M_i\}, \{M_i, R_i\}\}$ and $\{M_i\} \in Y \cap X \ne \varnothing$, is a contradiction. If possibility 3 holds, $M_i \in Y \cap X \ne \varnothing$, is a contradiction. These contradictions show that S_0 cannot be a subsystem of S_n. ■

2.1.5 Environments

Both the concepts of components and structures reflect the internal organization of systems, while interactions between systems and their environments reveal the external impacts on the systems by the environments. Only when one comprehends both the internal organization and the external impacts can he or she acquire a relatively complete understanding of the behaviors of the systems.

Definition 2.6

For a given system S, the totality of all matters that are located outside the system S and that have certain kinds of associations to the system S constitutes the environment E of S. Symbolically, using the ordered-pair Definition 2.4 of systems, the environment E is simply another system that contains S as one of its elements, and/or some elements of S are also elements of E. ■

Each system is born, develops, and evolves within a certain environment. Its structure, state, attributes, and behaviors are closely related to and correlate with its environment. Hence, systems rely on their environments; interactions between environments and systems constitute the external condition for the development of systems. A system may possess different structures in different environments. That explains why in different environments, the same system may possess a varied set of behaviors and attributes. Therefore, the environment plays an important role in the emergence of the whole system. When the environment changes, the system has to develop a certain new set of holistic attributes in order to adjust itself to the changing external conditions. This end explains the fact that the development of complexities in the environment creates complexities in the system. Therefore, to investigate a system, one has to pay attention to the system's interaction with its environment.

Each environment has its objectiveness and relativeness.

Definition 2.7

The set of all objects that separate system S and its environment E is referred to as the boundary of the system S, denoted $B(S)$. ■

Proposition 2.1

If a given system S and its environment E can be written as $S = (M, R)$ and $E = (M_E, R_E)$, then $B(S) = M \cap M_E$.

Proof is straightforward and is omitted. ■

For each physical system, its boundary objectively exists. However, for some systems, their boundaries can be clearly defined, while for some other systems, their boundaries cannot be easily located. For example, when civilizations are studied as systems, their boundaries at some geological locations and during certain historical moments can be surely drawn, while at some other geological locations and times, separating two cultures becomes an impossible task. Thus, according to the specifics of the problem of concern, one can select his or her desirable boundary and system. In such situations, the system will be defined as the totality of relevant events or matters that involve some of the research objects so that this totality is isolated out of the entire collection of interactive events and matters of the given task.

When investigating a physical system, it is most likely that the system is artificially carved out from its environment. Thus, at the same time when one recognizes the classification certainty of the chosen system and its environment and the definite difference between the internality of the system and the externality of the environment, he or she also needs to acknowledge the varied degrees of this certainty and definiteness and the relativeness of the classification between the system and its environment.

Definition 2.8

When a system does not interact with any external entity, the system is referred to as a closed system. Any system that is not closed is referred to as an open system. Symbolically, a system $S = (M,R)$ is closed if and only if its environment E is simply the trivial system $(\varnothing,\varnothing)$ or $(\varnothing, \{\varnothing\})$. ■

For input–output systems, which are systems that take matters in from their environments and give matters off into their environments, openness helps make the systems viable and evolve. For physical systems, their openness can be either voluntary or forced. This end has been empirically evidenced by the events of recent world history. On the other hand, closeness is a necessary condition instead of a negative factor for general systems to possess their individual identities. However, Lin and Forrest (2010) show that an expanded period of closeness is the sufficient condition for a civilization to experience internal turmoil and to suffer from external aggressions. Thus, for an input–output system, its viable and stable existence is guaranteed by a balance between its openness and closeness.

Using the relationship between systems and their respective environments, the totality of all systems can be classified into two classes: open systems and closed systems.

2.2 PROPERTIES OF SYSTEMS

2.2.1 Additive and Combinatorial Wholeness

Simple-minded emphasis on systems science as being a science of wholeness can be misleading. There are at least two classes of wholes and wholeness. One is the additive wholeness, where the whole is the sum of individually isolated elements. This situation is analogous to that of weights of physical object systems. Another class is the combinatorial wholeness. When speaking at the level of elements, the so-called combinatorial characteristic means that the same elements possess different attributes depending on whether they are positioned within the inside of the system or not. For instance, the characteristics of human organs are different depending on whether they are still within the body of a live person or separated from their carrying body. If speaking in terms of the systems' wholes, the characteristics of combinatorial wholeness are those that rely on specific relations between systems' components. To comprehend a system, one needs to know not only the system's components but also the relations between the components.

What needs to be pointed out is that any system possesses the attribute of additive wholeness. As long as such concepts as mass, energy, and others are involved, due to indestructibility of mass and conservation of energy, the whole has to be greater than or equal to the sum of parts. For such additive wholeness, natural science has done very thorough investigations so that it is no longer a problem systems science is interested in. On the other hand, combinatorial wholeness, of which a more scientific term is *emergence*, has been the focus of systems science.

To help the reader further understand the meanings of additive and combinatorial wholeness, let us now look at the structure of connectedness of a general system.

Let $S = (M,R)$ be a system and $r \in R$ a relation. The support of r, denoted $\text{Supp}(r)$, is defined by

$$\text{Supp}(r) = \{m \in M: \exists\, x \in r\, \exists\, \beta < n(r)\ (x(\beta) = m)\} \tag{2.6}$$

which is the set of all elements of M that appear in the relation r. The system S is said to be connected if it cannot be represented in the form $S_1 \oplus S_2 = (M_1 \cup M_2, R_1 \cup R_2)$, known as the free sum of the systems S_1 and S_2, where $S_1 = (M_1,R_1)$ and $S_2 = (M_2,R_2)$ are nontrivial subsystems of S such that $M_1 \cap M_2 = \varnothing$. That is, when the system S is connected, it is a combinatorial whole instead of an additive whole of some parts.

Theorem 2.4

For every system $S = (M, R)$, the following conditions are equivalent (Lin 1990):

1. The system S is connected.
2. For any two objects x and $y \in M$, there exists a natural number $n > 0$ and n relations $r_i \in R$ such that

$$x \in \text{Supp}(r_1) \text{ and } y \in \text{Supp}(r_n) \tag{2.7}$$

and

$$\text{Supp}(r_i) \cap \text{Supp}(r_{i+1}) \neq \varnothing, \text{ for each } i = 1,2,\ldots, n-1. \tag{2.8}$$

Proof

(1) → (2). We prove by contradiction. Suppose that the system S is connected and there exists two objects x and $y \in M$ such that relations do not exist $r_i \in R$, $i = 1,2,\ldots, n$, for any natural number $n \geq 1$, such that

$$x \in \text{Supp}(r_1) \text{ and } y \in \text{Supp}(r_n)$$

and

$$\text{Supp}(r_i) \cap \text{Supp}(r_{i+1}) \neq \varnothing, \text{ for each } i = 1, 2,\ldots, n - 1.$$

From the hypothesis that S is connected, it follows that there must be relations $r_1, s_1 \in R$ such that $x \in \text{Supp}(r_1)$ and $y \in \text{Supp}(s_1)$. Then our hypothesis implies that

$$\text{Supp}(r_1) \cap \text{Supp}(s_1) = \varnothing.$$

Let $U_0 = \text{Supp}(r_1)$ and $V_0 = \text{Supp}(s_1)$, and for each natural number n, let

$$U_n = \cup\{\text{Supp}(r)\text{: } r \in R \text{ and } \text{Supp}(r) \cap U_{n-1} \neq \varnothing\}$$

and

$$V_n = \cup\{\text{Supp}(s)\text{: } s \in R \text{ and } \text{Supp}(s) \cap V_{n-1} \neq \varnothing\}$$

Then, $U_0 \subseteq U_1 \subseteq \ldots \subseteq U_n \subseteq \ldots$, $V_0 \subseteq V_1 \subseteq \ldots \subseteq V_n \subseteq \ldots$, and $U_n \cap V_m = \varnothing$, for all natural numbers n and m.

We now define two subsystems $S_i = (M_i, R_i)$, $i = 1, 2$, of S such that

$$M_1 = U_0 \cup U_1 \cup \ldots \cup U_n \cup \ldots$$

$$M_2 = M - M_1$$

$$R_1 = \{r \in R\text{: } \text{Supp}(r) \cap M_1 \neq \varnothing\}$$

and

$$R_2 = \{r \in R\text{: } \text{Supp}(r) \subseteq M_2\}.$$

Then, $R_1 \cup R_2 = R$ and $R_1 \cap R_2 = \varnothing$. In fact, for each relation $r \in R$, if $r \notin R_1$, $\text{Supp}(r) \cap M_1 = \varnothing$ and so $\text{Supp}(r) \subseteq M_2$; thus, $r \in R_2$. Therefore, $S = (M,R) = (M_1 \cup M_2, R_1 \cup R_2) = S_1 \oplus S_2$, a contradiction.

(2) → (1). The proof is again by contradiction. Suppose condition 2 holds and S is disconnected. Thus, there exists nontrivial subsystems S_1 and S_2 of S such that $S = S_1 \oplus S_2$. Suppose that $S_i = (M_i, R_i)$, $i = 1, 2$. Pick an object $m_i \in M_i$, $i = 1,2$. Then there are no relations $r_j \in R$, $j = 1, 2,\ldots, n$, for any fixed natural number n, such that

$$m_1 \in \text{Supp}(r_1) \text{ and } m_2 \in \text{Supp}(r_n)$$

and

$$\text{Supp}(r_j) \cap \text{Supp}(r_{j+1}) \neq \varnothing$$

for each $i = 1,2,\ldots, n - 1$, a contradiction. ■

2.2.2 Holistic Emergence of Systems

2.2.2.1 Holistic Emergence

Those attributes possessed by the whole of a system and not shared by the elements or any part of the system are referred to as the emergence or whole of the system. In other words, the attributes that appear as soon as parts are combined together into a system and that will disappear as soon as the system is decomposed into independent parts are referred to as the (holistic) emergence or the whole.

The systems' emergence can also be intuitively and simply expressed as $1 + 1 > 2$. That is, the whole is greater than the sum of its parts. This statement should be understood not only as a quantitative but also as a qualitative characteristic of systems.

Systems' emergence is originated in three effects: the size effect, the structural effect, and the environmental effect. Speaking generally, the whole emergence is a systemic effect. Here, the basic quality and attributes of the parts are the material foundation for the systemic whole and its attributes to appear. When parts are given, the overall range of all possible systemic attributes of wholeness will be determined, even though in practice, such a range is not possible to specify.

Systems science is the science that deals with the appearance of wholeness and systemic emergence. It attempts to address such questions as, What is the wholeness emergence? What is the origin for causing the appearance of certain wholeness? What are the mechanism and the underlying laws that govern the appearance of systemic emergence? What are the main expressions of wholeness emergence? How can wholeness emergence be described? How can systemic emergence be practically applied to produce tangible benefits?

2.2.2.2 Scales and Emergence

Size first means the quantitative amount of the parts involved in the system. It can also mean the size of space the system occupies or the geological territory the system covers. If the system is positioned in the time dimension, the word "size" generally means the length of time the systemic process lasts.

Differences in size may lead to drastically different effects on the system's attributes and behaviors. Such effects are referred to as the size effect of systems. For example, new group behaviors of either ants or bees are created dependent on the size of the group.

2.2.2.3 Structures, Levels, and Emergence

Parts are the realistic foundation for systemic wholeness to appear. However, a complete list of parts is only a necessary prerequisite for wholeness emergence to form, because an unorganized pile of the parts cannot lead to the creation of the wholeness emergence. Each realistic wholeness emergence is established through connections, interactions, constraints, and excitation of many parts. That is, it is a result developed out of a system's structural excitations. In other words, the wholeness emergence is formed on the basis of parts with the guidance of the systemic structure. It is analogous to the situation that out of the same construction materials, different designs and implementations lead to completely dissimilar buildings. Out of the same set of words, different combinations lead to literature works of diverse qualities and styles.

One important aspect of structural effects is their consequent levels. Another illustration of systemic emergence is the attributes of the system's level that are not shared by any of the lower levels of the system. The appearance of each new level is originated in the appearance of a new wholeness emergence. As a concept for addressing systems' structures, what levels reflect in general are the evolutionary stages the wholeness emergence of the system experiences through systemic integration and organization. In any multileveled system, the jump from the attributes of the elements' level to those of the system's level is not accomplished in one step. Instead, it is gradually accomplished through many intermediate steps starting from the bottom level all the way to the overall system's level. For complex systems, it is impossible for the system's wholeness to emerge from the attributes of the elements of the bottom level with just one step jump. Instead, the system's wholeness emerges step by step through a series of intermediate stages, each of which represents a level; the totality of the attributes of a lower level supports the attributes of the next higher level, with the higher level dominating over the lower levels. The leveled structures are one of the origins of the complexities of systems. At the same time, levels provide a reference frame for investigating systems.

The appearance of new wholeness emergences may not necessarily lead to the creation of new levels. When a system $S = (M,R)$ is jointly acted upon by some internal and external factors, its structure, as expressed by the relations in R, generally experiences alteration, which inevitably forces the original systemic structure to be altered, leading to another kind of emergence. In such situations, no new levels are formed.

When seeing from the viewpoint of levels, each wholeness emergence is excited out of the mutual interactions of the objects and matters of a lower level. It is analogous to how springs belch from underground. For self-organizing systems, their wholeness emergences can also be referred to as self-emergences.

2.2.2.4 Environments and Emergence

The systemic emergence is determined by not only the system's internal components and structure but also the system's environment. First of all, the formation, maintenance, and evolutionary development of the system are made possible only with acquisitions of resources and conditions from the environment. Second, the formation, maintenance, growth, and role-play of each part require support from the environment. The organization of national sport teams is a good example. Each system generally has limited resources so that it has to acquire large amounts of supplies from the outside environment. On the other hand, the processing of assembling the different parts together, that is, the establishment of the system's structure, has to use the environment as its reference frame so that the assembled whole will adapt to the environment and be able to make use of the available conveniences of the environment as much as possible. That is, the determination of the system's structure cannot be separated from the environment. The desired structure is established in the process of interactions between the system and its environment. The system reconstructs and remakes the raw materials from the environment to form its parts; at the same time, it also develops the specific ways of how these parts interact with each other in order to ensure the resultant whole can continue to gain the necessary resources and supports from the environment.

Additionally, the effect of the environment on the system is not only to provide it with resources and supports; it also imposes constraints and limitations on the system. Although the constraints and limitations have their negative aspects in terms of the system's formation and growth, they also play constructive roles in helping the system to evolve. For the system to emerge out of the potentially voluminous environment to become a definite viable entity, the system has to evolve within its necessary boundaries. The constraints cannot be replaced by supports in terms of shaping the system. For instance, school regulations are indeed some kinds of limitation artificially imposed on the development of the students. However, they are indispensable in the healthy growth of these young minds. In theory, each specific system's part and structure are not only dependent on the particular supplies and supports of the environment but also greatly determined by the particular constraints and limitations the environment imposes on the system. For example, children with the same level of intelligence might live drastically differently in terms of lifestyles and might enjoy dramatically different levels of success, if they grow up in different environments. Additionally, the competition provided by the environment also cultivates and molds the structure of the system, because other than competition, the opponents also create mutually beneficial ways of survival. For instance, through scholastic debates, different thoughts and opinions often provoke one to reason differently. In the environment, there are often hostile forces that oppress and destroy the system. However, it is these hostile forces that play an important role of reshaping the system. For instance, although the severe coldness of the Arctic Ocean creates a harsh living environment for the region's biological species, it also cultivates the systemic adaptability of the region's organisms.

In short, regardless of whether the environment provides the system with resources and supports or imposes constraints and pressures, it creates on the system an environmental effect that is indispensable in the formation of the system's emergence.

Besides, cultivation is mutual; the system also exerts its very own effects on the environment. The environment is made up of and shaped by all the systems and nonsystems within the environment. From the stage of nonexistence to the stage of actual existence, the system has to bring about changes in the environment. When an intruder enters the environment, he or she surely creates alterations to the environment, which in turn causes changes in the system. As long as the system is evolving, it will sooner or later create specific effects on the environment so that the environment will either gradually or drastically change reactively. That is one fundamental way a system could shape the environment. It is analogous to how the evolution of the human system has been adversely affecting the geo-environment. The cultivating effect of systems on their environments also has both positive and negative directions. Positive cultivation means the constructive effects systems' behaviors have on the environments, while negative cultivation stands for the damages and pollutions systems' behaviors cause in the environments.

To gain additional understanding on the wholeness emergence of systems, let us look at the following result.

Theorem 2.5

If a subsystem C of a system S is connected, and for a pair S_1, S_2 of subsystems of S with disjoint object sets such that C is a subsystem of $S_1 \oplus S_2$, then C can only be a subsystem of either S_1 or S_2 but not both (Lin 1999, p. 218).

Intuitively, the connectedness of C makes it a true part of the system S. Then, what this theorem says is that if the system S is a combinatorial whole of its parts, then S possesses some wholeness properties, such as it being a partial free sum $S_1 \oplus S_2$ of some parts, that are not shared by any part.

Proof

We show the result by contradiction. Suppose that C is a subsystem neither of S_1 nor of S_2. Let $C = (M_C, R_C)$ and $S_i = (M_i, R_i)$, $i = 1, 2$. Then

$$M_C \cap M_1 \neq \varnothing \neq M_C \cap M_2.$$

Pick elements $x \in M_C \cap M_1$ and $y \in M_C \cap M_2$. Since C is connected and from Theorem 2.4, it follows that there exists a natural number $n > 0$ and n relations $r_i \in R_C$, for $i = 1, 2, \ldots, n$, such that

$$x \in \mathrm{Supp}(r_1) \text{ and } y \in \mathrm{Supp}(r_n)$$

and

$$\mathrm{Supp}(r_i) \cap \mathrm{Supp}(r_{i+1}) \neq \varnothing, \text{ for each } i = 1, 2, \ldots, n-1.$$

Therefore, there must be an i satisfying $0 < i \leq n$ such that

$$\mathrm{Supp}(r_i) \cap M_1 \neq \varnothing \neq \mathrm{Supp}(r_i) \cap M_2.$$

From the definition of subsystems, it follows that there exists a relation r in the free sum $S_1 \oplus S_2 = (M_1 \cup M_2, R_1 \cup R_2)$ such that $r_i \subseteq r|M_C$. Thus, r is contained either in R_1 or in R_2 but not both. Therefore, if $r \in R_1$, $\mathrm{Supp}(r_i) \cap M_2 = \varnothing$; if $r \in R_2$, $\mathrm{Supp}(r_i) \cap M_1 = \varnothing$, a contradiction. This implies that C is a subsystem of either S_1 or S_2 but not both. ■

2.2.3 Other Properties of Systems

The characteristics of systems can be summarized as follows:

1. *The wholeness.* As discussed above, each system consists of mutually reliant parts, among which there are organic connections. It is these parts and connections that make the totality of the parts and connections a comprehensive whole with additional attributes and functionalities. This end indicates that each system possesses all the properties of sets. That is, although the parts that make up the system may have respective attributes and functionalities, they are organically assembled together to materialize the functionalities of the wholeness. Hence, systems are not simple piles of their parts; instead they possess the oneness and wholeness. When designing practically usable systems, one needs to pay sufficient attention to the coordination and coupling of individual parts and levels in order to obtain the systems' orderliness and the desired operational effects of the wholeness.
2. *Relevance.* The collection of mutually associated parts constitutes the element set of the system. The characteristics and behaviors of the parts in the element set constrain and mutually affect each other. This kind of relevance underlies the attributes and states of the system.
3. *Purposefulness and functionalities.* Most systems' developments and behaviors can accomplish certain functionalities. However, not all systems, such as the solar system and some biological systems, are purposeful. All artificial systems or composite systems are designed purposefully to actualize certain specific goals. Such systems are the main focus of research of systems engineering. For example, managing administrative systems needs to optimize the allocation of the available resources in order to achieve the most desirable economic results. To provide safety for one's own people while eliminating the enemy, military systems need to employ the knowledge of modern science and technology in the organization of battlefields and weapon designs and production.
4. *Environmental adaptability.* There are exchanges of materials, energies, and information between a system and its environment. That is why changes in the external environment can alter the characteristics of the system, leading to changes in the association and functionalities of the parts within the system. To maintain and to resume the original characteristics, the system must possess its adaptability to the changing environment. Such systems include, but are not limited to, feedback systems, self-adaptive systems, self-learning systems, etc.
5. *The dynamic nature.* Materials and movements are inseparable. Each characteristic, state, structure, functionality, and regularity of materials is manifested through motions. Thus, to comprehend materials, one has to first study the movements of materials. The dynamic nature makes systems have life cycles. Open systems exchange materials, energies, and information with the environments, causing changes to the internal structure of the systems. Generally speaking, systems' developments are dynamic directional processes.
6. *Orderliness.* Due to the directionality in the dynamic evolutions of systems' structures, functionalities, and levels, systems reveal the characteristics of orderliness. One result of the general systems theory is considered important, if it connects the orderliness and purposefulness of biological organisms and the phenomena of life with the structural stability of systems. In other words, orderliness makes systems stable, while purposefulness helps systems evolve toward the expected stable structures.

2.2.4 Functionalities of Systems

For a given system, the effect of the system's behaviors that is beneficial in terms of sustainability and continued development to some parts or the entire environment of the system is referred to as the function of the system. It stands for the system's contribution to the sustainability and growth of the objects. In terms of functionalities, the wholeness emergence of systems stands for the systems' functions that are possessed by neither the individual parts nor the totality of the parts.

When the concept of systems' functions is employed on a subsystem, it means the responsibility and contribution of the subsystem to the sustained development of the whole system. Functional structure of a system means the classification of subsystems based on functions and their associations.

The concept of functions is different from that of capability. The latter stands for the specific attribute of the system expressed in its internal coherence and communication with the outside world. Capability generally is not a function, while functions are specific capabilities. Capability is the foundation of functions; it stands for the objective basis for the system to play out its functions. Functions are externalizations of capabilities and can only be expressed, observed, and evaluated through systems' exerted behaviors. For example, the fluidity of water is a capability, while transportations and generations of electricity using the fluidity of water are some of the functions of water. Capabilities can be observed and evaluated under the condition that the systems and the objects the systems exert their functions on are separate.

The multiplicity in systems' capabilities determines the great variety in systems' functions.

In short, for each system, its elements, structures, and environment jointly determine the functions of the system. Only when the environment is given can one say that the system's functions are determined by the system's structure. If one needs to construct a system with some predetermined functions, he or she will need to select those elements that have the necessary capabilities, choose the optimal design of the structure, and create the appropriate environment.

2.2.5 Attributes of Systems

The attribute of a thing, which is either a system or an object, represents the prescription, possessed by the thing itself and shown through the existence and movement of the things. When a system is investigated, the focus is on the system's attributes. Even for class attributes, a particular system may still show its own characteristics and individuality.

Systems' attributes are of differences in levels. When investigating systems, one needs to distinguish between shallow layer and deep layer attributes, and apparent and intrinsic attributes. These kinds of attributes can be separated by checking whether or not they can be directly observed, measured, and tested from outside the system. However, these kinds of attributes are also correlated with each other; apparent attributes can be employed to help understand those of deep layers.

Systems' attributes take many different forms, including multiplicity, relevance, wholeness, locality, openness, closeness, nonadditiveness, reliability, orderliness, process character, phases, dynamics, complexity, etc.

There are two basic perspectives in the investigation of systems: the synchronic and diachronic perspectives. When the former is concerned with, where either a particular time moment is chosen or the time factor is completely ignored, the researcher studies the characteristics of the spatial distribution of his or her system. When the diachronic perspective is dealt with, the system is studied from the time dimension, where the system's attributes are observed at different time moments and changes in the attributes are analyzed.

One important classification of systems is based on systems' qualitative and quantitative characteristics. Systems' attributes are of qualitative and quantitative aspects. Any system that describes a certain object has both qualitative and quantitative characteristics. To effectively describe a system quantitatively, the key is to find the system's characteristic quantities. The numerical value obtained from measuring a property of the system is referred to as a parameter or a parametric quantity of the system. Such parameters quantitatively represent the system's quantitative properties and are either constants or variables. In dealing with practical problems, the distinction of constants and variables are relative.

In terms of the description, investigation, and construction of systems, quantifying qualitative characteristics is generally the key step. If there is no quantitative expression with sufficient data support corresponding to the qualitative description of a system, then this qualitative description is not complete. However, there are large differences in the practicability of different quantitative descriptions. For instance, if physical systems are considered relatively easier to quantify, then the possibility for quantifying social systems is much less.

Fathoming the qualitative attributes of a system is also the foundation for acquiring a deeper understanding of the quantitative attributes of the system. Any system's quantity stands for the measurement of a certain qualitative attribute. Also, any qualitative attribute possesses its quantitative expression.

In the quantitative descriptions of systems, there are four kinds of commonly used variables: environmental quantities, input quantities, output quantities, and state quantities. Any quantity related to the system's input from the environment has something to do with the characteristics of the system; conversely, any quantity that relates to the system's output into the environment is also associated with the environment. Quantities of the input, output, and state of the system are the three fundamental classes of systems variables. A basic task of quantified systems theories is to select the necessary systems variables and to establish the appropriate models to describe the relationship of these variables.

State means those observable and recognizable conditions, trends, and characteristics of the system. It is a concept that characterizes the system's qualitative properties. However, the state of a system can be expressed using one of the several quantitative characteristics, known as state quantities, of the system. Because the system's state quantities can take different numerical values, they are referred to as state variables. The following are the requirements used for selecting state variables: objectivity (each selected state variable should possess specified systemic significance and is able to reflect the fundamental characteristics and behaviors of the system), completeness (enough state variables are chosen so that they can completely describe the state of the system), and independence (each chosen state variable is not a function of other selected state variables).

Sustained existence and evolutionism stand for another class of important attributes of systems. Each realistic system exists in space and evolves with time, while relying on some particular conditions. If a change occurs to the space, time, or sustaining conditions, the system will more or less change accordingly. However, if the external environment changes only slightly, the system will maintain its basic characteristics. This property of systems is referred to as the system's viability. Systems' viability depends on that of the parts, that of internal structures, and that of the environments. In the physical world, systems' viability is relative.

Changes a system might experience are also known as the evolution of the system. Evolution reflects such changes that occur in the system's structure, state, characteristics, behaviors, functionalities, etc., with time. Evolution is a generally existing character of systems. As long as the time scale is sufficiently large, any system can be treated as evolutionary; systems evolve under joint dynamic forces both internal and external to the systems. There are two basic forms for systems' evolution: special evolutions and general evolutions. The former stands for the transformation of the system from one structure to another; the latter includes the entire process of systems' appearance, growth, changes in states and structures, age, degeneration, and disappearance.

In the process of a general systems' evolution, the appearance of a system stands for the moment when the system initial becomes being. The growth represents the system's ability to continuously sustain itself. Changes in states and structures indicate that the viable existence of the system is limited. The disappearance implies that the lifespan of the system is bounded.

There are two basic directions along which a system evolves: one is the evolution from elementary to advanced stages, from primitive to sophisticated structures, and the other the degeneration from advanced to elementary stages, from sophisticated to primitive structures. These two directions complement each other. Each system goes through birth, growth, change, and death as a process, which is an important topic in systems science. That is, as long as a system is identified, it needs to be studied as a process as long as the time scale is sufficiently large.

2.3 DYNAMICS AND TIME SYSTEMS

Considering the static nature of the concepts introduced earlier, and considering the fact that each system should be studied as a process, in this section, we establish the concept of α-type hierarchies of systems with time systems as special cases.

Let $S_i = (M_i, R_i)$, $i = 1, 2$, be two systems and $h: M_1 \to M_2$ a mapping. Define two classes $\hat{M}_i$, $i = 1$, 2, and a class mapping $\hat{h}: \hat{M}_1 \to \hat{M}_2$ to satisfy the following properties:

$$\hat{M}_i = \bigcup_{n \in \text{Ord}} \hat{M}_i^n,\ i = 1,2 \tag{2.9}$$

where Ord stands for the totality of all ordinal numbers, and for each $x = (x_0, x_1, \ldots, x_\alpha, \ldots) \in \hat{M}_1$

$$\hat{h}(x) = \left(h(x_0), h(x_1), \ldots, h(x_\alpha), \ldots\right). \tag{2.10}$$

For each relation $r \in R_1$, $\hat{h}(r) = \{\hat{h}(r) : x \in r\}$ is a relation on M_2 with length $n(r)$. Without confusion, h will also be used to indicate the class mapping $\hat{h}$, and h is a mapping from S_1 into S_2, denoted $h: S_1 \to S_2$. When $h: M_1 \to M_2$ is surjective, injective, or bijective, $h: S_1 \to S_2$ is also called surjective, injective, or bijective, respectively. Let X be a subsystem of S_1. Then $h|X$ indicates the restriction of the mapping on X.

Let $(T, \leq)$ be a partially ordered set with order type α. An α-type hierarchy S of systems over the partially ordered set $(T, \leq)$ is a function defined on T such that for each $t \in T$, $S(t) = S_t = (M_t, R_t)$ is a system, called the state of the α-type hierarchy S at the moment t. Without causing confusion, we omit the words "over the partially ordered set $(T, \leq)$."

For an α-type hierarchy S of systems, let $\ell_{tr}: S_r \to S_t$ be a mapping from the system S_r into the system S_t, for any $r, t \in T$, with $r \geq t$ such that

$$\ell_{ts} = \ell_{tr} o \ell_{rs} \text{ and } \ell_{tt} = \text{id}_{S_t} \tag{2.11}$$

where r, s, t are arbitrary elements in T satisfying $s \geq r \leq t$, and $\text{id}_{S_t} = \text{id}_{M_t}$ is the identity mapping on the set M_t, defined by $\text{id}_{S_t}(m) = \text{id}_{M_t}(m) = m$, for any $m \in M_t$. The family $\{\ell_{ts}: t, S \in T, s \geq t\}$ is a family of linkage mappings of the α-type hierarchy S, and each mapping ℓ_{ts} a linkage mapping from S_S into S_t.

An α-type hierarchy of systems S, denoted $\{S, \ell_{ts}, T\}$ or $\{S(t), \ell_{tS}, T\}$, is referred to as a linked α-type hierarchy (of systems) if a family $\{\ell_{ts}: t, s \in T, s \geq t\}$ of linkage mappings is given.

Theorem 2.6

Let S be a nontrivial α-type hierarchy of systems; i.e., each state S_t is a nontrivial system. Then there exists a family $\{\ell_{ts}: t,s \leq T, s \geq t\}$ of linkage mappings of S [ZFC (Lin 1999, p. 224)].

Intuitively, what this theorem says is that as long as a hierarchy of systems is given, some kind of thread connecting the systems can be imagined, no matter whether such a thread is real or not.

Proof

Suppose $S_t = (M_t, R_t)$, for each $t \in T$. From the axiom of choice of the axiomatic set theory, which states that for any set X of nonempty sets, there exists a choice function f defined on X such that $f(s) \in s$, for any $s \in X$, it follows that there exists a choice function $C: T \to \cup\{M_t: t \in T\}$ such that $C(t) \in M_t$ for each $t \in T$. A family $\{\ell_{ts}: t, s \in T, s \geq t\}$ of linkage mappings of S can now be defined. For any $s, t \in$ T with $s > t$, let

$$\ell_{tS}(X) = C(t) \tag{2.12}$$

for all $x \in M_s$ and $\ell_{tt} = \text{id}_{M_t}$. ∎

Now, if the partially ordered set $(T, \leq)$ studied above is time, the set of all real numbers that is a linearly ordered set, then the concept of α-type hierarchies of systems becomes that of time systems, where each α-type hierarchy is referred to as a time system.

2.4 CLASSIFICATION OF SYSTEMS

Systems take different forms. Systems science studies general systems and classes of systems so that each particular class corresponds to a conventional scientific discipline.

2.4.1 Nonsystems Scientific Classification of Systems

All the myriad of things in the world exist in the form of systems. Each object studied in any discipline appears as a system. That is, different scientific disciplines focus on the investigation of systems of particular characteristics. As a matter of fact, modern science is compartmented according to the differences in the object systems. For instance, studies of natural systems are categorized into natural science; investigations of social systems are considered parts of social science. Natural science is further compartmented based on the attributes of the objects studied, leading to the creation of finer disciplines. Such method of classification emphasizes the basic qualitative characteristics of parts without considering the systemicality and wholeness. Thus, it is a nonsystems scientific classification.

2.4.2 Systems Scientific Classification of Systems

When systems are classified and studied according to their characteristics of wholeness without considering any particular properties of parts, what results are the disciplinary branches and theoretical structure of systems science.

In the following, let us briefly introduce several methods of systems classification.

General systems and particular systems: This is the classification method Bertalanffy (1968) established. He believes that there are models, principles, and laws that are applicable to general systems; these models, principles, and laws have nothing to with the classification of specific systems, properties of parts, and characteristics of the "forces" or relationships between the key factors. Here the phrase "specific systems" stands for different classes of general systems. It does not involve the basic properties of systems' parts. As systems scientific classification of systems is based on particular systemic characteristics, such as scales, structures, behaviors, functionalities, etc., the main classes of systems include

- Open and closed systems
- Elementary and complex systems
- Systems with and without control
- Self-organizing and hetero-organizing systems
- Physical and conduct systems
- Process and nonprocess systems
- Deterministic and nondeterministic systems
- Adaptive and nonadaptive systems
- Soft and hard systems

In the following, let us look at examples of several classification methods of systems.

2.4.2.1 Natural and Artificial Systems

Primitive systems, such as celestial bodies, oceans, ecological systems, etc., are natural. Artificial systems, such as manmade satellites, ships designed for sea transportation, machineries, etc., all exist within the inside of some natural systems. There are clearly boundaries between artificial and natural systems that affect each other. In recent years, adverse effects of artificial systems on natural systems, such as nuclear armaments, chemical weapons, environmental pollutions, etc., have caught people's attention. Natural systems, such as the weather system in which the seasons repeat themselves without any foreseeable end, the system of food chains, the water recycling system, etc., represent highly sophisticated equilibrium systems. In each natural system, organic matters, vegetations, and environments sustain a balanced state. In nature, what are most important are the circulation and evolutionary change of material flows. The natural environment system does not seem to have any beginning and abolishment. It only circulates and reciprocates, and evolves from one stage of development to another. Primitive man did not affect the natural system much. However, in the last 100 years, along with the benefits the fast-evolving science and technology have created for man, science and technology have also brought along with them crises and even disasters, capturing major attention from around the world. For example, the Aswan high dam in Egypt is a manmade system. Although it resolves the flood problem of the Nile River for Egyptians, it also creates some unexpected challenges. In particular, the existing food chain of the east area is destroyed, leading to huge economic losses to the fishery industry. Soil salinization in Nile River basin accelerates, which, together with newly appearing periodic droughts, greatly affects the agricultural production. Because of the river pollution, the health status of the nearby residents is adversely affected. If the methods of systems engineering were employed in the planning stage of the hydraulic project, there might be a better solution that could have resolved the flood problem as well as reduced the relevant losses.

The subjects of study of systems engineering tend to be such complex systems that include manmade and natural systems. Speaking from the point of view of systems, when analyzing systems, one should use a bottom-up instead of any down-top approach. For example, when investigating a system and its environment, one should first look at the effects of the environment on the system, because the environment is a higher level. After that, he or she would continue with the study of the system itself and then the various parts or factors that make up the system. The natural system involved in the study often stands for the highest level of the complex system of concern.

2.4.2.2 Physical and Abstract (Conceptual) Systems

A system is physical if it contains physically existing entities or factors as its parts. These physical entities, such as minerals and biological beings in nature, machineries and raw materials of a production department, etc., occupy certain space. Corresponding to physical systems are abstract conceptual systems, which are made up of such intangible entities as concepts, principles, hypotheses, methods, plans, regulations, procedures, etc. Management systems, legislation, education, cultures, etc., are examples of abstract systems. In recent years, conceptual systems have gradually been referred to as soft science systems, attracting an increasing amount of attention.

These two classes of systems are often combined in practice in order to realize some desired functionality. Physical systems are the foundations of conceptual systems, while conceptual systems provide guidance for and services to physical systems. For example, to materialize the construction of a certain engineering project, one needs to have the plan, design proposal, and goal reconciliation. If what is involved is a complex system, one needs to employ mathematical or other theoretical models and numerical simulations to extract the main factors of the system, and analyze multiple proposals before his eventual action. In this process, plans, designs, simulations, analysis of proposals, etc., all belong to the class of conceptual systems.

2.4.2.3 Static and Dynamic Systems

Systems' statics and dynamics are relative. Roughly speaking, all structures and systems that do not have any moving parts as well as those that are at rest, such as bridges, houses, roads, etc., are considered as static systems. As for dynamic systems, it means such entities that contain both static and moving parts. For example, each school is a dynamic system, which contains not only buildings but also teachers and students. Before the middle ages, people once believed that the cosmos is eternal and invariant and liked to treat matters as constant and static. Such a world point of view is idealistic or mechanic materialistic in philosophy. Along with the advances of science, man gradually recognized that the physical world is not a collection of invariant matters; instead it is a collection of dynamic processes, and only movement is eternal. The cosmos is a dynamic system, in which rest is only a relative concept.

2.4.2.4 Open and Closed Systems

A closed system is one that does not have any connection with the outside. No matter how the outside world changes, the closed system maintains its characteristics of equilibrium and internal stability. The chemical reaction that is taking place in a well-sealed container is an example of a closed system; under certain initial conditions, the reaction of the chemicals within the container reaches its equilibrium. Each open system exchanges information, materials, and energies with its environment. For example, business systems, production systems, or ecological systems are all open systems. When the environment undergoes changes, the open system stays in a dynamically stable state through mutual interactions of its parts and the environment and through its own capability of adjustment. Hence, open systems are generally self-adjusting or self-adapting systems.

Since scale, meaning the amount of parts, is an important attribute of systems, systems' scales can be relatively classified as small and large. Differences in scales surely create differences in systems' characteristics. In systems science, systems are classified into small, medium, large, giant, and extra-giant systems.

When scales and complexities are combined, systems can be classified into classes of simple systems, simple giant systems, and complex giant systems. Additionally, complex giant systems are studied as general complex giant systems and special complex giant systems.

2.4.3 Classifications of Systems Based on Mathematical Models

To a great degree, systems science is a quantified science. When the situation permits, quantitative models are established, and the system's characteristics and evolution are analyzed by using available mathematical tools. Due to this reason, systems science also pays attention to the classification of systems according to the mathematical models involved. The following are some of the most employed classes of systems.

1. *Continuous and discrete systems.* When the basic variables are continuous, the system is treated as continuous. When the basic variables take discontinuous values, the system is seen as discrete.
2. *Time-varying and time-invariant systems.* If in a mathematical model that describes the system of concern the parameters vary with time, then the system is time-varying; otherwise, the system is seen as time-invariant.
3. *Linear and nonlinear systems.* Systems that can be written by using linear mathematical models are known as linear systems, while those systems that have to be described by using nonlinear mathematical models are referred to as nonlinear systems. All published research seems to suggest that all realistic, physical systems are nonlinear, involving different degrees of nonlinearity. Different methods, such as that of local linearization and that of linearization with perturbations, have been employed to the research of nonlinear systems.
4. *Static and dynamic systems.* Systems whose state variables do not vary with time are treated as static; otherwise, they are seen as dynamic.

2.5 SYSTEMS DEFINED ON DYNAMIC SETS

Based on the previous description of static structures of general systems and investigations of systems dynamics using the concept of time systems without clearly pointing out where systems attributes would come into play, in this section, let us look at how to resolve this problem. We introduce the concept of dynamic sets so that the evolutions and changing structures of systems can be adequately described by using object sets, relation sets, attribute sets, and environments. On the basis of this background, we revisit some of the fundamental properties of systems, including systems emergence, stability, etc.

According to Definition 2.4, the objects of the general system are highly abstract; the only difference between sets of objects will be their index sets. That is why such a general theory of systems can be powerfully employed to investigate systems with parts of identical properties. That is, the general system, as defined in Definition 2.4, grasps the commonality of objects by ignoring their individual specifics. However, in terms of a realistic system, the objects' specifics might play important roles in the operations of the system. They influence not only the composite of the system but also the structure and functionality of the system, producing the relevant dynamic behaviors of evolution of the system. For example, a system that is made up of pure oxygen gas or of pure hydrogen gas is fundamentally different from that consisting of both oxygen and hydrogen gases. In particular, the interactions of the objects of the former system are mainly the repulsive and attractive forces between the gas molecules, while in the latter system, other than the similar repulsive and attractive forces, there are also following chemical reactions when the environment provides the needed condition:

$$2H_2 + O_2 \rightarrow 2H_2O. \tag{2.13}$$

The resultant system after the chemical reactions is different from the system that existed before the reactions in terms of the object set and also the relation set. What is more important is that the property of further chemical reactions in the system that resulted from the first round of chemical reactions is essentially different from that of the system that existed before the first round of chemical reactions. If the system that existed before the chemical reactions is written as $S_0 = (M_0,R_0)$ and the system that existed after as $S_1 = (M_1,R_1)$, then it is obvious that $M_0 \neq M_1$ and $R_0 \neq R_1$. Systems S_0 and S_1 are very different due to the objects in S_0 having undergone substantial changes with $2H_2O$ molecules added. Additionally, the interactions between the objects of S_1 are totally different from those in S_0. Other than repulsive and attractive reactions between the gas molecules, the interactions of the objects of S_1 also include those between liquid molecules and gas molecules, and those between liquid molecules.

From this example, it follows that for a certain circumstance, it is not enough to employ only object and relation sets to describe the systems and their behaviors, especially if changes and evolutionary behaviors of the systems are concerned. It is because both H_2 and O_2 are treated as abstract objects in S_0, while ignoring their differences in other aspects. When such differences do not affect the evolution of the systems much, the description of systems in Definition 2.4 will most likely be sufficient. However, the undeniable fact is that there are many such systems where the interactions between the objects of specific attributes greatly affect the structures, functionalities, and evolutions of the systems, just like the differences described by the two gases in Equation 2.13.

Therefore, there is a need to develop such a systems theory to deal with this situation, where the object and relation sets of systems need to be further deliberated so that the consequent evolutions of systems can be adequately investigated. The presentation of this section is mainly based on Duan and Lin (to appear).

2.5.1 Attributes of Systems

Because the usage of rigorous mathematical language in the discussion of properties of general systems is very advantageous, we will continue to employ this approach. In this section, we will not consider the case when a set is empty; all sets are assumed to be well defined. That is, we do not consider such paradoxical situations as the set containing all sets as its elements. To this end, those readers who are interested in set theory and the rigorous treatment of general systems theory are advised to consult with Lin (1999).

As discussed earlier, our main focus here is the difference between various parts of a system, that is, the differences between the system's objects and between the relations of the objects. We will employ the concept of attributes to describe such differences. Attribute means a particular property of the objects and their interactions and the overall behavior and evolutionary characteristic of the system related to the property.

If speaking in the abstract language of mathematics, the attributes of a system are a series of propositions regarding the system's objects, relations between the objects, and the system itself. Due to the differences widely existing between systems' objects, between systems' relations, and between systems themselves, these propositions can vary greatly. For example, a proposition might hold true for some objects and become untrue for others. In this case, we say that these objects have the particular attribute, while the others do not. As another example, a proposition can be written as a mapping from the object set to the set of all real numbers so that different objects are mapped onto different real numbers. In this case, we say that the system's objects contain differences.

Additionally, the discussion above also indicates that each system evolves with time so that its objects, relations, and attributes should all change with time. Thus, when we consider the general description of systems, we must include the time factor. By doing so, the structure of a system at each fixed time moment is embodied in the object set, relation set, and attribute set, while the evolution of the system is shown in the changes of these sets with time.

Summarizing what is analyzed above, the system of our concern can be defined as follows:

Definition 2.9

Assume that T is a connected subset of the interval $[0, +\infty)$, on which a system S exists. Then, for $t \in T$, the system S is defined as the following order triplet:

$$S_t = (M_t, R_t, Q_t) \tag{2.14}$$

where M_t stands for the set of all objects of the system S at the time moment t, R_t is the set of the relations between the objects in M_t, and Q_t is the set of all attributes of the system S. For the sake of convenience of communication, T is referred to as the life span and S_t as the momentary system of the system S. ■

What needs to be emphasized is that we study not only the evolution of systems with time but also the evolution of the system along with continuous changes of some conditions, such as temperature, density, etc. In such cases, T will be understood as one of those external conditions.

The object set of the system of our concern is made up of the system's fundamental units. So, for each $t \in T$, the objects in M_t are fixed. Let us write

$$\widehat{M} = \{m_{t,a} : a \in I_t, t \in T\} = \bigcup_{t \in T} M_t \tag{2.15}$$

where $M_t = \{m_{t,a} : a \in I_t\}$ is the object set of the momentary system S_t, and I_t is the index set of M_t as a function of time t. If for any $t \in T$, $M_t = M$, for some set M, then we say that the object set of the system S is fixed.

One of the most elementary relations between objects is binary, relating each pair of objects. Such a relation can be written by using the 2-D Cartesian product of the system's object set:

$$R_{t,2}^{0} = \left\{ \left(m_{t,a}, m_{t,b} \right) \in M_t \times M_t : \Phi_{t,2}^{0}\left(m_{t,a}, m_{t,b} \right) \right\} \quad M_t^2 \tag{2.16}$$

for each $t \in T$, where $\Phi_{t,2}^{0}(,)$ is a proposition that defines $R_{t,2}^{0}$. Corresponding to different properties, the system S_t might contain different binary relations as subsets of the 2-D Cartesian product M_t^2 of the object set. The set of all the binary relations in S_t is written as follows:

$$R_{t,2} = \left\{ R_{t,2}^{0} : k \in K_2 \right\} \tag{2.17}$$

where K_2 is the index set of all binary relations of S_t. Similarly, the system S_t might contain tri-nary relations:

$$R_{t,3}^{0} = \left\{ \left(m_{t,a}, m_{t,b}, m_{t,c} \right) \in M_t^3 : \Phi_{t,3}^{0}\left(m_{t,a}, m_{t,b}, m_{t,c} \right) \right\} \quad M_t^3 \tag{2.18}$$

for each $t \in T$, where $\Phi_{t,3}^{0}(,,)$ is a proposition that defines $R_{t,3}^{0}$. The set of all tri-nary relations of S_t is written as follows:

$$R_{t,3} = \left\{ R_{t,3}^{k} : k \in K_3 \right\} \tag{2.19}$$

where K_3 is the index set of all tri-nary relations of S_t. Higher-order relations of S_t can be introduced similarly. For convenience, let us define unitary relations of S_t as follows:

$$R_{t,1}^{0} = \left\{ m_{t,a} \in M_t : \Phi_{t,1}^{0}\left(m_{t,a} \right) \right\} \quad M_t \tag{2.20}$$

for each $t \in T$, where $\Phi_{t,1}^{0}()$ is a proposition that defines $R_{t,1}^{0}$. The set of all unitary relations of S_t is written as follows:

$$R_{t,1} = \left\{ R_{t,1}^{k} : k \in K_1 \right\} \tag{2.21}$$

where K_1 is the index set of all unitary relations of S_t. Now, the set R_t of relations of S_t can be written as follows:

$$R_t = \bigcup_{\alpha \in \mathrm{Ord}} R_{t,\alpha} \tag{2.22}$$

where Ord stands for the set of all ordinal numbers. When Ord is taken to be N = the set of all natural numbers, Equation 2.22 stands for the set of all finite relations the momentary system S_t contains.

From the discussion above, it can be seen that all kinds of algebras and spaces studied in mathematics are special cases of Equation 2.22 with Ord replaced by N. Without loss of generality, let us assume that $M_t \in R_{t,1}$. With this convention, it can be seen that in the following, when we talk about the attributes of a system S_t, we only need to mean the attributes of the relations of S_t; because the object set is also considered a relation, a unitary relation and the attributes of objects are now also those of relations.

The attribute set Q_t of the momentary system S_t is a series of propositions about the relations. Symbolically, we can write

$$Q_t = \{q_t(r_1, r_2, \ldots, r_\alpha, \ldots): q_t(\ldots) \text{ is a proposition of } r_\alpha \in R_t, \alpha = 1, 2, \ldots\}. \tag{2.23}$$

Without any doubt, with this notation in place, these propositions in Q_t embody all aspects of the momentary system S_t. For example, the concept of the mass of an object in physics is a mapping

$$\mu_t : R_t \to R^+ \tag{2.24}$$

which assigns each element in a unitary relation the mass of the element, and each element $\vec{x} = (x_1, x_2, \ldots, x_\alpha, \ldots)$ in an n-nary relation the sum of the masses of the objects contained in the element, if the sum exists. That is, for any $r \in R_t$, and any $\vec{x} = (x_1, x_2, \ldots, x_\alpha, \ldots) \in r$,

$$\mu_t(\vec{x}) = \mu_t(x_1, x_2, \ldots, x_\alpha, \ldots) = \Sigma_\alpha \mu_t(x_\alpha) \tag{2.25}$$

assuming that the sum on the right-hand side converges.

Let us look at the network model of systems as an example, where only unitary and binary relations are considered. In this model, all objects of the system of concern are treated as nodes; each binary relation is modeled as the set of edges of the network. By doing so, each binary relation of the system corresponds to a network or a graph; different binary relations correspond to different sets of edges. Such correspondence can be seen as an attribute of the edges. Accordingly, different unitary relations of the system correspond to different attributes of nodes in the network, such as size of the nodes, flow intensities of the nodes, etc. If different types of edges are treated as identical, then the network can be expressed as $G = (V, E, Q)$, where V stands for the set of all nodes, E the set of edges, and Q some attributes of either the nodes or the edges or both, such as weights of the edges. The ordered pair (V,E) completely describes the topological structure of the network, which is sufficient for some applications. However, if the system we investigate is quite specific, for example, it is a network of railroads, a network of human relationships, etc., we may very well need to model multiple relations. In this case, the attribute set Q can be employed to describe the scales of the stations in the railroad network, the traffic conditions or transportation capabilities between stations, etc. If the system is a network of human relationships, then Q can be utilized to represent the social status of each individual person, the intensity of interaction between two chosen persons, etc. As a matter of fact, the concept of systems, as defined in Equations 2.14, is a generalization of that as defined in Definition 2.4 (Lin 1999). In other words, we can rewrite Definition 2.4 in the format of Equation 2.14 as follows.

Let all object sets be static. Thus, for any $t \in T$, we have $M_t = M$ and $R_t = R$. Also, what is interesting is how an attribute is introduced. To this end, we can introduce a proposition q_0 on the Cartesian product $\widehat{M} = \sum_{\alpha \in \mathrm{Ord}} M^\alpha$ of the object set M so that for any $r \in \widehat{M}$,

$$q_0(r) = \begin{cases} 1, & \text{if } r \in R \\ 0, & \text{otherwise.} \end{cases}$$

Then, we take $Q = \{q_0\}$. That is, the attribute set is a singleton. Now, each system written in the format of Definition 2.4 is rewritten in the format of Equation 2.14. Here, the attribute q_0 describes if an arbitrarily chosen relation in the Cartesian product $\widehat{M}$ belongs to the system's relation set R or not. In essence, it restates the membership relation to the relation set R from the angle of attributes. In particular, because the relation set R of the system S is a subset of $\widehat{M}$, now the membership in the relation set R is determined by a proposition q_0, while such a description is an attribute of the system S. That is, the general systems theory developed on set theory (Lin 1999) has already implicitly introduced the concept of attributes. What we do here is to make this fact explicit. Also, because the systems we are interested in can have multiple attributes, our contribution to the general systems theory is to make the concept of attributes more general as a set Q of attributes, including more than just the particular attribute q_0.

2.5.2 Subsystems

Just like each set has its own subsets, every system has subsystems, which can be constructed from the object set, relation set, and the attribute set of the system. In short, a system s is a subsystem of the system S, provided that the object set, relation set, and attribute set of s are corresponding subsets of those of S so that the restrictions of the attributes of S on s agree with the attributes of s. Symbolically, we have the following.

Definition 2.10

Let $s_t = (m_t, r_t, q_t)$ and $S_t = (M_t, R_t, Q_t)$ be two systems, for any $t \in T$. If the following hold true:

$$m_t \quad M_t, r_t \quad R_t\big|_{s_t} \text{ and } q_t = Q_t\big|_{s_t}, \forall t \in T \tag{2.26}$$

then s_t is known as a subsystem of S_t, denoted

$$s_t < S_t, \forall t \in T, \text{ or } s < S \tag{2.27}$$

where $R_t\big|_{s_t}$ and $Q_t\big|_{s_t}$ represent, respectively, the restrictions of the relations and attributes in R_t and Q_t on the system S_t and are defined as follows:

$$R_t\big|_{s_t} = \left\{ r\big|_m : r \in R_t \right\} \text{ and } Q_t\big|_{s_t} = \left\{ q\big|_{m_t \cup r_t} : q \in Q_t \right\}$$ ■

In this definition, the notation of "less than" of mathematics is employed for the relationship of subsystems, because the relation of subsystems can be seen as a partial ordering on the collection of all systems. Similarly, when Equation 2.26 does not hold true, we say that s is not a subsystem of S, denoted $s_t < S_t$, $\forall t \in T$, or $s < S$. Let the set of all subsystems of S be $\mathbb{S}$; then we have

$$\mathbb{S} = \{s : s < S\} = \{s_t : s_t < S_t, \forall t \in T\}. \tag{2.28}$$

For any given system $S = (M, R, Q)$, where $M = \{M_t: t \in T\}$, $R = \{R_t: t \in T\}$, and $Q = \{Q_t: t \in T\}$, as defined by Equation 2.14, take a subset set of its object set $A = \{A_t \subseteq M_t: t \in T\}$. Then by restricting the relation set R and the attribute set Q on A, we obtain the following subsystem of S induced by A:

$$S\big|_A = \left\{ s_t = \left(A_t, R_t\big|_{A_t}, Q_t\big|_{\left(A_t, R_t|_{A_t}\right)} \right) : t \in T \right\} \tag{2.29}$$

where the restrictions $R_t\big|_{A_t}$ and $Q_t\big|_{A_t}$ are assumed, respectively, to be

$$R_t\big|_{A_t} = \left\{ r\big|_{A_t} : \text{each element in } r\big|_{A_t} \text{ has the same length, } r \in R_t \right\} \tag{2.29a}$$

and

$$Q_t\Big|_{\left(A_t, R_t|_{A_t}\right)} = \left\{ q\Big|_{\left(A_t, R_t|_{A_t}\right)} : q\Big|_{\left(A_t, R_t|_{A_t}\right)} \text{ is a well-defined proposition on } s_t,\ q \in Q_t \right\}. \qquad (2.29b)$$

Proposition 2.2

The induced subsystem $S|_A$ is the maximum subsystem induced by A.

Proof

Let $s = \{(A_t, r_{t,A}, Q_{t,A}): t \in T\} < S$ be an arbitrary subsystem with the entire A as its object set. According to Equation 2.26, we have $r_{t,A} \quad R_t|_s$, and $q_{t,A} = Q_t|_s = Q_t|_{(A_t, r_{t,A})}$, $\forall t \in T$.
So, from Equation 2.29a, it follows that $\left(Q_t|_{\left(A_t, R_t|_{A_t}\right)} \right)\Big|_{(A_t, r_{t,A})} = q_{t,A}$, which leads to $s < S|_A$. ■

Proposition 2.3

Assume that A and B are subsets of M satisfying that $B \subseteq A \subseteq M$; then $S|_B \le S|_A$.

Proof

According to Equation 2.29, we have $S|_B = \left\{ S_t = \left(B_t, R_t|_{B_t}, Q_t|_{\left(B_t, R_t|_{B_t}\right)} \right) : t \in T \right\}$. $B \subseteq A$ implies that $B_t \subseteq A_t$, $\forall t \in T$. Thus, it follows that $R_t|_{B_t} \quad R_t|_{A_t}$ and, consequently, $R_t|_{B_t} \quad \left(R_t|_{A_t} \right)\Big|_{B_t}$ and $Q_t|_{\left(B_t, R_t|_{B_t}\right)} = \left(Q_t|_{\left(A_t, R_t|_{A_t}\right)} \right)\Big|_{\left(B_t, R_t|_{B_t}\right)}$. Therefore, $S|_B \le S|_A$. ■

Proposition 2.4

Let S be a system. Then the collection of all subsystems of S forms a partially ordered set by the subsystem relation "$<$."

Proof

This result is a straightforward consequence of Proposition 2.3. ■

Assume that S^1, S^2, and S are systems with the same time span such that $S^1 < S$ and $S^2 < S$. Let $\mathbb{S}^i, i = 1, 2$, denote the set of all subsystems of S^i. Then each element in the set $\mathbb{S}^1 - \mathbb{S}^2 = \{s : s < S^1, s \not< S^2\}$ is a subsystem of S^1 but not a subsystem of S^2. Similarly, each element in the set $\mathbb{S}^2 - \mathbb{S}^1 = \{s : s < S^2, s \not< S^1\}$ is a subsystem of S^2 but not a subsystem of S^1. Let $\mathbb{S}^1 \Delta \mathbb{S}^2 = \{s : s < S^1, s \not< S^2\} \cup \{s : s < S^2, s \not< S^1\}$ be the union of the previous two sets of subsystems of either S^1 or S^2, and $\mathbb{S}^1 \cap \mathbb{S}^2 = \{s : s < S^1, s < S^2\}$ the set of all subsystems of both S^1 and S^2.

Proposition 2.5

Assume that $S^1 < S$ and $S^2 < S$. Then $\mathbb{S}^1 \cup \mathbb{S}^2 = \{s : s < S^1\} \cup \{s : s < S^2\}$ is a subset of $\mathbb{S}|_{M^1 \cup M^2}$.

Proof

$S|_{M^1 \cup M^2}$ is a maximal element in the partially ordered set $\left(\mathbb{S}|_{M^1 \cup M^2}, \right)$, satisfying $\forall s \in S|_{M^1 \cup M^2}$, $s < S|_{M^1 \cup M^2}$. On the contrary, $\forall s < S|_{M^1 \cup M^2}$, we have $s \in \mathbb{S}|_{M^1 \cup M^2}$. Thus, $\forall s \in \mathbb{S}^1 \cup \mathbb{S}^2$, we have $s \in \mathbb{S}|_{M^1 \cup M^2}$. ■

What this result indicates is that the union $\mathbb{S}^1 \cup \mathbb{S}^2$ of the sets of subsystems of two subsystems S^1 and S^2 is a subset of the $\mathbb{S}|_{M^1 \cup M^2}$ of the subsystems of the induced system on the union $M^1 \cup M^2$.

Proposition 2.6

Given two arbitrary systems S^1 and S^2, there is always a system S^{12} such that $S^1 < S^{12}$ and $S^2 < S^{12}$.

Proof

Without loss of generality, assume that $S^1 = \left\{\left(M_t^1, R_t^1, Q_t^1\right) : t \in T^1\right\}$ and $S^2 = \left\{\left(M_t^2, R_t^2, Q_t^2\right) : t \in T^2\right\}$. To construct the system $S^{12} = \left\{\left(M_t^{12}, R_t^{12}, Q_t^{12}\right) : t \in T^{12}\right\}$, $T^{12} = T^1 \cup T^2$, we first assume that the object sets M_t^1 and M_t^2 are disjoint, that is, $M_t^1 \cap M_t^2 = \varnothing$, for any $t \in T^1 \cap T^2$. Then, the desired system S^{12} is defined as follows: for $t \in T^1 \cap T^2$,

$$\begin{cases} M_t^{12} = M_t^1 \cup M_t^2 \\ R_t^{12} = R_t^1 \cup R_t^2 \\ Q_t^{12} = Q_t^1 \cup Q_t^2 \end{cases} \tag{2.30}$$

for $t \in T^1 - T^2$, define $M_t^{12} = M_t^1$, $R_t^{12} = R_t^1$, and $Q_t^{12} = Q_t^1$, and for $t \in T^2 - T^1$, define $M_t^{12} = M_t^2$, $R_t^{12} = R_t^2$, and $Q_t^{12} = Q_t^2$, and is denoted as $S^{12} = S^1 \oplus S^2$. Now, if $M_t^1 \cap M_t^2 \neq \varnothing$, for some $t \in T^1 \cap T^2$, we simply take two systems $*S^1 = \left\{\left(*M_t^1, *R_t^1, *Q_t^1\right) : t \in T^1\right\}$ and $*S^2 = \left\{\left(*M_t^2, *R_t^2, *Q_t^2\right) : t \in T^2\right\}$ with $M_t^1 \cap M_t^2 = \varnothing$, for any $t \in T^1 \cap T^2$, such that $^*S^i$ is similar to S^i, $i = 1, 2$, where similar systems are defined in the same fashion as in the work of Lin (1999, p. 201). Then, we define $S^{12} = {}^*S^1 \oplus {}^*S^2$. Up to a similarity, the system S^{12} is uniquely defined. Therefore, it can be seen as well constructed such that $S^1 < S^{12}$ and $S^2 < S^{12}$, where the time spans T^1 and T^2 are seen as the same as T^{12} such that when $t \in T^1 - T^2$ (respectively, $t \in T^2 - T^1$), we treat S_t^2 (respectively, S_t^1) as a system with an empty object set. ■

From this proposition, it follows that when the interactions of some given systems are considered, these systems can always be seen as subsystems of a larger system. On the other hand, this proposition also shows that there is always some kind of interaction between two given systems, which is embodied in the fact that they are subsystems of a certain system.

2.5.3 Interactions between Systems

When there is an interaction between two objects of a system $S = (M, R, Q)$ (of time span T), where $M = \{M_t : t \in T\}$, $R = \{R_t : t \in T\}$, and $Q = \{Q_t : t \in T\}$, it can be described by using a binary relation of the system. We say that objects m_1 and $m_2 \in M$, which means either $m_i = (m_t)_{t \in T}$ such that $m_t \in M_t$, for each $t \in T$, or $m_i = m_t \in M_t$, for a particular $t \in T$, $i = 1, 2$, interact with respect to an attribute $q \in Q$, provided that the proposition q holds true for a binary relation $r \in R$ that contains either (m_1, m_2) or (m_2, m_1) or both. The idea of interactions between systems is a natural generalization of that between two objects. Based on the discussion of the previous section, we will discuss interactions of systems in the framework of subsystems. Assume that $S_i < S$, $i = 1, 2$. Now, let us look at how these subsystems could interact with each other.

Definition 2.11

The system S_1 is said to have a weak effect on the system S_2 with respect to an attribute $q \in Q$ provided that for any $r_2 \in R_2$, there is $r_1 \in R_1$ such that the proposition q holds true for the ordered pair (r_1, r_2). When no confusion is caused, we simply say that the system S_1 affects the system S_2 weakly without mentioning q. If the system S_2 also exerts a weak effect on S_1, then we say that these systems interact with each other weakly. ■

Definition 2.12

The system S_1 is said to have a strong effect on the system S_2 with respect to an attribute $q \in Q$, provided that for any $r_2 \in R_2$ and any $r_1 \in R_1$, the proposition q holds true for the ordered pair (r_1, r_2). When no confusion is caused, we simply say that the system S_1 affects the system S_2 strongly without mentioning q. If the system S_2 also exerts a strong effect on S_1, then we say that these systems interact with each other strongly. ■

From these definitions, it follows that strong interaction requires interactions between every ordered pair of objects, which is a more rigorous requirement than that of weak interactions. Also, to maintain the intuition behind the concepts of interactions, in Definitions 2.11 and 2.12, we look at only two relations, $r_i \in R_i$, $i = 1, 2$. In order to capture the general spirit, these individual relations should be replaced by subsets $\{r_i \in R_i : \Phi_i(r_i)\}$, where $\Phi_i()$ stands for the proposition that defines the set, for $i = 1, 2$. By doing so, what are discussed in Definitions 2.11 and 2.12 become special cases.

Definition 2.13

Given a subsystem $s = (m, r, q)$ of a system $S = (M, R, Q)$, the totality of all objects in $M - m$, each of which interacts weakly with at least one object in m, is known as the environment of the subsystem s in S, denoted E^S. ■

2.5.4 Systems Properties Based on Dynamic Set Theory

2.5.4.1 Basic Properties

For two given systems $S^1 = \left\{S_t^1 = \left(M_t^1, R_t^1, Q_t^1\right) : t \in T^1\right\}$ and $S^2 = \left\{S_t^2 = \left(M_t^2, R_t^2, Q_t^2\right) : t \in T^2\right\}$, let us consider the following.

Definition 2.14

These systems S^1 and S^2 are equal, provided that

$$M_t^1 = M_t^2, R_t^1 = R_t^2, Q_t^1 = Q_t^2 \text{ and } T^1 = T^2 \tag{2.31}$$

fo reach $t \in T^1 = T^2$.

The systems S^1 and S^2 are said to be identical on the time period $T \subseteq T^1 \cap T^2$, which means that

$$M_t^1 = M_t^2, R_t^1 = R_t^2, Q_t^1 = Q_t^2 \text{ for each } t \in T. \tag{2.32}$$

■

Definition 2.15

The system S^1 is said to be homomorphically embeddable into the system S^2 provided that there is a non-decreasing mapping f: $T^1 \to T^2$ such that for any $t_1 \in T^1$, if $t_2 = f(t_1) \in T^2$, then $S_{t_1}^1 = S_{t_2}^2$, or equivalently

$$M_{t_1}^1 = M_{t_2}^2, R_{t_1}^1 = R_{t_2}^2, \text{ and } Q_{t_1}^1 = Q_{t_2}^2. \tag{2.33}$$

The mapping f is referred to as an embedding mapping from S^1 into S^2. If the system S^1 can be homomorphically embeddable into S^2 and S^2 into S^1, then the systems S^1 and S^2 are said to be homophorhically equivalent. Evidently, equal systems are homomorphically equivalent with the identity mapping on the time set as the canonical embedding mapping. ■

Proposition 2.7

If the embedding mapping f: $T^1 \to T^2$ from the system S^1 into S^2 is bijective, then the systems are homomorphically equivalent.

Proof

It suffices to show that the inverse mapping f^{-1}: $T^2 \to T^1$ is an embedding mapping from S^2 into S^1.

Because f is bijective, it is strictly increasing from T^1 into T^2; its inverse f^{-1} is also a strictly increasing mapping from T^2 into T^1, satisfying that for any $t_2 \in T^2$, if $t_1 = f^{-1}(t_2) \in T^1$, then $M_{t_2}^2 = M_{t_1}^1, R_{t_2}^2 = R_{t_1}^1$, and $Q_{t_2}^2 = Q_{t_1}^1$. Therefore, f^{-1}: $T^2 \to T^1$ is an embedding mapping from S^2 into S^1. ■

Definition 2.16

A system $S = \{S_t = (M_t, R_t, Q_t): t \in T\}$ is said to be cyclic or periodic provided that there is time $t_c > 0$ such that $S_{t+t_c} = S_t$, for any $t \in T$. Evidently, in this case, for any natural number $n \in N$, nt_c is also a period of the system S. The minimum period is named the period of S, denoted T_c. ■

2.5.4.2 Systemic Emergence

Systemic emergence means the properties of the whole system that parts of the system do not have. Such a property might suddenly appear at a particular time moment. In terms of the system, the holistic emergence is mainly created and excited by the system's specific organization of its parts and how these parts interact, supplement, and constrain on each other. It is a kind of effect of relevance, the organizational effect, and the structural effect.

To be specific, let P represent such a property. It is defined on the entire system S. Because of the system's dynamic characteristics, the interactions between the system's objects, relations, attributes, and the environment change constantly. Thus, the value of P also varies accordingly. Define

$$P(S_t) = \begin{cases} 1, & \text{if } S \text{ has this property} \\ 0, & \text{otherwise} \end{cases}, \ \forall t \in T. \tag{2.34}$$

It satisfies the following properties:

$$P(s_t) = 1 \rightarrow P(S_t) = 1, \forall s < S, \forall t \in T \tag{2.35}$$

That is, as long as a subsystem s has this property, the overall system S also has the property. That is another way to say that the whole is greater than the sum of parts. Of course, there are also such properties that some subsystems have that the overall system does not have. For instance, let P be a property the system S does not have. Define $P^{-1}(1) = \{s < S: s \neq S\}$ be the collection of all proper subsystems of S, where the superscript (–1) stands for the inverse operation. Then, for any $s \in P^{-1}(1)$, $s < S$ and $P(s) = 1$. However, $P(S) = 0$ does not satisfy Equation 2.35.

When a property P is said to be system S's holistic emergence, provided that for any subsystem $s < S$, $P(s_t) = 0$, $\forall t \in T$, and $P(S_{t_0}) = 1$ for at least one $t_0 \in T$.

In the traditional static description of systems, there is another definition of systemic emergence. In particular, consider a monotonically increasing sequence of subsystems $\varnothing \neq s^1 < s^2 < \ldots < s^k < \ldots < S$; there is a k_0 such that

$$P(s^k) = 0, \ \forall k < k_0, \ P(s^{k_0}) = 1. \tag{2.36}$$

In this case, the system S is said to have the emergent property P, while its parts s^k, $k < k_0$, do not share this property until they reach the whole s^{k_0} of certain scale. From a detailed analysis, it follows readily that what was just presented is a special case of the systemic emergence of dynamic systems, where we can surely treat the collection of all the superscript k as a subset of the time index T so that the sequence $\{s^k: k = 1, 2, 3,\ldots\}$ of subsystems is a subsystem of a dynamic system by letting $S_t = s^t$, for $t = 1, 2, 3, \ldots$ From Equation 2.36, it follows that

$$P(S_t) = 0, \ \text{for } t = 1, 2, 3, \ldots < k_0, \text{ and } P(S_{k_0}) = 1.$$

That is, the property P emerges at the time moment k_0.

2.5.4.3 Stability of Systems

Stability means the maintenance or continuity of a certain measure of the dynamic system on a certain time scale. That is, under small disturbances, the measure does not undergo noticeable changes. Let us look at the trajectory system of a single point, where the focus is how the point moves under the influence of an external force. If a disturbance is given to a portion of the trajectory that is not at a threshold point, then the disturbed trajectory will not differ from the original trajectory much. However, if the same disturbance is given to the trajectory at an extremely unstable extreme point, then a minor change in the disturbed value could cause major deviations in the following portion of the trajectory. Thus, only the trajectory of motion at stable critical extrema is stable, while the trajectory systems with unstable critical points, such as a saddle point, are unstable.

In general, assume that an attribute $q \in Q$ of the system S satisfies that q is a real-valued function defined for each momentary system S_t, for any $t \in T$. Let $t_0 \in T$. If for any $\varepsilon > 0$, there is a $\delta_{t_0,\varepsilon} > 0$ such that

$$\left|q\left(S_{t_0}\right)-q\left(S_t\right)\right|<\varepsilon,\ \forall t \in\left(t_0-\delta_{t_0,\varepsilon}, t_0+\delta_{t_0,\varepsilon}\right), \tag{2.37}$$

then the attribute q of the system S is regionally stable over time at $t_0 \in T$.

If for any $\varepsilon > 0$, there is $\delta = \delta(\varepsilon) = \delta_\varepsilon > 0$ such that

$$\left|q\left(S_{t_1}\right)-q\left(S_{t_2}\right)\right|<\varepsilon,\ \forall t_1,\ t_2 \in T \text{ such that } \left|t_1-t_2\right|<\delta_\varepsilon \tag{2.38}$$

then the attribute q is said to be holistically stable over time or uniformly stable over time. When no confusion can be caused, the previous concepts of stability of the attribute q are respectively referred to as that the system S is regionally stable at $t_0 \in T$ or uniformly stable over T.

In addition, the structural stability of systems can also be defined. In particular, Let $d \in Q$ be such that d: $\mathbb{S}_t \to R^+$ is a positively real-valued function defined for each subsystem of the momentary system S_t, $t \in T$, satisfying

1. $d(\varnothing) = 0$, where $\varnothing$ stands for the subsystems with the empty set as their object set
2. $\forall S^1, S^2 < S$, if $S^1 < S^2$, then $d(S^1) \leq d(S^2)$
3. $\forall S^1, S^2 < S$, if $S^1 \Delta S^2 = \varnothing$, then $d(S^1 \cup S^2) = d(S^1) + d(S^2)$.

It can be readily seen that such an attribute d can be employed to measure the difference between subsystems of S; for any chosen $S^1 < S$, and for any $S^2 \in$ S, the greater the attribute value $d(S^1 \Delta S^2)$ is, the more different the systems S^1 and S^2 are. By making use of such a $d \in Q$, which satisfies the previous properties, for any chosen $s \in$ S and any real number $\delta > 0$, we can define the δ-neighborhood $Nbrd_d(s,\delta)$ of s as follows:

$$Nbrd_d(s,\delta)=\left\{s' \in \mathbb{S}: d\left(s' \Delta s\right)<\delta\right\}. \tag{2.39}$$

The attribute $q \in Q$ of the system S is said to be structurally stable in the neighborhood of a subsystem $s < S$ with respect to attribute $d \in Q$, provided that for any $\varepsilon > 0$, there is $\delta = \delta_\varepsilon > 0$ such that

$$|q(s) - q(s')| < \varepsilon,\ \forall s' \in Nbrd_d(s,\delta). \tag{2.40}$$

Similarly, the concepts of regionally structural stability and uniformly structural stability of a system S at all of its subsystems can be defined.

If a system S is both uniformly stable over time and uniformly structurally stable, then the system is referred to as a uniformly stable system.

2.5.4.4 Evolution of Systems

As far as systems' evolution is concerned, the focus is how the system develops over time. Speaking rigorously, each system can be described by using the dynamic format in Equation 2.14. As for those systems, which do not seem to change with time, when seen from the angle of dynamics, they are either evolving with time extremely slowly or considered within a very short period of time. That is how they project a mistakenly incorrect sense of being static.

The evolution of systems takes two main forms: One is the transition of the system from one structure or form to another structure or form, and the other is that a system appears from its earlier state of nonexistence and grows from an immature state to a more mature state. Both of these forms of evolution can be described and illustrated by using the language of the dynamic systems in Equation 2.14. Assume that the system of our concern is indeed described in the form of Equation 2.14. Then for the first form of evolution, the relation set R_t and the attribute set Q_t change with time, while for the second form of evolution, the elements in the object set M_t develop over time together of course with changes of the relation and attributes of sets R_t and Q_t. They represent two specific cases of the evolution of dynamic systems.

2.5.4.5 Boundary of Systems

Boundary stands for the separation between a system and its environment. The boundary is a part of the system and interacts with the environment more closely than the interior of the system. By reviewing the definition of environments in Definition 2.13, we can define the boundary of a system as follows. Assume that $s < S$ is a subsystem of the system S over the time span T. Let us denote the subsystem s and the system S respectively as $s_t = (m_t^s, r_t^s, q_t^s)$ and $S = (M, R, Q)$, for $t \in T$. For the sake of convenience of communication, we write $s = (m_s, r_s, q_s)$ for a fixed time moment. Then, the (external) environment E^s of s within the system S is defined to be

$$E^s = \left\{ \left(M - m_s, r|_{m^E}, q|_{m_s^E, r^E} \right) : r \in R\left(r|_{m_s} \quad r_s \right), \right.$$
$$\left. q \in Q\left(q|_{m_s} \quad q_s \right) \right\} = \left(m^E, r^E, q^E \right)$$

where $m^E - M - m_s, r^E = r|_{m^E}$, and $q^E = q|_{m_s^E, r^E}$, for $r \in R\left(r|_{m^E} \quad r_s \right)$ and $q \in Q\left(q|_{m^E} \quad q_s \right)$;

Assume that there is an attribute $\mu \in Q_t$ of the system S such that it assigns each relation to a positive real number $\mu(r): R_t \rightarrow R^+$. Intuitively, this attribute μ is an index that measures the intensity of each relation of the objects of S. Now, let us fix a threshold value μ_0 for the relational intensity; then by combining with the concept of weak interactions between systems (Definition 2.11), we can obtain the set of all relations in S of intensity at least μ_0 that relate the subsystem s and its environment E^s as follows:

$$r_b = \left\{ r \in R : r|_{m_s} \in r_s, \mathrm{Supp}(r) \cap m_s \neq \varnothing \neq \mathrm{Supp}(r) \cap m^E, \mu(r) \geq \mu_0 \right\} \quad (2.41)$$

where Supp(r) stands for the support of the relation r, which is the set of all objects that appear in the relation r. The set of all objects of s that interact with the environment E^s of intensity of at least μ_0 is given by the following:

$$m_b = \bigcup_{r \in r_b} \mathrm{Supp}(r) - m^E \quad (2.42)$$

Then the system

$$\partial s = \left(m_b, r_b|_{m_b}, q_b|_{m_b, r_b|_{m_b}} \right) \quad (2.43)$$

satisfies that $\partial s < s$ and is referred to as the boundary (system) of s within the system S. The intensity of its interaction with the environment is no less than μ_0, while any other subsystem of s interacts with the environment E^s with strictly less intensity than μ_0.

If in the evolution of the system, ∂s has good stability, then we say that the boundary of the system is clear. Otherwise, we say that the boundary of the system is fuzzy. Symbolically, let $q \in Q$ be an attribute and $t_0 \in T$ chosen. If for any $\varepsilon > 0$, there is $\delta = \delta_{t_0,\varepsilon} > 0$ such that

$$\left|q\left(\partial s_{t_0}\right) - q\left(\partial s_t\right)\right| < \varepsilon,\ \forall t \in \left(t_0 - \delta_{t_0,\varepsilon}, t_0 + \delta_{t_0,\varepsilon}\right) \tag{2.44}$$

then we say that the boundary of the system s at time $t_0 \in T$ is definite. Otherwise, the boundary is said to be fuzzy at $t_0 \in T$.

It is not hard to see that the definiteness of a system's boundary is defined by using the stability of the boundary system ∂s. Therefore, the stability of the boundary system at one time moment corresponds to the definiteness of the system's boundary. Similarly, we can study the concepts of regional definiteness of boundaries over time and structural definiteness of boundaries.

2.6 OPEN-ENDED PROBLEMS

1. When are systems seen as one of the forms of existence of all matters and things? How can one understand systems and nonsystems?
2. In your own language, illustrate the wholeness emergence principle using real-life scenarios. Please check the existing literature and think about how to comprehend the microcosmic reductionism.
3. It is defined in set theory that an ordered pair $(x,y) = \{\{x\}, \{x,y\}\}$. Show that for two given systems $S_1 = (M_1,R_1)$ and $= (M_2,R_2)$, $S_1 = S_2$ if and only if $M_1 = M_2$ and $R_1 = R_2$.
4. Write out a proof for Proposition 2.1.
5. Given a system $S = (M,R)$ and a subset $M^* \subseteq M$, define the star neighborhood of the subset M^*, denoted Star(M^*), as follows:

   $$\text{Star}(M^*) = \bigcup \{\text{Supp}(r): r \in R \text{ and } \text{Supp}(r) \cap M^* \neq \varnothing\}.$$

   By applying mathematical induction on natural number n, we can define

   $$\text{Star}^1(M^*) = \text{Star}(M^*) \text{ and } \text{Star}^{n+1}(M^*) = \text{Star}(\text{Star}^n(M^*)).$$

   Then prove that for every system $S = (M,R)$, the following conditions are equivalent: (i) The system S is connected, and (ii) for each object $x \in M$, $\text{Star}^\infty(\{x\}) = \bigcup_{n=1}^{\infty} \text{Star}^n(\{x\}) = M$.
6. A well-known Song dynasty poem reads as follows: "If you say the music is from the instrument, how does the instrument not sound when it is packed in a box? If you say the music is from the fingers of the musician, why don't you just listen to the musician's fingers?" Please explain the effect of holistic emergence by using this poem.
7. Please analyze systems emergence from the angle of both collaboration and competition.
8. Please list and analyze examples of systems emergence from the angle of dynamic equilibrium.
9. The classical set-theoretic definition of systems $S = (M,R)$ does not explicitly consider the expressions of time and systems' emergence. Please compare what are presented in Sections 2.3 and 2.5 and gain a deeper understanding of how abstract concepts are generalized.
10. Conduct an analysis of general systems using input–output dynamic processes.

11. Explore in your own words the separation and connection between a system's structure and its environment.
12. Emergence could be potentially caused by changes first in the environment and then in the correlations between the parts of the system. Try to analyze this possibility using water and its changes between different forms of appearance.
13. Use examples of nonliving systems to illustrate the emerging effect of scales, say, the threshold effect of self-organization of sand dunes.
14. Please conduct a literature search and look for the current interests of research in systems science.
15. How would you define a system's boundary that separates the system from its environment?
16. Try to gain a deeper understanding about the concept of systems and about the characteristics of systems. Then see what methods have been employed to investigate systems. Can you find a better method for analyzing systems?

3 Nonlinear Systems Dynamics and Catastrophe

3.1 MATHEMATICAL DESCRIPTION OF SYSTEMS DYNAMICS

3.1.1 Linear Systems and Nonlinear Systems

Systems are classified as linear and nonlinear by the characteristics of systems in terms of the available mathematical tools. This classification is applicable to dynamical and static systems. The theory of linear systems is well studied, while that of the nonlinear systems is more widely applicable in systems science. It is because when studying practical systems, one is often faced with nonlinear mathematical models.

First, a linear system is defined as follows.

Definition 3.1

Let A be a field, X and Y linear spaces over A, and S an input–output system such that

(1) $\varnothing \neq S \subset X \times Y$, where X is known as the input space and Y the output space
(2) $s \in S$ and $s' \in S$ imply $s + s' \in S$
(3) $s \in S$ and $\alpha \in A$ imply $\alpha \cdot s \in S$

Here the symbols + and · stand for the addition and scalar multiplication in the product space $X \times Y$, respectively, and are defined as follows: for any $(x_1, y_1), (x_2, y_2) \in X \times Y$ and any $\alpha \in A$,

$$(x_1, y_1) + (x_2, y_2) = (x_1 + x_2, y_1 + y_2) \tag{3.1}$$

$$\alpha(x_1, y_1) = (\alpha x_1, \alpha y_1) \tag{3.2}$$

The input–output system S is then called a linear system. ■

Theorem 3.1

Suppose that X and Y are linear spaces over the same field A. Then the following statements are equivalent:

1. $S \subset X \times Y$ is a linear system.
2. There exists a linear space C over A and a linear mapping $\rho: C \times D(S) \to Y$ such that $(x, y) \in S$ if and only if there exists $c \in C$ such that $\rho(c,x) = y$.
3. There exists a linear space C over A and linear mappings $R_1: C \to Y$ and $R_2: D(S) \to Y$ such that the mapping ρ in (2) satisfies that

$$\rho(c, x) = R_1(c) + R_2(x), \text{ for every } (c, x) \in C \times D(S).$$

Here, $D(S) = \{x \in X : \exists y \in Y((x,y) \in S)\}$ is the domain of the input–output system S. ■

The detailed proof can be found in the work of Lin (1999, p. 263).

The concept of linear systems includes those concepts well studied in the names of linear transform, linear operation, linear function, linear functional, linear equation, and so on. The basic distinction between linear systems and nonlinear systems is whether the systems satisfy the additional principle given in Equations 3.1 and 3.2.

Example 3.1

Let us look at two examples and show that the concept of linear systems is really an abstraction of some well-known different mathematical structures.

a. Suppose that S is a system described by the following linear equations:

$$\begin{cases} a_{11}x_1 + a_{12}x_2 + \cdots + a_{1n}x_n = b_1 \\ a_{21}x_1 + a_{22}x_2 + \cdots + a_{2n}x_n = b_2 \\ \vdots \\ a_{m1}x_1 + a_{m2}x_2 + \cdots + a_{mn}x_n = b_m \end{cases}$$

where x_i, $i = 1, 2, \ldots, n$, are n unknown variables, m is the number of equations, a_{ij}, $i = 1, 2, \ldots, m$; $j = 1, 2, \ldots, n$, are the coefficients of the system, and b_j, $j = 1, 2, \ldots, m$, are the constraints of the system. The system S can then be rewritten as a linear system $(M, \{r\})$, where the object set M is the set of all real numbers, and

$$r = \{(x_1, x_2, \ldots, x_n) \in M^n : a_{i1}(x_1 + y_1) + \ldots + a_{in}(x_n + y_n) = b_i, \text{ for } i = 1, \ldots m\}$$

where $(y_1, y_2, \ldots, y_n) \in M^n$ is fixed such that $a_{i1}x_1 + a_{i2}x_2 + \ldots + a_{in}x_n = b_m$ for each $i = 1, 2, \ldots, m$.

b. Suppose that a given system S is described by the differential equation

$$\frac{d^n x}{dt^n} + a_1(t)\frac{d^{n-1}x}{dt^{n-1}} + \cdots + a_{n-1}(t)\frac{dx}{dt} + a_n(t)x = f(t)$$

where $a_i(t)$, $i = 1, 2, \ldots, n$, and $f(t)$ are continuous functions defined on the interval $[a, b]$. The system S can then be rewritten as a linear system $(M, \{r\})$ in the following way: the object set M is the set of all continuous functions defined on $[a, b]$, and the relation r is defined by

$$r = \left\{x \in M : \frac{d^n(x+y)}{dt^n} + a_1(t)\frac{d^{n-1}(x+y)}{dt^{n-1}} + \cdots \right.$$
$$\left. + a_{n-1}(t)\frac{d(x+y)}{dt} + a_n(t)(x+y) = f(t)\right\}$$

where $y \in M$ is a fixed function such that

$$\frac{d^n y}{dt^n} + a_1(t)\frac{d^{n-1}y}{dt^{n-1}} + \cdots + a_{n-1}(t)\frac{dy}{dt} + a_n(t)y = f(t).$$

From the examples above, we can see that linear and differential equations both can be described by using the language of set theory. Additionally, each continuous mapping can also be written in terms of the set theory.

Definition 3.2

Let $S_1 = (M_1, R_1)$ and $S_2 = (M_2, R_2)$ be two systems, f a mapping from M_1 to M_2, written as $S_1 = (M_1, R_1) \to S_2 = (M_2, R_2)$. The mapping f is said to be continuous, short for systemic continuous, from S_1 to S_2, if $\forall r_2 \in R_2, f^{-1}(r_2) \in R_1$, where

$$f^{-1}(r_2) = \left\{ (x_1, \ldots, x_\alpha, \ldots) \in M_1^{n(r_2)} : f(x_1, \ldots, x_\alpha, \ldots) = (f(x_1), \ldots, f(x_\alpha), \ldots) \in r_2 \right\}. \quad \blacksquare$$

A continuous system, which is defined as such that its inputs and outputs are capable of changing continuously with time, can be seen as the images in different time slots by a continuous mapping.

Definition 3.3

A dynamical system is an ordered tuple (T, M, f), where T is a monoid, written additively, M is a set, and f is a function satisfying that

$$f : U \subset T \times M \to M$$

and that

$$I(x) = \{t \in T : (t, x) \in U, f(0, x) = x,$$

$$f(t_2, f(t_1, x)) = f(t_1 + t_2, x), \text{ for } t_1, t_2, t_1 + t_2 \in I(x).$$

The function $f(t,x)$ is referred to as the evolution function of the dynamical system: it associates to every point in the set M a unique image, depending on the variable t, known as the evolution parameter. M is referred to as the phase space or state space, while the variable x is the initial state of the system. $f(t,x){:}I(x) \to M$ is known as a flow through x, and its graph is a trajectory through x. The set $\gamma_x{:} = \{f(t,x){:}\ t \in I(x)\}$ is referred to as the orbit through x. ■

Also, a continuous dynamic system can be modeled by a differential equation, which can describe the dependent relation between different state variables. If the state variables are functions of time, it would be represented by an ordinary differential equation. If the state variables are functions of time and space, it would be represented by a partial differential equation.

When the state variables of a system are written as a vector $X = (x_1, \ldots, x_n)^T$ of differentiable functions $x_1, x_2, \ldots,$ and x_n of time t, then if the linear system is linear, it can be generally written or modeled symbolically as follows:

$$\frac{dX}{dt} = AX \tag{3.3}$$

where A is a constant matrix.

This model may also be employed to describe two kinds of nonlinear systems. One is for systems involving weak nonlinearity; the other stands for linear approximations of continuous nonlinear systems in local areas.

If Equation 3.3 represents the underlying physical system well, the system's characteristics would be described by the structure and parameters of the equation. In this case, the development characteristics of the system would be clearly expressed by the solution of Equation 3.3.

If Equation 3.3 has k real characteristic values, $\lambda_1,\ldots, \lambda_k$ of multiplicities n_i, $i = 1, 2,\ldots,k$, and $2l$ complex characteristic values $\lambda_j = \alpha_j \pm i\beta_j$ of multiplicities m_j, $j = 1,\ldots,l$, respectively, such that $\sum_{i=1}^{k} n_i + 2\sum_{j=1}^{l} m_j = n$, then the solution of Equation 3.3 would be

$$x(t) = \sum_{i=1}^{k} r_{n_i-1}(t)e^{\lambda_i t} + \sum_{j=1}^{l} (p_{m_j-1}^{(t)} \cos\beta_j t + q_{m_j-1}^{(t)} \sin\beta_j t)e^{\alpha_j t}$$

where $r_{n_i-1}, p_{m_{j-1}}, q_{m_{j-1}}$ are polynomials in t of the largest orders $n_i - 1$, $m_j - 1$, and $m_j - 1$, respectively. For example,

$$r_{n_i-1} = C_{n,r}^{(0)} + C_{n,r}^{(1)} t + \cdots + C_{n,r}^{(n_i-1)} t^{n_i-1}$$

where $C_{n,r}^{(s)}$ is an n-dimensional vector, $s = 0, 1, \ldots, n_i - 1$. These coefficients can be determined by the given initial conditions, and the parameters $r_{n_i-1}, p_{m_{j-1}}$, and $q_{m_{j-1}}$ represent the dynamical characteristics of the underlying linear system.

If an input–output system S does not satisfy conditions 3.1 and 3.2 in Definition 3.1, it is then referred to as a nonlinear system.

The general nonlinear system is much more complex than any linear system. It could embody the diversity, variation, and complexity of the realistic, physical systems of concern.

The general form of the dynamic equation of a nonlinear system is

$$\frac{dX}{dt} = F(X,C), \tag{3.4}$$

where $X = (x_1,\ldots,x_n)^T$ is the state variable vector and $C = (c_1,\ldots,c_m)^T$ is a vector of control parameters.

The available methods for solving the dynamic system 3.4 include (Xu 2000) the following:

1. *Analytical methods:* The equation can be solved analytically when the function F has an appropriate form.
2. *Geometrical methods:* Generally, geometric analyses can be employed to extract useful qualitative information from equations and parameters without the need to solve the equations. Sometimes, this approach is simple and very effective. For instance, Poincaré invented a theory that can be utilized to qualitatively analyze differential equations. This theory provides a good mathematical tool for conducting system analysis.

3. *Numerical computation:* Numerical analysis could produce approximate solutions. This approach is important, when the symbolic model cannot be analytically solved, for analyzing the system's characteristics and producing qualitative and quantitative result.
4. *Method of linearization:* The local behaviors and characteristics of a nonlinear system can be generally investigated by using the method of linearization, if the system satisfies the conditions of continuity and differentiability at the locality of concern. In this case, at the said location, the nonlinear system can be approximated by a linear system. This method is known as a local linearization of the nonlinear system and has been widely applied. Additionally, it should be noted that sometimes, appropriate nonlinear terms could be added as modifications to the original model. This method is referred to as "linearization and small perturbations." It is also a main approach used to study nonlinear systems.

However, the methods of linearization and small perturbations belong to the theory of linear systems; they cannot be employed to deal with nonlinear systems involving discontinuity, and they are applicable only to analyzing local characteristics of the underlying system. As a matter of fact, nonlinearity is the origin of diversity, singularity, and complexity of systems. Thus, it takes novel methods for one to effectively research a nonlinear system; for details, please consult with the work of Lin and OuYang (2010).

In short, even though the applicability of the theory of linear systems is limited, this theory and relevant methods are still important in the development of systems science. It is because they are relatively mature and form a basis on which one investigates nonlinear systems. To a degree, one can understand nonlinearity better when he compares nonlinear systems with linear ones.

3.1.2 Kinetic Mechanism of Dynamical Systems

When the symbolic model describes the underlying system, each of its solutions represents a behavioral process of a system. If the solution is plotted in the phase space, namely, a state variable space, the set of all the plotted points is referred to as an orbit of the system. The movement of a point along the orbit as time goes by exhibits the evolvement of the system. It embodies the dependence relationship of the different state variables. Each point of the phase space would have at least one orbit.

On the other hand, the parameter space of the control parameter is known as a control space. When the control parameter is given, the state space can then be constructed. Any change in the control parameter may deduce a quantitative or qualitative change in the system's characteristics. The control space focuses on the family of systems with the same evolvement or evolution structure.

The states of a dynamical system may be classified into two kinds: the *transient state*, which can be reached by the system at certain time moments, but cannot be held on to or returned to by the system without the act of some external forces; and the *steady state*, which the system can hold on to or return to without any act of external forces. The steady state includes the equilibrium states, such as such fixed points (e.g., centers, nodes, foci, and saddle points), periodical states, pseudo-periodical states, chaos states, and so on.

The qualitative characteristics of the system are determined by the steady states. Each change from one steady state to another embodies a change of the system's qualitative characteristics. Steady states are closely related to the system's dimensionality, which is an important parameter that determines the system's characteristics.

3.1.3 Stability and Instability

Each system could be disturbed by either the environment or its own internal evolution, leading to changes in its structure, state, and behaviors. The magnitude of change reflects the physical maintenance of the system.

Definition 3.4

Stability of the system is an attribute of the system with which the system's structure, state, and pattern of behaviors are maintained with relative stability when the system is disturbed by either the environment or internal changes of the system itself. ■

If one state of the system is not stable, it can either be difficult to reach or occur only at one time moment in the dynamic process of the system's evolvement. Without any steady state, the system cannot operate normally and cannot materialize its functionality. On the other hand, from the viewpoint of evolvement, if a system were maintained stably throughout its entire existence, there would be no change, development, and innovation occurring in the system. Therefore, only when the original mode of the state, structure, and pattern of behaviors loses its stability will there be an emergence of new wholeness for the system.

Stability is a precondition of development. If a new state or structure is not stable, it would not replace the old state and structure. Stability is the ultimate problem in the theory of dynamical systems. The stability of each dynamical system always focuses on the concepts of states and orbit that are made up of states. These concepts can express indirectly the way the components of the system are related to each other and whether the overall pattern of behaviors is stable.

Generally, the concept of stability is relative. As for the stability of a system, it means the stability of the solution of the system's evolvement equation, because each solution of the evolvement equation represents an evolution process of the system. There are many different ways to define the concept of stability. In the following, we will look at only two of them (Xu 2000).

Definition 3.5

(Lyapunov stability). Let $\Phi(t)$ be a chosen solution of the vector differential equation $\frac{dX}{dt} = F(X,C)$ of the underlying system of concern, and $X(t)$ the solution derived from the initial value $X_0 = X(t_0)$. The solution $\Phi(t)$ is said to be Lyapunov stable if for any $\varepsilon > 0$ (small enough), there exists $\delta = \delta(\varepsilon) > 0$, so that if at $t = t_0$,

$$|X_0 - \Phi(t_0)| < \delta$$

holds true, then

$$|X(t) - \Phi(t)| < \varepsilon$$

holds for all $t \geq t_0$. ■

Definition 3.6

(Lyapunov asymptotical stability). Let $\Phi(t)$ be a chosen solution of the vector differential equation $\frac{dX}{dt} = F(X,C)$ of the underlying system of concern, and $X(t)$ the solution derived from the initial value $X_0 = X(t_0)$. If $\Phi(t)$ is Lyapunov stable and satisfies $\lim_{t\to\infty}|X(t) - \Phi(t)| = 0$, then $\Phi(t)$ is known as being Lyapunov asymptotically stable. ■

Remark 1

As for the general solution stability of the nonlinear system 3.4, it could be transformed into the zero solution stability. In particular, if we assume $X(t)$ and $\Phi(t)$ the same as in Definition 3.6 above, then by letting $Y(t) = X(t) - \Phi(t)$, we can then transform Equation 3.4 into the following:

$$\frac{dY(t)}{dt} = G(X,C) \tag{3.5}$$

where $G(X,C) = F(X,C) - F(\Phi,C) = F(Y + \Phi,C) - F(\Phi,C)$, with $G(0,C) = 0$. Thus, the arbitrary solution $X(t) = \Phi(t)$ in Equation 3.4 corresponds to the particular solution of $Y = 0$ of Equation 3.5. ■

Remark 2

The concept of Lyapunov asymptotical stability is stronger than that of Lyapunov stability. If a solution is Lyapunov stable, it requires only that the solution's deviation is sufficiently small, when the initial disturbance is small enough, without the need for them to approach the same value as the time approaches infinity. However, if the solution is Lyapunov asymptotically stable, it must satisfy the condition that the solution's deviation would disappear when the system evolves into its final state. Figure 3.1 illustrates the Lyapunov stability of a solution. ■

Example 3.2

Given the system

$$\begin{cases} \dfrac{dx}{dt} = ax - y^2 \\ \dfrac{dy}{dt} = 2x^3 y \end{cases}$$

where a is a parameter, prove that (1) when $a < 0$, the solution zero is Lyapunov asymptotically stable; (2) when $a = 0$, the solution zero is Lyapunov stable but not Lyapunov asymptotically stable; and (3) when $a > 0$, the solution zero is not stable.

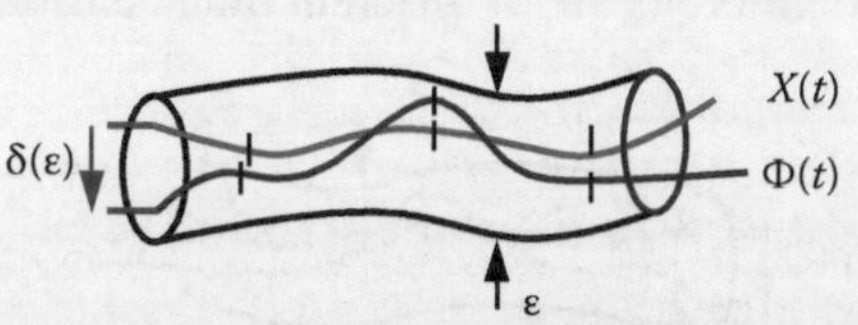

FIGURE 3.1 Illustration of the Lyapunov stability of a solution.

Proof

Construct a potential function $V(x, y) = x^4 + y^2$, which is positive definite. Thus, we have

$$\dot{V} = 4ax^4 \begin{cases} > 0, a > 0 \\ = 0, a = 0 \\ < 0, a < 0 \end{cases}. \quad (3.6)$$

Now, the needed conclusions can be readily derived from Equation 3.6. ■

Lyapunov stability has a very strict requirement for two solutions: they must be close to each other at any time moment. In order to loosen the restriction, let us look at a different kind of stability.

Definition 3.7

(Poincaré stability or orbital stability). Let Γ be the phase orbit of the solution $\Phi(t)$ of Equation 3.4, and Γ' the orbit of another solution $X(t)$ defined for all t. Γ is said to be Poincaré stable or orbital stable if for any $\varepsilon > 0$ (small enough), there exists $\delta(\varepsilon) > 0$ so that

$$|\Phi(t) - X(\tau)| < \delta(\varepsilon)$$

for some τ, then there exists t' so that $|\Phi(t) - X(t')| < \varepsilon$ holds for any $t > 0$. ■

Definition 3.8

(Poincaré asymptotical stability or orbital asymptotical stability). Assume the same as in Definition 3.7. The solution $\Phi(t)$ is said to be Poincaré asymptotically stable (or orbital asymptotically stable) if $\Gamma' \to \Gamma$ when $t \to \infty$. ■

Figure 3.2 illustrates the concept of orbital stability of the solution $\Phi(t)$.

Remark 3

The meaning of orbital stability is that two solutions share the same history but are given on different time scales. Additionally, their orbits are of different time densities. ■

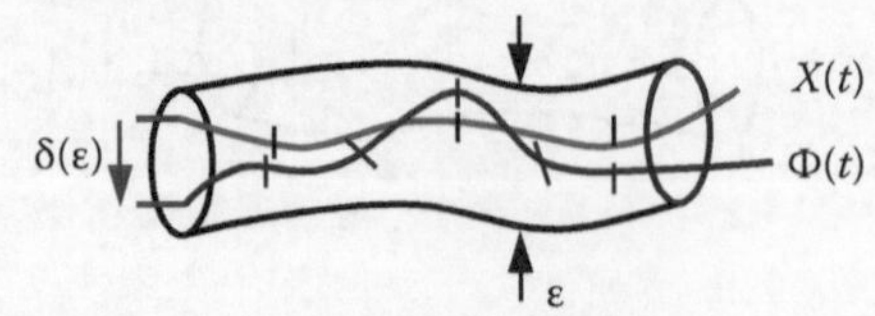

FIGURE 3.2 Illustration of the orbital stability of the solution $\Phi(t)$.

Remark 4

What has been discussed above concerns only local stabilities. We will not look at the concept of global stability for now. ■

When looking from the angle of phase spaces, the stability of a steady state is actually the stability of nearby orbits. Because the system's stability means whether or not the system could eliminate its behavioral deviations when it is disturbed, the stability of a particular steady state of the system can be determined by the final states of all the nearby orbits (Xu 2000).

Definition 3.9

(The stability of fixed point).

1. *Focal fixed point:* The neighborhood is filled with spiraling phase orbits. The orbit that starts from an arbitrarily chosen initial state in the neighborhood is a spiral path with the fixed point as its limiting point. The foci are classified as either stable or unstable (see Figure 3.3a).
2. *Fixed point of the node type:* The neighborhood is filled with nonspiraling phase orbits. Under normal conditions, the phase orbits are straight lines pointing to the fixed point. The nodes can also be classified into stable and unstable (see Figure 3.3b).

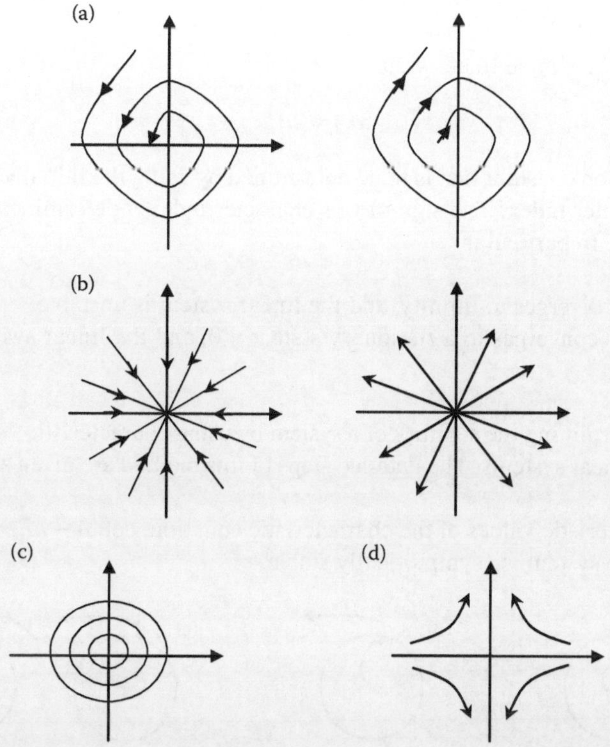

FIGURE 3.3 Phase orbits of fixed points of different types. (a) Phase orbits of stable and unstable foci. (b) Phase orbits of stable and unstable nodes. (c) Phase orbit of center. (d) Phase orbit of saddle.

3. *Fixed point of the center type:* The neighborhood of the fixed point is filled with closed orbits of different periodicities. From an arbitrarily chosen initial state in the neighborhood, the underlying system shows a periodic movement surrounding the fixed point. The central fixed point is stable but not asymptotically stable (see Figure 3.3c).
4. *Fixed point of the saddle point type:* Two phase orbits converge to the fixed point from two opposite directions, while the respective fixed points of the phase orbits diverge along opposite directions. The saddle point is globally unstable (see Figure 3.3d). ■

Definition 3.10

(Stability of limit circles or periodical orbits).

1. *Stable limit circle:* All nearby orbits converge to the limit circle in a helix fashion. In particular, all the orbits located outside the limit circle involve to the limit circle, and all the orbits inside the limit circle convolve to the limit circle (see Figure 3.4a).
2. *Unstable limit circle:* All orbits spiral (diverge) away from the limit circle. In particular, all orbits outside the limit circle convolve away from the limit circle, and all orbits inside the limit circle involve away from the limit circle (see Figure 3.4b).
3. *One-side stable limit circle:* The orbits outside the limit circle involve to the limit circle, while the orbits inside the limit circle diverge from the limit circle; or the orbits outside the limit circle diverge from the limit circle, while the orbits inside the limit circle convolve to the limit circle (see Figure 3.4c). ■

3.1.3.1 Stability of Linear System

The general solution of a simple linear system $\dot{x} = ax$ is

$$x(t) = ce^{at}$$

where c is the integration constant that is to be determined by using the initial value. The coefficient a is known as a character index. The sign of this character index a determines whether or not this linear system is stable. In particular

- When $a > 0$, $x(t)$ diverges to infinity, and the linear system is unstable.
- When $a < 0$, $x(t)$ converges to a stationary state $x = 0$, and the linear system is stable (see Figure 3.5).

The method of determining the stability of a system by using characteristic values is widely applicable to the study of linear systems. The detailed steps of this method are given as follows (Xu 2000):

1. If all the characteristic values of the characteristic equation, $\det|A - \lambda E| = 0$, of the system are negative, the system is asymptotically stable.

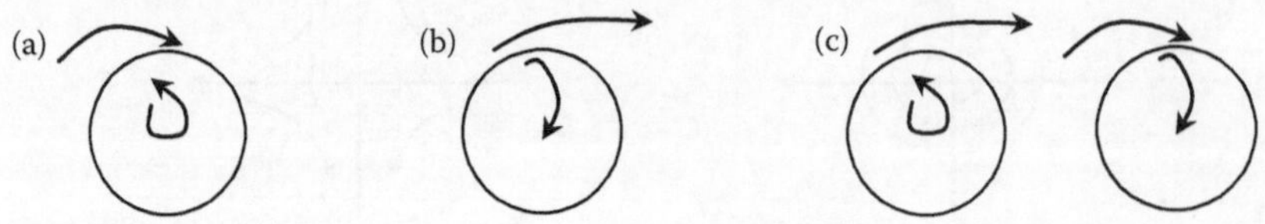

FIGURE 3.4 Phase orbit of different limit circle stabilities. (a) Stable limit circle. (b) Unstable limit circle. (c) Semistable limit circles.

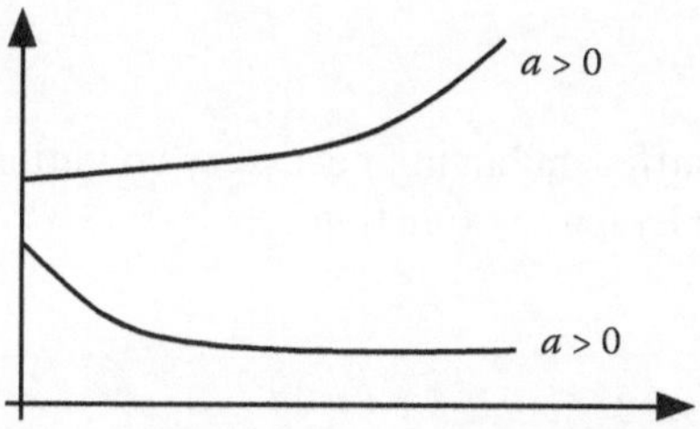

FIGURE 3.5 Stability of a linear system with different parameter cases.

2. If at least the real part of one characteristic value is positive, the system is unstable.
3. If all the characteristic values of the characteristic equation of the system are nonnegative with at least one being zero, then the system may be stable or unstable. In these instances, additional deliberate criteria are needed.

3.1.3.2 Stability of Nonlinear System

Different from the case of linear systems, the stability of one solution of the nonlinear system is unrelated to the stability of another solution. When one is interested in the local stability, if the evolution equation of the system satisfies the requirements of continuity and smoothness, one commonly employed method is to expand the system of nonlinear equations at the specific point; by ignoring the high-order terms, one obtains a system of linear equation of the same dimension. Then, the stability of this linear system represents the stability of the original nonlinear system at this specified locality. This method is known as the linear stability analysis of nonlinear systems or Lyapunov's first method. It has been widely employed in systems science.

If the established dynamical system of equations reflects the kinetic law that governs the underlying system correctly, the system will contain the information for one to distinguish whether the system is stable or unstable. The methods of extracting the desired information include either solving the equation directly or acquiring the information from the structure and parameters of the equation system directly.

3.1.3.3 Lyapunov's Theorem of Stability (Lyapunov's Second Method)

For a given nonlinear system $\frac{dX}{dt} = F(X,C)$, if there exists a continuous, differential, and positive-definite function $V(x)$, known as the Lyapunov function, then we have the following:

1. The total derivative along the orbit of the function is nonpositive, namely,

$$\frac{dV}{dt} = \sum_{i=1}^{n} \frac{\partial V}{\partial x_i} \frac{\partial x_i}{\partial t} = \sum_{i=1}^{n} \frac{\partial V}{\partial x_i} f_i \le 0,$$

then the zero solution of the system is stable.
2. The total derivative along the orbit of the function is negative definite, namely,

$$\frac{dV}{dt} = \sum_{i=1}^{n} \frac{\partial V}{\partial x_i} f_i < 0,$$

while the total derivative as a function in x is nonnegative, then the zero solution of the system is asymptotically stable.

Remark 5

Lyapunov's second method is a sufficient but not necessary condition. If the zero solution is asymptotically stable, there will exist a Lyapunov function. ■

Remark 6

The construction of the desired Lyapunov function is the key to distinguish whether the system is stable or not. For the general nonlinear system, there is no general method available for the construction of $V(x)$. ■

Example 3.3

Verify that the following systems have Lyapunov functions of the form $V(x,y) = ax^2 + by^2$. Then, judge whether or not the zero solutions of the systems are stable.

1. $$\begin{cases} \dfrac{dx}{dt} = -xy^2 \\ \dfrac{dy}{dt} = -yx^2 \end{cases}$$

2. $$\begin{cases} \dfrac{dx}{dt} = -x + xy^2 \\ \dfrac{dy}{dt} = -2x^2y - y^3. \end{cases}$$

Solution

1. If the system has a Lyapunov function of the form $V(x,y) = ax^2 + by^2$, then
$$\frac{dV}{dt} = 2ax\frac{dx}{dt} + 2by\frac{dy}{dt} = -2ax^2y^2 - 2bx^2y^2.$$
By taking $a = b = 1$, we have $\dfrac{dV}{dt} = -4x^2y^2$. Thus, it is easy to see that the total derivative along the orbit of V is nonpositive. Therefore, the zero solution of this system is stable.
2. If this system has a Lyapunov function of the form $V(x,y) = ax^2 + by^2$, then
$$\begin{aligned}\frac{dV}{dt} = 2ax\frac{dx}{dt} + 2by\frac{dy}{dt} &= 2ax(-x + xy^2) + 2by(-2x^2y - y^3) \\ &= -2ax^2 + 2ax^2y^2 - 4bx^2y^2 - 2by^4.\end{aligned}$$
By taking $a = 2$, $b = 1$, we have $\dfrac{dV}{dt} = -4x^2 - 2y^4$. Thus, it is easy to see that the total derivative along the orbit of the function V is negative. Therefore, the zero solution of this system is stable.

3.1.4 Fixed Point Stability of 2-D Linear System: A Case Study

Consider the following 2-D linear system:

$$\frac{d}{dt}\begin{pmatrix} x \\ y \end{pmatrix} = A\begin{pmatrix} x \\ y \end{pmatrix} \tag{3.7}$$

where $A = \begin{pmatrix} a & b \\ c & d \end{pmatrix}$. When $\det(A) = \begin{vmatrix} a & b \\ c & d \end{vmatrix} \neq 0$, it can readily be seen that (0,0) is the unique primary singular point (equilibrium point) of system 3.7. The rest of the discussion of the (0,0) stability is based on this assumption.

Using the invertible linear transformation

$$\begin{pmatrix} x \\ y \end{pmatrix} = T\begin{pmatrix} \xi \\ \eta \end{pmatrix},$$

where $\det(T) \neq 0$, changes system 3.7 into the following:

$$\frac{d}{dt}\begin{pmatrix} \xi \\ \eta \end{pmatrix} = T^{-1}AT\begin{pmatrix} \xi \\ \eta \end{pmatrix}. \tag{3.8}$$

Now, according to linear algebra, it follows that each quadratic form defined on the field of real numbers can be transformed through a nondegenerate linear transformation into a standard form of quadratic forms. Thus, by choosing an appropriate T, one can make $T^{-1}AT$ be the Jordan canonical form of A so that the phase diagram of system 3.8 can be produced on the (ξ,η) plane. Now, by applying the inverse T^{-1}, one returns to the (x,y) plane and obtains the phase diagram of system 3.7.

For the sake of convenience, let us consider only the phase diagram of system 3.8. Assume that $T^{-1}AT$ is already in the standard form. Then $T^{-1}AT$ should possess one of the following structures:

$$\begin{pmatrix} \lambda & 0 \\ 0 & \mu \end{pmatrix}, \begin{pmatrix} \lambda & 0 \\ 1 & \lambda \end{pmatrix}, \text{ and } \begin{pmatrix} \alpha & -\beta \\ \beta & \alpha \end{pmatrix},$$

where λ, μ, and β are nonzero. Now, let us look at each of these cases.

Case 1: $T^{-1}AT = \begin{pmatrix} \lambda & 0 \\ 0 & \mu \end{pmatrix}$ such that $\lambda\,\mu \neq 0$.

In this case, the system looks as follows:

$$\begin{cases} \dfrac{d\xi}{dt} = \lambda\xi \\ \dfrac{d\eta}{dt} = \mu\eta \end{cases} \begin{cases} \xi = A_1 e^{\lambda t}, \\ \eta = A_2 e^{\mu t}, \end{cases} \eta = C|\xi|^{\mu/\lambda}. \tag{3.9}$$

Our discussion can be divided into three possibilities:

Case 1.1: $\lambda = \mu$. Thus, both ξ and η are reduced into $\eta = c|\xi|$, and the node (0,0) of the system is known as stellar (Figure 3.6).

The characteristics of the trajectory are as follows:

1. Each trajectory is a ray, which either approaches or moves away from the equilibrium point.
2. When $\lambda = \mu > 0$, (0,0) is an unstable stellar node.
3. When $\lambda = \mu < 0$, (0,0) is a stable stellar node.

Case 1.2: $\lambda\ \mu > 0$. In this case, other than the ξ-axis and the η-axis, each curve in class 3.9 is a "parabola" with (0,0) as its vertex. When $|\mu/\lambda| > 1$, these parabolas are tangent with the ξ-axis, and when $|\mu/\lambda| < 1$, they are tangent with the η-axis. Each parabola is cut into two trajectories of the system by the point (0,0). Thus, in this case, (0,0) is known as a bidirectional node of the system. See Figure 3.7 for more details.

The trajectories have the following characteristics:

1. When $\mu, \lambda < 0$, the equilibrium point is asymptotically stable; when $\mu, \lambda > 0$, the equilibrium point is unstable.
2. In the process when the trajectories approach or leave far away from the equilibrium point, other than some individual trajectories, they all share a common tangent line with the equilibrium point as their tangent point. A point satisfying such a property is known as a node.

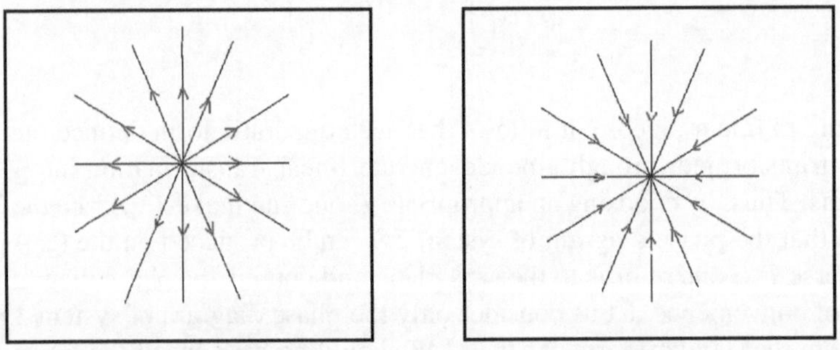

$\lambda = \mu > 0$
(Unstable stellar node)

$\lambda = \mu < 0$
(Stable star-shaped node)

FIGURE 3.6 Stellar nodes.

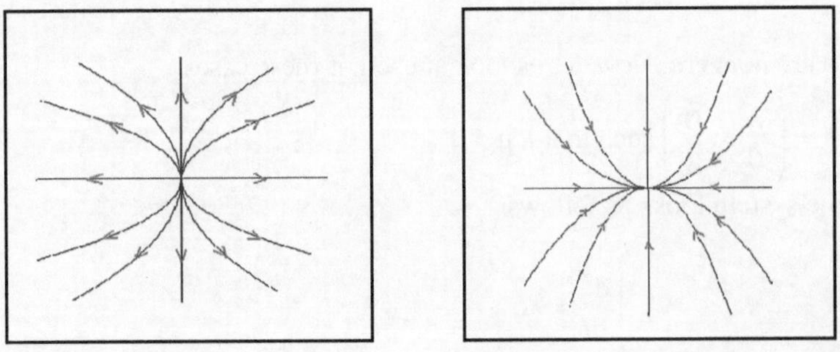

$\lambda > \mu > 0$
(Unstable, tangent to η-axis)

$\lambda < \mu < 0$
(Stable, tangent to ξ-axis)

FIGURE 3.7 Parabolic trajectories.

Case 1.3: $\lambda\mu < 0$. In this case, other than $\xi = 0$ and $\eta = 0$, the class of trajectories contains hyperbolas, to each of which $\xi = 0$ and $\eta = 0$ are the asymptotes. The trajectories of the system consist of the positive and negative ξ- and η-axes and these hyperbolic curves; along each of the hyperbolic curves, when $t \to \infty$, $(\xi(t), \eta(t))$ moves far away from (0,0). See Figure 3.8 for more details.

The characteristics of the trajectories are as follows:

1. There are always two phase trajectories that approach the equilibrium point from the positive direction and two phase trajectories that approach the equilibrium point from the negative direction.
2. All other trajectories first approach the equilibrium point; and when getting close to the equilibrium point within a certain distance, they start to move away from the equilibrium point. In this case, the equilibrium point is unstable and is known as a saddle point.

Case 2: $T^{-1}AT = \begin{pmatrix} \lambda & 0 \\ 1 & \lambda \end{pmatrix}$ such that $\lambda \neq 0$. That is, this matrix has a nonzero real characteristic root of multiplicity 2, and the corresponding Jordan block is of second order. Solving

$$\begin{cases} \dfrac{d\xi}{dt} = \lambda\xi \\ \dfrac{d\eta}{dt} = \xi + \mu\eta \end{cases}$$

produces

$$\begin{cases} \xi = Ae^{\lambda t}, \\ \eta = (At + B)e^{\lambda t}. \end{cases} \tag{3.10}$$

Therefore, we have $\eta = C\xi + \frac{\xi}{\lambda}\ln|\xi|$ and $\xi = 0$. That is,

$$\lim_{\xi\to 0}\eta = 0;\ \lim_{\xi\to 0}\frac{d\eta}{d\xi}\begin{cases} +\infty, \lambda < 0 \\ -\infty, \lambda > 0. \end{cases} \tag{3.11}$$

See Figure 3.9 for more details.

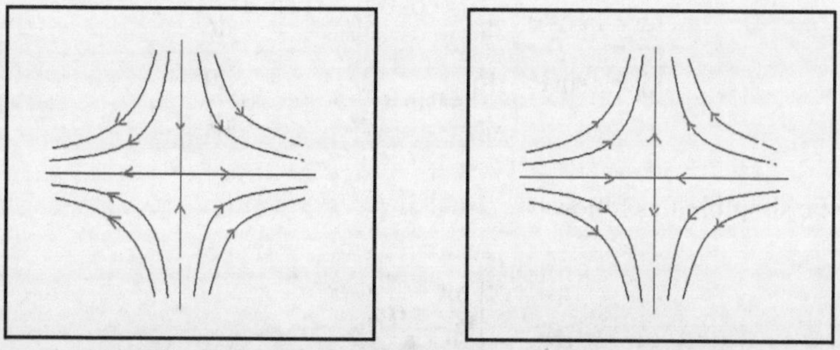

When $\lambda > 0$, $\mu < 0$ When $\lambda < 0$, $\mu > 0$

FIGURE 3.8 Hyperbolic curves.

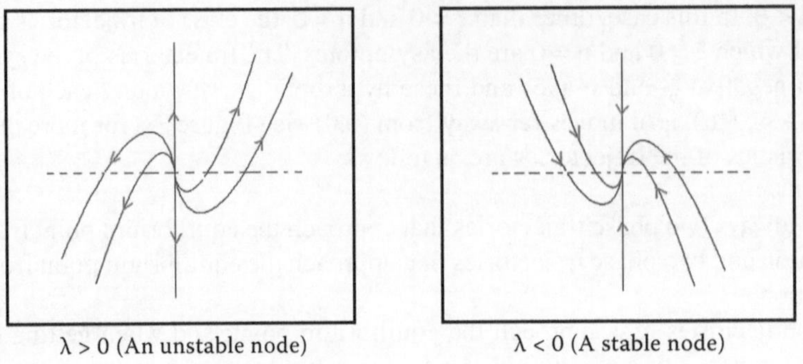

FIGURE 3.9 Trajectories of the system for case 2.

The characteristics of the trajectories are as follows:

1. Both the positive and negative η-axes are trajectories.
2. All other trajectories cut across the ξ-axis and are tangent to the η-axis.
3. All trajectories either approach or move away from the equilibrium point in the same direction.
4. When $\lambda < 0$, the equilibrium point is asymptotically stable; when $\lambda > 0$, the equilibrium point is unstable. In this case, the equilibrium point is known as an irregular node.

Case 3: $T^{-1}AT = \begin{pmatrix} \alpha & -\beta \\ \beta & \alpha \end{pmatrix}$ such that $\beta \neq 0$. That is, this matrix has a pair of conjugate complex characteristic roots $\alpha \pm \beta i$. In this case, the given system can be written as follows:

$$\begin{cases} \dfrac{d\xi}{dt} = \alpha\xi - \beta\eta \\ \dfrac{d\eta}{dt} = \beta\xi + \alpha\eta. \end{cases} \tag{3.12}$$

By using the polar coordinate transformation,

$$\begin{cases} x = r\cos\theta \\ y = r\sin\theta \end{cases} \begin{cases} \dfrac{d\xi}{dt} = \cos\theta\dfrac{dr}{dt} - r\sin\theta\dfrac{d\theta}{dt} \\ \dfrac{d\eta}{dt} = \sin\theta\dfrac{dr}{dt} + r\cos\theta\dfrac{d\theta}{dt}. \end{cases} \tag{3.13}$$

System 3.12 can be simplified as follows:

$$\begin{cases} \dfrac{dr}{dt} = \alpha r, \\ \dfrac{d\theta}{dt} = \beta. \end{cases}$$

Thus, we have

$$\begin{cases} r = Ae^{\alpha t}; \\ \theta = \beta t + C_0; \end{cases} \quad r = Ce^{(\alpha/\beta)\theta}. \tag{3.14}$$

where $C \geq 0$. The sign of β determines the direction of rotation of the trajectory. When $\beta > 0$, the trajectory spins counterclockwise. When $\beta < 0$, the trajectory spins clockwise. Based on the different signs of α, the phase plot can be considered in three possibilities.

Case 3.1: $\alpha < 0$ and $\beta > 0$ when the focus is stable. In this case, the phase plot is given in Figure 3.10.

Case 3.2: $\alpha = 0$ and $\beta > 0$, where a center point exists. In particular, in this case, the following holds true:

$$\begin{cases} \dfrac{d\xi}{dt} = -\beta\eta \\ \dfrac{d\eta}{dt} = \beta\xi. \end{cases}$$

Therefore, $\dfrac{d\xi}{dt}\xi + \dfrac{d\eta}{dt}\eta = 0$, namely, $\xi^2 + \eta^2 = C$, where C is a constant.

The phase plot for this case is given in Figure 3.11.

Case 3.3: $\alpha > 0$ and $\beta > 0$ when the focus is unstable. In this case, the phase plot is given in Figure 3.12.

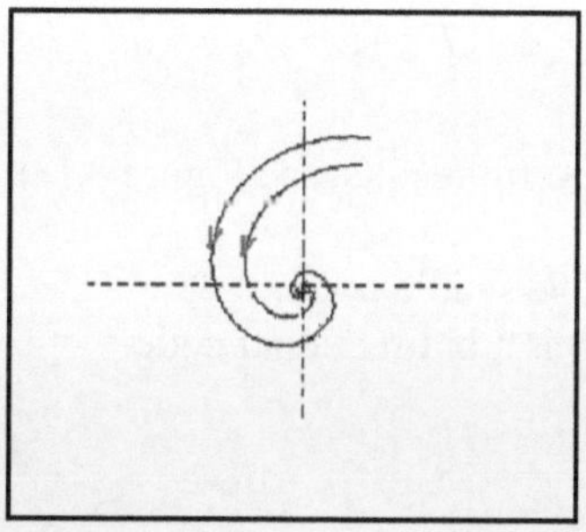

FIGURE 3.10 Phase plot when $\alpha < 0$ and $\beta > 0$.

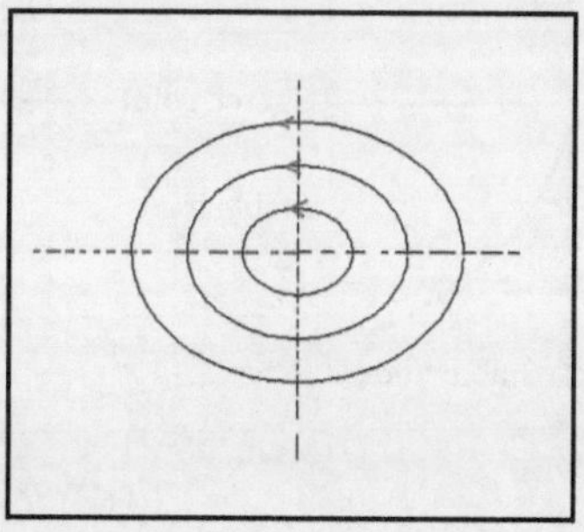

FIGURE 3.11 Phase plot when $\alpha = 0$ and $\beta > 0$.

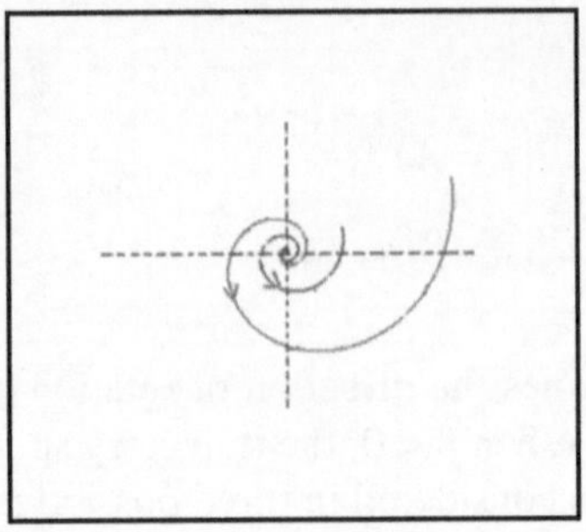

FIGURE 3.12 Phase plot when $\alpha > 0$ and $\beta > 0$.

The characteristics of the trajectories are given as follows:

1. When $\alpha = 0$, $\beta > 0$, the trajectories are a class of concentric circles. The point (0,0) is stable but not asymptotically stable. In this case, the point (0,0) is known as the center. For all other cases, all the trajectories are a class of logarithm helical curves.
2. When $\alpha < 0$, the trajectories asymptotically approach the equilibrium point. In this case, (0,0) is asymptotically stable and is known as a focus.
3. When $\alpha > 0$, the trajectories evolve away from the equilibrium point. In this case, (0,0) is unstable and also known as a focus.

To summarize, based on the discussion above, for system 3.7, denote

$$\begin{cases} p = tr\left(T^{-1}AT\right) = (\lambda + \mu) = (a + d); \\ q = \det\left(T^{-1}AT\right) = \lambda\mu = (ad - bc). \end{cases}$$

Then we can obtain the following useful results (see Figure 3.13):

1. When $q < 0$, (0,0) is an unstable saddle point.
2. When $q > 0$ and $p^2 > 4q$, (0,0) is a bidirectional node.

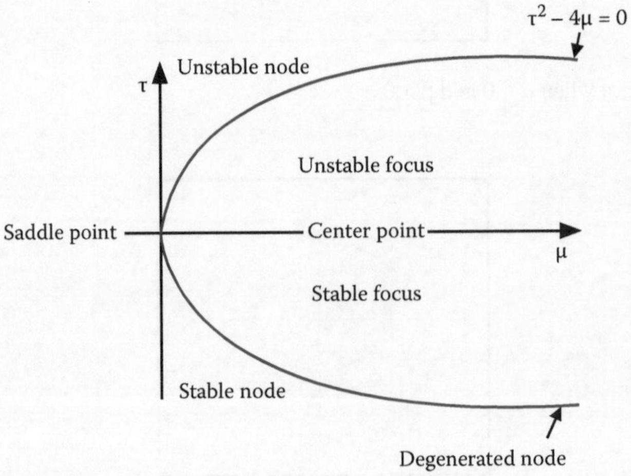

FIGURE 3.13 Distribution of saddle points, nodes, foci, and center points.

3. When $q > 0$ and $p^2 = 4q$, (0,0) is either a unidirectional or a stellar node.
4. When $q > 0$ and $0 < p^2 < 4q$, (0,0) is a focus.
5. When $q > 0$ and $p = 0$, (0,0) is a center point.

3.2 DYNAMIC SYSTEMS THEORY OF NONLINEAR SYSTEMS

3.2.1 Attractor and Purposefulness

The definition of attractors has many varied versions and meanings in different research fields. As for that of dynamic systems, let us look at the following.

Definition 3.11

(Attractor of a dynamic system). A point set is referred to as an attractor or a sink of the dynamic system if the phase orbit of the dynamic system satisfies the following three properties:

1. *Ultimacy:* The point set represents the ultimate and final states of the system's evolution. In these states, there is no longer any motivation to change the states. Thus, the set represents the system's steady states.
2. *Stability:* In these states, the system possesses the ability to resist disturbance and to maintain its characteristics, reflecting the prescriptiveness of the system's quality. All the elements in the set represent stable steady states.
3. *Attractability:* The states in the set attract other nearby states or orbits. ■

Let us in the following describe the concept of attractability symbolically.

Definition 3.12

(Attractability). Let $\Phi(t)$ be an arbitrary solution of the vector differential equation $\frac{dX}{dt} = F(X,C)$, and $X(t)$ the solution derived from the initial value condition $X(t_0) = x_0$. If there exists $\delta_0 > 0$ such that when $|X_0 - \Phi(t_0)| < \delta_0$, $\lim_{t\to\infty} |X(t) - \Phi(t)| = 0$, then $X(t)$ is referred to as having attractability. ■

Remark 7

The concepts of attractability and stability represent two different attributes of the system. For instance, the center point of an autonomous system is stable but not an attractor. Also, there exists steady states that possess attractability but do not satisfy the condition of Lyapunov stability. In particular, in the system $\theta' = 1 - \cos\theta$, $\theta = 0$ is a fixed point. It possesses attractability for all orbits, but it is not Lyapunov stable (see Figure 3.14). ■

Remark 8

The following are some of the well-studied attractors: (1) foci and nodes, which represent equilibrium movements of the system; (2) limit circles, which represent periodical movements of the

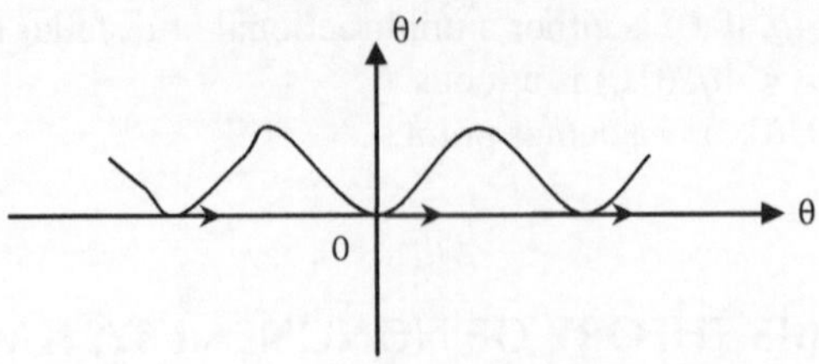

FIGURE 3.14 Illustration for why (0,0) possesses attractability for all orbits but is not Lyapunov stable.

system; (3) anchor rings, which represent quasi-periodical movements of the system; and (4) edges of chaos, which represent the movements of the system between order and chaos. ■

Definition 3.13

(Attractors defined by sets and mappings).

1. Let f be a given mapping. If a set X^* from within the domain of f satisfies $f(X^*) = X^*$, then X^* is known as an invariant set.
2. Let set X^* be a nonempty invariant set of the mapping f: $R^n \to R^n$. If there is $x \in R^n$ satisfying that $f^i(x) \to x^* \in X^*$, for some natural number i, then x is known as an attraction point of the invariant set X^*.
3. The set of all attraction points of the invariant set X^* is known as the attraction domain or field of X^*, denoted by $A(X^*)$. If there exists an open set U such that $A(X^*) \supset U \supset X^*$, then X^* is an attractor of the mapping f: $R^n \to R^n$. Here, $f^i(x) = f(f^{i-1}(x))$. ■

The purpose of introducing the concept of attractors is to describe the destination of the system, which is a dynamic characteristic of the system. As a matter of fact, each transformation from a transient state to a stable steady state represents an evolutionary process of the system's search for its destination. However, the destination of a system is determined by not only the system itself but also the environment.

A figurative description of the attraction domain is that each attractor defines a sphere of influence around itself in the phase space. Each orbit that starts with one of the points within the sphere tends to approach that attractor. The set of all the points in this sphere of influence of the phase space is known as the attraction domain of the attractor.

Definition 3.14

A point is referred to as a repeller or source if it repels all the orbits around it. Each orbit that starts with a nearby point departs from the point as time elapses. Such a point or the set of all such points is known as a repeller or source. ■

Definition 3.15

The number of characteristic values of the system with positive real parts is known as the system's index. ■

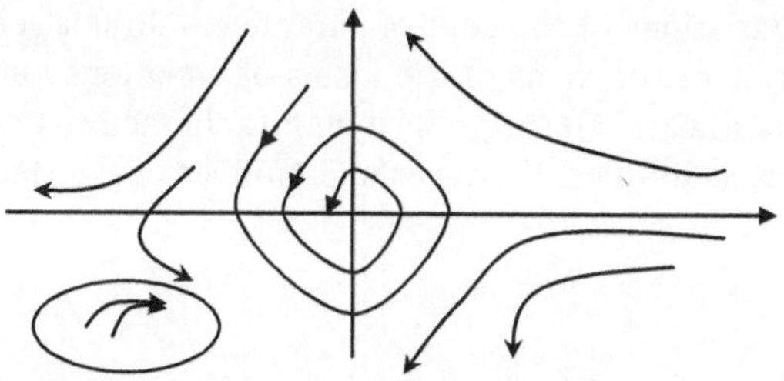

FIGURE 3.15 Illustration of a phase portrait.

The following facts can be shown: each steady state with index zero is an attractor. Each steady state with index n, the dimensionality of the system, is a repeller. Each steady state with index between 0 and n stands for a saddle point.

Definition 3.16

For each set of given values of the control parameters, a geometric figure can be constructed in the phase space with all steady states, including their types, numbers, and distributions, and the characteristics and directions of the orbits near each steady state, intuitively labeled. Such a figure is known as the phase portrait of the system. See Figure 3.15 for more details. ■

Each linear system contains only fixed-point-type steady states. It is impossible for a linear system to have any limit circle or steady state of more complex type. Additionally, each linear system can have at most an attractor. In this case, the entire phase space of the linear system is an attraction domain; the attractor depicts the behavioral characteristics of the system in the entire phase space.

As for a nonlinear system, its phase portrait is much richer than that of any linear system and can contain all kinds of fixed points, planar and spatial limit circles, strange attractors, and edges of chaos.

It is often the situation that in a nonlinear system, there are multiple types of steady states; the coexisting multiple attractors divide the entire phase space into various attraction domains.

When a particular system has multiple attractors, it implies that the system will have different evolutionary outcomes (final states). In this case, the region where the initial state falls in signals the specific attractor to be the ultimate evolutionary state of the system. Hence, the choice of the initial state of the system determines its ultimate evolutionary results.

After having established the evolution equations for the system of concern, the theory of attractors, a part of the dynamic systems research, mainly discusses the following problems of systems:

1. Whether or not there exists any attractor
2. How many attractors there are
3. What types of attractors they are
4. How the attractors are distributed in the phase space and how the attraction domains are divided
5. What the relationships are between the attractors and the control parameters

3.2.2 Bifurcation

For each given value of the control parameter, a specified system is determined, and its state space can be constructed. In each parameter space, the family of systems with the same symbolic structure is investigated. Variations of the control parameters do not generally change the mathematical structure of the systems in terms of the forms of equations but can essentially alter the system, especially if it is nonlinear, in terms of its dynamic characteristics, systems' portrait structures, systems' steady states, stabilities, the types and number of the steady states, the phase portraits, and so on.

Definition 3.17

(Stability of system structure). Given a small perturbation to the control parameters, if the qualitative properties of the system's phase portrait, including the types, number, and stability of the steady states, do not change accordingly, then the system structure is said to be stable; otherwise, the structure is said to be unstable. ■

Remark 9

The concepts of system's stability and system's structural stability are different. In particular, a system's stability refers to the stability of the system's solution or state that evolves with time in the state space. In contrast, the stability of a system's structure refers to the potential changes of the qualitative properties of the system's phase portrait, including the types, number, stability of the steady states, and so on. ■

Definition 3.18

(Bifurcation). For the nonlinear system $\frac{dx}{dt} = F(x,c)$, if a small perturbation of the control parameter c around a fixed value c_0 would lead to one of the following changes in the system, then this phenomenon is referred to as a bifurcation, and the point of change is known as a bifurcation point of the system:

1. Creation or disappearance of a steady state.
2. A change of the system's stability.
3. An originally stable steady state loses its stability, while giving birth to one or more new steady states.
4. One type of steady state is transformed into other type of steady state.
5. A change in the distribution of steady states in the phase portrait.

The bifurcation theory studies and classifies phenomena characterized by sudden shifts in behaviors arising from small changes in circumstances. It also analyzes how the qualitative nature of the model's solutions depends on the parameters that appear in the model. For example, the system described by the Van der Pol equation behaves very differently for varied values of the parameter c. In particular, $c = 0$ is the bifurcation point, and the system changes totally from when $c > 0$ to when $c < 0$:

- There exists one steady limit circle when $c > 0$.
- There exists one center (0,0) when $c = 0$.
- There exists only one equilibrium point (0,0) when $c < 0$.

■

Remark 10

The phenomenon of bifurcation may also appear to a linear system [only in the form of (1) and (2) in Definition 3.16]. However, it is a common phenomenon for nonlinear systems. ■

Studying when bifurcation occurs and the type and number of bifurcation points and determining the stability of bifurcation solutions have been among the important contents of dynamic systems. In the following, we study several simple bifurcation types (Xu 2000).

3.2.2.1 Saddle–Node Bifurcation

Let us look at the system $\dot{x} = a + x^2$ with parameter a. When $a > 0$, the system has no steady state; when $a = 0$, the system has only one steady state; and when $a < 0$, the system has two steady states, where $x = -\sqrt{-a}$ is stable and $x = \sqrt{-a}$ is unstable. For further details, see Figure 3.16.

With a change in the control parameter, an existing system's steady state could possibly vanish, and a nonexisting steady state could potentially be created; the created steady state turns out to be semistable. This bifurcation is referred to as a saddle–node bifurcation.

3.2.2.2 Transcritical Bifurcation

For the system $\dot{x} = ax - x^2$, the fixed points are $x_1 = 0$ and $x_2 = a$. When $a > 0$, $x_2 = a$ is stable, while $x_1 = 0$ is not stable. When $a < 0$, $x_1 = 0$ is stable, while $x_2 = a$ is not stable. Thus, $a = 0$ is a bifurcation point, and the stability of the system changes at this point. This kind of bifurcation is referred to as a transcritical bifurcation.

3.2.2.3 Pitchfork Bifurcation

Consider the system $\dot{x} = ax - x^3$. When $a < 0$, there is only one steady point $x = 0$. When $a > 0$, there are three steady points

$$x_1 = 0,\ x_2 = \sqrt{a},\ \text{and}\ x_3 = -\sqrt{a},$$

where the first solution is not stable, and the latter two are stable.

Here, $a = 0$ is a bifurcation point. When the value of a transits from being negative to the bifurcation point, a new steady state is created in the system; also, the system experiences a change of stability. The creation or disappearance of fixed points always occurs in pairs. This kind of bifurcation is known as a pitchfork bifurcation.

There are two kinds of pitchfork bifurcations. One is known as supercritical pitchfork, which happens when the control parameter increases. The other kind of pitchfork bifurcation is known as subcritical pitchfork, which occurs when the control parameter decreases.

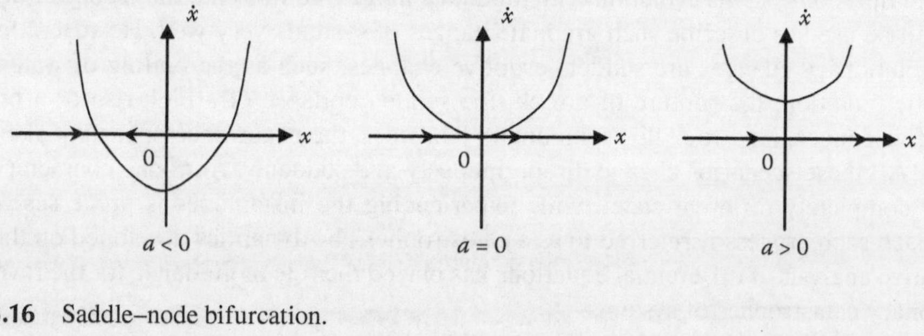

FIGURE 3.16 Saddle–node bifurcation.

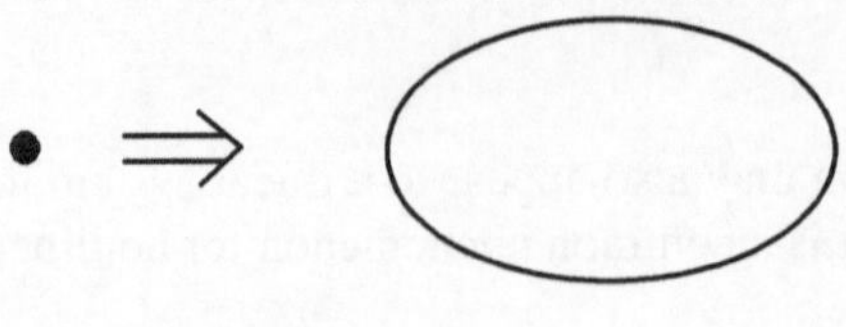

FIGURE 3.17 One-focus-to-limit-circle bifurcation.

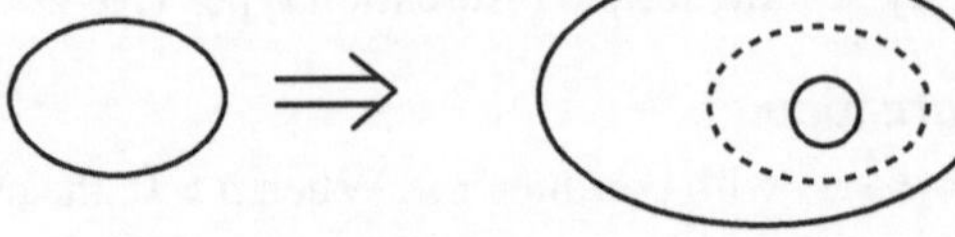

FIGURE 3.18 Bifurcation of one limit circle to two new steady limit circles.

3.2.2.4 One-Focus-to-Limit-Circle Bifurcation

In a continuous dynamic system, the bifurcation of a fixed point appears in the real part of a characteristic value λ. If there are two complex conjugated characteristic values

$$\lambda_1 = \lambda' + i\omega \text{ and } \lambda_2 = \lambda' - i\omega,$$

when the real part λ' increases from being negative to positive, $\lambda' = 0$ stands for a bifurcation point, and one steady focus will change to a steady limit circle. This phenomenon is known as a Hopf bifurcation. See Figure 3.17 for more details.

In some cases, one limit circle could change to two new steady limit circles (see Figure 3.18).

What has been discussed above is only about simple attractors. Sometimes, a change in the control parameter could create a strange attractor, such as chaos.

3.3 CATASTROPHE THEORY

Catastrophe theory was initiated by René Thom in 1972. Although the classical physics studies the gradual change of continuous processes, there exists instantaneous processes involving sudden and eruptive changes. The catastrophe theory deals with the mathematical modeling for the research of this kind of discontinuous change.

3.3.1 Catastrophe Phenomena and Catastrophe Theory

In systems evolutions, there exists both gradual and sudden changes. The growth of an organism, the circumrotation of earth around the sun, and the continuous flow of water in a river are just some of many examples of systems evolution with gradual changes. To this end, the theory of differential calculus can be used to describe such gradual changes of systems very well. However, opposite to gradual evolutionary changes are sudden, eruptive changes, such as the boiling of water, cellular split, genetic mutation, the rupture of a rock, the sudden collapse of a high rise or a bridge, the eruption of volcanoes, mud-rock flows, mountain landsides, the occurrence of earthquakes and tsunamis, etc. All these scenarios change discontinuously and suddenly, from one characteristic state to another completely different state, while experiencing the instantaneous processes of sudden changes. Each such process is referred to as a catastrophe. The dynamics developed on the basis of the qualitative analysis of differential equations has played the role of guidance for the investigation of evolutionary catastrophes of systems.

The concept of catastrophe is defined differently from one field to another. Below is an example.

Definition 3.19

(Catastrophe). In biological genetics, each mutation a genetic substance experiences that alters the substance's biological form and behavior is known as a catastrophe. In systems science, each sudden change of the system's qualitative behavior is referred to as a catastrophe. ■

Distinguished from the concept of catastrophe is that of gradual changes of systems. The latter stands for indistinguishable changes of systems' qualitative characteristics along with the elapses of time and variations of the control parameters.

Because the theory of calculus could not deal with discontinuous processes of sudden changes, Thom (1989) initiated the concept of catastrophe and published a book on structural stability with emphasis placed on discontinuous sudden changes by using topology and singular point theory. His original works have been considered the start of the catastrophe theory.

One characteristic of catastrophe is that a tiny vibration of the environment induces a sudden change of the system at a macroscopic scale. Such a phenomenon could appear only in nonlinear systems, because for linear systems, each continuous change of the environment could cause only another continuous change. In the realm of mathematics, catastrophe theory is considered a branch of the bifurcation theory in the study of dynamical systems. It provides a theory for handling discontinuous changes in systems' states occurring along with varying control parameters. It also stands for a particular case of a more general singularity theory of geometry, which is totally different from the case that the long-lasting stable equilibrium can be identified with the minimum of a smooth, well-defined potential function.

Catastrophe represents a particular way for a system's state to vary. It is a widely existing objective phenomenon. Its tragic consequences need to be controlled or minimized, while its constructive consequences can be positively utilized.

From the viewpoint of systems science, gradual changes lead to sudden changes. Such association corresponds to the mutual transformations between structural stability and instability. Each system with stable structure everywhere would not experience catastrophe and transitional change, and systems that are unstable everywhere would not exist. That is, each objectively existing system contains regions of both stable and unstable structures. Only such systems can experience transformations from gradual changes to sudden changes.

3.3.2 Singular Points and Topological Equivalence

Catastrophe theory is based on the following mathematical theory: group theory, manifold, singularity theory in mappings, topology, and so on (Li 2006).

Each singular point is defined relative to regular points. Generally, there are a large number of regular points, while the number of singular points is small.

Catastrophe theory mainly considers such systems transformations that evolve from one stable state to another, where each system is described by using equations involving parameters. In a stable state, some functions, such as the energy function or the entropy function, of the system's state would reach their extrema. In the point of view of mathematics, the stability of the system would be generally maintained by the extrema of some of the functions, when the derivatives of the functions are zero. Each such point represents the simplest singularity, known as a critical point.

Assume that an equation $F_{uv}(x) = 0$ is given, where u and v are parameters. Then, finding the critical points of the equation $F_{uv}(x)$ is equivalent to solving the following differential equation with given u and v values:

$$\frac{\mathrm{d}F_{uv}(x)}{\mathrm{d}x} = 0.$$

Each critical point x can be regarded as a function of the parameter u and v, denoted $x = l(u,v)$. Evidently, this function determines a curve in the (u,v,x) space, known as a critical curve. The points that make this function take extreme values are known as stable. Thus, critical points may not be stable. That is, at a particular critical point, the system can be either stable or unstable.

Definition 3.20

(Topology equivalence). If two geometric objects can be transformed continuously to each other without tearing, and adhesion of different points with a one-to-one mapping, then they are seen as topologically equivalent. That is, a geometric object can keep its topological structure unchanged with topologically equivalent transformations. ■

Catastrophe theory researches the topological invariability of the structures of systems.

3.3.3 Thom's Principle (Potential Function and Subdivision Lemma)

The stability of a singular point can be determined by the second-order derivatives of the potential function V (Thom 1989). The minimum of the potential function is an attractor, while the maximum of the function represents a source. By forcing the gradient ∇V to be 0, one can find the singular points, and the characteristics of the singularity can be determined by the matrix of second-order partial derivatives:

$$\begin{bmatrix} \frac{\partial^2 V}{\partial x_1^2} & \frac{\partial^2 V}{\partial x_1 \partial x_2} & \frac{\partial^2 V}{\partial x_1 \partial x_3} & \cdots & \frac{\partial^2 V}{\partial x_1 \partial x_n} \\ \frac{\partial^2 V}{\partial x_2 \partial x_1} & \frac{\partial^2 V}{\partial x_2^2} & \frac{\partial^2 V}{\partial x_2 \partial x_3} & \cdots & \frac{\partial^2 V}{\partial x_2 \partial x_n} \\ \vdots & \vdots & \vdots & \ddots & \vdots \\ \frac{\partial^2 V}{\partial x_n \partial x_1} & \frac{\partial^2 V}{\partial x_n \partial x_2} & \frac{\partial^2 V}{\partial x_n \partial x_3} & \cdots & \frac{\partial^2 V}{\partial x_n^2} \end{bmatrix},$$

which is known as a Hessen matrix.

If the determinant of the Hessen matrix satisfies $\det V_{n\times n} \neq 0$, then the singular points determined by $\nabla V = 0$ are referred to as isolated or Mose singular points. Each Hessen matrix is symmetric and can be transformed by using a diagonal matrix of a linear transformation, where the diagonal entries $\omega_1, \omega_2, \ldots$, and ω_n are the characteristic values of the Hessen matrix.

Each singular point is related to the control parameters. Thus, the characteristic values are also functions of the control parameters. If the control parameters $u_1, u_2, \ldots, u_m$ are zero when taking certain given values ω_i, $i = 1,\ldots,l$, then the Hessen matrix is not full rank, namely, $\det V_{n\times n} = 0$. In this case, the singular points determined by the equation $\nabla V = 0$ are known as nonisolated or non-Mose singular point. Here, l is the arithmetic complement of the rank of the Hessen matrix, while the rank of the Hessen matrix is $n - l$.

The potential function $V(\{x_i\}, \{u_a\})$ can be expanded into a Taylor series in a neighbor of a singular point. Assume that the singular point is the origin (0,0) of the phase space. Then the constant term of the expansion can become zero. From the definition of singular points, we know $\nabla V = 0$. Thus, we have

$$V(\{x_i\},\{u_a\}) = \sum_{i=1}^{n} \omega_i x_i^2 + O(x_i^3).$$

Assume that this equation has changed the Hessen matrix of rank n into a diagonal matrix through a linear transformation, then the properties of the singular points are completely determined by the characteristic values ω_i of the Hessen matrix. Because the higher-order terms do not really play any role, the potential function is known as Morse potential, whose structure is stable.

If the rank of the Hessen matrix is $n - l$, then ω_i is 0, for $i = 1,\ldots,l$, and the second-order partial derivatives cannot determine the singularity of the state variables $x_1, x_2, \ldots, x_l$. In this case, the third-order partial derivatives of the potential function need to be considered. Now, the potential function can be divided into Morse part and non-Morse part, namely,

$$V(\{x_i\},\{u_a\}) \approx \sum_{i=1}^{n} \omega_i x_i^2 + V_{NM}(x_i^3)$$

where $\sum_{i=1}^{n} \omega_i x_i^2$ is the Morse part corresponding to the isolated singular points, while $V_{NM}(x_i^3)$ is the non-Morse part corresponding to the nonisolated singular points with the third-order partial derivative terms.

At the same time, the state variables can also be grouped into those that are relevant to structural stability and those that are irrelevant to structure stability. That is the so-called subdivision lemma. When analyzing the type of a catastrophe, the catastrophe is examined not by the state variable number but by l, the arithmetic complement of the rank of the Hessen matrix.

3.3.4 Main Characteristics of Catastrophe

The basic characteristics of the catastrophe phenomenon are as follows:

1. *Multiple stable states.* In general, there are multiple, two or more, stable steady states in each catastrophe system. Because each point of the parametric system can correspond to multiple stable steady states, it makes it possible for the underlying system to jump from one gradual stable steady state solution to another so that a catastrophe appears in the evolution of the system. The root reason for the existence of multiple steady-state solutions is the nonlinearity of the system. Thus, catastrophes can occur only to nonlinear systems.

   For the folding catastrophe of a one-dimensional system involving one parameter, there is only one stable steady state; for the cusp catastrophe of a one-dimensional system involving two control parameters, there are two stable steady states.
2. *Unreachability.* There are unstable steady states, such as the maximal points and others, that exist in between different stable steady states. These unstable states cannot be reached in reality.
3. *Sudden jump.* At each cusp of the bifurcation curve, the system abruptly transforms from one stable steady state to another.
4. *Delay.* At noncusp locations, a system's catastrophes do not immediately appear when the system is transformed from one stable state to another. Also, this phenomenon of lagging varies with the direction of change of the control parameters.
5. *Divergence.* At around a bifurcation point, the sensitivity of the system's terminal states to the particular path of change of the system's parameters is known as divergence. When seen from the angle of the parametric space, the ultimate direction of development of the system is sensitive only in the region near the bifurcation curve.

3.3.5 Basic Catastrophe Types

Suppose that the dynamics of a system could be derived from a smooth potential function. Then Thom proves by using topological means that the number of discontinuous structures of different characteristics is determined not by the number of state variables but by the number of control variables. According to the classification theorem of Thom's catastrophe theory, a large number of discontinuous phenomena existing in natural and social scenarios can be expressed by using certain specified geometric shapes. As long as the number of control parameters is not more than five, there are a total of 11 kinds of catastrophes in terms of a certain equivalence relation. For the elementary catastrophes, occurring in the three-dimensional space plus the additional time dimension, that are controlled with four factors, there are only seven basic types of different properties (see Table 3.1 for more details).

A famous suggestion (Catastrophe Theory, Wikipedia, accessed on October 14, 2010) is that the cusp catastrophe can be used to model the behavior of a stressed dog, which may respond by becoming cowed or becoming angry. The suggestion is that at a moderate stress level ($a > 0$), the dog will exhibit a smooth transition of response from crowed to angry, depending on how it is provoked. However, higher stress levels correspond to moving to the region ($a < 0$). In particular, if the dog starts to be cowed, it will remain cowed as it is irritated more and further until it reaches the folding point, when it will suddenly, discontinuously snap through to angry mode. Once in the angry mode, it will remain to be angry, even if the direct irritation parameter is considerably reduced.

Example 3.4

(Xu 2000). Consider a gradient system $x' = f(x)$ with the potential function

$$V(x) = x^4 + ax^2 + bx$$

where a potential function is a scalar function satisfying $\frac{dV(x)}{dx} = -f(x)$. The fixed point of the system satisfies the following equation, because the derivative of the potential function at the fixed point should be zero:

$$4x^3 + 2ax + b = 0 \tag{3.15}$$

In the a-b-x product space of the phase and the parametric spaces, let the surface of this system be denoted by M. Then, in the half-space of $a \le 0$, the surface M contains a gradually opening folded region of three leaves, where the top and the bottom leaves stand for the minimal points

TABLE 3.1
Basic Catastrophe Type

Catastrophe Type	No. of Control Parameter	Potential Parameter
Fold	1	$V(x) = x^3 + ux$
Cusp	2	$V(x) = x^4 + ux^2 + vx$
Swallowtail	3	$V(x) = x^5 + ux^3 + vx^2 + wx$
Butterfly	4	$V(x) = x^6 + tx^4 + ux^3 + vx^2 + wx$
Hyperbolic umbilic	3	$V(x,y) = x^3 + y^3 + wxy + ux + vy$
Ellipse umbilic	3	$V(x,y) = x^3 - xy^2 + w(x^2 + y^2) + ux + vy$
Parabolic umbilic	4	$V(x,y) = y^4 + x^2y + wx^2 + ty^2 + ux + vy$

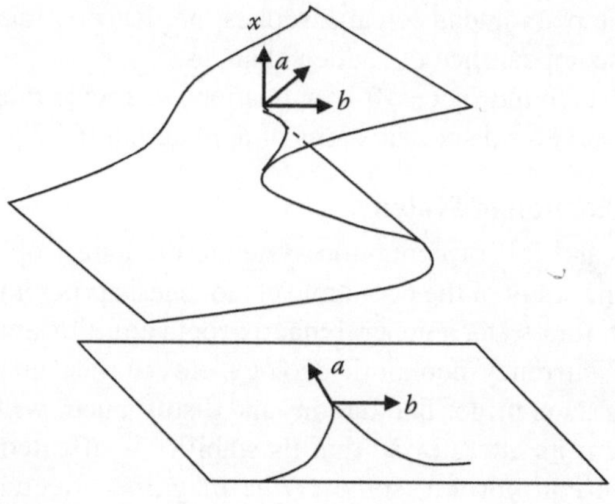

FIGURE 3.19 Bifurcation curve.

of $V(x)$, while the middle leaf, maximal points of $V(x)$. The two folding seams of the surface, which exist between the top and middle leaves and between the middle and bottom leaves, are of the following dynamic significance: on the top and bottom leaves, $V(x)$ is stable, while on the middle leaf, $V(x)$ is unstable. These two seams satisfy Equation 3.15 and the following equation, because at the transition point, the derivative of Equation 3.15 also holds true:

$$12x^2 + 2a = 0.$$

Therefore, we have $x^2 = -\frac{a}{b}$, and $b^2 = 4x^2 (a + 2x^2)^2$. That is,

$$8a^3 + 27b^2 = 0.$$

It is a cusp curve in the a–b plane crossing over the point (0,0). Each point of the curve stands for a bifurcation point of the system (see Figure 3.19 for more details).

3.3.6 Typical Applications

For the study of dynamic systems that can be quantitatively described, one can employ the catastrophe theory. For those systems, such as economic systems, social systems, etc., that are difficult to describe using dynamics, the catastrophe theory can be utilized qualitatively. In each of such qualitative applications, other than that the dynamic equations of the system is unknown, the state variables that describe the problem of concern and the total number of these variables cannot be accurately determined, either. Only relevant conjectures can be made based on the perceptual knowledge of the problem and related experiments. To this end, the catastrophe theory, developed on the foundation of topology, provides a tool for establishing these conjectures, while simplifying the problem by grouping the state variables into two classes. One class contains those variables that are relevant to the system's structural stability, while the other class is irrelevant to the stability. The former variables are known as influential, only with which catastrophes are associated. Generally, there are only one or two influential state variables. Thus, selecting those influential state variables that represent catastrophes becomes a relatively easy task.

After having selected the influential state variables, the catastrophe theory can be applied to further determine the control parameters and to describe and interpret the theoretical outcomes.

The catastrophe theory is a mathematical model of the phenomena of sudden changes. It investigates these changes widely existing in nature and social activities. By employing the catastrophe theory, one can interpret the phenomena of catastrophes, predict possible sudden changes that are forthcoming, and control the appearance of sudden changes.

The catastrophe theory has found successful applications in a wide range of human activities. In the following, we will look at several specific cases of application (Li 2006).

3.3.6.1 Applications in Economic Systems

In the research of the rise and fall of economic systems, the catastrophe theory has found very important applications. In the study of the economic cladogenesis problems, when an economic system is situated in an equilibrium state, a general macroscopic equilibrium, that is, when the overall system in terms of supplies, currency, economic ecology, etc., in their material forms, is in an equilibrium or near-equilibrium state, minor fluctuations and disturbances will not be sufficient to make the system evolve away from its attractor so that its stability is affected. When a system evolves a good distance away from equilibrium, some of the magnified fluctuations could be suddenly changed to various states. The economic development of any nation stands for processes of branching and combination of state catastrophes caused by the economic rises and falls of time periods. If each deviational rise and fall from the equilibrium leads to a sustained growing catastrophe in the phase space, then such a historical process stands for a long-lasting economic growth; otherwise, economic fluctuations appear. If the fluctuations show gaps of time, they represent periodic cycles. The key for the study of these cycles is to determine the factors that cause economic rises and falls and predict the appearance of those factors that could signal the next economic rise and fall.

The Keynesian equilibrium theory mainly utilizes the aggregated demand as the fluctuation factor and price level as the state variable. In any realistic economic system, there are many factors that cause economic fluctuations. Thus, in comparison, Keynesian equilibrium theory employs only a very limited number of fluctuation factors. When this theory is applied to analyze a realistic economic system, one needs to consider many additional conditions. If the theories of catastrophe and dissipative structures are employed to analyze nonlinear regions of an economic system that are far away from the equilibrium, inflation might be caused either by the attractions of the regions with stable price levels or by some sudden changes. Therefore, in order to provide additional grounds for economic decision-making, there is a need to investigate the future economic evolution using the concepts of catastrophes and dissipative structures.

For instance, bimodal distributions can be employed to represent opposing information, such as the approvals and oppositions of exploiting a certain natural resource. In such a situation, there are two opposing or mutually conflicting control factors, denoted respectively by u,v. For the simplest possible scenario, the decision-making process involving these two parameters (u,v) can be expressed using the following potential function: $V(x) = x^4 + ux^2 + vx$.

When the profit is high while the cost is low, it might be easy for the decision-maker to pursue after the decision of additional investment with the hope of acquiring high levels of profits. As the market condition changes, higher and higher prices need to be paid to create the desired profits. When the level of investment reaches a certain level, the corresponding profits start to fall from the top leaf of a cuspidal folding curve toward its edge. When the level of investment reaches the edge, it experiences a jump and then enters into the bottom leaf of the curve, indicating the transition from active investment to a complete stop of investment.

When the cost is very high while the profit is low, the decision-maker would refuse to make any initial investment. Along with the development of the economic conditions, when a transition of lowering cost and rising profit appears, the decision-making process of the investor moves the level of investment from the bottom leaf of the folding curve toward the edge. When it reaches the edge, a jump to the top leaf occurs, indicating the decisional transition from refusing to invest to actively investing.

When neither the cost nor the profit shows any clear advantage, the investor takes the neutral position of not making any decision and spends more time to wait for the further development of the market conditions. In this case, one needs to first modify his or her model and then derive the consequent results.

3.3.6.2 Applications in Military Affairs

Cusp catastrophe has been applied in the works of military operations. For instance, when two forces confront each other under a certain war environment, the outcome will be either that one side wins while the other side loses or that the forces hold up to each other, a stalemate. The defense line could be either unbreakable or instantly broken. By employing the theory of catastrophe, one can investigate the condition under which one side is victorious while the other side is defeated. He can also study how the offensive side takes actions in order to defeat the defensive side and how the defensive side arranges its defense into order to strengthen its chance of survival, etc.

In this area of applications, catastrophe models and the Monte Carlo method have been combined to simulate combat tactics. On the basis of the computer programs developed to simulate cusp catastrophes, scholars have used the Monte Carlo method to simulate face-to-face air-strikes for achieving the purpose of breaking into the air corridor of the enemy. Through simulation programs, the computer terminal can figuratively display the changes of the combat situations under various conditions.

3.3.6.3 Applications in Social Sciences

Throughout the history of China, the rise and fall of a dynasty can be described by using a cusp catastrophe model. The history has been well described as "divide after a long period of unity, and unite after a long period of separation." In short, the state of China has been historically reciprocating between the state of unity and that of separation. Each such reciprocating change can be seen as a sudden change, which occurs within a relative short period of time of turmoil. Evidently, such processes of evolution can be appropriately described by employing cusp catastrophe models, where the state variable represents the social condition with separation corresponding to a minimal value, decline a medium value, and prosperity a maximal value.

When two control variables are introduced, one can be used to represent the adjusting force for the maintenance of a centralized government. This variable measures the capability of decision-making and the consequent implementation of the decisions. The other variable stands for the centrifugal force leading to separation. It is jointly determined by the annexation degree of the land, level of corruption of the bureaucracy, and changes in the societal ideologies, etc. When the magnitudes of the centrifugal force and the unifying force fluctuate in comparison with each other, the states of centralism and feudal separatism alternate in the form of catastrophes.

3.4 OPEN-ENDED PROBLEMS

1. Explain the principles and methods of the linear science that can be employed to deal with nonlinear problems.
2. What are the characteristics of nonlinearity? In your own language, illustrate the fact that the essence of the physical world is nonlinear by considering nonlinear scenarios existing in nature, society, the mind, and engineering processes.
3. Please depict the dialectical relationship between linearity and nonlinearity, while describing the role of linear methods in the study of nonlinearity.
4. Nonlinearity is the origin of the appearance of the world's infinite diversity, abundance, strangeness, and complexity. Try to explain why using your own language.
5. Explain the concept of bifurcation using examples.
6. What are the characteristics of catastrophe? What are the differences and connections between the concepts of bifurcation and catastrophe?

7. Show that each of the following systems of equations contains Lyapunov function of the form $V(x,y) = ax^2 + by^2$:

a. $$\begin{cases} \frac{dx}{dt} = -x + 2y^3 \\ \frac{dy}{dt} = -2xy^2 \end{cases}$$

b. $$\begin{cases} \frac{dx}{dt} = x^3 - 2y^3 \\ \frac{dy}{dt} = xy^2 + x^2y + \frac{1}{2}y^3. \end{cases}$$

8. Solve each of the following systems for their respective Lyapunov functions:

a. $$\begin{cases} \frac{dx}{dt} = -x - y + (x - y)(x^2 + y^2) \\ \frac{dy}{dt} = x - y + (x + y)(x^2 + y^2) \end{cases}$$

b. $$\begin{cases} \frac{dx}{dt} = -xy^6 \\ \frac{dy}{dt} = x^4y^3 \end{cases}$$

c. $$\begin{cases} \frac{dx}{dt} = ax - xy^2 \\ \frac{dy}{dt} = 2x^4y \end{cases}$$

d. $$\begin{cases} \frac{dx}{dt} = y - xf(x,y) \\ \frac{dy}{dt} = -x - yf(x,y). \end{cases}$$

9. Analyze Figure 3.16 for its transcritical bifurcation and pitch fork bifurcation.
10. Based on Figure 3.13, analyze the phase graph distribution of fixed points of the two-dimensional linear system.

4 Self-Organization: Dissipative Structures and Synergetics

In order to explain explicitly the general and fundamental postulates of different systems, be they natural or social, one needs theories to describe the intrinsic mechanism underlying the systems and the interactions between these individual systems and their outside environments. In this chapter, self-organization systems and relevant theories are introduced. In particular, the theory of dissipative structures focuses on the relationship between a system and its outside environment, while the theory, named synergetics, aims at the investigation of the intrinsic operational mechanism of systems.

4.1 ORGANIZATION AND SELF-ORGANIZATION

Organization means to organize according to a certain goal, a predetermined task, and form. It stands for the evolution of the system of concern. The structure formed during the process of organization is also known as organization. Speaking generally, comparing to the state of the system before organization takes place, the organizational structure shows an increased degree of order and a lowered degree of symmetry. Each organizational process represents a qualitative change of the system with increasing degree of order. There are many different forms of organization, such as various kinds of societal organizations and economic entities of people, ants, bees, biological chains, etc. Based on the characteristics, organizations are classified into two categories: self-organization and planned organization.

4.1.1 General Definitions of Organization and Self-Organization

The term "self-organizing" was initially introduced into contemporary science by William Ross Ashby (1947), a psychiatrist and engineer. It was soon taken up by such well-known cyberneticians as Heinz von Foerster, Gordon Pask, Stafford Beer, and Norbert Wiener, where Wiener discussed this concept in the second edition of his book *Cybernetics: or Control and Communication in the Animal and the Machine* (MIT Press 1961).

Self-organization as a concept was used by those associated with general systems theory in the 1960s but did not become commonly accepted in the scientific literature until its adoption by physicists and researchers in the field of complex systems in the 1970s and 1980s. After the award of a Nobel Prize to Ilya Prigogine in 1977, the *thermodynamic concept of self-organization* received the attention of the public, and scientific researchers started to migrate from the point of view of cybernetics to that of thermodynamics.

Definition 4.1

Self-organization stands for such a process that a structure or pattern appears in a system without being influenced by a central authority or through an external element that imposes its planning. This globally coherent pattern appears from local interactions of the elements that make up the system. Thus, the organization is achieved in a way that is parallel (all the elements act at the same time) and distributed (no element is a coordinator) (Glansdorff and Prigogine 1971). ■

Self-organization and planned organization are relatively classified, without an absolute borderline. When analyzing specific systems, the theory of self-organization or the theory of planned organization is adopted based on the needs of investigation. The following provides some of the widely employed methods of classification.

1. In terms of the formation of system's organization, there is a coordinator that exists outside the system and is responsible for the appearance and establishment of the planned organization. By setting a goal, the coordinator has a predetermined plan and scheme to reach the eventual goal.

   A system is referred to as a self-organization system, if the system acquires its time, space, and/or functional structure without being influenced by any imposing external element.
2. In terms of the relationship between the external control and system response, for a planned organization, this relationship is explicit and can be studied analytically, such as that of a pilot who controls the movement of his or her plane. When that relationship is not explicit, the system can be regarded as a self-organization, such as the organization of a family of ants.
3. In terms of the relationship of input and output, in a planned organization system, the relationship between input and output is determined ahead of time. For instance, a guided missile system can hit the target according to the information input beforehand. Otherwise, the system can be regarded as a self-organization system, such as the Rayleigh–Bénard convection (for details see Section 4.1.3); the location of the design could not be determined beforehand.

Self-organization is a natural phenomenon that includes the evolvement of complex systems. The theory of self-organization investigates the appearance and development of various different phenomena of self-organization that exist in the natural world. Because there is not any universal method available to deal with nonlinear systems, the theory of self-organization is far from being mature when seen as a discipline. Generally, dissipative structures and synergetics are considered two theories about self-organization systems. The former focuses on the study of the relationship between the system of concern and its external environment, while the latter investigates the intrinsic operational mechanism of the system.

At the most generality, there are four forms of self-organization, including self-creation, self-replication, self-growth, and self-adaptation.

Self-creation embodies the increase in order of each new state of the self-organization process. It analyzes the relationship between the different states before and after the self-organization.

Self-replication is a description of the interaction of subsystems in a self-organization process. It guarantees that the system can reach a steady state of new order. In most cases, it describes and is concerned with subsystems.

Self-growth describes the state evolution of the system with time in the self-organization process from the viewpoint of systems at the unitary level. The system's characteristics stay unchanged, while its volume becomes larger.

Finally, self-adaptation is a description of the self-organization in terms of the relationship between the system and its environment.

4.1.2 Definition of Organization and Self-Organization by Set Theory

What has been discussed above is a basic description of the phenomenon of self-organization and its corresponding concepts. In the following, we will represent these concepts by using the language of set theory.

Definition 4.2

(Organization). For a given time system $S_t = (M_t, R_t)$, $t \in T$, if there exists $t_1, t_2 \in T$ such that if $t_1 < t_2$ implies that S_{t_1} is a proper subsystem of S_{t_2}, then the evolvement process of the given time system from time moment t_1 to t_2 is referred to as an organization or an organizational process. In this case, the fact that the system S_{t_1} is a proper subsystem of S_{t_2} is seen as that the degree of order of the time system $S_t = (M_t, R_t)$ has been increased from that at the time moment t_1 to that at t_2. ■

Definition 4.3

(Self-organization and planned organization). Assume that the evolutional process of the time system $S_t = (M_t, R_t)$, $t_1 \in T$, from t_1 to t_2 is an organization. If for each $t \in T$, the moment system $S_t = (M_t, R_t)$ has an environment E_t, and the organizational change of S_t from t_1 to t_2 depends only on that of E_t from t_1 to t_2, that is, when the law of change of E_t from t_1 to t_2 is given, the corresponding evolution of S_t plays out automatically, then the organization of S_t from t_1 to t_2 is known as a self-organization. Otherwise, the organization is known as planned. ■

In other words, the key that separates a self-organization from any planned organization is whether or not the evolution of the system S_t is affected by any elements from outside the object set M_t, when the law of change of the environment E_t is given.

Definition 4.4

(Self-creation and self-adaptation). In a self-organizational process, assume the system $S_{t_1} = \left(M_{t_1}, R_{t_1}\right)$ evolves into $S_{t_2} = \left(M_{t_2}, R_{t_2}\right)$. If either $m \in M_{t_2}$ and $m \quad M_{t_1}$ or $r \in R_{t_2}$ and $r \quad R_{t_1}$ such that the particular m or r cannot be obtained through any combination of components and/or relations of S_{t_1}, then the element m or the relation r or both are referred to as self-created by system S_t; otherwise, either m or r or both are referred to as self-adaption(s). ■

Definition 4.5

(Self-replication). In a self-organizational process, assume that the system $S_{t_1} = \left(M_{t_1}, R_{t_1}\right)$ evolves into $S_{t_2} = \left(M_{t_2}, R_{t_2}\right)$. If there are $M_{t_1} \quad M^*_{t_2} \quad M_{t_2}$, $R^*_{t_2} \quad R_{t_2}$, and a mapping $f\colon M^*_{t_2} \to M_{t_1}$ satisfying that $f(m) = m$, for all $m \in M_{t_1}$, and $f\left(M^*_{t_2}, R^*_{t_2}\middle| M^*_{t_2}\right) = \left(M_{t_1}, R_{t_1}\right) = S_{t_1}$, then each object $m \in M_{t_1}$ such that $|f^{-1}(m)| > 1$ is said to have self-replicated in the self-organizational process from S_{t_1} to S_{t_2}. ■

In terms of the daily language, self-replication of an object m means that identical copies of m are introduced during the evolution of the time system. For instance, the replication of cells is a kind of self-replication.

Definition 4.6

(Self-growth). In a self-organizational process, assume that the system $S_{t_1} = \left(M_{t_1}, R_{t_1}\right)$ evolves into $S_{t_2} = \left(M_{t_2}, R_{t_2}\right)$ and that there are $M_{t_1} \quad M^*_{t_2} \quad M_{t_2}$ and a mapping $f\colon M^*_{t_2} \to M_{t_1}$ satisfying that $f(m) = m$, for all $m \in M_{t_1}$. If there are $m \in M_{t_1}$, $(x_1, x_2, ..., x_\alpha, ...) \in r_{t_1} \in R_{t_1}$, and $r_{t_2} \in R_{t_2}\big|M^*_{t_2}$ such that $|f^{-1}(m)| = |\{m_1, m_2, \ldots, m_\beta, \ldots\}| > 1$, $m = x_i$, $i \in I_m$, for a given nonempty index set I_m, and $(y_1, y_2, ..., y_\gamma, ...) \in r_{t_2}$, where $(y_1, y_2, \ldots, y_\gamma, \ldots)$ is obtained from $(x_1, x_2, \ldots, x_\alpha, \ldots)$ by replacing each $m = x_i$, $i \in I_m$, with an ordered set of $\{m_1, m_2, \ldots, m_\beta, \ldots\}$, then each such object $m \in M_{t_1}$ is said to have self-grown in the self-organizational process from S_{t_1} to S_{t_2}. ■

Intuitively, an object $m \in M_{t_1}$ is self-grown in the self-organizational process from the state S_{t_1} to the state S_{t_2} if the object m has been split into a set X_m of additional copies at state S_{t_2} such that some of the relations originally satisfied by m at state S_{t_1} now are collectively satisfied at state S_{t_2} by the entire family X_m. In other words, the importance and functionality of the object m at state S_{t_1} become those of the set X_m. In many widely seen scenarios, self-replications of subsystems are the reason for the systems to self-grow.

As a matter of fact, in the self-organization process of many systems, self-creation, self-adaptation, self-replication, and self-growth can all appear either collectively or simultaneously. That creates and accomplishes each and every process of self-organization.

4.1.3 Some Examples of Self-Organization

4.1.3.1 Rayleigh–Bénard Convection

Rayleigh–Bénard convection is a type of natural convection that occurs in a fluid that fills the space between two parallel planes when heated from below, where the distance between the planes is small compared to the horizontal dimension. The fluid when heated from below develops a regular pattern of convection cells, known as Bénard cells, with the initial upwelling of warmer liquid from the heated bottom layer. Rayleigh–Bénard convection is one of the most commonly studied convection phenomena because of its analytical and experimental accessibility. The convection patterns are the most carefully examined example of self-organizing nonlinear systems.

In such fluid placed between two parallel planes, as described above, when the temperature difference between the top and the bottom is small, the heat is transported throughout in the form of thermal conduction so that the fluid is in a static state. When the temperature gradient passes a threshold value, the fluid located on the bottom starts to float to the top and then returns to the bottom along the side of the cell in a visible macroscopic motion (Figure 4.1). What is surprising is that a pattern that consists of quite regular right hexagonal prisms appears within the fluid. When the temperature gradient continues to increase, each of the rotational hexagonal prisms vibrates along its axis.

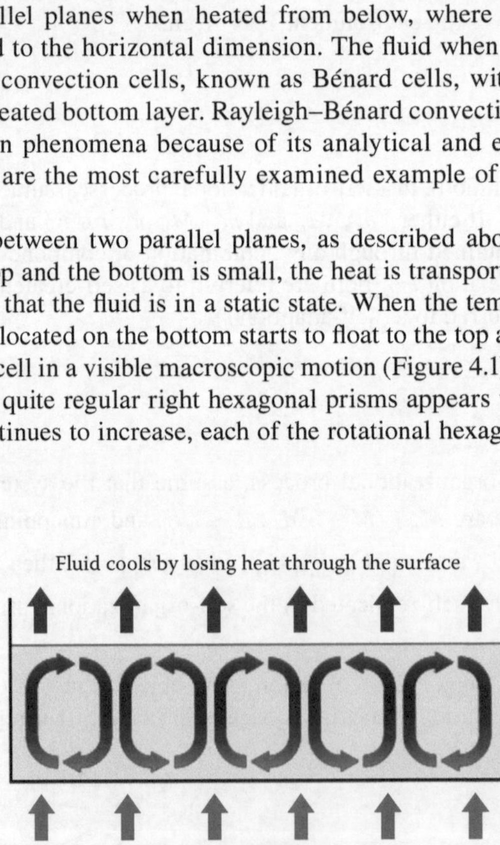

FIGURE 4.1 Rayleigh–Bénard convection. (From Getling, A.V. *Rayleigh–Bénard Convection: Structures and Dynamics*. Singapore: World Scientific, 1998.)

We know that the heat merely increases the energy of each molecule of the fluid, while the distances among the molecules are much larger than the diameter of each molecule. However, the molecules form the regular macroscopic pattern of cells. That indicates that the range of effect of the molecules is way greater than the scale of the molecules, leading to the formation of the macroscopic structure. The appearance of this spatial structure evidently is not a product of the properties and characteristics of the molecules that make up the system of interest. Instead, it is produced out of the interaction and joint effect of the molecules.

4.1.3.2 Belousov–Zhabotinsky Reaction

A Belousov–Zhabotinsky reaction, or BZ reaction, is one of the many reactions that serve as classical examples of nonequilibrium thermodynamics. Resulting from these examples is the establishment of a nonlinear chemical oscillator. The only common element in these oscillating systems is the inclusion of bromine and an acid. The reactions are theoretically important because they show that chemical reactions do not have to be dominated by equilibrium thermodynamic behavior. These reactions are far from equilibrium and remain so for a significant length of time. In this sense, they provide an interesting chemical model of nonequilibrium biological phenomena, and the mathematical models of the BZ reaction themselves are of theoretical interest.

An essential aspect of the BZ reaction is its so-called "excitability"—under the influence of stimuli, patterns develop in what would otherwise be a perfectly quiescent medium.

The discovery of the phenomenon is credited to Boris Belousov. He noted sometime in the 1950s (the dates change depending on the source one references to but range from 1951 to 1958) that in a mix of potassium bromate, cerium(IV) sulfate, propanedioic acid, and citric acid in diluted sulfuric acid, the ratio of concentration of the cerium(IV) and cerium(III) ions oscillates, causing the color of the solution to vary between a yellowish solution and a colorless solution. That is due to the cerium(IV) ions being reduced by propanedioic acid to cerium(III) ions, which are then oxidized back to cerium(IV) ions by bromate(V) ions. See Figure 4.2 for more details.

In the BZ reaction, the color "structure" is once again caused not by any external coordinator but by the interaction of the molecules.

FIGURE 4.2 Stirred BZ reaction mixture showing changes in color over time. (From Zhabotinsky, A.M. *Biophysics*, 9, 306–311, 1964.)

4.1.3.3 Laser

A laser is a device that emits light, more specifically, electromagnetic radiation, through a process of optical amplification based on the stimulated emission of photons. The emitted laser light is notable for its high degree of spatial and temporal coherence that is unattainable by using any other currently available technology. Spatial coherence is typically expressed as the output being a narrow beam that is diffraction-limited and is often a so-called "pencil beam." The temporal (or longitudinal) coherence implies a polarized wave at a single frequency whose phase is correlated over a relatively large distance (the coherence length) along the beam. This is in contrast to thermal or incoherent light emitted by ordinary sources whose instantaneous amplitudes and phases vary randomly with respect to time and position. Although temporal coherence implies monochromatic emission, there are such lasers that emit a broad spectrum of light or emit different wavelengths of light simultaneously.

Laser is not a natural light. Instead, it is made specially by using the method of self-organization. Without any external command or coordinator, in the coherence system, the self-organization of electrons produces the laser beam. To establish the concept of self-organization of general systems, H. Haken studied the underlying mechanism of laser for over 20 years. It is due to his work that the mechanism underlying the process of self-organization becomes clear through massive mathematical computations.

4.1.3.4 Continental Drift

In 1915, the German geologist and meteorologist Alfred Wegener first proposed the theory of continental drift in his book *On the Origin of Continents and Oceans*. It states that parts of the earth's crust slowly drift atop a liquid core. Together with the theory of plate tectonics, the theory of continental drift has been well supported by fossil records.

Wegener hypothesized that there was a gigantic supercontinent 200 million years ago, which he named Pangaea, meaning "all-earth" in Greek. Then, Pangaea started to break up into two smaller supercontinents, named Laurasia and Gondwanaland, respectively, during the Jurassic period. By the end of the Cretaceous period, the continents further broke up into land masses that look like our modern-day continents (Figure 4.3). For more details, please consult with *Continental Drift, Dinosaur and Paleontology Dictionary.*

The cause of continental drift can also be explained by using the analysis and results of the Rayleigh–Bénard convection. It is because there is a great temperature difference for the fluid that exists in between the crust and core of the earth, and that temperature difference moves the crust, although the movement is very slow.

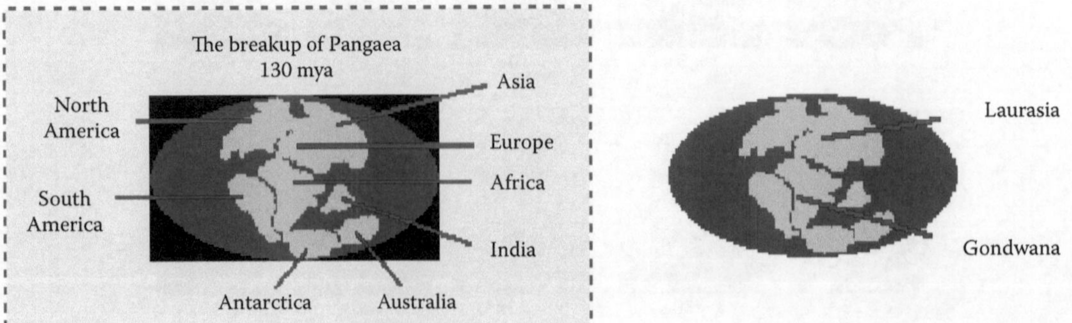

FIGURE 4.3 Breakup of Pangaea.

4.2 DISSIPATIVE STRUCTURES

4.2.1 Classical and Nonequilibrium Thermodynamics

In the physical world, the motion of any matter or object takes place in a determinate time and space. In the theory of dynamics, the concept of time is regarded as a fourth-dimensional scale that describes the related physical process similarly as the three-dimensional coordinates do in terms of the spatial position of the object. In Newtonian physics, quantum mechanics, and relativity theory, time is a geometry and numerical parameter introduced to describe reversible motions. What needs to be emphasized is that in these theories, time is reversible. Therefore, in the classical dynamics, the concept of evolution does not truly exist. That is, the concept of time is not intrinsically related to the characteristics of movement of objects. The classical dynamics presents a reversible and symmetric image of the physical world.

4.2.1.1 Classical Thermodynamics

In the nineteenth century, the invention and application of steam engines promoted the introduction and development of thermodynamics and statistical physics. The former theory is a part of physics that deals with the relationships between heat energy and other forms of energy.

The first law of thermodynamics, established, respectively, by Mayer, Joule, and Hermann von Helmholtz during 1842–1848, is an expression of the conservation law of energy. It states that energy can be transformed or changed from one form to another but cannot be created or destroyed. It is usually formulated as that the change in the internal energy of a system is equal to the amount of heat supplied to the system minus the amount of work performed by the system on its surroundings.

The second law of thermodynamics, whose first formulation is credited to the German scientist Rudolf Clausius, states that over time, differences in temperature, pressure, and chemical potential tend to equilibrate in an isolated physical system. In the classical thermodynamics, this second law is a basic postulate applicable to any system involving measurable heat transfer, while in statistical thermodynamics, this law is a fundamental postulate of the assumed randomness of molecular chaos. It is also detrimental in the development of the concept of thermodynamic entropy.

What is explained by the second law of thermodynamics is the widely existing phenomenon of irreversibility in nature; it posts the physically existing irreversibility of natural evolution on the concept of time. For example, if the distribution of the initial temperature of a metal stick is asymmetric, and if the stick is also adiabatic from the environment, then the stick would experience an irreversible change in terms of its temperature. In particular, the higher-temperature part of the stick would pass heat to the low-temperature part, spreading out the uneven distribution of temperature in the stick. This process of temperature redistribution will not be reversed forever unless an external influence is imposed. This heat transportation process can be represented by using a Fourier equation. Similarly, if one pours some water-based ink into a container that holds some clear water, the ink will spread and mix with the clear water into some well-proportioned liquid. Once again, the ink and the clear water in this eventual mixture of fluid will not separate from each other without an external force.

In other words, the second law of thermodynamics implies that each closed system that is isolated from the environment would evolve in the direction of becoming even with any initially existing difference being eliminated. What it describes is actually a kind of degradation of the initial condition. Rudolf Clausius generalizes this theory and concludes that the whole universe is obviously isolated; therefore, its entropy should increase until it runs into such a state of no thermodynamic free energy that no motion or life could be sustained, that is, the heat death.

On the other hand, Darwinism claims that monads emerge from the hungry earth and that natural selection takes place based on the principle of "survival of the fittest." Gradually speaking, all advanced biological beings have been evolving with time, and it is through evolution that human beings eventually come into existence. The development of creatures forms the evolutionary direction of biological evolvement. This evolutionary direction is stridently contradictory to that of thermodynamic degradation.

The aforementioned contradiction of two images of nature embodies an implicit relationship between dynamics and thermodynamics or a relationship between physics and biology. Reversibility or irreversibility, the direction of time arrow reveals the existence of a gap between the dynamics and thermodynamics. Additionally, the direction of time arrow constitutes a basic problem for or difference between the research in the lifeless sciences and life sciences.

Ilya Prigogine, a Russian-Belgian physical chemist, whose theory of dissipative systems will be introduced in Section 4.2.2, divided the study of physics into two parts: one for the research of existence, such as the classical mechanics and quantum mechanics, in which time is reversible, and the other for the research of evolvement or irreversible phenomena from simple heat conducts to complex self-organizations of biological beings. By using microcosmic theories and based on the fact that entropy changes irreversibly, Prigogine materialized the transition from existence to evolvement and unified the two parts of physics.

4.2.1.2 Nonequilibrium Thermodynamics

Definition 4.7

(System state). State of a system $S_t = (M_t, R_t)$ means the characteristics of the system at the fixed time moment t. The state contains a microcosmic or macroscopic exhibition of the sets M_t and R_t, which sometimes may be described by a function of M_t and R_t. ■

For example, two macroscopic parameters for gas are temperature and intensity of pressure, while microcosmic parameters for a gas molecule may be the molecule's moving speed, direction, and kinetic energy.

Definition 4.8

(Equilibrium state). Equilibrium state means such a state of the system that does not change with time, or there does not exist any macroscopic physical quantity that flows within the system. An equilibrium state often sustains the uniformity of the space. If an isolated or closed system reaches an equilibrium state, it would no longer change any further. Near-equilibrium state means such a state of the system that lies in a linear region very close to an equilibrium state. Exit-from-equilibrium state means that after reaching an equilibrium state, the interior of the system becomes terribly uneven so that there appears a macroscopic physical quantity that flows within the system, causing the system to change again with time. ■

For an isolated system, its steady state is also an equilibrium state. However, for open systems, steady and equilibrium states are essentially different. In other words, an open system would not evolve into a steady state, or its steady state might not be an equilibrium state.

Definition 4.9

(Symmetry). A characteristic of the system is said to be symmetric if the attributes of the system stay unchanged with some transition of the system's state. ■

The concept of symmetry is roughly related to that of order. The better the symmetry is, the lower the order becomes. That is, order predicates a reduction of symmetry. Therefore, a system at an equilibrium state possesses more symmetry than when it is in a nonequilibrium state. Both symmetry and order are important concepts in the theory of self-organization.

Entropy is one of the most important concepts introduced in thermodynamics. It could measure how much a system is out of order. Therefore, the concept of entropy is also very important for the research of complex systems.

Definition 4.10

(Entropy). The arrangement number of each macroscopic distribution in a thermodynamic system W is referred to as thermodynamic odds. Based on the thermodynamic odds, the entropy of the system is defined as

$$S = k\ln W \tag{4.1}$$

where $k = 1.381 \times 10^{-23}$ J/K is the Boltzmann constant, and W is the thermodynamic odds (Li 2006). ■

Different macroscopic distributions have different thermodynamic odds, while the uniform distribution corresponds to the largest thermodynamic odds W_M with the largest entropy:

$$S_M = k\ln W_M. \tag{4.2}$$

When an isolated system is in a steady state, the particles of the system will always be uniformly distributed. At this time, the system is out of order and in chaos to the largest degree; hence, the entropy is the largest. Thus, the concept of entropy could be used to measure how much a system is out of order.

In 1856, the German physicist Rudolf Clausius stated his "second fundamental theorem in the mechanical theory of heat" in the following form (Clausius 1865):

$$\int \frac{\delta Q}{T} = -N$$

where Q stands for heat, T temperature, and N the "equivalence value" of all uncompensated transformations involved in the cyclical process. Later in 1865, Clausius defined the "equivalence value" as entropy. On the basis of this definition, in that same year, the most famous version of the second law was read in a presentation at the Philosophical Society of Zurich on April 24, in which at the end of his presentation, Clausius concluded:

> The entropy of the universe tends to maximize.

This statement is the best-known phrasing of the second law.

As a matter of fact, the heat conduction from high-temperature areas to low-temperature areas and the simplex transformation from work to heat are both irreversible. In other words, the process through which the entropy of an isolated system increases is also irreversible. Namely, an isolated system could be developed into an equilibrium state from any nonequilibrium state, while the reverse process is impossible.

As for an open system, the entropy dS of the system should be divided into two parts:

$$dS = dSe + dSi \tag{4.3}$$

where the entropy dS includes two parts: the system's entropy flow dSe and the interior, intrinsic entropy dSi. The entropy flow dSe reflects the entropy exchange between the system and environment. It could be positive, negative, or zero. The interior, intrinsic entropy dSi represents the entropy of the system and could not be negative, namely,

$$dSi \geq 0 \text{ or } dS \geq dSe. \tag{4.4}$$

Similar to the concept of energy, entropy is also a physical quantity that can be transformed. Distinct from the energy of one isolated system that has to stay unchanged, the entropy of the system increases in an irreversible process.

In an irreversible process, the entropy increases. Within a unit time, the increase is referred to as the entropy production rate, namely,

$$P = dSi/dt. \tag{4.5}$$

When the initial temperature of a metal stick that is adiabatic with the environment is asymmetric, the stick will experience an irreversible change. In particular, the part of higher temperature will pass the heat to that of lower temperature; eventually, the temperature distribution in the metal stick will become even.

This flow of temperature is irreversible. The cause that produces the irreversible flow is referred to as an irreversible force. As for the example of the metal stick, the uneven temperature distribution is the essential irreversible force.

Evidently, in all kinds of irreversible processes, the entropy production rate depends on the magnitude of flows and the magnitude of forces. In thermodynamics, the entropy production rate is defined to be the product of the magnitude of flow and that of force in the irreversible process, symbolically,

$$p = \sum_i y_i x_i \tag{4.6}$$

where y_i and x_i represent the magnitudes of the irreversible flow and force. Equation 4.6 is applicable to any irreversible process.

The intensity of the irreversible flow Y_i depends on the irreversible force X_i. Generally, the flow is a complex function of the force. When the magnitudes of irreversible flow and force are not large, there exists a nearly linear relationship between Y_i and X_i shown as

$$Y_i = L_{ii} X_i$$

where L_{ii} is a self-phenomenological coefficient. It is a constant independent of X_i. Generally, in a linear region, the relationship between the flow and force takes the following form:

$$Y_i = \sum_j L_{ij} X_j \tag{4.7}$$

where L_{ij} is a linear phenomenological coefficient. When $i \neq j$, L_{ij} is referred to as a cross-phenomenological coefficient.

Lars Onsager (1931) found that the phenomenological coefficients satisfy

$$L_{ij} = L_{ji}, \tag{4.8}$$

which is named as the Onsager reciprocal relation. That is, the influence of the *i*th force on the *j*th flow equals that of the *j*th force to the *i*th flow.

The equality of certain ratios between flows and forces of thermodynamic systems that are out of equilibrium is not dependent on the specific types of the flows and forces. However, "reciprocal relations" occur between different pairs of forces and flows in a variety of physical and chemical systems. For example, let us consider fluid systems described in terms of temperature, matter density, and pressure. In any of such systems, it is known that temperature differences lead to heat flows from warmer areas to colder parts of the system. Similarly, pressure differences lead to matter flow from high-pressure regions to low-pressure fields. What is truly remarkable is the observation that when both pressure and temperature vary, temperature differences at constant pressure can cause matter flow (as in convection), and pressure differences at constant temperature can cause heat flow. It is perhaps more surprising to know that the heat flow per unit of pressure difference is identical to the density (matter) flow per unit of temperature difference. This equality was shown to be necessary by Onsager by using statistical mechanics as a consequence of the time reversibility of microscopic dynamics. The theory developed by Onsager is of course much more general than what this example shows and is capable of treating more than two thermodynamic forces.

For his discovery of the reciprocal relations, Onsager was awarded the 1968 Nobel Prize in chemistry. In his award presentation speech, Onsager referred to the three laws of thermodynamics and then added (Lars Onsager), "It can be said that Onsager's reciprocal relations represent a further law making possible a thermodynamic study of irreversible processes." Wendt (1974) has even described Onsager's reciprocal relations as the "fourth law of thermodynamics."

The second law of thermodynamics indicates that the entropy production rate could not be negative. Thus, it is easy to see that

$$p = \sum_i L_{ij} X_i Y_j \geq 0. \tag{4.9}$$

That requires the coefficient matrix L be a positive definite.

Prigogine proved Equation 4.9. It is a universal principle for any irreversible process that when a system is in a nonequilibrium state, the entropy production rate of the system takes the minimum value. If a nonlinear condition holds true in a linear region, then when the system reaches a steady state, the entropy production rate of the system would be smaller than that when the system is at a nonsteady state. This principle is known as the minimum entropy postulate, which is derived by Prigogine on the basis of the Onsager reciprocal relations.

An isolated system would evolve in a direction of increasing entropy until it reaches an equilibrium state when the entropy reaches its maximum and the entropy production rate becomes zero. If there is an external restriction on the system, making it nonequilibrium, the system will have a positive entropy production rate. The minimum entropy postulate indicates that the entropy production rate continues to diminish until reaching its minimum value, while at the same time, the system reaches a steady state. Once the restriction of uniform temperature is broken, the system will evolve toward a new steady state. Namely, a thermodynamic system would evolve to an equilibrium state as its destination, and the destination is mostly a uniform and out-of-order state.

Therefore, the minimum entropy postulate requires the system of concern to stay near an equilibrium state or within the linear region of the nonlinear conditions of a nearly equilibrium state. Also, it implies that within the linear thermodynamic region of a nonequilibrium state, each nonequilibrium steady state is stable. When the system is far away from any equilibrium state, there is no longer any linear relationship between flows and forces. In this case, the minimum entropy postulate becomes invalid. It was in his investigation of systems that are far away from any equilibrium state that Prigogine established his theory of dissipative structures.

4.2.2 Dissipative Structures

Such a structure that is characterized by spontaneous appearance of symmetry breaking (anisotropy) and formation of complex, sometimes chaotic, structures, where interacting particles exhibit long-range correlations, is known as a dissipative structure. The term *dissipative structure* was coined together by Ilya Prigogine, a Russian-Belgian physical chemist. He was awarded the Nobel Prize in Chemistry in 1977 for his pioneering work on these structures. The dissipative structures considered by Prigogine have dynamical régimes that can be regarded as thermodynamically steady states and sometimes can at least be described by some suitable extremal principles of nonequilibrium thermodynamics. Simple examples include convection, cyclones, and hurricanes. More complex examples include lasers, Bénard cells, and the Belousov-Zhabotinsky reaction. At the most sophisticated level, the form of life itself is included.

Definition 4.11

A dissipative system is such a system that is far from thermodynamic equilibrium, and hence efficiently dissipates the heat needed to sustain it and has the capacity of changing to higher levels of orderliness (Brogliato et al. 2007). ■

According to Prigogine, dissipative systems contain subsystems that continuously fluctuate. At times, a single fluctuation or a combination of fluctuations may become so magnified by feedback that the system shatters its preexisting organization. At such revolutionary moments or "bifurcation points," it is impossible to determine in advance whether the system will disintegrate into "chaos" or leap to a new, more differentiated, higher level of "order." The latter case defines dissipative structures, so termed because they need more energy to sustain them than the simpler structures they replace and are limited in growth by the amount of heat they are able to disperse. A dissipative system is a thermodynamically open system that operates out of thermodynamic equilibrium in an environment with which it exchanges energy and matter.

Definition 4.12

(Web Dictionary of Cybernetics and Systems). A dissipative structure is a dissipative system that has a dynamical regime and is, in some sense, in a reproducible steady state. This reproducible steady state may be reached by natural evolution of the system or by human artifice or by a combination of these two. ■

Definition 4.13

Fluctuation stands for the phenomenon wherein the system of concern deviates from a steady state. ■

In general, each fluctuation can lead to disturbance and convection. The further a fluctuation point deviates from the steady point, the weaker the disturbance and convection caused by the fluctuation is. At such a point, the magnitude of the fluctuation will strengthen. To guarantee the dynamic stability, either there will be exchanges of matter and energy or a new structure will be formed to materialize a new level of stability. Whether or not a new structure will be formed is a stochastic process, where the ultimate state of the system, the threshold value of transition, the magnitude of fluctuation, and the relevant probability are related to each other.

A dissipative system would demonstrate some important characteristics:

1. The unilateralism of time. The system's behavior with time is an irreversible process. For example, a chicken being hatched could not be returned back to the original egg.
2. The spatial asymmetry in an utter sense. The imbalance existing in the system's structure is the motivation for the system to further evolve.
3. The system stays steady and creates order from the increasing negative entropy. The negative entropy comes from the environment by exchanging matter, energy, and so on. Let us look at a biological being as an example. As a basic principle, it absorbs nutrients to maintain its viable form of life, while the theory of dissipative systems provides an explanation of this dependence on nutrients from the viewpoint of energy flow and entropy change.
4. Fluctuation always exists. Fluctuations close to steady points would be minor, while at any bifurcation point, fluctuations would be magnified and force the system to evolve into a new state. A small fluctuation of the initial state would lead to huge differences in the evolution of the system. The dissipative structure could be regarded as a large fluctuation that is sustained by exchanging matter and energy with the external environment. The evolvement of the system is determined by three factors: which fluctuation happens first, the magnitude of the fluctuation, and the underlying driving force. The occurrence of fluctuation is random. In a particular system structure, fluctuation might very well determine how the system would evolve under some specific circumstances.

Each biological organism stands for a typical example of dissipative systems, although each such organism also contains mechanical, physical, and chemical movements of the primitive form.

Although the theory of dissipative structures is originated from the study of nonequilibrium physics, thermodynamics, and inorganic systems, it has also been employed in the investigations of problems of various organic or quasi-organic systems. Therefore, the theory of dissipative structures can be seen as a bridge connecting both organic and inorganic systems.

4.2.3 Conditions for Existence of Dissipative System

1. The system should be an open system.

   The second law of thermodynamics indicates that the process of the entropy increase of an isolated system is irreversible. Hence, only an open system could have the chance to decrease its total entropy by exchanging matter, energy, etc., with the external environment. Therefore, an isolated system definitely cannot be dissipative.

   A biological organism is both a typical open system and a typical dissipative structure. It needs metabolism, acquires nutrients from the environment, and gives off its waste of metabolism. Otherwise, no biological organism can survive and be living.

   It should be pointed out that the system is different from the environment. The system could not open entirely to the external world, meaning that the system has to be identifiable from the environment. Otherwise, the system would be a counterpart of the environment or other system without the necessary relatively verifiable independence of itself.
2. The system should exist far away from any equilibrium state.

   The system has to be driven somehow to a state that is far from being equilibrium. Only when it is far from any equilibrium state can an orderly structure exist within the system, because any equilibrium state possesses a steady structure, whose maintenance does not need any input of outside energy. The dissipative structure is a "living" structure, which could possibly exist only when the system is far away from being in equilibrium and when it is an open system. Prigogine even claimed that nonequilibrium is the origin of order.

3. There is a threshold over which mutation and self-organization begin to appear within the system.

   When the evolution of the system is near a critical point or threshold, a tiny change of the control parameter or initial value condition can alter the characteristics and behavior of the system drastically.
4. There is nonlinearity in the interaction of the constituents of the system. Through nonlinear interactions, the constituents could cooperate with each other and make the whole (the system) evolve into a state with more order from its previously more out-of-order states. It has been well documented that linear systems cannot evolve into a steady structure from the previously unstable states.
5. Fluctuation deduces the order. It is observed that macroscopic quantities of a system, such as temperature, pressure, energy, entropy, and so on, reflect the statistical mean effect of the microcosmic particles. However, these quantities measured at each exact time moment may not be equal to their respective mean values. There would be deviations, namely, fluctuations. When the system is in a steady state, its fluctuation is known as a disturbance, and the system has to resist the disturbance in order to maintain its steadiness. When the system is in a critical unsteady state, any fluctuation could induce the system to transit into another state of different orderliness.

Prigogine believes that the macroscopic order of a dissipative system is dominantly determined by the fluctuation that increases the order of the system the fastest.

4.2.4 Theoretical Analysis of Dissipative Systems

In this section, we will see how dissipative systems can be analytically investigated. For a further in-depth discussion, please consult the works of Li (2006) and Xu (2000).

In the Bénard convection experiment, if we control the outside temperature and make the temperature difference between the top and the bottom plate be ΔT, then the system could experience a transition from an initial equilibrium state to a nonequilibrium state, which is named as thermodynamic embranchment. When the nonequilibrium state deviates further from the initial equilibrium state, it would experience another transition into a dissipative structure, which is referred to as a nonequilibrium phase change. The nonequilibrium phase change occurs in an equilibrium state; the macroscopic structure of the system experiences a mutation.

As for the nonequilibrium phase change, suppose that the system could be represented by a group of ordinary equations as follows:

$$\begin{cases} \dfrac{dx_1}{dt} = f_1(A, x_1, \ldots, x_n) \\ \dfrac{dx_2}{dt} = f_2(A, x_1, \ldots, x_n) \\ \cdots \\ \dfrac{dx_n}{dt} = f_n(A, x_1, \ldots, x_n) \end{cases} \tag{4.10}$$

where $x_1, \ldots, x_n$ are n state variables that describe the degree of orderliness of the system. Their values for out-of-order states would be 0. If Equation 4.10 has a steady nonzero stable solution, it means that the system is a dissipative structure.

The parameter A represents the control imposed on the system from the outside environment, such as the temperature difference ΔT of the two plates in the Bénard convection experiment. This control parameter would not change in the system's process of evolution, while changing this parameter would, in general, mean that the evolution of the system is altered. That is, the control parameter has a detrimental influence on how the system would evolve.

First, let us consider the following single-variable linear parameter equation:

$$\frac{dx}{dt} = (A - A_c)x \tag{4.11}$$

where A_c is some fixed value, A is a control parameter, and x is the state variable.

The stationary solution of Equation 4.11 is $(A - A_c)\,x = 0$. That is, $x = 0$ is a stationary solution.

To determine the stability of the thermodynamics embranchment, let us give a small disturbance a to the state x at initial time moment. If the system returns to its original state when the disturbance vanishes, the thermodynamics embranchment is considered stable. If when the disturbance strengthens, the system exits from the thermodynamics embranchment, there then is a need to investigate the system's stability. With a given initial condition, the solution of Equation 4.11 is given below:

$$x = ae^{(A-A_0)t}. \tag{4.12}$$

Thus, the stability of $x = 0$ depends on the value of the control parameter A.

1. When $A < A_c$, the exponent of Equation 4.12 is negative, and the thermodynamic embranchment is stable.
2. When $A > A_c$, the exponent of Equation 4.12 is positive, and the thermodynamic embranchment is unstable.

Thus, A_c stands for a critical value, and $A = A_c$ is a bifurcation point. In the region near the bifurcation point, the characteristics of the solution would experience a primary change.

This discussion on the case of a single variable could be generalized to a multivariable system. For simplicity, let us consider a two-variable system:

$$\begin{cases} \dfrac{dx}{dt} = B_{11}x + B_{12}y \\ \dfrac{dy}{dt} = B_{21}x + B_{22}y \end{cases} \tag{4.13}$$

where B_{11}, B_{12}, B_{21}, B_{22} are constants that depend on a control parameter A. Here, $x = y = 0$ is the stationary solution of the system.

When $t = 0$, a disturbance could make the initial values of the parameters x, y deviate from the thermodynamic embranchment, namely, $x = a$, $y = b$. The solution of Equation 4.13 at time t is

$$\begin{cases} x = \dfrac{(B_{11} - \lambda_2)a + B_{12}b}{\lambda_1 - \lambda_2} e^{\lambda_1 t} + \dfrac{(\lambda_1 - B_{11})a + B_{12}b}{\lambda_1 - \lambda_2} e^{\lambda_2 t} \\ y = \dfrac{B_{21}a + (B_{22} - \lambda_2)b}{\lambda_1 - \lambda_2} e^{\lambda_1 t} + \dfrac{(\lambda_1 - B_{22})b - B_{21}a}{\lambda_1 - \lambda_2} e^{\lambda_2 t} \end{cases} \tag{4.14}$$

where λ_1, λ_2 are the solutions of algebraic equation (latent equation)

$$\lambda^2 - \omega\lambda + T = 0 \tag{4.15}$$

known as the latent roots. In the latent equation 4.15,

$$\begin{cases} \omega = B_{11} + B_{22} \\ T = B_{11}B_{22} - B_{12}B_{21} \end{cases} \tag{4.16}$$

When $\omega^2 = 4T$, $\lambda_1 = \lambda_2 = \lambda$. The solution of Equation 4.14 could be written as

$$\begin{cases} x = \{a + [(\lambda - B_{11})a + B_{12}b]t\}e^{\lambda t} \\ y = \{b + [(B_{22} - \lambda)b + B_{21}a]t\}e^{\lambda t}. \end{cases} \tag{4.17}$$

We are going to see that the stability of the steady thermodynamic embranchment $x = y = 0$ of the system depends on the real part of the latent roots. Only when the real part is negative is the thermodynamic embranchment steady. By looking at the roots of Equation 4.15, the relationship between different values of ω and T and the signs of the latent roots could be determined.

There are five possible situations for different characteristic values with different solutions.

1. If ω, $T > 0$, $\omega^2 > 4T$, and λ_1, $\lambda_2 < 0$, then the moving point representing the state variable would approach $x = y = 0$ directly, the characteristic value is real, the original point is a stationary node, and the thermodynamic embranchment is steady.
2. If ω, $T > 0$, $\omega^2 < 4T$, and λ_1, λ_2 are conjugate complex roots with negative real parts, then the moving point representing the state variable would approach $x = y = 0$ along an orbital trajectory in an eddy form. Thus, $x = y = 0$ is a steady focus, and the thermodynamic embranchment is also steady.
3. If $T < 0$ and $\lambda_1 > 0 > \lambda_2$, then the moving point representing the state variable would approach $x = y = 0$ in a specific direction and exit far from the original point in all other directions. In this case, $x = y = 0$ stands for a saddle point, and the thermodynamic embranchment is not steady.
4. If $\omega < 0$, $T > 0$, $\omega^2 > 4T$, and λ_1, $\lambda_2 > 0$, then the moving point representing the state variable would stay away from the original point. The original point is an unstable node, and the thermodynamic embranchment is not steady.
5. If $\omega < 0$, $T > 0$, $\omega^2 > 4T$, and λ_1, λ_2 are conjugate complex roots with positive real parts, the moving point representing the state variable would exit from the original point in an eddy fashion. The original point is an unsteady focus, and the thermodynamic embranchment is not steady.

Thus, in a linear system of either one or more variables, the evolution of the state variables would approach either zero with stable thermodynamic embranchment or infinity with unstable thermodynamic embranchment.

Under some particular conditions, if the system could break away from the thermodynamic embranchment and develop into a nonlinear dissipative structure, then the nonlinear system of the state variables begets detailed investigation. To this end, let us take such a system in a single variable as an example:

$$\frac{dx}{dt} = (A - A_c)x - x^3. \tag{4.18}$$

The solutions of its stationary equation $(A - A_c)x - x^3 = 0$ are $x = 0$ and $x = \pm\sqrt{A - A_c}$. Next, let us look at the relevant scenarios.

1. When $A < A_c$, the solution $x = \pm\sqrt{A - A_c}$ is an imaginary number. Since the state variable is a physical quantity, there is no practical significance for this imaginary number. Thus, the stationary solution $x = 0$ is a thermodynamic embranchment solution. Within a region close to $x = 0$, Equation 4.18 could be linearized as $\frac{d\Delta x}{dt} = (A - A_c)\Delta x$. Its solution is $\Delta x = \Delta x_0 e^{(A-A_c)t}$. Because $A - A_c < 0$, $x = 0$ is steady.
2. When $A > A_c$, $x = 0$ is still the stationary solution of Equation 4.18. However, it is no longer stable. Because $A - A_c$ as the exponent of e is positive, any deviation from $x = 0$ would make the system exit far away from the thermodynamic embranchment. However, the nonlinear term $-x^3$ would confine the state variable to a limited but nonzero value.
3. When $A > A_c$, $x = \pm\sqrt{A - A_c}$ are real numbers. As for the stationary solution $x_0 = \sqrt{A - A_c}$, in the neighborhood of x_0, the nonlinear equation (Equation 4.17) could be linearized as $\frac{d\Delta x}{dt} = -2(A - A_c)\Delta x$. Its solution is $\Delta x = \Delta x_0 e^{-2(A-A_c)t}$. Since $A - A_c > 0$, $x_0 = \sqrt{A - A_c}$ is steady. Similarly, $x_0 = -\sqrt{A - A_c}$ can also be shown to be steady.

From this example, we can readily see that the nonlinearity of the system of our concern does matter and does make a difference. In particular, it makes the thermodynamic embranchment not diverge to infinity after losing stability while converging to a dissipative structure with a nonzero state variable.

Therefore, the evolution of nonlinear systems, in general, is not as monotonous as that of linear systems. Nonlinear systems could represent biological diversity. When nonlinear restrictions drive a system into a state exiting far from any equilibrium state, the nonlinear terms would become dominant and correspond to multistationary solutions, some of which are stable, while others, unstable.

4.3 SYNERGETICS

4.3.1 Basic Concepts

In Chinese society since antiquity, the philosophical thought that all worldly things give birth while in opposition to each other and mutually support and react to each other (wan wu xiang sheng xiang ke, xiang gan xiang ying) has been well accepted and widely recognized. However, this philosophical thought has not been rigorously described using scientific language. Also, it could not explain the ultimate whys and hows of things in terms of analytically describing interactions. In other words, there has not been a formal theory developed on the particular philosophical thought that can be analytically employed to explain interactions of world things, to promote desirable changes and creations, and to prescribe attributes.

Inspired by the study of lasers, Hermann Haken (1977, 1991) established the theory of synergetics in 1977. Along with the successes of this theory, the task of explaining the ultimate reason underlying systems' evolution is considered finished by some.

The word "synergetics" is rooted in Greek. It means "coordination and cooperation." It attempts to represent the fundamental principle that underlies structural changes of systems from totally different scientific fields.

Definition 4.14

Synergetics is an interdisciplinary science that studies the formation and self-organization of patterns and structures appearing in open systems that are far from thermodynamic equilibrium (Haken 1982). ■

Essential in synergetics is the concept of order parameters, which was originally introduced in the Ginzburg–Landau theory in order to describe phase transitions in thermodynamics. The major idea of synergetics is that the macroscopic and holistic characteristics and behaviors of a system are determined through cooperation and combination by the characteristics of and interactions among the subsystems, instead of by simple additions of the subsystems. Such a holistic effect of cooperation and combination is the key mechanism that turns a system from an out-of-order state to an orderly state. It also stands for the ability of evolution and self-organization of the system.

The research synergetics include all kinds of complex systems, such as liquid systems, chemical systems, ecosystems, the earth system, celestial systems, economical systems, management systems, etc. This theory studies how the competition and cooperation mechanisms among the parts of a complex system contribute to the holistic self-organization behavior of the system. Therefore, the method employed in synergetics is mainly based on the idea of synthesis. Since the diversity of the world originates from nonlinearity, the equations and systems considered in synergetics are nonlinear. Although systems of concern might be very different from each other, the interaction of subsystems is synergetic, represented by competition, cooperation, and/or feedback. Such interaction makes the subsystems form a macroscopic structure of order through self-organization.

The objective of synergetics is to research the qualitative change of the macroscopic characteristics of different complex systems and to investigate the general principle that enslaves subsystems to cooperate. The qualitative change of the macroscopic characteristics means the creation of structures of higher order or a structural transition from one level of order to another. Here, general means such a principle that is irrelevant to specific characters of particular subsystems.

The outstanding contribution Haken made to the scientific world is that he established the "enslaving principle." It states that in the process of a phase transition, the dynamics of fast-relaxing modes is completely determined by the "slow" dynamics of, as a rule, only a few "order parameters."

4.3.2 Order Parameter and Enslaving Principle

In a system that is able to self-organize, there are many state variables. In its process of self-organization, these state variables would interact and affect each other, making the system experience qualitative changes. In his theory, Haken promotes the concept of order parameters, which is a very good tool useful for describing the self-organization behavior of the system.

Definition 4.15

For a system that is in an out-of-order state, some of its state variables take the value zero. When the system transits from the out-of-order state to an orderly state, each of the previously zero-valued variables changes its value to a positive number. This positive value can be seen as an expression of the orderliness of the system and is referred to as an order parameter. ■

When compared to other variables, if one variable changes very slowly, then this variable is also known as a slow-change parameter. On the contrary, if a variable changes faster than other variables, it is known as a fast-change variable.

While an order parameter can describe the orderliness of the system, the number of order parameters is much fewer than that of fast-change parameters. At the same time, when order parameters describe the orderliness, change, and characteristics of the system, they also dominate over other fast parameters. Because of this reason, each order parameter can also be named as a "command parameter."

Owing to the great variety of different systems, the method of determining order parameters can take different forms depending on the particular system involved. For some simple systems, there is one slow-change variable among all variables so that this variable can be regarded as an order parameter. As for complex systems, it might be very difficult for one to distinguish slow-change from fast-change variables. However, for certain particular scenarios, by using an appropriate coordinate transformation, an order parameter of a complex system could be determined easily. Unfortunately, there are such complex systems wherein the idea of coordinate transformation fails to work successfully. When this happens, new variables of a different level may be chosen in order to find a desired order parameter. For example, for an ideal aerothermodynamic system, the location, momentum, energy, and angle momentum could not be taken as the order parameter, while such macroscopic parameters as temperature and/or pressure could be taken as the order parameter in order to analyze the evolution of the system.

Once a desirable order parameter is determined, the research of the system's evolution could focus mainly on the order parameter. The order parameter extracts information of the whole system and offers a way to recognize the system readily. However, as mentioned before, determining a desirable order parameter is not an easy job. It requires an in-depth understanding of the system. Generally, although the principle on how to select an order parameter can be given, particular criteria and procedures of finding an order parameter for any given system do not exist.

For many systems, their order parameters are formed during their processes of self-organization. Thus, one can say that when a system self-organizes, some order parameters are formed on the basis of all the system's variables. On the other hand, order parameters, in general, are composed of other state variables of the system and dominate and enslave the changes of other state variables. Each process of phase change of systems consists of the formation of order parameters on the basis of the systems' state variables and the process of these order parameters enslaving other variables of the systems. Haken named the enslaving and obeying relationship of the order parameters and fast-change variables, which appears in the process of phase changes of systems, as the enslaving principle. More specifically, this principle says (Haken 1982) that in the process of a phase transition, the dynamics of fast-relaxing (stable) modes is completely determined by the slow dynamics of, as a rule, only a few order parameters (unstable modes). The order parameters can be interpreted as the amplitudes of the unstable modes that determine the macroscopic pattern of the system.

In the following, let us use an example to explain the enslaving principle.

Example 4.1

Given a system of two variables X and Y, assume that the system of evolution is given as follows:

$$\begin{cases} \dfrac{dX}{dt} = \alpha X - XY \\ \dfrac{dY}{dt} = -\beta Y + X^2 \end{cases} \tag{4.19}$$

Without the nonlinear term, the two equations would be independent. Given parameter $\beta > 0$, Y is an attenuation variable and approaches zero with time. With $\alpha > 0$, X diverges to infinity. When $\alpha < 0$, the pattern of change of both X and Y is similar. Y is named as a stable module, while X an unstable module; its stability depends on the sign of α (Li 2006; Xu 2000).

Let us now look at the nonlinear term. Suppose that the relation between X and time t is known, and the solution of the second equation in Equation 4.19 is

$$Y = \int_{-\infty}^{t} e^{-\beta(t-\tau)} X^2(\tau) d\tau \tag{4.20}$$

where the initial condition $Y(-\infty) = 0$ is applied. To comprehend the enslaving principle, by integrating Equation 4.20, we have

$$Y = \frac{X^2}{\beta} - \frac{2}{\beta}\int_{-\infty}^{t} e^{-\beta(t-\tau)} X\dot{X} d\tau. \tag{4.21}$$

The second term on the right-hand side of Equation 4.21 could be computed as follows:

$$\begin{aligned}\frac{2}{\beta}\int_{-\infty}^{t} e^{-\beta(t-\tau)} X\dot{X}\, d\tau &\le \frac{2}{\beta}\left|X\dot{X}\right|_{max}\int_{-\infty}^{t} e^{-\beta t}e^{\beta\tau}\, d\tau \\ &= \frac{2}{\beta}\left|X\dot{X}\right|_{max} e^{-\beta t}e^{\beta t} = \frac{2}{\beta}\left|X\dot{X}\right|_{max}.\end{aligned} \tag{4.22}$$

The condition under which the second term on the right-hand side of Equation 4.21 can be neglected is

$$|X| >> \frac{2}{\beta}\left|\dot{X}\right|_{max} \tag{4.23}$$

where the rate of change $\dot{X}$ is very small. Thus, X is seen as a slow-change variable.

Correspondingly, if X is a slow-change variable satisfying Equation 4.23, then the solution of the second equation in Equation 4.19 is $Y = \frac{X^2}{\beta}$, which can be regarded as the relation between X and Y. This solution can also be obtained by solving Equation 4.19. Substituting $Y = \frac{X^2}{\beta}$ into the first equation, the differential equation of X is

$$\dot{X} = \alpha X - \beta X^3. \tag{4.24}$$

Now, X can be solved for readily. By using $Y = \frac{X^2}{\beta}$, we can also solve for Y.

From this example, we see that for a system in two state variables, by letting the derivative of the fast-change variable be zero and by substituting the result into another equation to remove the fast-change variable, the system is reduced into a single equation in only the slow-change variable. By solving this resultant equation for the slow-change variable and then by using back substitution, the given system is completely solved.

The basis of this method is that the order parameter, or the slow-change variable, determines the system's evolution. The situation could be regarded as that the fast-change variable evolves faster than the other variable, reaches the bifurcation point first, and then does not change any further. Thus, by letting the derivative of the fast-change variable be zero, we can derive the needed relation between the fast-change and slow-change variables. Consequently, the system is simplified to a single equation containing only the slow-change variable. Because the slow-change variable evolves much slower than the fast-change variable, the process of solving the system is greatly simplified.

4.3.3 Hypercycle Theory

A hypercycle is a new level of organization wherein the self-replication units are connected in a cyclic, autocatalytic manner. The self-replication units are themselves (auto-)catalytic cycles. The concept of hypercycle is a specific model established for the chemical origin of life. It was pioneered by Eigen and Schuster. From random distributions of chemicals, the hypercycle model seeks to find and grow sets of chemical transformations that include self-reinforcing loops. Hypercycles are similar to autocatalytic systems in the sense that both represent a cyclic arrangement of catalysts, which themselves are cycles of reactions. One difference between the two concepts is that the catalysts that constitute hypercycles are themselves self-replicative (Eigen and Schuster 1978; Padgett et al. 2003).

Generally, transformation of motion takes place only conditionally. Other than the condition that the corresponding physical parameters, such as temperature, pressure, density, chemical potential, etc., have to reach their respective threshold values, the appearance of a new organizational form (or structure of interaction) or disintegration of the existing organization is also one of the main conditions. To this end, Manfred Eigen established the theory of hypercycles.

From the angle of prebiotic evolutions, Eigen classified reaction networks into three classes: reactive cycles, catalytic cycles, and hypercycles. A reactive cycle stands for such a mutually related reaction where the products of a reaction step are exactly the reactants of the previous step. For instance, in the carbon–nitrogen cycle of the solar fusion of hydrogen into helium, the carbon atoms are both reactants and reaction products. The orderly structure of an entire reactive cycle functions like a catalyst of activities.

Within a reactive cycle, if there is an intermediate that also plays the role of a catalyst to the reactive cycle itself, then this reactive cycle is known as a catalytic cycle. The semiconservative replication of DNAs is such a catalytic cycle where a single chain plays the role of a template that "molds" the substrates into another complementary single chain according to particular requirements. The orderly activity structure of an entire catalytic cycle behaves like a unit of self-replication.

If several units of self-replication are coupled into a new cycle, then the form of organization of hypercycles appears. Manfred Eigen pointed out clearly that a catalytic hypercycle stands for such a system that is made up of several units of self-catalysts and self-replication connected by a cyclic relationship. Each cycle that is made up of several nucleic acid molecules and protein molecules is an example of catalytic hypercycles.

It can be seen that although reactive cycles do not alter the attributes of the reaction products, they indeed change the speed of reaction. That is very important for many chemical reactions. In practice, some chemical reactions that do not easily take place can happen as soon as reactive cycles are established. In this sense, it can be seen that the organizational form of reactive cycles provides a condition to transform physical motions into chemical motions. Catalytic cycles can create "communications" between parental and offspring generations so that certain information can be invariantly passed on through generations. Furthermore, hypercycles not only can sustain units' information unaltered but also can maintain the connections of the various units of different attributes. That is extremely important for the evolution of life. No matter whether it is a simple or complicated system of life, the system stands for an organic whole composed of various units of different functionalities. In order to replicate these units with their attributes unaltered and to maintain the coupling relations between these units of different functionalities through metabolic processes, the organizational form of hypercycles is probably unavoidable. This form of organization provides a necessary condition to transform physical and chemical motions into movements of life.

4.4 SOME PHILOSOPHICAL THOUGHTS

4.4.1 On Dissipative Structures

1. Time: reversible and irreversible, symmetry and asymmetry. The theory of dissipative systems introduces a new meaning to the concept of time. The classical thermodynamics regards that time is reversible, while the study of dissipative systems is based on such a time that is irreversible and focuses on irreversible processes existing far from an equilibrium state. At the same time, time is an intrinsic evolutionary variable of the systems.

The theory of dissipative systems also explores the essence of science research. From the point of view of this theory, when attempting to understand the physical world, humans are also participants of the natural evolution. That is, the world cannot be analyzed in any absolutely objective means.

2. Structure: equilibrium and nonequilibrium, stable and unstable. Equilibrium and nonequilibrium are a pair of contradictions. They restrict and transform into each other under different sets of conditions. A macroscopic equilibrium state may not necessarily be equilibrium at the microcosmic level. It is because at the microscopic level, there are always fluctuations caused by the Brownian motion of molecules.

   To resolve this pair of contradictions, both the classical mechanics and statistical physics employ the concept of statistical means. By using statistical means, it can be shown that nonequilibria of the microcosmic level can possibly exist in equilibrium states of the macroscopic level.

   Prigogine divides the system of concern into components, which are small when seen from the macroscopic level, while sufficiently large from the microcosmic point of view. These components could be approximately seen as even and in equilibrium states during a very short time interval. Thinking this way creates a method to deal with the contradiction of macroscopic equilibria and microcosmic nonequilibria and transforms a nonequilibrium question into many local equilibrium questions.

   The theory of dissipative systems investigates the rules of transformation between equilibrium and nonequilibrium, order and out of order, stable and unstable, and so on. It can be and has been applied to explain the phenomenon of life. At present, the research on structural stability constitutes a fundamental basis of the evolution of ecosystems.

3. System: simplicity and complexity, locality and integrity. A system is complex because it cannot be treated as a simple combination of its parts due to the reason that there exists intertwining relations and interactive restrictions among the parts. In the point of view of systems theory, the whole can be greater than the sum of its parts; parts of low reliability can be combined organically into a whole of high-level reliability. Therefore, if a system is far away from any equilibrium state, the interaction of its parts is nonlinear; the complexity and integrity of the system should be considered. That is, the system should be investigated as a whole. The interaction of the internal parts of a dissipative structure possesses the characteristics of nonlinearity and coherence, which are very significant in the research of biology and human societies. For example, the joint human visual acuity of both eyes is about 6–10 times higher than that of just one eye. For anyone, it is readily seen that the collective function of both eyes cannot be explained by simply using linear additions of the functionalities of the individual eyes.

4. Rule: determinism and indeterminism, dynamics and thermodynamics. From the point of view of time reversibility and system simplicity, given an initial state, all individual states a system evolves through can be derived in thermodynamics. That is historically known as determinism. Prigogine correctly recognizes that deterministic theories can be employed to analyze simple systems and become invalid when facing complex systems. For instance, chaos theory claims that long-term weather prediction is impossible. Therefore, reversibility and determinism are applicable only to elementary and local situations. On the other hand, irreversibility and indeterminism are the primary rule under which the physical world evolves. Additionally, it should be noted that determinism and indeterminism, dynamics and thermodynamics are not pairs of contradictions. Instead, they mutually complement each other.

   Most importantly, the investigation of dissipative systems implies that the current focus of scientific research has changed from the studies of existence into the explorations on evolution.

4.4.2 On Synergetics

Einstein once pointed out the main function of developing basic concepts in the construction of the theories of physics. Haken is a life-long theoretical physicist, and throughout his theory of synergetics, from fundamental concepts to basic theory to the relevant methods, dialectic thinking has been widely employed.

4.4.2.1 Synergy

The word "synergy" means cooperation and collaboration. The rule of change common to different systems prescribes the oneness of relevant matters. Although different systems may possess different characteristics, the qualitative structural change of various systems is similar or even the same. The concept of synergies comes from the systemic oneness of the world; it parallels with the dialectic law of unity of opposites.

4.4.2.2 Orderliness and Order Parameters

Each system has to face the contradiction of order and out of order. Orderliness is a concept for the whole system at the macroscopic level. It is not applicable to particles or elements in the system of concern unless a particle or element is also treated as a system.

Each scientific discipline applies its own individual set of physical measurements in its description of the macroscopic orderliness of the systems of interest. In the theory of dissipative structures, the concept of entropy is employed, while in synergetics, the concept of order parameters is introduced to represent the orderliness of systems.

The values of the order parameter represent the macroscopic orderliness of the system. If the system is out of order, it means that the relationship between subsystems is weak; their independence dominates, and the value of the order parameter is zero. Along with a change of the environment, the value of the order parameter also varies accordingly. Here, the relatively independent, irregular movements of subsystems bring changes to the whole system through the dynamic interaction and coupling of the subsystems. These changes in the subsystems and the whole system affect each other mutually. When the whole system evolves into a critical state, the cooperation of the subsystems, established on the basis of their interaction and coupling, dominates the next stage of the whole system's evolution. In this case, the order parameters change and reach their peak values fast. At such a time moment, mutation occurs within the critical state; an orderly structure along with its corresponding functionality appears within the system. Thom employs his catastrophe theory to describe such phenomena of critical changes.

4.4.2.3 Slow-Change and Fast-Change Parameters

Synergetics analyzes the varied functions of different parameters. It distinguishes the essential factors from those unessential, temporary acting factors from those that are long-term acting, and occasional factors from those that are necessary. In the process of change, the rates of change of different parameters are different. According to the rates, parameters are classified into two classes, slow-change and fast-change parameters. Fast-change variables experience greater dumping effects with short relaxation time, while slow-change variables do not experience any dumping effect in the system's evolution. It is those slow-change variables that are selected as order parameters that measure the system's degree of orderliness and eventually dominate the formation of the ultimate structure and functionality of the system. It is through competition with the fast-change variables that the slow-change variables become the order parameters by winning over the fast-change variables. Through the competition among the order parameters, a form of evolution is rigorously selected for the system.

4.4.2.4 Analogy: Main Research Method of Synergetics

Similarity is the foundation for making comparisons. The existence of similarity between different forms of motion and between varied systems implicitly articulates the uniformity of the world. In synergetics, comparisons are done in two different ways. On one hand, different subsystems of the whole behave similarly in the process that the whole system evolves from a state of no order to one with order. The behaviors of the subsystems can be described by using mathematical equations of a same class. As pointed out by Haken, the advantage of analogy is evident, because when a problem is resolved successfully in one discipline, the results can be generalized to another discipline. A particular system well studied in one area can be treated as a simulated computer in another discipline. On the other hand, synergetics compares the phase changes or behaviors, which are similar to phase changes, of an equilibrium system at a critical point with those phase changes of the system situated in a nonequilibrium state. As a result, it is found that these phase changes are similar and satisfy mathematical equations of the same class.

In short, in the theory of synergetics, cooperation, orderliness, order parameters, and slow-change and fast-change variables are all products of competition. Through conflicts, opposites are unified; within the unification, there are still opposites. This fundamental law of dialectics plays out vividly in synergetics.

4.4.3 On Structure, Organization, and Chance

For any chosen natural phenomenon, the characteristics of the initial state could be classified into two categories. One is simplicity, and the other is out of order. The latter we name as chaos; for details, please see Chapter 5. That is to say, the structural elements and their relationship are elementary and straightforward, while the overall state of existence and the form of motion are chaotic with neither order nor visible structure. The pervasive nebula, the floating atmosphere, the turbulent oceans, the mighty rivers, and even a seed, an egg, and a fetus are just some of the specific examples of such natural phenomena. Such a situation emerges out of an out-of-order state and will in turn evolve into another state of order. For example, a rising orderly column of smoke starts to circle around and then returns to its original state of rise into the sky. The flag hanging at the tip of its pole sometimes crackles loudly and sometimes flaps gracefully in the wind. Along the heavily traveled streets, there are always jams somewhere with the traffic flow completely blocked; however, in a little while, everything returns to its normal state of smooth operation. The current in rivers rushes downstream unstoppably; however, at different locations along its path, there might be densely distributed whirlpools and turbulences. All these are quite elementary, ordinarily seen phenomena. However, for mathematicians and physicists, they represent some of the most difficult problems of the world; they are some of the most unpredictable mazes. The German physicist Werner Heisenberg, one of the founders of quantum physics, commented on his death bed that there are two problems he has been thinking about and for which he cannot find any resolution; now he can go to ask God about these problems. The first problem is why there is relativity, and the second problem is why there is turbulence. Heisenberg believed that God could answer the first problem; as for the second problem of turbulence, it was about the occurrence and transformation between orderliness and orderlessness. Even God could not answer this problem.

Thus, what is the mystery underneath all those quite elementary, ordinarily seen phenomena? The key is how their structures or organizations are preserved and embodied; how a minor factor or epitome is magnified into a major macroscopic effect; and whether the connection between the minor cause and the magnificent result constitute a rigid must or a probabilistic accidence.

4.4.3.1 Structures

Structure determines attributes. Although it exists within the inside of the system of concern, or maybe it cannot be totally understood and controlled, or there is no real need to control it, it represents the essence of the system. When designing an artificial system, one has to establish the structure. When it is seen statically, it is known as structure; when it is seen dynamically, it is known as organization. For example, each societal system can be called a social structure, and each relatively fixed network of people is referred to as social organization. For practical purposes, when a system is seen as a black box, the structure of the system can be temporarily ignored, creating a mystery feeling about the structure. The most unthinkable is the periodic appearance and disappearance of structures in the alteration of generations. For example, how can an apple tree grow out of a small seed? How can a tiny fertilized egg, invisible to the naked eye, grow into a huge elephant or camel? Is there a tiny apple tree or a tiny camel inside the originally tiny seed? Discoveries of modern science of course tell us that in the tiny seed, instead of an apple tree or a camel, there is only the genetic information of the future tree or future camel. What happens if we replace the metaphor of tree or camel in the previous discussion with nebula, atmosphere, light smoke, or turbulence? Does it imply that what is hidden in the tiny "seed" is the future panorama of the entire universe? The eternally changing crimson clouds at sunrise and sunset? The naturally coiling and unrestrained rising smokes? Or the gigantic mighty cyclones? Does it mean that all future happenings have been included and germinated at the very beginning, similar to the situation of embryos? Evidently, that is impossible. If so, then how is the magnificent universe constructed? How can a madly spinning tornado be originated at a location that does not seem any different from other nearby locations? These are some of the difficult questions awaiting the scientific mind to explore.

4.4.3.2 Self-Organization

When one looks at the mechanism for chaotic and random phenomena to appear and to evolve from the angle of behavioral forms, he or she can define these phenomena as such of self-organization. Or in the language of Ervin Laszlo, self-organization means such systems' behaviors that are shown during the time when the systems' internal structure and complexity go through a phase change. It is the foundation for all evolutionary processes to take place, be they natural or historical, physical or chemical, biological or human or societal. No matter what properties the systems' components possess, and no matter which level they are situated at, these processes are similar, and their attributes are all self-organization. That is to say, due to the interaction between a system's internal factors, the degree of orderliness and organization of the system grows automatically with time. Such systems are self-organization systems; such phenomena are self-organization phenomena. Evidently, each such system is exactly the so-called autopoiesis system, so named by Humberxo Marurana. Biological cells represent some of the most elementary entities of autopoiesis systems, while organs, human bodies, and societies consisting of humans are also such systems. They are considered as process networks of self-production consisting of mutually dependent internal parts. They produce such networks, while they themselves are products of such networks. They sustain themselves within matter and energy flows, replicate themselves, and when the condition is right, they expand, grow, and become more advanced with complexity. Of course, such systems have two forms: one form contains such systems that have the capability of self-repairing and self-control. The other form has the tendency to grow indefinitely like cancer cells, and the growth can be controlled only with external constraints.

From molecules to the universe, from cells to societies, all the objects that are so seemingly different possess one common attribute without a single exception: they evolve in the direction of an increasing degree of orderliness. Such evolutions are spontaneous. They require only the appropriate environmental conditions without any need for external commands. If we use Hermann Haken's theory, we have the following metaphor. Assume that there is a group of workers, each of whom acts perfectly according to the commands of the leader. Such a group is known as an organization. If the workers work together as a team to produce their products based on some sort of unspoken agreement without any external instruction, then such a process of operation is known as a self-organization. The phenomena of self-organization appear not only in social systems but also in physical, chemical, and biological systems. Evidently, self-organization systems and behaviors are widely seen in natural and human-related matters and events. When one gains sufficient understanding of this concept, he or she will undoubtedly become efficient with this valuable scientific method.

4.4.3.3 Chance and Inevitability

Stepwise magnification and bifurcative departure are very attractive because they make people believe that by using a tiny weight, one can balance a heavy load; by putting in nearly no effort, one can receive disproportional return and acquire great fortune. However, if he or she did not control the situation right or was out of luck, he or she would feel regretful for the rest of his or her life. For such an initial difference as thin as a piece of hair, the consequence could be either a happy ending or a devastating disaster. With such sensitivity, how can the universe still evolve and advance? The answer might be found within the theories of chaos and self-organization. If one looks at the ebb and flow of randomness and the phenomena of constant evolution as "chaos" where no clear order can be seen, then hidden behind these disorderly phenomena are some unexpected regularity and inevitability. Originally, the concept of chaos is defined on the basis of irregular movements produced out of deterministic equations, because the nondeterministic equations of molecular movements are derived out of the deterministic equations of Newtonian mechanics. The thoughts of probabilisticity and randomness that arose from quantum mechanics and neutral mutation theory can also find their footholds in chaos theory. Such a situation of three contenders has almost entirely reshaped the conception of science of the twentieth century.

4.5 OPEN-ENDED PROBLEMS

1. What is the characteristic of self-organization? How does the concept of self-organization unify the concepts of inevitability and randomness, determinism, and indeterminism?
2. What is the meaning of spontaneity in the evolution of matters and events?
3. From the viewpoint of information transformation, classify the difference between organization and self-organization.
4. Explain why societal phenomena could be explained scientifically by jointly employing the theories of self-organization and organization. Please try to explain the relationship and interaction between market mechanism and macrocontrol in economics.
5. Please explain the relationship between a system and its outside environment, and the intrinsic mechanism developed in synergetics.
6. Please use your own words to explain the enslaving principle.
7. Provide five examples of self-organization that you observe either in your personal life or in the natural environment.
8. Conduct an analysis on the origin of life using the concept of self-organization based on your literature search.
9. How would you analyze the phenomena of self-organization on different scales, such as the ripples in a calm pond caused by a rock thrown into the pond, the hurricane eye on Jupiter, etc., and their dynamic stabilities?
10. Considering the harmonic coexistence of man and nature, what should be done to achieve the optimal state of self-organization?
11. What are the characteristics of self-organizing dissipative structures?

5 Chaos

The study of nonlinear science has touched on reconsiderations of many important concepts and theories, such as determinism and randomness, orderliness and disorderliness, accidentalness and inevitability, qualitative and quantitative changes, whole and parts, etc. These reconsiderations are expected to affect the accustomed ways of how people think and reason and have been involved in addressing some of the most fundamental problems of the logic system that underlies the entire edifice of modern science. It is commonly believed that nonlinear science consists mainly of the chaos and fractal theories, the latter of which will be studied in the next chapter. Discrete cases of chaos are often seen in chaotic time series, while chaotic time series contains rich information on the dynamics of the underlying system. How to extract this information and how to practically employ the extracted information represents one important aspect of applications of chaos theory. In this chapter, we will study some of the fundamental concepts of the chaos theory and related methods.

5.1 PHENOMENA OF CHAOS

Bertalanffy (1968) once pointed out that the principles of general systems, although they are originated from different places, are clearly similar to those of dialectical materialism. What is described in this statement is equally applicable to the chaos theory, a new research area. The development of the chaos theory has led to many dialectical thoughts.

The phenomena of chaos were initially discovered by E. N. Lorenz (1963) in his study of a three-dimensional autonomous dynamic system. To investigate the prediction of weather, he simplified the equation system of atmospheric dynamics. Then, in his numerical simulation of the resultant system, he found that the repeated simulations depart drastically from each other, leading to totally different results, as the computing time increased, even though the initial conditions were roughly the same (they are the same up to many places after the decimal point). This observation implies that short-term weather forecast is possible, while long-term prediction is impossible.

Chaos is a seemingly random phenomenon emerging out of deterministic nonlinear dynamic systems. Such uncertain results out of deterministic systems stand for an inherent randomness of nonlinear dynamic systems. Coinciding with the current belief of a probabilistic universe, chaos has become a hot topic in modern science. As of this writing, this concept has been widely employed in the investigations of such physical phenomena as turbulence, climate, earthquake, life, threshold state of condensation, economies, growth, etc.

Historically, what has been believed is that deterministic systems produce deterministic results. However, what has been found in chaos theory indicates that indeterminate results can also be obtained. In the past, it was known that iterations of a discrete dynamic system can converge to some fixed points, while today, they can also "converge" to chaos attractors. Of course, according to the past point of view, the latter "convergence" is not really the defined convergence. Because of this, the concept of chaos has greatly furthered our understanding of nature.

5.2 DEFINITION OF CHAOS

The iteration formula $x_{n+1} = f(x_n)$ is deterministic. However, for some sensitive function f, when n is large, $\{x_n\}$ becomes unpredictable so that it is "random." One well-known example is Ulam–von Neumann mapping f:(−1,1) → (−1,1), defined by

$$f(x_n) = 1 - 2x^2 \tag{5.1}$$

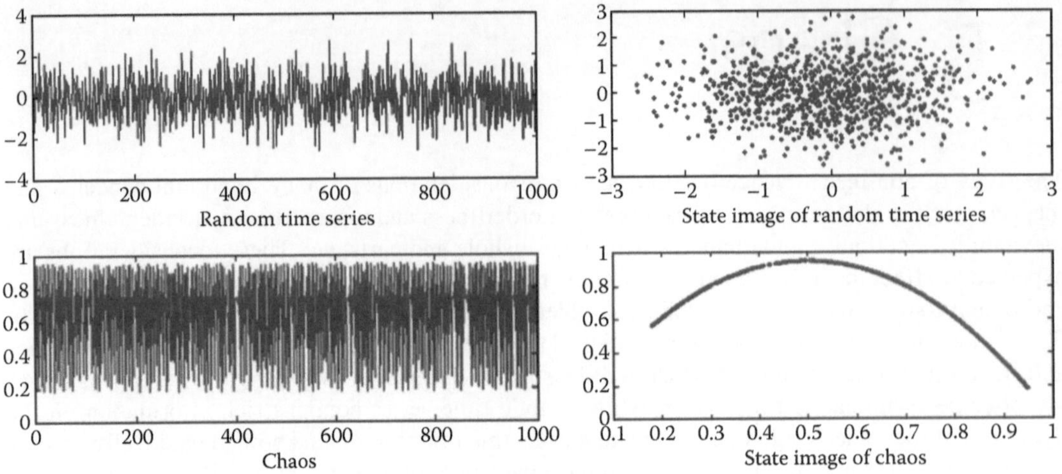

FIGURE 5.1 Comparison between random and chaotic time series.

whose iteration formula is

$$x_{n+1} = 1 - 2x_n^2 \tag{5.2}$$

When an initial value is given, a pseudorandom sequence is produced. Figure 5.1 compares the time series plots and phase space plots of a random time series and a chaotic time series.

5.2.1 Pseudorandomness

Pseudorandomness is really not the same as randomness. When an event A is said to be random under condition C, it means that when experiments are repeated under C, A might occur or might not occur. Even if A occurred during the last experiment run, it still may not occur for the next experiment run. For example, when a fair coin is tossed, although a head is obtained for the first toss, the head may not show up again for the second toss. That is, observing a head is a random event. The fact that when the coin is tossed three times, person X observes "head, head, tail" does not mean that if person Y tossed the coin three times, he would also obtain the same results, because the probability for such a coincidence to occur is 1/8.

Assume that F is a given probability distribution. Pick m independent numbers $x_1, x_2,\ldots, x_m$ from the F population. Then these are referred to as F-random numbers. Assume that this sequence satisfies some properties $F_1, F_2,\ldots,F_l$ that are related to F, say, $(x_1 + x_2 + \ldots + x_m)/m$. When m is sufficiently large, this value should be roughly equal to F's mean. One obvious characteristic of this F-random number sequence is that if person X randomly picks $x_1, x_2,\ldots,x_m$ from F and person Y $y_1, y_2,\ldots,y_m$, then, in general, $\{x_i\}$ is different from $\{y_i\}$, although both sequences satisfy the properties $F_1, F_2,\ldots,F_l$. Such unpredictability is exactly the essence of "randomness."

Now, assume that we also obtain a sequence $x_1, x_2,\ldots,x_m$ by employing a certain deterministic method, say Equation 5.2. Although by using statistical testing, we might find that $\{x_i\}$ also satisfies the properties $F_1, F_2,\ldots,F_l$, the previously described essence has been lost. That is, from a chosen starting value b_0, using the predetermined method, both X and Y obtain two sequences $\{x_i\}$ and $\{y_i\}$. Because the method is deterministic, we know $x_i = y_i (i = 1, 2,\ldots,m)$. Evidently, $\{x_i\}$ cannot be seen as F-random numbers. Instead, it can be referred to only as an F-pseudorandom number sequence. This analysis implies that through deterministic iteration, one can, at most, obtain pseudorandom numbers. That is, we can state that chaos stands for the pseudorandomness of deterministic systems.

For a true random system, the particular value of any future moment cannot be determined by using the value of the current moment. That is, the system's behavior is not predictable even for the short term. On the other hand, for any deterministic system, its short-term behaviors are completely determined except that, due to its sensitivity to the initial value, its exact long-term movement is unpredictable. That is caused by its inherent internal randomness. Also, all nonlinear systems do suffer from this phenomenon.

For two- or lower-dimensional nonlinear dynamic systems, no matter how their control parameters change, the new mechanism of the systems' movement, caused by bifurcations and sudden changes, can only be either an equilibrium or a periodic state, that is, an elementary orderly motion.

Nonlinear dynamic systems of higher dimensions (>2), along with the changes in the control parameters and all the possible states the systems might take, such as the equilibrium states, periodic states, and quasiperiodic states, after the systems have gone through a series of bifurcations and sudden changes, can lose their stability and become chaotic. That is an extremely irregular and extraordinarily complicated form of motion.

The states of nonlinear dissipative dynamical systems change constantly (local-unstably), while their overall states are constrained within bounded spaces (the global stability). Such inconsistencies between local and global state behaviors have to lead to multiple time–space scales of greatly varied magnitudes, where many physical quantities change with the scales.

One most important aspect in studying these dynamic systems is the search for such quantities that are invariant with scale, known as scale invariance. Fractal dimensions in the research of fractals are one such invariance. Hence, chaos and fractals are naturally related.

5.2.2 Definition of Chaos

The following is the mathematical definition of chaos (Devaney 1989):

Definition 5.1

Assume that (V,d) is a metric space. A mapping $f: V \to V$ stands for a chaos if it satisfies the following conditions:

1. *Sensitivity to the initial value:* There is $\delta > 0$ such that for any $\varepsilon > 0$ and any $x \in V$, there are y in the ε-neighborhood of x and a natural number n such that $d(f^n(x), f^n(y)) > \delta$.
2. *Topological transitivity:* For any open sets $X, Y \subseteq V$, there is $k > 0$ such that $f^k(X) \cap Y \neq \varnothing$.
3. *Density of periodic points:* The set of all periodic points of f is dense in V. ■

The following are some remarks on this definition of chaos:

1. No matter how close the points x and y can be, their trajectories under the effect of f can depart beyond a large distance. Also, in each neighborhood of any chosen point x in V, there is another point y that will eventually travel away from x under the effect of f. If one computes the trajectory of such function f, any minor initial noise can lead to failed computations after a certain number of iterations.
2. The property of topological transitivity implies that for any chosen point x, each of its neighborhoods under the effect of f will "scatter" throughout the entire metric space V. It implies that the function f cannot be decomposed into two subsystems that do not interact with each other.
3. These two items, in general, are properties of random systems. However, condition 3, density of periodic points, on the other hand, indicates that the system evolves with definite

certainty and regularity instead of being completely chaotic. The property of realistic orderliness that looks chaotic is exactly what attracts the academic attention to chaos.

In Section 5.2.3, another definition of chaos (Li and Yorke 1975) will be introduced.

5.2.3 Period Three Implies Chaos

5.2.3.1 Definition of Periodic Point

Let $F:J \to J$ be a mapping such that $F^0(x) = x$ and $F^{n+1}(x) = F(F^n(x))$, for any $n = 0,1,2,...$

1. If there is $p \in J$ such that $p = F^n(p)$ and $p \neq F^k(p)$, for any $1 \leq k < n$, then p is called a periodic point of period n.
2. If $\exists n \in N$ such that p satisfies condition 1 above, then p is called a periodic point.
3. If $\exists m \in N$ such that $p = F^m(q)$, where p is a periodic point, then q is an eventually periodic point.

As an example, let us look at period 3. Suppose that $x_{n+1} = f(x_n)$ is an iteration defined on the unit interval $[0,1] \to [0,1]$. If $x_0 \in [0,1]$ is a three-periodic point of this iteration, then the following hold true:

$$x_1 = f(x_0) \neq x_0$$

$$x_2 = f(x_1) \neq x_1 \neq x_0$$

$$x_3 = f(x_2) = x_1.$$

That is, after one iteration, x_0 is mapped to x_1. After another iteration, x_1 is mapped to x_2. When the third iteration is applied, x_2 is mapped back to the original point x_0. Because three iterations map x_0 back to its original place, x_0 is called a periodic point with period 3.

Evidently, it can be noted that x_1 is also a periodic point with period 3 because x_1 can return back to itself after three iterations. The same holds true for x_2. Therefore, the function f with period 3 would have at least three periodic points with period 3. The so-called fixed point is actually a periodic point with period 1.

5.2.3.2 Sarkovskii Theorem

Ex-Soviet mathematician Sarkovskii arranges all the natural numbers in accordance with the following order:

$$\begin{cases} 3 & 5 & 7 & 9 & 11 & \cdots & (2n+1)\cdot 2^0 & \cdots \\ 3\cdot 2 & 5\cdot 2 & 7\cdot 2 & 9\cdot 2 & 11\cdot 2 & \cdots & (2n+1)\cdot 2^1 & \cdots \\ 3\cdot 2^2 & 5\cdot 2^2 & 7\cdot 2^2 & 9\cdot 2^2 & 11\cdot 2^2 & \cdots & (2n+1)\cdot 2^2 & \cdots \\ 3\cdot 2^3 & 5\cdot 2^3 & 7\cdot 2^3 & 9\cdot 2^3 & 11\cdot 2^3 & \cdots & (2n+1)\cdot 2^3 & \cdots \\ & \vdots & & & & & & \\ \cdots & 2^n & \cdots & 2^4 & 2^3 & 2^2 & 2^1 & 2^0 \end{cases}$$

This sequence is now known as the "Sarkovskii order." For iterations of a continuous interval, Sarkovskii proved the following: Assume that in the Sarkovskii order M is arranged in front of N. Then if there is a periodic point of period M, then there must be a periodic point of period N. That is the Sarkovskii theorem. According to this theorem, it follows that if a function has a periodic point of period 3, then this function has periodic points of any natural number period because 3 is listed at the first place in the Sarkovskii order.

5.2.3.3 Li–Yorke Theorem

The following is the main content of the Li–Yorke theorem. For details please, consult Li and Yorke (1975).

Theorem 5.1

Let J be an interval of real numbers and let $F:J \to J$ be a continuous mapping. Assume that there is a point $a \in J$ such that $b = F(a)$, $c = F^2(a)$, $d = F^3(a)$, and $d \le a < b < c$ or $d \ge a > b > c$. Then

1. For every $k = 1,2,\ldots$, there is a periodic point $p_k \in J$ of period k.
2. There is an uncountable $S \subset J$ that contains no periodic point such that
   2.1. $\forall p,q \in S$ satisfying $p \neq q$, the following is true:

$$\limsup_{n\to\infty} \left|F^n(p) - F^n(q)\right| > 0 \text{ and } \liminf_{n\to\infty} \left|F^n(p) - F^n(q)\right| = 0.$$

   2.2. $\forall p \in S$ and for any periodic point $q \in J$, the following holds true:

$$\limsup_{n\to\infty} \left|F^n(p) - F^n(q)\right| > 0.$$

Note: When $F(x)$ has periodic points of period 3, the conditions of Li–Yorke Theorem will be satisfied. Thus, when speaking in this sense, it indeed means that period 3 causes chaos. ■

Based on the Li–Yorker theorem, Li and Yorker introduce a definition of chaos as follows: Let J be an interval of real numbers and $F:J \to J$ a continuous mapping. The mapping $F(x)$ would have demonstrated chaos if it satisfies the following:

1. The periods of the periodic points of $F(x)$ do not have an upper bound.
2. There is an uncountable subset $S \subset J$, on which $F(x)$ is defined, such that
   2.1. $\forall p,q \in S$, $p \neq q$ implies

$$\limsup_{n\to\infty} \left|F^n(p) - F^n(q)\right| > 0 \text{ and } \liminf_{n\to\infty} \left|F^n(p) - F^n(q)\right| = 0.$$

   2.2. $\forall p \in S$, if $q \in J$ is a periodic point, then

$$\limsup_{n\to\infty} \left|F^n(p) - F^n(q)\right| > 0.$$

5.2.4 Characteristics of Chaos

1. *Nonperiodicity.* Chaos is one possible state of each nonlinear dynamic system. The trajectory of the system changes neither monotonically nor periodically. Instead, it represents a change that is nonperiodic and consists of zigzag rises and falls.

2. *Dynamic behaviors over strange attractors.* The chaos of dissipative systems appears only with the bifurcation structures, known as strange attractors. This kind of complex motion can be described only when the system has a set of fractal points of a certain self-similarity in the phase space.
3. *Stability and instability.* Each strange attractor attracts external trajectories; after entering the strange attractor, no trajectory can leave again. Therefore, the attractor is globally stable. However, trajectories inside the attractor repel each other, showing a level of extreme instability.
4. *Determinacy and randomness.* As soon as the trajectory of a chaos system enters into a strange attractor, its irregularity in principle is not any different from that of random motion. However, this irregularity is caused by the internal nonlinear factor of the deterministic system and has nothing to do with any external conditions.
5. *The unpredictability of long-term behaviors.* Because the system is deterministic, its short-term behaviors can be predicted. However, due to its sensitivity to the initial values, its long-term behaviors cannot be predicted. (For simple orderly motions, both of their short-term and long-term behaviors can be forecasted, while for random motions, neither their short-term nor long-term development can be predicted.)

5.3 LOGISTIC MAPPING—CHAOS MODEL

The purpose of investigating a discrete dynamic system is to answer the following question: For any given initial state $x(0)$, what is the system's final or asymptotic state?

In particular, for the dynamic system $f{:}X \to X$, $x(0) = x_0 \in X$, $x(n) = f(x(n-1)) = f^n(x(0))$, its final or asymptotic state is the behavioral state of $x(n)$ when $n \to \infty$. That is,

1. Will $x(n)$ converge to a point $x^* \in X$? In other words, will $x^* = f(x^*)$ hold true? That is, is x^* a fixed point (equilibrium point) of f?
2. Is x^* an asymptotic stable point?

If both (1) and (2) hold true, then the system is known as asymptotically stable.

Evidently, when f is a contraction mapping, the system has a unique fixed point so that the fixed point is asymptotically stable. Specifically, the linear system

$$x(k+1) = Ax(k)$$

is globally asymptotically stable, if and only if all the characteristic values $\lambda_1, \ldots, \lambda_n$ of A satisfy $|\lambda_i| < 1$, $\forall i$. If $\exists i$ such that $|\lambda_i| > 1$, then the system is unstable; when $\exists i$ such that $|\lambda_i| = 1$, then the system is critically stable.

Each deterministic system has deterministic results and maybe indeterminate conclusions. By employing simple models, one can produce complicated nonperiodic results. The logistic mapping is such a typical example of simple models that illustrate the aforementioned situation.

5.3.1 LOGISTIC MAPPING

In particular, when considering an ecological system, the following is a model that describes the population size of a specific species:

$$x_{n+1} = \mu x_n(1 - x_n) = f(x_n) \tag{5.3}$$

where x_n and x_{n+1}, respectively, stand for the population sizes of the nth and $(n+1)$th generations and μ is a control parameter that is related to the living condition of the species. The function f is referred to as a one-dimensional logistic mapping.

When the μ value of the control parameter falls in the interval 0–4, the effect of the logistic mapping f is to map any value $x_n \in [0,1]$ back into this interval, that is, $x_{n+1} = f(x_n) \in [0,1]$. Because the function f is nonlinear, one cannot define its inverse mapping $x_n = f^{-1}(x_{n+1})$ so that all nonlinear mappings in this class do not have inverses. Under certain conditions, inversability can be seen as dissipation. Thus, one-dimensional nonlinear mappings can be seen as simple dissipative systems.

In the following, let us analyze the effect of the control parameter μ on the long-term behaviors of the specific species.

Let us use the method of iteration to solve the algebraic Equation 5.3. If the iteration converges to a fixed point, then it is essentially a fixed point of the discrete dynamic system and satisfies asymptotic stability. From the stability principle of discrete dynamic systems, each fixed point of Equation 5.3 satisfies

$$x^* = f(x^*) = \mu x^*(1 - x^*). \tag{5.4}$$

1. *The situation when there is only one fixed point:* When $\mu < 1$, the system has the unique fixed point $x^* = 0$, $|f'(0)| = \mu < 1, f(1) = 0, f^k(1) = 0, k \geq 1, \forall x_0 \in (0,1), 0 < x_1 = f(x_0) = \mu x_0(1 - x_0) < \mu x_0 \leq x_0 < 1, x_2 = \mu x_1(1 - x_1) < \mu x_1 \leq x_1$. Hence, $\{x_n\}\downarrow$ has a lower bound and converges to 0. That is, $x_n \to 0$. Therefore, $x^* = 0$ is the system's unique attraction fixed point with [0,1] being its attraction domain.

   When $\mu = 1$, $|f'(0)| = \mu = 1$. The system is situated at a critical state, and its stability needs to be checked using other methods.
2. *The situation when there are two fixed points:* When $\mu \in (1,3)$ increases from the initial value $\mu_0 = 1$, the system loses its stability at the equilibrium point $x^* = 0$ and reaches a new fixed point, which can be obtained as $x_1^* = 1 - 1/\mu$ from Equation 5.4. The corresponding stable critical value satisfies $|f'(x_1^*)| = 1$. Thus, μ_1 can be determined by the stable boundary

$$f'(x^*) = \mu - 2\mu x^* = -1. \tag{5.5}$$

   Substituting $x_1^* = 1 - 1/\mu$ into Equation 5.5 produces $\mu_1 = 3$. That is, at x_1^*, $|f'(x_1^*)| < 1$, and the system is asymptotically stable.
3. *Two-periodic points:* When $\mu_1 < \mu < \mu_2$ (it is assumed to be the next bifurcation point), the fixed point $x_1^* = 1 - 1/\mu$ will be linearly destabilized and become a two-periodic point:

$$f^2(x^*) = ff(x^*) = \mu[\mu x^*(1 - x^*)][1 - \mu x^*(1 - x^*)] = x^*. \tag{5.6}$$

   Canceling one x^* produces

$$\mu^2 x^*(1 - x^*)[1 - \mu x^*(1 - x^*)] - 1 = 0,$$

   which is simplified into

$$x^{*3} - 2x^{*2} + \left(1 + \frac{1}{\mu}\right)x^* - \frac{\mu^2 - 1}{\mu^3} = 0.$$

   Factoring this expression gives

$$\left(x^* - \frac{\mu - 1}{\mu}\right)\left[x^{*2} - \left(1 + \frac{1}{\mu}\right)x^* + \frac{\mu + 1}{\mu^2}\right] = 0.$$

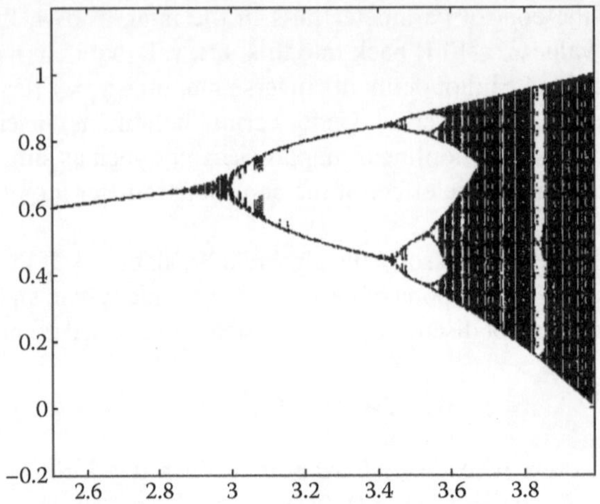

FIGURE 5.2 Period-doubling bifurcation of the logistic equation.

Because the first root $x_1^* = 1 - 1/\mu$ has been linearly destabilized, new fixed points are given below:

$$x_{2,3}^* = \frac{1}{2\mu}\left[(1+\mu) \pm \sqrt{(1+\mu)(\mu-3)}\right]. \tag{5.7}$$

4. *From period-doubling bifurcation to chaos:* Based on the reasoning above, it can be seen that the bifurcations of the logistic equation are realized by period-doubling (Figure 5.2, where the horizontal axis stands for the parametric dimension μ and the vertical axis stands for x^*); that is, 1 bifurcates into 2, 2 into 4, 4 into 8,.... The bifurcation values can be computed. Corresponding to $\mu = \mu_\infty = 3.569945672$, a chaos area appears.

Example 5.1

For the given population model

$x_{n+1} = \mu(1 - x_n)x_n,\ x_n \in (0,1),\ \mu > 0,$

prove that when $\mu = \frac{5}{3}$, the system has a unique attractor. After that, compute the attractor and analyze its field of attraction.

Proof

When $\mu = \frac{5}{3}$, denote $f(x) = \frac{5}{3}x(1-x)$. Solving the equation $f(x) = x$ leads to the fixed points $x_1 = 0$ and $x_2 = \frac{2}{5}$, where x_1 is not stable, because $|f'(x)| = \left|\frac{5}{3}(1-2x)\right|$ so that $|f'(x_1)| = \left|\frac{5}{3}(1-2x)\right|_{x_1=0} = \frac{5}{3} > 1$. That means that the field of attraction of the fixed point x_1 consists of two isolated points $\{0,1\}$. At the same time, x_2 is a stable fixed point, because $|f'(x)| = \left|\frac{5}{3}(1-2x)\right|$ so that $|f'(x_2)| = \left|\frac{5}{3}(1-2x)\right|_{x_2=\frac{2}{5}} = \frac{1}{3} < 1$, and the field of attraction of this fixed point is the entire open interval (0,1). For any $x_n \in (0,1)$, we have

$$\left|f(x_n) - \frac{2}{5}\right| = \left|\frac{5}{3}x_n(1-x_n) - \frac{2}{5}\right| = \left|\frac{5}{3}x_n(1-x_n) - \frac{5}{3}x_2(1-x_2)\right|$$

$$= \left|\left(x_n - \frac{2}{5}\right)\left(1 - \frac{5}{3}x_n\right)\right| < r\left|\left(x_n - \frac{2}{5}\right)\right|, \text{ for } r \in (0,1).$$

Thus, we can derive that $\left|f^k(x_n) - \frac{2}{5}\right| < r^k \left|x_n - \frac{2}{5}\right|$. ■

5.3.2 Feigenbaum's Constants

In the 1970s, American physicist M.J. Feigenbaum carefully analyzed the period-doubling sequence $\mu_1, \mu_2, \ldots$ and discovered that

$$S_n = \frac{\mu_n - \mu_{n-1}}{\mu_{n+1} - \mu_n} \to \delta = 4.669201609\ldots. \tag{5.8}$$

That is the first Feigenbaum's constant. At the same time, he also discovered the following second Feigenbaum's constant:

$$\lim_n \frac{\Delta_n}{\Delta_{n+1}} = 2.502907875\ldots = \alpha, \tag{5.9}$$

which is an irrational number, where Δ_n stands for the distance between the nth intersection point of the horizontal line $x^* = \frac{1}{2}$ in Figure 5.2 and the other branching curve of the same doubling bifurcation (for details, see Figure 5.3). The parameter μ^* that corresponds to Δ_n also satisfies

$$\lim_n \frac{\mu_n^* - \mu_{n-1}^*}{\mu_{n+1}^* - \mu_n^*} = \delta. \tag{5.10}$$

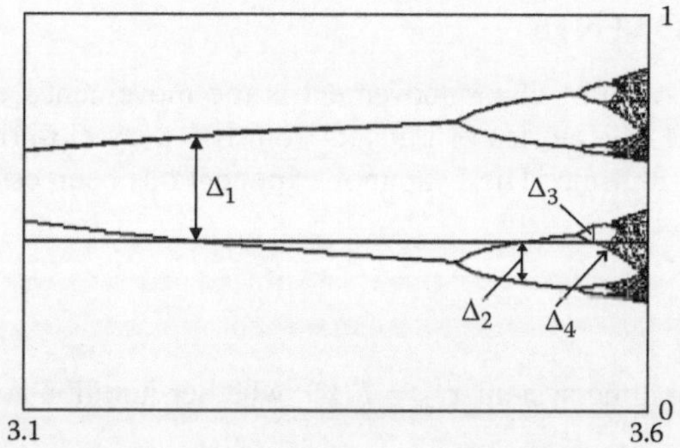

FIGURE 5.3 Feigenbaum's constant α.

5.4 PHASE SPACE RECONSTRUCTION

The basic idea for reconstructing the phase space is the following: the evolution of any particular component of the dynamic system of concern is determined by the other interacting components. So, the information of these relevant components is implicitly contained in the development of the said component. To construct an "equivalent" state space, one can observe a component and treat its measurements at certain fixed, delayed time moments as new dimensions to generate a multidimensional state space. Takens (1981) proves the possibility of finding an appropriate embedding dimension. That is, if the dimension m of the delay coordinate system satisfies $m \geq 2d + 1$, where d is the dimension of the dynamic system, then in this embedding space, regular trajectories (attractors) can be recovered.

Assume that the control equation of the dynamic system is

$$\frac{\mathrm{d}x_i}{\mathrm{d}t} = f_i\left(x_1, x_2, \ldots, x_n\right), i = 1,2,\ldots,n \tag{5.11}$$

and that the system's evolution with time is described by the trajectory $X(t) = [x_1(t), x_2(t), \ldots, x_n(t)]^T$ of the n-dimensional space generated by the variables $(x_1, x_2, \ldots, x_n)$.

For any problem with which only a time series of a single variable is available, one needs to transform the first-order differential equations of multivariables in Equation 5.11 into a higher-order differential equation of the single variable by using the method of elimination. For a continuous variable, the following nth-order differential equation can be obtained:

$$x^{(n)} = F(x, x', x'', \ldots, x^{(n-1)}). \tag{5.12}$$

The trajectory of the new dynamic system, which resulted from the transformation, is

$$X(t) = [x(t), x'(t), \ldots, x^{(n-1)}(t)]^{\mathrm{T}}. \tag{5.13}$$

Both Equations 5.13 and 5.12 reflect the characteristics of the same dynamic system except that it now evolves in the phase space generated by the axes $x(t)$ and its derivatives $x^{(1)}(t), \ldots, x^{(n+1)}(t)$.

For situations involving a discrete time variable, the previous $x(t)$, $x^{(1)}(t)$, $\ldots, x^{(n+1)}(t)$ can be replaced with drifts up to the $(n - 1)$th order. In particular, the trajectory of the dynamic system in the phase space is

$$X(t) = [x(t), x(t + \tau), x(t + 2\tau), \ldots, x(t + (n - 1)\tau)]^T$$

where τ stands for the time step of the drifts, also known as delay time.

5.5 LYAPUNOV EXPONENTS

The fundamental characteristic of chaos movement is the movement's extreme sensitivity to the initial value condition. The trajectories originated from two very close initial values depart from each other exponentially with time. The Lyapunov exponent has been established to quantitatively describe this phenomenon.

5.5.1 The Concept

In the one-dimensional dynamic system $x_{n+1} = F(x_n)$, whether iterations of two given initial values are departing from each other or not is determined by the value of the derivative $\left|\frac{\mathrm{d}F}{\mathrm{d}x}\right|$. If $\left|\frac{\mathrm{d}F}{\mathrm{d}x}\right| > 1$, then the iterations depart from each other. If $\left|\frac{\mathrm{d}F}{\mathrm{d}x}\right| < 1$, the iterations move the two points closer. However, in the process of iteration, the value of $\left|\frac{\mathrm{d}F}{\mathrm{d}x}\right|$ changes constantly so that the iterated initial values might get closer to or depart from each other from one step to the next. To look at the overall state situation of two neighboring points, one can take the average over the number of iterations. Thus, without loss of generality, assume that the average exponential departure of each iteration is λ. Thus, the distance of the starting points that are originally of a distance ε apart after the nth iteration is given below:

$$\varepsilon e^{n\lambda(x_0)} = \left|F^n(x_0+\varepsilon) - F^n(x_0)\right|. \tag{5.14}$$

Taking the limit by letting $\varepsilon \to 0$, $n \to \infty$ makes Equation 5.14 become

$$\lambda(x_0) = \lim_{n\to\infty}\lim_{\varepsilon\to 0}\frac{1}{n}\ln\left|\frac{F^n(x_0+\varepsilon) - F^n(x_0)}{\varepsilon}\right| = \lim_{n\to\infty}\frac{1}{n}\ln\left|\frac{\mathrm{d}F^n(x)}{\mathrm{d}x}\right|_{x=x_0}.$$

This equation can be simplified into

$$\lambda = \lim_{n\to\infty}\frac{1}{n}\sum_{i=0}^{n-1}\ln\left|\frac{\mathrm{d}F(x)}{\mathrm{d}x}\right|_{x=x_i}$$

which is referred to as the Lyapunov exponent of the dynamic system. This index indicates the average degree of exponential departure caused by one step of iteration.

In particular, if we look at the logistic mapping

$$x_{n+1} = f(x_n) = \mu x_n(1 - x_n), f: [0,1] \to [0,1].$$

Then, its Lyapunov exponent is

$$\lambda = \lim_{n\to\infty}\frac{1}{n}\sum_{i=0}^{n-1}\ln\left|\mu - 2\mu x_i\right|$$

where $\lambda > 0$ corresponds to a chaos movement. When $\mu \in [3.7,4]$, the corresponding Lyapunov exponents are shown in Figure 5.4.

If $\lambda < 0$, it means that the neighboring points will eventually combine into one single point, and their trajectories contract. If $\lambda > 0$, it means that the originally neighboring points will eventually depart from each other, which corresponds to unstable trajectories. Thus, $\lambda > 0$ can be employed as one criterion of chaos behaviors of the system.

For the general n-dimensional dynamic system, the Lyapunov exponent is defined as follows.

Assume that a mapping $F{:}R^n \to R^n$ determines an n-dimensional discrete dynamic system $x_{n+1} = F(x_n)$. Let the initial condition of the system be an infinitesimally small n-dimensional ball. Due to the natural deformation occurring in the evolution process of the system, the ball will become an ellipsoid. After ordering the principal axes of these ellipsoids according to their magnitudes,

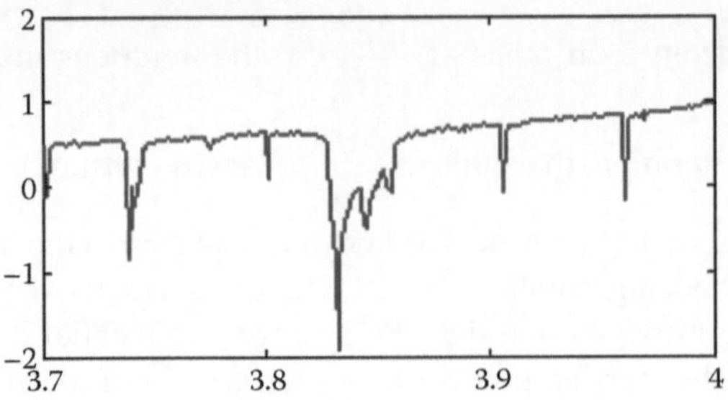

FIGURE 5.4 Lyapunov exponents of the logistic mapping.

the acceleration of the ith Lyapunov exponent based on the length of the ith principal axis $p_i(n)$ is defined as follows:

$$\lambda_i = \lim_{n\to\infty} \frac{1}{n} \ln\left[\frac{p_i(n)}{p_i(0)}\right], \; i = 1,2,....$$

Therefore, the Lyapunov exponents are closely related to the contraction or expansion of the trajectory in the phase space. In the negative direction of the Lyapunov exponents, the trajectories contract, and the motion is stable; in the positive direction of the Lyapunov exponents, the trajectories depart rapidly.

Note that the lengths of the principal axes of the ellipsoids increase in the order of e^{λ_1}; the area of the region defined by the previous two principal axes grows in the order of $e^{(\lambda_1+\lambda_2)}$; the volume of the body defined by the previous three principal axes enlarges in the order of $e^{(\lambda_1+\lambda_2+\lambda_3)}$; etc. That is, the sum of the previous j Lyapunov exponents is determined by the average long-term acceleration of the volume of the j-dimensional body defined by the previous j principal axes.

5.5.2 Selection of Parameters

First, let us look at the choice of the mean period p.

For a given time series $\{x_1, x_2, \ldots, x_N\}$ and the period condition $x_{N+j} = x_j$, compute the autocorrelation function (the discrete convolution):

$$C_j = \frac{1}{N}\sum_{i=1}^{N} x_i x_{i+j}.$$

Then, apply the discrete Fourier transform (DFT) on C_j by computing the Fourier coefficients:

$$P_k = \sum_{j=1}^{N} C_j e^{\frac{i2\pi kj}{N}}.$$

Cooley and Tukey (1965) proposed their fast Fourier transform (FFT) algorithm when they tried to compute DFTs. By applying FFT, one can directly do an FFT using x_i, producing the coefficients

$$a_k = \frac{1}{N}\sum_{i=1}^{N} x_i \cos\frac{2\pi ki}{N} \text{ and } b_k = \frac{1}{n}\sum_{i=1}^{N} x_i \sin\frac{2\pi ki}{N}.$$

Then, compute $P_k' = a_k^2 + b_k^2$. From many groups of $\{x_i\}$, one obtains a group of $\{P_k'\}$ by computing the mean of which one approximates the power spectrum P_k.

Second, let us look at the choice of the delay time τ.

For an infinitely long data sequence without any noise, there is, in principle, no constraint on the choice of the delay time τ. However, numerical experiments indicate that the characteristic quantities of the phase space are dependent on the choice of τ. Choosing an appropriate τ naturally requires linear independence, which is choosing the first zero of the autocorrelation function. Later, Rosenstein et al. (1994) proposed in their experiments to apply the time lag needed for the autocorrelation function value to drop to $1-\frac{1}{e}$ times of the initial value as the choice of the delay time τ.

As for the methods for numerical computation, other than the definition, there currently are many numerical methods for computing the Lyapunov exponents. They can be roughly categorized into two groups: the Wolf and Jacobian methods, where the Wolf methods are appropriate for no-noise time series whose spatial evolutions of small vectors are highly nonlinear, and where the Jacobian methods are appropriate for noisy time series whose spatial evolutions of small vectors are nearly linear. Barana and Tsuda (1993) once introduced a new p-norm method, which bridges the Wolf and Jacobian methods. However, due to complications in the selection of the p-norm, practical application of this method has been very difficult.

On the other hand, the method proposed by Rosenstein et al. (1994) for dealing with small samples possesses operational convenience with the following advantages: (1) it is relatively reliable for data samples; (2) the amount of computation is relatively small; and (3) it is easy to operate. In this chapter, we improve on this method, apply the resultant new method to resolve one practical problem, and develop some interesting conclusions.

5.5.3 Method for Small Samples of Data

This small-sample method computes the largest Lyapunov exponent of chaos time series. In practical applications, there might not be any need to compute the entire spectrum of the Lyapunov exponents of the given time series. Instead, all that is needed is to calculate the largest Lyapunov exponent. For instance, to check whether or not a time series represents a chaos system, one needs only to see if the largest Lyapunov exponent is greater than 0. That is, such prediction problems are resolved by obtaining the largest Lyapunov exponent. Hence, the computation of the largest Lyapunov exponent is very important in the spectrum of Lyapunov exponents.

Assume that the given chaos time series is $\{x_1, x_2, \ldots, x_N\}$ with embedding dimension m and time delay τ. Then the reconstructed phase space is

$$Y_i = (x_i, x_{i+\tau}, x_{i+2\tau}, \ldots, x_{i+(m-1)\tau}) \in R^m,\ i = 1,2,\ldots,M$$

where $M = N - (m-1)\tau$. The embedding dimension m is chosen based on Takens theorem (Takens 1981) and the G-P algorithm (Sivanandam and Deepa 2007), and the time delay τ is selected using the time lag, as proposed by Rosenstein et al. (1994). It takes the autocorrelation function value to drop to $1-\frac{1}{e}$ times of the initial value.

After the phase space is reconstructed, locate the closest proximal points for the points on the trajectory. That is,

$$d_j(0) = \min_{x_j} \left\| Y_j - Y_l \right\|, \left| j - l \right| > p$$

where p stands for the mean period of the time series, which can be estimated by using the reciprocal of the mean frequency of the energy spectrum. Then the largest Lyapunov exponent can be estimated by using the mean diverging speed of the proximal points nearest the points of the trajectory.

Sato et al. (1985) estimate the largest Lyapunov exponent as follows:

$$\lambda_1(i) = \frac{1}{i\Delta t}\frac{1}{(M-i)}\sum_{j=1}^{M-i}\ln\frac{d_j(i)}{d_j(0)}$$

where Δt stands for the sample space and $d_j(i)$ is the distance after i discrete time steps of the jth pair of the nearest proximal points on the basic trajectory.

Later, Sato et al. (1987) improved their estimation to the following expression:

$$\lambda_1(i,k) = \frac{1}{k\Delta t}\frac{1}{(M-k)}\sum_{j=1}^{M-k}\ln\frac{d_j(i+k)}{d_j(i)}$$

where k is a constant and $d_j(i)$ is the same as above.

The geometry of the largest Lyapunov exponent represents a quantification of the exponential divergence of the initialized closed trajectory and a quantity that estimates the overall level of the system's chaos. Thus, combining with Sato et al.'s estimation formula, we have

$$d_j(i) = C_j e^{\lambda_1(\Delta t)}, C_j = d_j(0).$$

Taking the logarithm of both sides of this equation produces

$$\ln d_j(i) = \ln C_j + \lambda_1(i\Delta t), j = 1,2,\ldots,M.$$

Evidently, the largest Lyapunov exponent is roughly equal to the slope of these straight lines. It can be estimated by using the least squares method to approximate this system of linear equations. That is,

$$\lambda_1(i) = \frac{1}{\Delta t} < \ln d_j(i) >$$

where $<.>$ stands for all the means about j.

Below are the specific computational steps of using the small sample method:

1. For the given time series $\{x(t_i), i = 1, 2, \ldots, N\}$, compute its time delay τ and the mean period P.
2. Compute the correlation dimension d using the method described in Section 6.1.4 and the embedding dimension m using $m \geq 2d + 1$.
3. Reconstruct the space $\{Y_j, j = 1, 2, \ldots, M\}$ based on the time delay τ and embedding dimension m.
4. Locate the nearest proximal point Y_k for each point in the phase space Y_j, satisfying

$$d_j(0) = \min_j \left\| Y_j - Y_k \right\|, \left| j-k \right| > p.$$

5. For each point Y_j in the phase space, compute the distance $d_j(i)$ of the ith discrete time step after the pair (Y_j, Y_k):

$$d_j(i) = |Y_{j+i} - Y_{k+i}|,\ i = 1, 2, \ldots, \min(M - j, M - k).$$

6. For each i, compute the mean $y(i)$ of all j-related $\ln d_j(i)$, that is,

$$y(i) = \frac{1}{q\Delta t} \sum_{j=1}^{q} \ln d_j(i)$$

where q is the number of nonzero $d_j(i)$'s.

7. Construct the $y(i){:}i$ regression line using the least squares method. The slope of this line is the maximum Lyapunov exponent λ_1.

Example 5.2

The data analyzed in this example are from the closing prices of the 792 stocks traded on the Shanghai Stock Exchange from January 5, 1998, to March 29, 2004, including a total of 1508 trading days. We compute the Lyapunov exponents of the closing prices of the exchange index and five randomly selected stocks. The results are listed in the following table:

Stock Code	Index	600000	600368	600466	600873	600759
Lyapunov	0.0035	−0.0005	−0.0002	0.0083	0.0252	0.04121

The changes in the index's Lyapunov exponent for the amounts of data from 800 to 1500 are listed in the table below and Figure 5.5:

800	900	1000	1100	1200	1300	1400	1500
0.0085	0.0063	0.0021	0.0018	0.0049	0.0064	0.0075	0.0034

Divide the 1500 data points of the index into groups of 150 points each and compute the respective Lyapunov exponents, producing the results listed in the table below.

1	2	3	4	5	6	7	8	9	10
0.0024	0.0065	0.0047	0.0049	0.0030	0.0036	0.0025	0.0054	0.0065	0.0020

The table provides a mean value of 0.0036, while the actual value is 0.0035, and the standard deviation is 0.0018. This result indicates that the Lyapunov exponent of Shanghai Stock Index is reliably greater than zero.

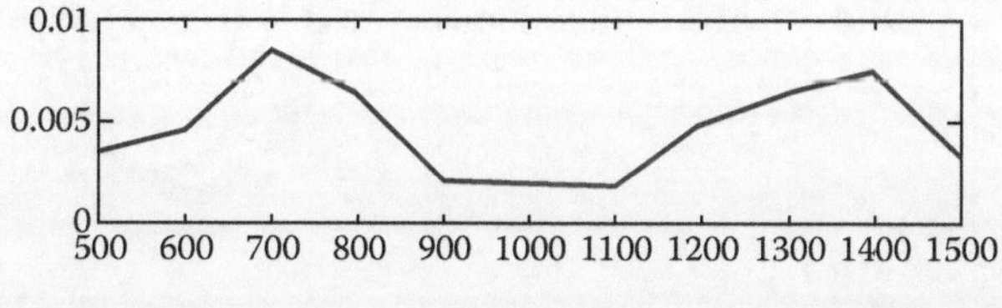

FIGURE 5.5 Changes in Lyapunov exponent for the amounts of data from 500 to 1500.

What is discussed above implies that Shanghai Stock Exchange contains chaos. (Although the Lyapunov exponents of two of the five randomly selected stocks are less than zero, from the global consideration, we can still see that there is chaos in the Shanghai Stock Exchange.) Stock markets used to be investigated using randomness and linear points of view. Doing so has led to mistakes and loss of much rich information. Chaos, on one hand, expands on the collection of objects that can be predicted. (If those behaviors that were studied as randomness before are chaos, then one can employ the technique of phase space reconstruction to produce much richer and deeper results than those of mean value predictions.) On the other hand, the sensitivity of chaos to initial values limits the time span over which predictions can be meaningfully made. That is, long-term chaos behaviors cannot be accurately predicted.

However, it is also difficult to make use of the short-term prediction capability of chaos because of the following.

1. Takens theorem only guarantees to recover the stock market, seen as a dynamic system, to an underlying dynamic system that is topologically equivalent to the market. Such a recovered system is still not the same as the true system of the market.
2. Although the dimensionality, as a topologically invariant quantity, is maintained, one has lost the opportunity to accurately locate appropriate variables to quantify and to model the underlying system. Also, because of the fractal dimensions, there is not any elementary relation between the number of variables that should be used and the dimensionality.
3. The technique of phase space reconstruction requires the available data to be noiseless. However, each stock market has not only observational noise but also system noise.
4. The effect of inflation is not considered.
5. With the assumption of randomness, the conventional statistics requires only a certain amount of data. On the other hand, the technique of phase space reconstruction requires that the available data cover a sufficiently long period of time.

5.6 CASE STUDIES AND FIELDS OF APPLICATIONS

In this section, we will look at several classical case studies of chaos and related graphs.

Example 5.3

The phase graph of the differential equation

$$y'''(x) + 2y''(x) + 2y'(x) = \sin[4x] + \cos[5x]$$

is given in Figure 5.6 if initial values are $y(0) = 0$, $y'(0) = 0$, $y''(0) = 0$. If the initial values are $y(0) = 0$, $y'(0) = 0$, $y''(0) = 5$, the phase graph is shown in Figure 5.7.

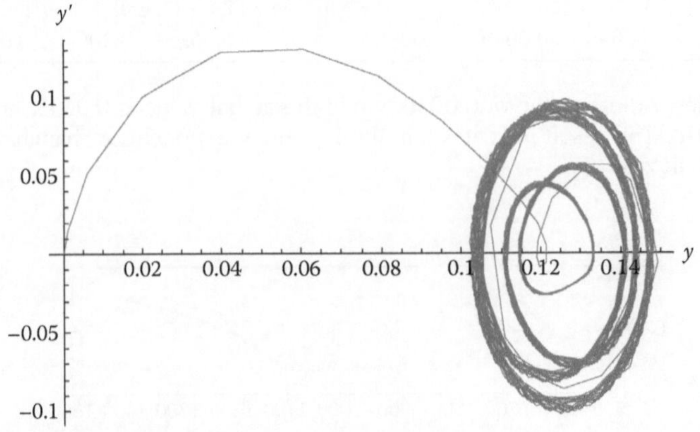

FIGURE 5.6 Phase graph of a third-order initial value problem.

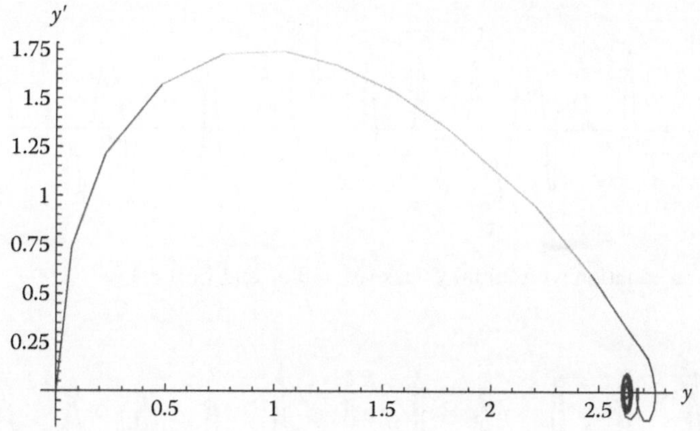

FIGURE 5.7 Phase graph of the same system as in Figure 5.6 with a different set of initial values.

Example 5.4

The Duffing equation

$$y''(x) + 0.05y'(x) + y(x)^3 = 7.5\cos[x]$$

has a unique chaos attractor. When the initial values are $y(0) = 0$, $y'(0) = 0$, its graph is given in Figure 5.8. When the initial values are $y(0) = 3$, $y'(0) = 4$, its graph is shown in Figure 5.9. When the initial values are $y(0) = 3.01$, $y'(0) = 4.01$, the graph is depicted in Figure 5.10.

From comparing Figures 5.9 and 5.10, it can be seen that with the initial values "$y(0) = 3$ and $y'(0) = 4$" and "$y(0) = 3.01$ and $y'(0) = 4.01$," the figures are quite close for $x \leq 3$ and then start to depart from each other as x increases further beyond 3. Figure 5.11 shows another comparison between these figures for $x \in [5.6, 6.2]$, where the overlapping curves show some clear differences.

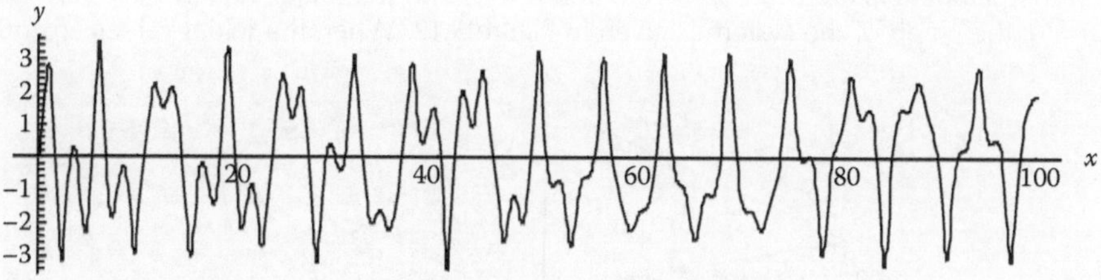

FIGURE 5.8 Duffing equation with initial values $y(0) = 0$ and $y'(0) = 0$.

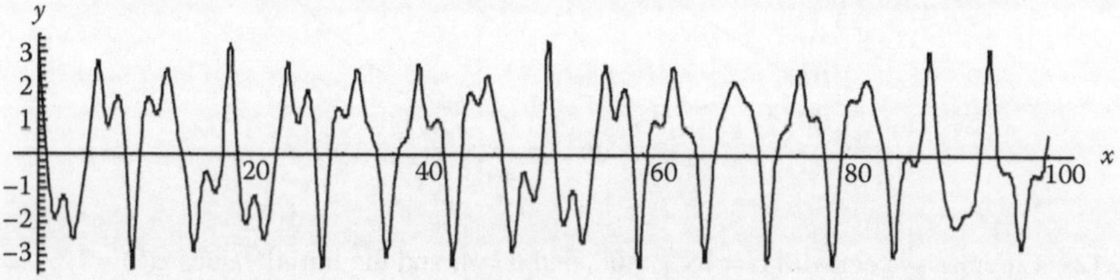

FIGURE 5.9 Duffing equation with initial values $y(0) = 3$ and $y'(0) = 4$.

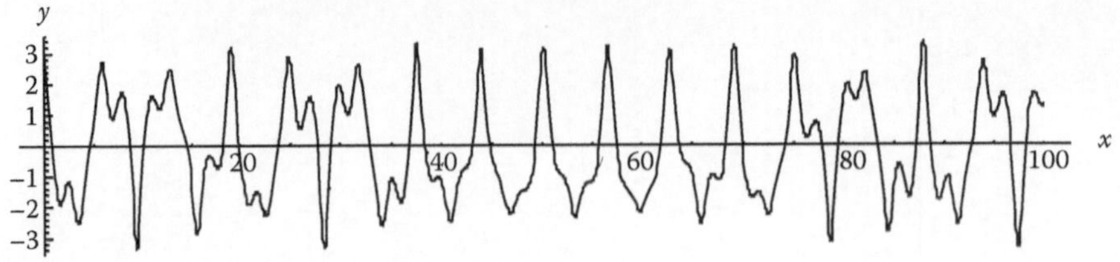

FIGURE 5.10 Duffing equation with initial values $y(0) = 3.01$ and $y'(0) = 4.01$.

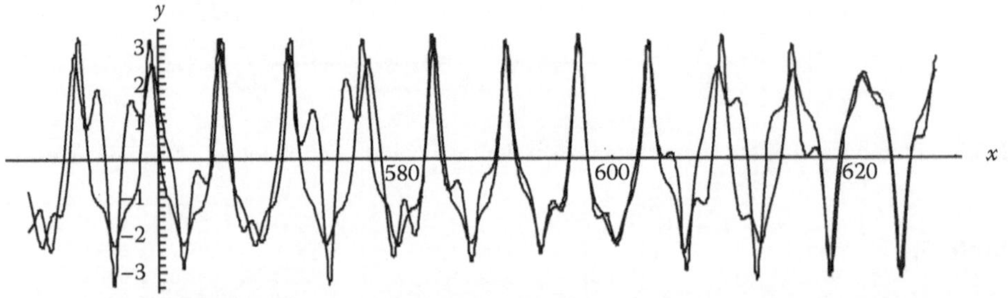

FIGURE 5.11 Overlapping curves of Duffing equation in interval [5.2, 6.2] with initial values $y(0) = 3$, $y'(0) = 4$ and $y(0) = 3.01$, $y'(0) = 4.01$.

Example 5.5

The Lorenz differential equation system is

$$\begin{cases} x' = -\sigma(x-y) \\ y' = rx - y - xz. \\ z' = xy - bz \end{cases}$$

When the parameters are $\sigma = 16$, $r = 60$, and $b = 4$, and the initial values $x(0) = 12$, $y(0) = 4$, and $z(0) = 0$, the graph of the system is given in Figure 5.12. When the initial values are $x(0) = 6$,

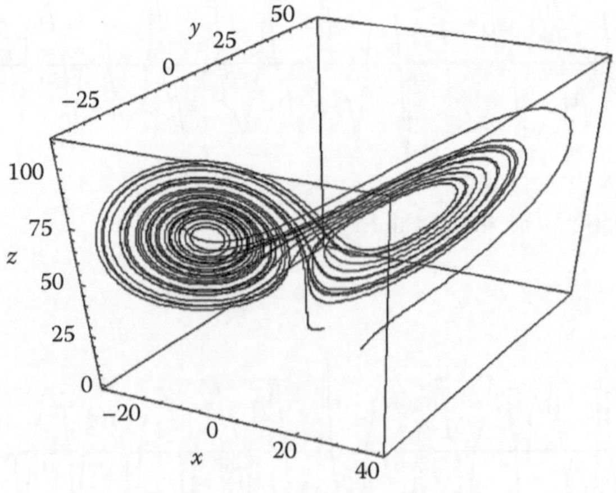

FIGURE 5.12 Lorenz's system with $\sigma = 16$, $r = 60$, and $b = 4$, and the initial values $x(0) = 12$, $y(0) = 4$, and $z(0) = 0$.

$y(0) = 10$, and $z(0) = 10$, the graph of the system is given in Figure 5.13. Now, if the parameters are changed to $\sigma = 10$, $r = 10$, and $b = 8/3$, then the graph of the system is presented in Figure 5.14. When the parameters are changed to $\sigma = 10$, $r = 24.5$, and $b = 8/3$, the graph of the system is depicted in Figure 5.15. When the parameters are changed to $\sigma = 10$, $r = 99.5$, and $b = 8/3$, the system's graph is given in Figure 5.16.

Example 5.6

The graph of the one-dimensional logistic equation

$$x_{n+1} = f(x_n) = \mu x_n(1 - x_n)$$

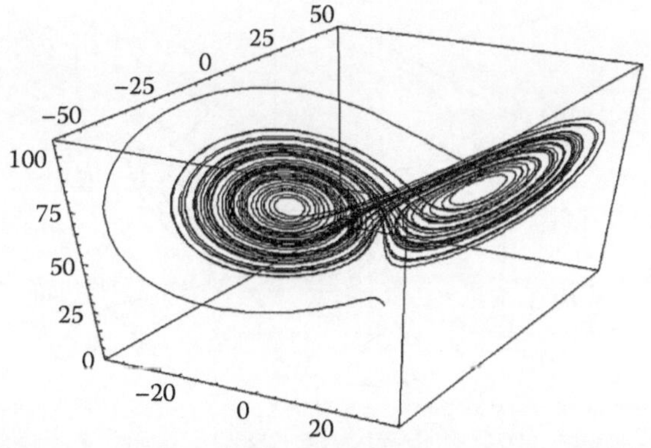

FIGURE 5.13 Lorenz's system with the initial values changed to $x(0) = 6$, $y(0) = 10$, and $z(0) = 10$.

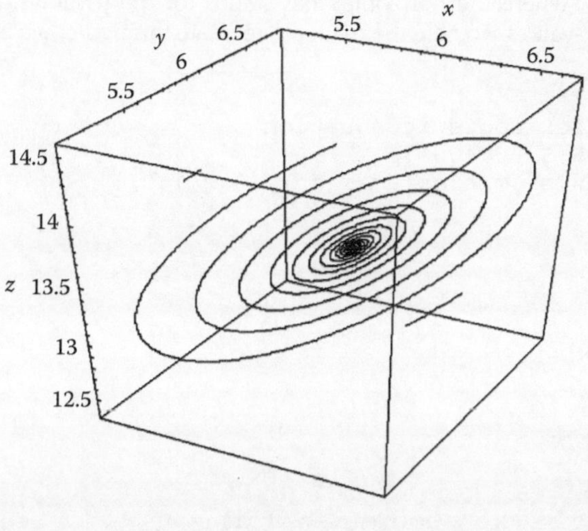

FIGURE 5.14 Lorenz's system with the parameters changed to $\sigma = 10$, $r = 10$, and $b = 8/3$.

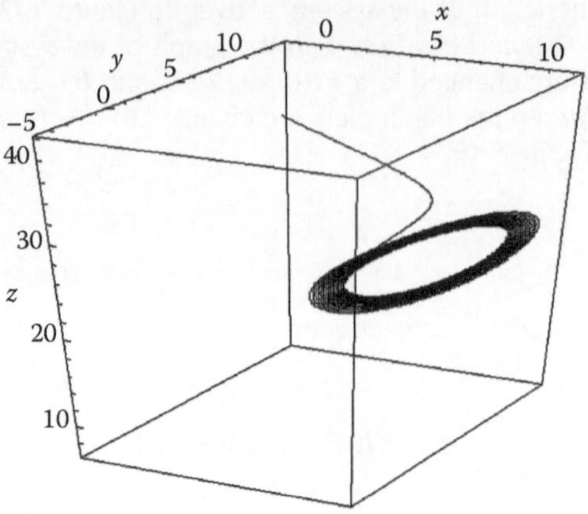

FIGURE 5.15 Lorenz's system with the parameters changed to $\sigma = 10$, $r = 24.5$, and $b = 8/3$.

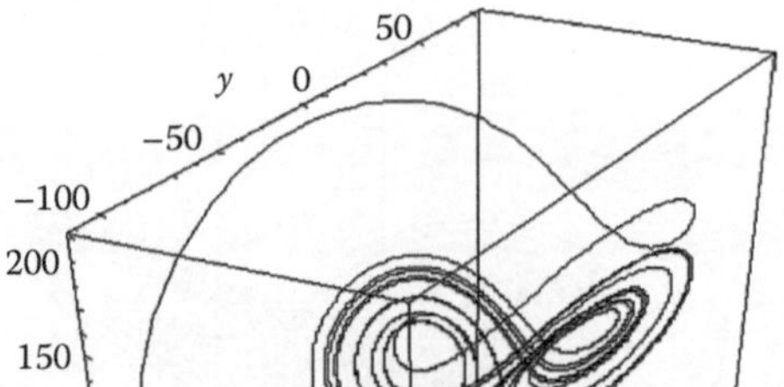

FIGURE 5.16 Lorenz's system with the parameters changed to $\sigma = 10$, $r = 99.5$, and $b = 8/3$.

is given in Figure 5.17, where the horizontal axis stands for the μ-values, and the vertical axis stands for the iterated x-values. To help the reader to recreate this figure, we provide the MATLAB® source code below:

```
******************
%%% Matlab code for logistic equation
an = linspace(2.5,3.99,400);
hold on;box on;axis([min(an),max(an),-0.2,1.2])
N = 1000;
xn = zeros(1,N);
for a = an;
x = rand;
for k = 1:20;
    x = a*x*(1-x);
end
for k = 1:N;
    x = a*x*(1-x);
    xn(k)  = x;
end
plot(a*ones(1,N),xn,'k.','markersize',1);
end
******************
```

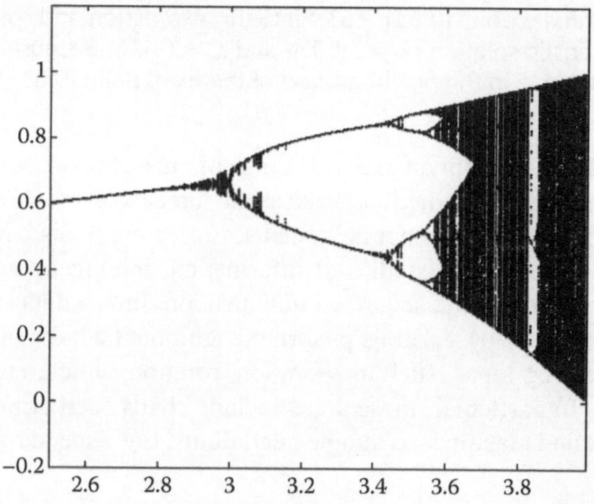

FIGURE 5.17 Graph of the one-dimensional logistic equation.

Example 5.7

The general form of the two-dimensional Henon iteration is given as follows:

$$:\begin{cases} x(n+1) = f(x(n),y(n)) \\ y(n+1) = g(x(n),y(n)). \end{cases}$$

As a particular example, let us use

$$:\begin{cases} x(n+1) = 1+0.3y(n)-1.4x^2(n) \\ y(n+1) = x(n) \end{cases}$$

to compute (1) the fixed point(s) of this system and (2) the Jacobian matrix of this iteration, and then analyze the system's stability in the neighborhood of the fixed point(s) in the first quadrant.

Solution. (1) From the characteristics of fixed points $\xi = f(\xi,\eta)$ and $\eta = g(\xi,\eta)$, it follows that

$$\begin{cases} \xi = 1+0.3\eta-1.4\xi^2 \\ \eta = \xi. \end{cases}$$

The following two fixed points can be readily computed:

(0.631,0.631) and (−1.131,−1.131).

(2) The Jacobian matrix of this iteration is

$$J = \left(\frac{\partial(x(n+1),y(n+1))}{\partial(x(n),y(n))}\right) = \begin{pmatrix} -2.8x(n) & 0.3 \\ 1 & 0 \end{pmatrix}.$$

If we substitute the first fixed point (0.631,0.631) into this expression and solve $\det(J-\lambda I) = 0$, then we obtain two characteristic solutions $\lambda_1 = -1.924$ and $\lambda_2 = 0.156$. Because $|\lambda_1| > 1$, it means that the given system is not stable in the neighborhood of the fixed point (0.631,0.631).

The concept of chaos has been applied to a wide range of areas. In particular, chaos time series has been well employed in the study of predictions, such as the forecast on the short-term loads of electricity systems, dynamic predictions of economic systems, time predictions based on neural networks, etc. Also, chaos has been seen in the studies of information, intelligent automatic control, including intelligent information processing, secure communication, image data compression, high-speed retrieval, chaos optimization, chaos learning, pattern recognition, fault diagnosis, laser chaos, etc.

In mechanical engineering areas, such as vibration, rotation, shock, etc., the concept of chaos has also been well cited. In particular, these areas include chaos oscillation of satellites and chaos vibration of such devices and machines as simple pendulums, high-speed rotors, high-speed cutting tools, printers, etc.

5.7 SOME PHILOSOPHICAL THOUGHTS ON CHAOS

5.7.1 Unification of Order and Disorder

Nonlinear dynamic systems enter chaos through doubling periodic bifurcations so that the periodic states of motion of the original orderly structures are destroyed, while forming infinitely many structural nonperiodic movements and chaotic states. However, in the region of chaos, there are infinitely many nested self-similar structures. Any small window in the chaos region, if magnified to the right scale, possesses the same structure of the original chaos area and has a structure that is similar to that of the whole. This self-similar structure contains the part of orderliness. In the chaos region, the stability of the whole is organically combined with local instabilities. That is, the chaos phenomena possess the disorderly aspect of chaos as well as the orderliness of regularity. There is order in disorder, while order also contains disorder.

The studies of chaos reveal a unification of order and disorder. The orderliness is not absolute but relative to a certain degree, including the internal conditions and factors necessary for producing chaos. Chaos is neither absolute disorder nor simple mess, either. It contains various complicated factors of order. The order and disorder oppose against each other and transform into each other. They coexist within the inside of chaos attractors. Chaos contains rich contents and various kinds of information. Therefore, chaos is an origin for order and an origin of information.

5.7.2 Unity of Such Opposites as Determinacy and Randomness

The laws of motion established by Newton and others are elementary; deterministic and definite conclusions can be derived by using deterministic equations. Statistical mechanics and probability theory, developed since the nineteenth century, looked at randomness and attempted to uncover statistical facts out of large amounts of random events. Randomness, in terms of deterministic equations, is a kind of external interference, rise and fall, or noise. The movements of individual objects obey Newton's laws, while the motions of multiobject groups follow statistical laws. Definite conclusions are produced out of deterministic equations, while statistical results are drawn out of equations of randomness. This implies that the relationships between determinacy and randomness, and inevitability and occasionality, are exogenous and paratactic. Along with the discovery of chaos phenomena, determinacy and randomness, and determinism and indeterminism, are unified at a higher level.

Because each chaos movement is highly sensitive to the initial state, when the initial value experiences small changes, its short-term behaviors can still be predicted. That is different from totally stochastic processes. However, after a long period of evolution, its state becomes unpredictable.

Therefore, out of deterministic equations, some uncertain results are produced. Such phenomena depict the connotative behaviors of deterministic nonlinear equations instead of interferences of the external world. That is why such randomness is referred to as connotative.

There is such connotative randomness in large amounts of conservative and dissipative systems. Such connotative randomness is different from randomness, and completely deterministic equations without involving any factor of randomness can show certain kinds of random behaviors, leading to results of chaos. In any chaos system, periodic and chaos solutions can be organically combined to materialize a higher-level unification of determinacy and randomness, and determinism and indeterminism.

5.7.3 Unification of Universality and Complexity

Feigenbaum discovers that all processes from doubling periodic bifurcation to chaos evolve at a constant convergence speed. The Feigenbaum's constants have a relationship neither with the parabolicity of the logistic equation nor with exponential mapping, sine mapping, and other forms of mappings. Although they have different forms of strange attractors, they all have the self-similarity structure of infinite nestings with the same scale factor.

Such universality has nothing to do with the specific forms of equations, the dimension of the phase spaces, and any specific scientific field, but it has something to do with the complexity of the problem. It is a universal law complexity. These Feigenbaum's constants indicate that there is a unification between the universality and the complexity existing in the evolutionary process from doubling periodicity to chaos. These Feigenbaum's constants reveal the laws inherent to the complicated chaos movements. Elementary iterations lead to behaviors of complex systems, while within complex systems, there are fundamental laws. Thus, chaos can be seen as a unification of simplicity and complexity.

5.7.4 Unification of Driving and Dissipative Forces

In the population model, the linear term stands for a driving force, and the nonlinear term stands for a dissipative force. Changes in the state of the underlying system are completely caused by the competition of these forces. When the driving force is relatively weak, the state can only be constant. When the dissipative force is roughly equivalent to the driving force, various periodic states appear, from which the system evolves into chaos.

The theory that explains how the competition of these two forces causes the state of the nonlinear logistic mapping changes can also be employed to study two-dimensional nonlinear mappings. It is readily seen that in the process of iterations, and that of evolution of nonlinear mappings from doubling periodic bifurcations to chaos, the opposition of the driving forces and dissipative forces are united.

5.8 CHAOS OF GENERAL SINGLE-RELATION SYSTEMS

In this section, we study the concepts of chaos and attractors of general systems of single relations. That is, we consider such general systems $S = (M, R)$, each of which contains only one relation. In particular, $R = \{r\}$.

5.8.1 Chaos of Single-Relation General Systems

A single-relation system $S = (M, \{r\})$ is referred to as an input–output system if there are nonzero ordinal numbers n and m such that

$$\varnothing \neq r \subset M^n \times M^m. \tag{5.15}$$

Without loss of generality, we let $X = M^n$ and $Y = M^m$; instead of using the ordered pair $S = (M, \{r\})$, we will simply think of S as the binary relation r such that

$$\varnothing \neq S \subset X \times Y \tag{5.16}$$

where the sets X and Y are, respectively, referred to as the input space and the output space of S. In the real world, most systems we see are input–output systems. For example, each human being and each factory are input–output systems.

If we let $Z = X \cup Y$, the input–output system S in Equation 5.16 is a binary relation on Z. Let $D \subset Z$ be an arbitrary subset. If $D^2 \cap S = \varnothing$, then D is referred to as a chaos of S (Zhu and Wu 1987). Intuitively, the reason why D is known as a chaos is because the system S has no control over the elements in D.

Theorem 5.2

A necessary and sufficient condition under which an input–output system S over a set Z has a chaos subset $D \neq \varnothing$ is that

$$I \text{ is not a subset of } S, \tag{5.17}$$

where I is the diagonal of the set Z defined by $I = \{(x,x): x \in Z\}$ (Lin 1999, p. 256).

Proof

Necessity. Suppose that the input–output system S over the set Z has a nonempty chaos D. Then $D^2 \cap S = \varnothing$. Therefore, for each $d \in D$, $(d, d) \notin S$. This implies that $I \not\subset S$.

Sufficiency. Suppose that Equation 5.17 holds true. Then there exists $d \in Z$ such that $(d, d) \notin S$. Let $D = \{d\}$. Then the nonempty subset D is a chaos of S. ■

Given $S \subset Z^2$, define $S * D = \{x \in Z : \forall y \in D, (x,y) \notin S\}$ for any $D \subset Z$. Then, we have the following result.

Theorem 5.3

Let $D \subset Z$. Then D is a chaos of the system S iff $D \subset S*D$ (Lin 1999, p. 256).

Proof

Necessity. Suppose that D is a chaos of S. Then for each object $d \in D$ and any $y \in D$, $(d,y) \notin S$. Therefore, $d \in S*D$. That is, $D \subset S*D$.

Sufficiency. Suppose that D satisfies $D \subset S*D$. Then $D^2 \cap S = \varnothing$. Therefore, D is a chaos of S. ■

Let COS(S) denote the set of all chaos subsets on Z of the input–output system S. If $S \not\subset I$, then

$$S_I = \{x \in Z : (x,x) \in S\} \neq Z. \tag{5.18}$$

We denote the complement of S_I by $S_I^- = Z - S_I$. Then Theorem 5.2 implies that each chaos of S is a subset of S_I^-. Therefore, we have

$$\left|\mathrm{COS}(S)\right| \le 2^{|S_I^-|}. \tag{5.19}$$

The following theorem is concerned with determining how many chaos subsets an input–output system has.

Theorem 5.4

Suppose that an input–output system $S \subset Z^2$ satisfies the following conditions (Zhu and Wu 1987):

1. S is symmetric, that is, $(x,y) \in S$ implies $(y,x) \in S$.
2. S is not a subset of I.
3. $\left|S_I^-\right| = m$ is finite.
4. For any $x \in S_I^-$, there exists at most one $y \in S_I^-$ such that either $(x,y) \in S$ or $(y,x) \in S$.

Let $n = \left|\left\{x \in S_I^- : \ y \in S_I^-((x,y) \in S)\right\}\right|$. Then,

$$\left|\mathrm{COS}(S)\right| = 2^m \times \left(\frac{3}{4}\right)^k \tag{5.20}$$

where $k = n/2$.

Proof

According to the discussion prior to the theorem, it is known that every chaos subset of Z is a subset of S_I^-. The total number of all subsets of S_I^- is 2^m. Under the assumption of this theorem, a subset of S_I^- is a chaos of S if the subset does not contain two elements of S_I^- that have S-relations. Among the elements of S_I^-, there are $n/2$ pairs of elements with S-relations, and every two pairs do not have common elements. Let $\{x, y\}$ be a pair of elements in S_I^- with an S-relation; then every subset of $S_I^- - \{x, y\} \cup \{x, y\}$ forms a subset of S_I^- not belonging to COS(S). There are a total of 2^{m-2} of such subsets. There are $n/2$ pairs of elements in S_I^- with S-relations; we subtract these $(n/2) \times 2^{m-2}$ subsets from the 2^m subsets of S_I^-. However, some of the $(n/2) \times 2^{m-2}$ subsets there are in fact the same ones. Naturally, every subset of S_I^-, containing two pairs of elements in S_I^- with S-relations, has been subtracted twice. There are $\binom{k}{2} \times 2^{m-4}$ such sets. We add the number of subsets that have been subtracted twice and obtain

$$2^m - k \times 2^{m-2} + \binom{k}{2} \times 2^{m-4}. \tag{5.21}$$

However, when we add $\binom{k}{2} \times 2^{m-4}$ subsets, the subsets of S_I^- containing three pairs of elements of S_I^- with S-relations have been added once more than they should have. Therefore, this number should be subtracted. Continuing this process, we have

$$\begin{aligned}
|\text{COS}(S)| &= 2^m - k \times 2^{m-2} + \binom{k}{2} \times 2^{m-4} - \binom{k}{3} \times 2^{m-6} + \ldots \\
&= \sum_{i=0}^{k} \binom{k}{i} \times 2^{m-2i} \times (-1)^i \\
&= 2^{m-n} \sum_{i=0}^{k} \binom{k}{i} \times 2^{n-2i} \times (-1)^i \\
&= 2^{m-n} \sum_{i=0}^{k} (-1)^i \binom{k}{i} \times 2^{2k-2i} \\
&= 2^{m-n} \sum_{i=0}^{k} (-1)^i \binom{k}{i} \times 4^{k-i} \\
&= 2^{m-n} (4-1)^k \\
&= 2^{m-n} \times 3^k \\
&= 2^m \left(\frac{3}{4}\right)^k .
\end{aligned}$$

■

5.8.2 Attractors of Single-Relation General Systems

Let S be an input–output system on a set Z. A subset $D \subset Z$ is referred to as an attractor of S (Zhu and Wu 1987), provided for each $x \in Z - D$, $S(x) \cap D \neq \varnothing$, where $S(x) = \{y \in Z\text{: } (x,y) \in S\}$. Figures 5.18 and 5.19 show the geometric meaning of the concept of attractors. When S is not a function, Figure 5.18 shows that the graph of S outside the vertical bar $D \times Z$ overlaps the horizontal bar $Z \times D$. When S is a function, Figure 5.19 shows that the graph of S outside the vertical bar $D \times Z$ must be contained in the horizontal region $Z \times D$.

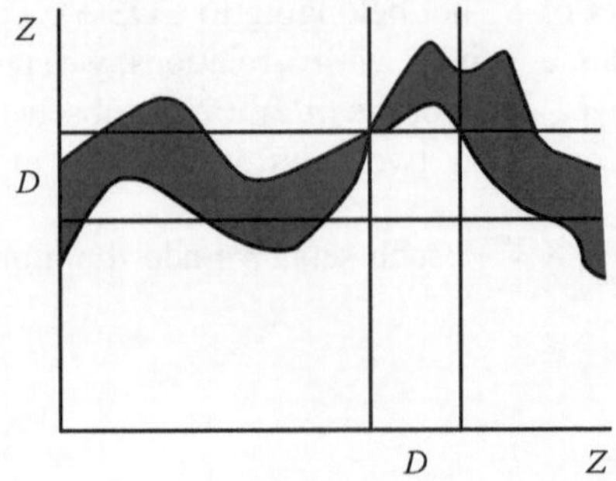

FIGURE 5.18 D is an attractor of S, and $S(x)$ contains at least one element for each $x \in Z - D$.

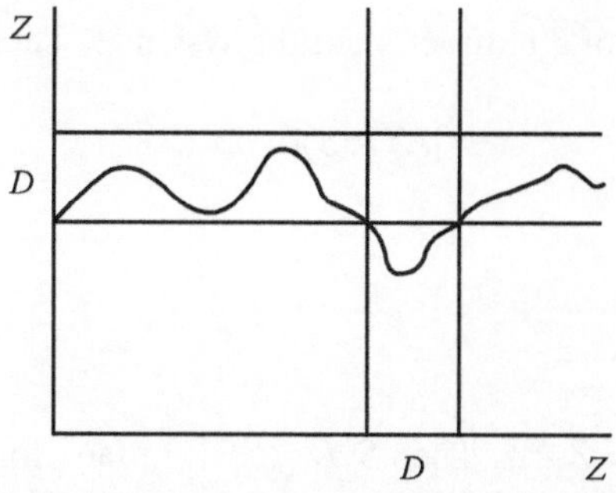

FIGURE 5.19 *D* is an attractor of *S*, and *S* is a function from *Z* to *Z*.

Theorem 5.5

Suppose that $S \subset Z^2$ is an input–output system over the set Z and $D \subset Z$. Then D is an attractor of S, iff $S^*D \subset D$ (Lin 1999, p. 258).

Proof

Necessity. Suppose that D is an attractor of S. Let $d \in S^*D$ be an arbitrary element. From the definition of S^*D, it follows that for every $x \in D$, $(d,x) \notin S$. Therefore, $d \in D$. That is, $S^*D \subset D$.

Sufficiency. Suppose that D satisfies $S^*D \subset D$. It follows from the definition of S^*D that for each object $x \in Z - D$, $x \in S^*D$. Thus, there exists at least one object $d \in D$ such that $(d,x) \in S$. That is, $S(d) \cap D \neq \varnothing$. This implies that D is an attractor of S. ■

Theorem 5.6

A necessary and sufficient condition under which an input–output system $S \subset Z^2$ *has an attractor* $D \neq Z$ is that S is not a subset of I (Lin 1999, p. 259).

Proof

Necessity. Suppose S has an attractor $D \neq Z$. By contradiction, suppose $S \subset I$. Then for each $D \subset Z$, $Z - D \subset S^*D$. Therefore, when $Z - D \neq \varnothing$, $D \not\subset S^*D$; that is, S, according to Theorem 5.5, does not have any attractor not equal to Z. This tells us that the hypothesis that S has an attractor D such that $D \neq Z$ implies that $S \not\subset I$.

Sufficiency. Suppose $S \not\subset I$. Choose distinct objects x, $y \in Z$ such that $(x,y) \in S$, and define $D = Z - \{x\}$. Then, according to the definition of S^*D, $S^*D \subset D$. Applying Theorem 5.5, it follows that S has attractors. ■

Theorem 5.7

Suppose that an input–output system $S \subset Z^2$ satisfies the condition that if $(x,y) \in S$, then for any $z \in Z$, $(y,z) \notin S$ (Zhu and Wu 1987). Let

$$S_{I^-} = \{x \in Z: \ y \in Z(y \neq x \text{ and } (x,y) \in S)\} \tag{5.22}$$

$n = |S_{I^-}|$, and ATR(S) be the set of all attractors of the system S. Then

$$|\text{ATR}(S)| = 2^n. \tag{5.23}$$

Proof

For each subset $A \subset S_{I^-}$, let $D_A = Z - A$. Then $S^*D_A \subset D_A$. In fact, for each object $d \in S^*D_A$, $(d,x) \notin S$ for every $x \in D_A$. Therefore, $d \notin A$. Otherwise, there exists an element $y \in Z$ such that $y \neq d$ and $(d,y) \in S$. The hypothesis stating that if $(x,y) \in S$ then for any $z \in Z$, $(y,z) \notin S$ implies that the element $y \in Z - A$, a contradiction; thus, $d \in Z - A = D_A$. This implies that $S^*D_A \subset D_A$. Applying Theorem 5.5, it follows that the subset D_A is an attractor of S. Now it can be seen that there are 2^n many such subsets D_A in Z. This completes the proof of the theorem. ■

5.9 OPEN-ENDED PROBLEMS

1. Is the existence of randomness an inherent characteristic of the universe? Or is it a product of the finiteness of human wisdom?
2. Is chaos the same as being chaotic? What can be known as "chaos order"?
3. How can you understand that "chaos stands for deterministic randomness"? How could the sensitivity of chaos to the initial values be produced?
4. Explore how the concept of chaos destroys Laplace's demon. How does chaos help expand human capability on prediction?
5. It is believed to be Henry Adams who said that chaos often breeds life, while order breeds habit. Please write down what you feel from reading this statement.
6. For the population model

$$x_{n+1} = \mu(1 - x_n)x_n,\ x_n \in (0,1),\ \mu > 0,$$

   prove the following: When $\mu = 2$, then the system has a unique attractor. Then, compute the attractor and analyze its field of attraction.
7. In the two-dimensional Henon iteration of Example 5.7, try to alter the parameter 1.4 to see the consequent changes accordingly.

6 Fractals

Continuing from the previous chapter, here we will study some of the fundamental concepts of the fractal theory and relevant applications. Speaking generally, all the research objects from areas like systems, structures, information, etc., that are of self-similarities and/or same order self-similarities are referred to as fractals. The creation of fractal sets makes people realize the existence of regularity, self-similarities without any characteristic scale, within seemingly chaotic phenomena, where the regularity cannot be described by using the conventional Euclidean geometry. However, not every complex scenario is a fractal phenomenon.

6.1 DIFFERENT ASPECTS OF FRACTALS

6.1.1 Fractal Geometry, Fractal Phenomena, and Multiscale Systems

The concept of fractals was initially proposed by B.B. Mandelbort in 1975, after he had spent a long period of time on several scientific areas that have been historically remote and had hardly anyone considered. Like other concepts in history, the groundwork for such a concept had been laid for quite some time. For example, in the early nineteenth century, French mathematician Poincare employed some innovative geometric methods in his studies of the three-body problem. However, due to its level of theoretical difficulty, hardly anyone noticed these methods. In 1875, German mathematician Weierstrass constructed a surprisingly continuous function that is not differentiable everywhere. After that, Cantor, the founder of set theory, constructed his well-known trichotomic Cantor set with a great many strange properties. In 1890, Italian mathematician Guiseppe Peano discovered an incredible curve that can theoretically fill a space. In 1904, Swedish mathematician Cohen designed a curve that looks like a snow flake and the edge of an island. Ten years later, Polish mathematician Sierpinski successfully drew such geometric figures that look like a carpet and sponge. However, due to their peculiarity, all these discoveries suffered from the same fate: they were laid aside without being investigated further. However, it is the commonalities of these peculiar discoveries that some rich mathematical thoughts were germinated. In the 1920s, German mathematician Hausdorff introduced the concept of fractal dimension in order to study the properties of peculiar sets. After that, several mathematicians employed the concept of fractal dimensions to resolve their individual research problems. However, these exotic and innovative mathematical thoughts and concepts were initially introduced to overthrow an accepted conclusion. It was not until 1975 that these incomprehensible small pieces of knowledge were assembled together to form a new field of scientific activities.

The basic idea behind fractal geometry is the self-similarity in the sense of statistics that exists in between the whole and parts in terms of form, functionality, information, time, space, and other aspects. For instance, any part of a magnet is like the whole, which has both a positive and negative polarity. No matter how the division continues, each resultant part has a magnetic field that is similar to that of the whole. The overall structure of such a layered self-similarity stays invariant no matter how it is magnified or reduced.

The fractal theory believes that dimensionality can also be a fraction. Such dimensions represent some of the important concepts established by physicists in their studies of chaotic attractors and other theories. To quantitatively describe the degree of "irregularity" of objective matters, in 1919, mathematicians established the concept of fractal dimensions, which generalizes the concept of dimensionality from whole numbers to fractions so that the whole-number boundary of general topological dimensions is broken.

The emergence of fractal geometry makes people realize the existence of regularity underneath the seemingly chaotic messes. That is, the self-similarities of no particular scales cannot be described by using the classical Euclidean geometry.

The following are two examples of fractals provided by the initiators of the fractal theory: (1) a set whose parts are similar to the whole in a certain way and (2) a set whose Hausdorff dimension is greater than its topological dimension.

There are also scholars who point out that fractal, in general, stands for a class of objects that possess symmetric expansions and contractions and whose states are difficult for Euclidean geometry to describe. Here, symmetric expansions and contractions mean that if a magnifying glass of multiple times is employed to make observations, one can see similar structures. No matter which definition is applied, the concept of similarity is mentioned. As a matter of fact, self-similarity is the core concept and the fundamental characteristics among all characteristics of the fractal theory. Self-similarity means that each regional form is similar to that of the whole. In other words, each chosen part of the whole can represent the basic spirit and main characteristics of the whole. Any figure of the fractal geometry represents a form that is composed of parts, each of which is similar to the whole in some specific way. Its essential characteristic is like an infinitely nested net.

A well-known figure once commented that the so-called pathological structures created by those mathematicians who broke loose from the cage of the nineteenth century naturalism are, in fact, inherent in the familiar matters in our surroundings. These inherent characteristics of nature are eliminated in the idealized research of mathematicians. Even so, the studies on the rest of the remaining objects still helped the man to create the enormous material wealth and to satisfy man's various demands that nature cannot provide. However, along with the introduction of the fractal theory, the world of learning once again returns to the investigation of those inherent characteristics. In nature, there are countless many examples of fractals, such as cloud clusters, range upon range of mountains, coastal lines, barks, lightning, etc. All these natural phenomena stand for some of the most straightforward expressions of the fractal geometry.

To further investigate the geometric properties of fractals, the concept of characteristic scales is established. The so-called characteristic scale of a matter stands for the specific magnitude order the matter possesses in either space or time. To measure a particular magnitude order, an appropriate ruler is needed. For instance, the magnitude order of meter is the characteristic scale of human heights. For typhoons, their characteristic scale is several thousand kilometers. If we treat each typhoon as a whirlpool and study it from the angle of geometric structures, then small whirlpools are nested inside the large whirlpool. This phenomenon occurs within the individual ranges of different scales. Systems that involve multiple different scales are referred to as multiscale systems or systems without any characteristic scale.

The chaos of dissipative systems appears with the mechanism of motion of these kinds of fractal structures of strange attractors. These kinds of complex motions can be described only by using certain kinds of fractal point sets in phase spaces. Both Lorenz and Rössler attractors are such point sets, where when any chosen portion is magnified, it looks just like the whole without much order but with infinitely fine structures and a certain kind of self-similarity.

6.1.2 Topological and Fractal Dimensions and Fractal Statistical Models

In Euclidean spaces, the dimensionality of a geometric object is equal to the number of independent coordinates needed to determine the location of any point in the object. The dimensionality defined in this way is known as the Euclidean or topological dimension, because domain is a core concept of topology. No matter how a geometric object is stretched, compressed, and twisted, as long as continuity is guaranteed, any two neighboring points in the object will maintain neighboring. It is because topological dimensions are an invariant quantity under topological transformations. For example, the topological dimension of a line or curve is 1; the topological dimension of a plane figure is 2; and the topological dimension of each spatial figure is 3.

First of all, let us start with the concept of integer dimension and then expand the concept later. Assume that a unit segment of the "radius" r (i.e., the length of the segment is $2r$) is used to measure a line of length L. Let the number of times the unit segment is used be $N(r)$. Then, we have

$$N(r) = \frac{L}{2r} = cr^{-D}$$

where $D = 1$ and $c = \frac{L}{2}$. That indicates that the one-dimensional line segment of length L contains $N(r)$ copies of the unit segment of "radius" r. The previous formula can be rewritten as follows:

$$N(r) = cr^{-D_f} \propto r^{-D_f}$$

where $D_f = 1$ stands for the dimension of the line segment of length L.

Similarly, assume that there is a two-dimensional circular disk and that we measure its area S by using the area of a small circular disk of radius r. If the area S is equal to $N(r)$ copies of the unit disk, then we have

$$N(r) = \frac{S}{\pi r^2} = cr^{-D_f}$$

where, $c = \frac{S}{\pi}$ and $D_f = 2$ represents the dimension of the circular disk to be measured.

Now, let us generalize the two dimensionality into the three dimensionality and measure the volume V of a three-dimensional ball. Assume that we now measure V using a unit ball of radius r; the resultant reading is that $N(r)$ times of the unit volume fills the volume V. Then we have

$$N(r) = \frac{V}{\frac{4}{3}\pi r^3} = cr^{-D_f}$$

where $c = \frac{V}{\frac{4}{3}\pi}$ and $D_f = 3$ is the dimension of the ball to be measured.

According to the previous cases as just discussed, the concept of dimension can be extended to the general case. If the unit that is used to make measurement has radius r, and the number $N(r)$ of copies of the unit contained in the object to be measured satisfies the relation

$$N(r) = cr^{-D_f},$$

then the object that is to be measured is of D_f dimensions. That is the classic fractal dimension, known as the Hausdorff dimension.

In the following, let us look at the concept of box dimension.

If we use a ruler of length ε to measure a unit square or a cube, we can obtain the number $N(\varepsilon)$ of small squares or small cubes needed to cover the unit square or cube. The following formula can be employed to calculate the topological dimension D of the unit square or cube:

$$D = \frac{\log N(\varepsilon)}{\log\left(\frac{1}{\varepsilon}\right)},$$

which can be treated as the definition of topological dimensions.

Each topological dimension is a whole number. Although the count $N(\varepsilon)$ of small boxes increases with the shortening length ε of the ruler, the length (or total area, or total volume) of the geometric object stays the same.

In terms of the length of a coastal line, it becomes longer and eventually approaches infinity as the length ε of the ruler shortens. In order to generalize the concept of topological dimensions to that of fractal dimensions, one has to break through the limitation that dimensionality has to be whole numbers. Additionally, he or she has to take limit to the defined topological dimensions. For instance, the following is a definition for fractal dimension:

$$D = \lim_{\varepsilon \to 0} \frac{\log N(\varepsilon)}{\log\left(\frac{1}{\varepsilon}\right)},$$

which is the definition Hausdorff introduced in 1919 and so is also referred to as the Hausdorff dimension.

A fractal statistical model stands for the exponential relationship, $P(\xi \le r) \propto r^D$, $r > 0$, between the probability of smaller than (or greater than) a particular characteristic scale and this characteristic scale, where D stands for the fractal dimension. Evidently, the distribution function is $F(x) = Cx^D$, $x > 0$, where C is a constant that makes this distribution function satisfy the condition of unity. Since for any $0 < x < x'$, $P(\xi \le x | \xi \le x') = (x/x')^D$ holds true, we have the following fractal property of invariant scales: $P(\xi \le x | \xi \le x') = P(\xi \le cx | \xi \le cx')$, where c is an arbitrary positive constant. By taking the logarithm of the basic expression of the fractal statistical model, one obtains $\log P(\xi \le r) = D \log r + \log C$. Therefore, by computing the linear regression of $\log P$ and $\log r$, one can obtain the fractal dimension D, and in the plot of $\log P/\log r$, one can observe a set of dots lined up in a line format, whose slope is equal to the fractal dimension.

Proposition 6.1

There is not any fractal dimension for any sample that is obtained from a normally $N(0, \sigma)$ distributed population.

Proof

We prove by contradiction. Assume that there is a fractal dimension; that is, there is D such as $P(\xi > r) = Cr^D$, $r > 0$. Then,

$$D = \frac{\log P(\xi > r)}{\log r}.$$

Hence, $\lim_{r \to \infty} D = \lim_{r \to \infty} \frac{\log P(\xi > r)}{\log r}$. By applying the L'Hospital's rule of finding limits, we differentiate the numerator and the denominator at the same time. By letting the probability density function of the normal distribution be $f(x)$ and the distribution function $F(x)$, we have

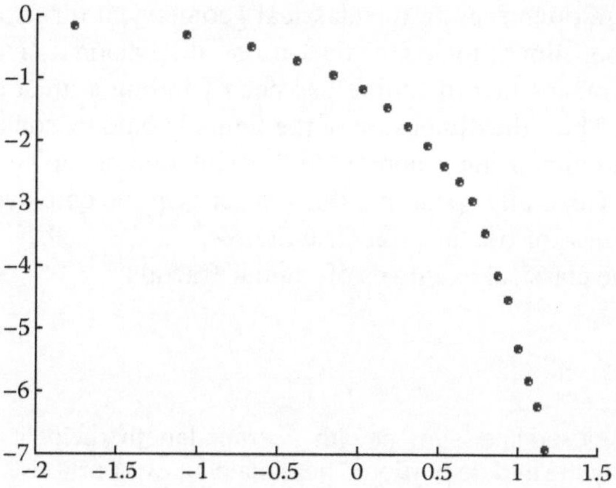

FIGURE 6.1 Ratio of log $P(\xi > r)$ to log r of random numbers satisfying $N(0,1)$ distribution.

$$\lim_{r\to\infty} D = \lim_{r\to\infty}\frac{\log P(\xi > r)}{\log r}$$

$$= \lim_{r\to\infty}\frac{-f(r)/\left[1-F(r)\right]}{1/r},$$

$$= \lim_{r\to\infty}\frac{rf(r)}{F(r)-1} = -\infty$$

which means that there is no stable slope so that no fractal dimension exists. ■

Let us take 2000 random numbers that satisfy the normal $N(0,1)$ distribution. For each r of the positive numbers, compute log $P(\zeta > r)$ and log r (for the negative numbers, the results are similar). Over the entire region $r \in \left[\frac{1}{100}, 30\right]$, namely, log $r \in [-2, 1.5]$, the ratio of log $P(\xi > r)$ to log r is in the state of a curve with slopes varying from 0 to negative infinity without any clearly defined straight segment (Figure 6.1), where the horizontal axis stands for log r and the vertical axis log $P(\xi > r)$. Thus, for normal distributions, there is no fractal dimension.

6.1.3 Regular Fractals and Similarity Dimensions

First of all, let us look at the concept of similarity dimensions.

For a complex geometric figure, the ordinary dimension might change with the scale. For example, a 10-cm-diameter ball of woolen yarn is made by winding a 1-mm-diameter woolen yarn. When this woolen ball is observed differently, its dimension will be different. In particular, if one moves from a distant zero-dimensional point to the three-dimensional ball, as he or she gets closer, he or she will in turn see a one-dimensional thread, a three-dimensional round pole, one-dimensional fiber, and eventually zero-dimensional molecules. These differences in dimensionality caused by differences in the distance scales are similar to the problem of coastal line in the previous discussions. Thus, these scenarios suggest that we need to establish a uniform standard for measuring dimensionality.

There are many different kinds of fractal dimensions, with similarity dimension being one that is relatively simple. When the concept of similarity dimensions is applied to points, lines, surfaces, and volumes, such as those often seen in the classical geometry, it turns out to be the same as the conventional whole-number dimensions. Consider a one-dimensional smooth curve. If the curve nearly fills up an area, it means that the curve has been winding a great deal and almost forms a two-dimensional surface. Thus, the dimension of the figure should be somewhere between 1 and 2. If the concept of similarity dimension is applied to this situation, it can be seen as a measure of the complexity of this curve. Generally speaking, this concept can be employed as a measurement for the complexity and roughness of fractal geometric figures.

The following are some classical examples of regular fractals.

Example 6.1

(The Cantor set). Take a closed line segment with a certain length, divide it into three equal-length portions, and throw away the middle portion. The remaining two portions are still closed intervals. For the next step, divide each of the remaining intervals into three equal-length portions, throw away the middle portion, and keep the remaining four closed intervals. Continuing this inductive process until the infinity leads to a remaining discrete set. According to the definition of similarity dimensions, we see that the dimension of this point set is between 0 and 1 exclusively. Different from the concept of points in the traditional geometry, the points in the Cantor set have their lengths and are not zero-dimensional (see Figure 6.2). The MATLAB® source codes are given below.

```
**********************
%%% Matlab code for The Cantor set
[za,zb] = cantor(50+110*i,750+110*i);
ylim([-50 120]); box on;
function [za zb] = cantor(za,zb)
if nargin = =0;
za = 50+110i;
zb = 750+110i;
end
c = 4;
d = 30;
hold on;
if abs(real(za)-real(zb))>c
plot(real([za,zb]),imag([za,zb]),'b','linewidth',3);
zc = za*2/3+zb*1/3;
zc = zc-d*i;
```

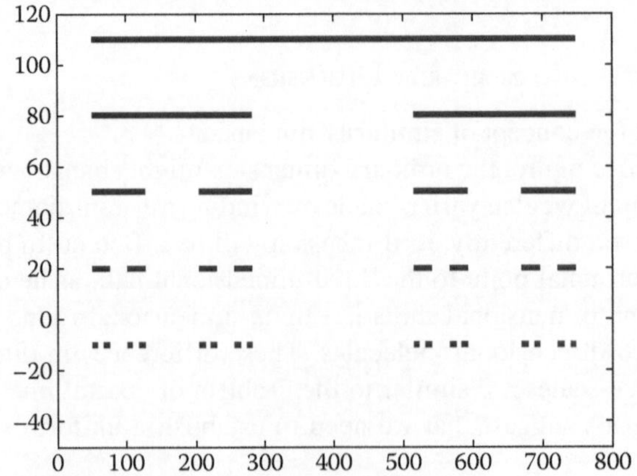

FIGURE 6.2 Cantor set.

```
za = za-d*i;
[za,zc] = cantor(za,zc);
za = za+d*i;
zd = za/3+zb*2/3;
zb = zb-d*i;
zd = zd-d*i;
[zd,zb] = cantor(zd,zb);
end
title('Cantor Set')
**********************
```

Example 6.2

(Koch curve and its generalization). Take a line segment of a certain length. After evenly dividing this segment into three pieces of equal length and keeping the two end segments, draw an equilateral triangle that has the middle segment from step 1 as its base and points upward, and then remove the line segment that is the base of the triangle. As the next step, perform the same operation as before on each of the new line segments. By repeating this procedure until infinity, we obtain a fractal geometric figure whose similarity dimension is between 1 and 2 exclusively (see Figure 6.3). The following are the MATLAB source codes.

```
******************************
%%% Matlab code for The Koch curve
M = [1/3,0,0,1/3,0,0;...
1/6,1/sqrt(12),-1/sqrt(12),1/6,1/3,0;...
1/6,-1/sqrt(12),1/sqrt(12),1/6,1/2,1/sqrt(12);...
1/3, 0,0, 1/3, 2/3,0 ];
p = ones(1,4)/4;
IFS_draw(M,p)
toc
title('Koch Curve')
function IFS_draw(M,p)
N = 30000;%number of iteration
for k = 1:length(p);
eval(['a',num2str(k),' = reshape(M(',num2str(k),',:),2,3);']);
end
xy = zeros(2,N);
pp = meshgrid(p);
pp = tril(pp);
pp = sum(pp,2);
for k = 1:N-1;
a = rand-pp;
d = find(a< = 0);
```

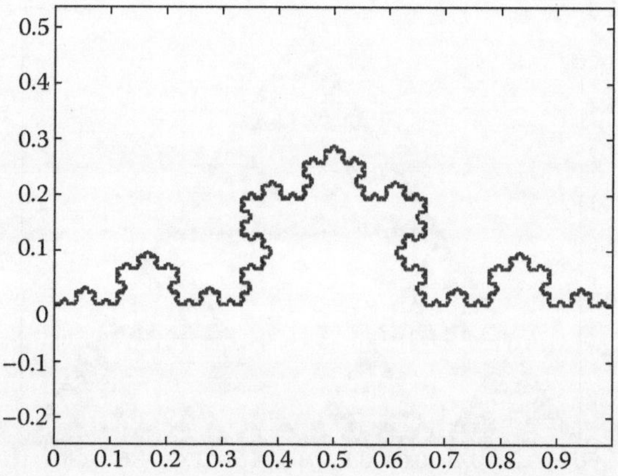

FIGURE 6.3 Koch curve.

```
xy(:,k+1) = eval(['a',num2str(d(1)),'(:,1:2)'])*xy(:,k)+eval(['a',num2str(d(1)),'(:,3)']);
end
P = complex(xy(1,:),xy(2,:));
plot(P,'b.','markersize',2);
axis equal
******************************
```

Now, let us replace the initial line segment from above with a plane area. Then we obtain a series of surfaces with uplifted regions. The result of continuing this procedure is a fractal geometric figure whose similarity dimension is between 2 and 3.

Example 6.3

(Sierpinski gasket). Let us image a continued "tunneling" operation on a regular cubic body. In particular, we divide every face of the cube into 9 squares, which subdivide the cube into 27 smaller cubes; then we remove the cube at the middle of every face and the one in the center, leaving 20 cubes. As the second step, we repeat this operation on each of the remaining small cubes. The limit of this process after an infinite number of iterations will be a fractal geometric figure that looks like a sponge (see Figure 6.4). Its similarity dimension is between 2 and 3 exclusively. The following are the MATLAB source codes.

```
*********************************
%% matlab code for Sierpinski gasket
N = 40000;
p = 1/3;
q = 2/3;
e = [1,0];
f = [0.5,sqrt(3)/2];
pxy = zeros(N,2);
r = rand(1,N-1);
for k = 1:N-1;
if r(k)<p
     pxy(k+1,:) = pxy(k,:)/2;
else if r(k)<q;
     pxy(k+1,:) = pxy(k,:)/2+e;
     else
     pxy(k+1,:) = pxy(k,:)/2+f;
     end
end
```

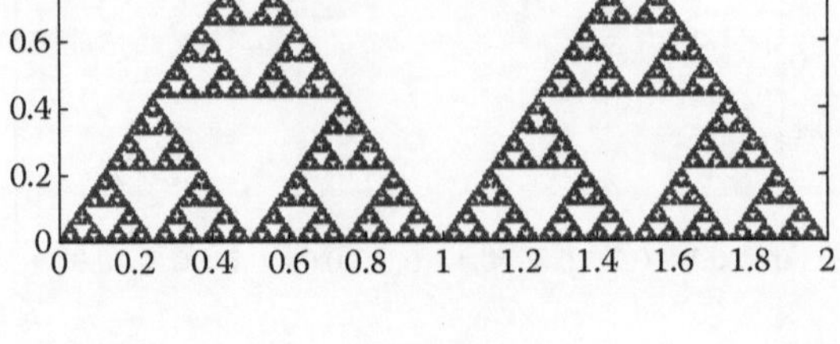

FIGURE 6.4 Sierpinski gasket.

```
end
P = complex(pxy(:,1),pxy(:,2));
plot(P,'b.','markersize',2)
title('Sierpinski gasket ')
*********************************
```

Example 6.4

(Julia set). A technique, often known as "hopalong" following an article published in *Scientific American* in 1986 by Barry Martin, is normally used to represent the strange attractor of a chaos system, for example, the well-known Julia set (Figure 6.5). For a complex number *c*, the filled-in Julia set of *c* is the set of all points *z* of the complex plane for which the iteration $z \rightarrow z^2 + c$ does not diverge to infinity. The Julia set is the boundary of the filled-in Julia set. For almost all complex numbers *c*, these sets are fractals.

Similar to the formation of the Julia set, the Mandelbrot set is the set of all *c* for which the iteration $z \rightarrow z^2 + c$, starting from $z = 0$, does not diverge to infinity. Julia sets are either connected (one piece) or a dust of infinitely many points. The Mandelbrot set contains those points *c* for which the Julia set is connected.

The Mandelbrot set (Figure 6.6) may well be the most familiar image produced by the mathematics of the last century.

All these examples provide scenarios of regular fractals. That is, there is a given construction principle. The similarity dimension of the resultant geometric figure depends on the particular operations.

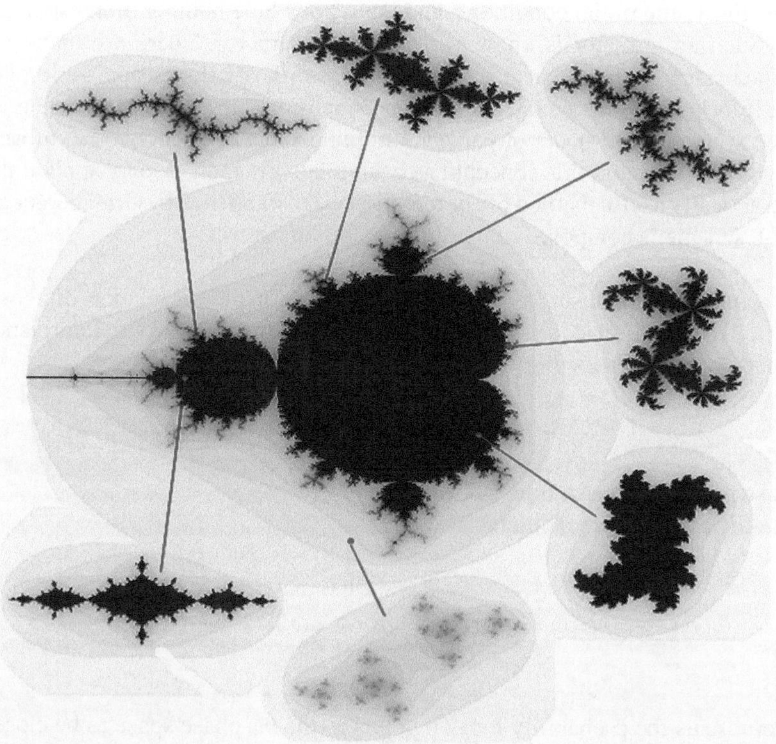

FIGURE 6.5 Julia set.

FIGURE 6.6 Mandelbrot set. (From http://graffiti.u-bordeaux.fr/MAPBX/roussel/fractals.html, accessed on October 5, 2010. With permission.)

6.1.4 Irregular Fractals and Correlation Dimensions

When the proportional operational details in the construction of regular fractals are replaced by random operations, what are produced are referred to as random fractals. In the investigation of Brownian motions, there were once people who traced the motion trajectory of a suspended particle. After connecting the observed trajectory points using line segments, one obtains a zigzagged line. If the time interval between consecutive observations is shortened, the zigzagged line obtained earlier will be replaced by a new one. When the length of the time interval shortens indefinitely to zero, then the motion trajectory of the particle will also be a fractal geometric figure, a random fractal.

6.1.4.1 Correlation Dimensionality

When a nonlinear system is in a chaos state, its motion trajectory in the phase space is very complicated, is sensitive to the initial conditions, and has a nonwhole number dimensionality—a fractal characteristic. Such an attractor is known as a strange attractor. There are many ways to compute fractal dimensions, such as Hausdorff dimension, similarity dimension, Kolmogorov capability dimension, information dimension, correlation dimension, etc. Because by using only one data sequence of the system, the concept of correlation dimensions can provide the information on the dimensionality of the attractor, this concept has been relatively more widely applied than others.

One quite practically useful method of computation was established by Grassberger and Procaccia (1983). The G-P scheme is composed of the following main steps:

1. Choose a relatively small m_0 as the initial estimate for the correlation dimension of the attractor of the time series $x_1, x_2, \ldots, x_{n-1}, x_n, \ldots$ Then, construct the corresponding m_0 phase space with the following elements:

$$Y(t_i) = [x(t_i), x(t_i + \tau), x(t_i + 2\tau), \ldots, x(t_i + (m_0 - 1)\tau)],\ i = 1,2,$$

   where τ stands for a chosen time delay.

2. Compute the correlation function

$$C(r) = \lim_{N\to 0} \frac{1}{N^2} \sum_{i,j=1}^{N} \theta\left(r - \left|Y(t_i) - Y(t_j)\right|\right),$$

   which represents the probability for two points in the m_0 phase space to be closer than the distance r, where $|Y(t_i) - Y(t_j)|$ stands for the distance between the phase points $Y(t_i)$ and $Y(t_j)$ and $\theta(z)$ is an indicator function (a symbolic function).

3. From the theoretical reasoning, it is known that the correlation function $C_n(r)$, which is defined as the function within the limit operation in step 2 with N replaced by n, satisfies the following relation with r as $r \to 0$:

$$\lim_{r \to 0} C_n(r) \propto r^{d(m)},$$

that is,

$$d(m) = \frac{\ln C_n(r)}{\ln r}$$

where $d(m)$ stands for the estimate for the correlation dimension of the attractor.
4. Increase the embedding dimension from m_0 to m_1 and repeat steps 1–3 until the corresponding dimension estimates $d(m_0)$, $d(m_1)$, $d(m_2)$, $d(m_3)$,... converge to d. Then this d is the correlation dimension of the attractor. However, if $d(m)$ increases with m without converging to any definite number, then this fact implies that the system of concern is a random time series.

Example 6.5

The data of this experiment represent the closing index prices of the Shanghai Stock Exchange from January 5, 1998 to March 29, 2004 for a total of 1503 trading days. The time series plot of these data is given in Figure 6.7.

As indicated in Figures 6.8(a) and (b), all the curves represent the relationship of $\ln C_n(r)$ versus $\ln r$ when the embedding dimension is taken from 4– to 18. Because these curves are roughly parallel to each other, it means that the slopes are stable. That is, the correlation dimension functions converge with the increase of in the embedding dimension. Through actual fitting computation, we obtain 2.63 as the correlation dimension for the Shanghai Stock Exchange; when the time delay τ is 5, the convergence of the correlation dimension is the best, and the reconstruction of the phase space is most ideal. When the time delay is less than 2, the correlation dimension has the tendency of increase, while the reconstruction of the phase space is not good. That end indicates that when the time delay τ is too small, the embedded vectors behave like random variables without much or any correlation so that the potential determinacy is lost.

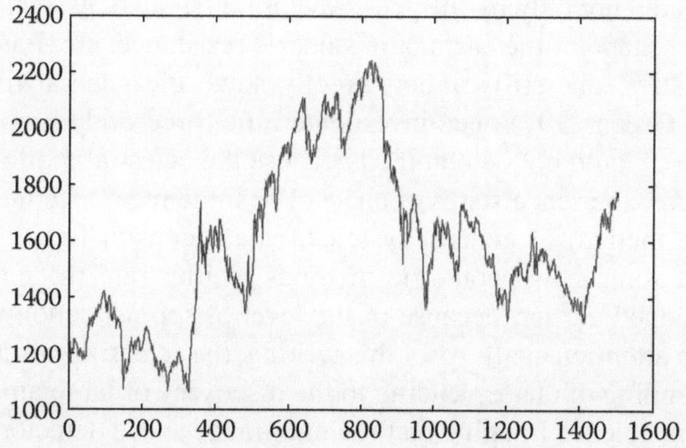

FIGURE 6.7 Time series plot of Shanghai Stock Exchange Index (January 5, 1998–March 29, 2004).

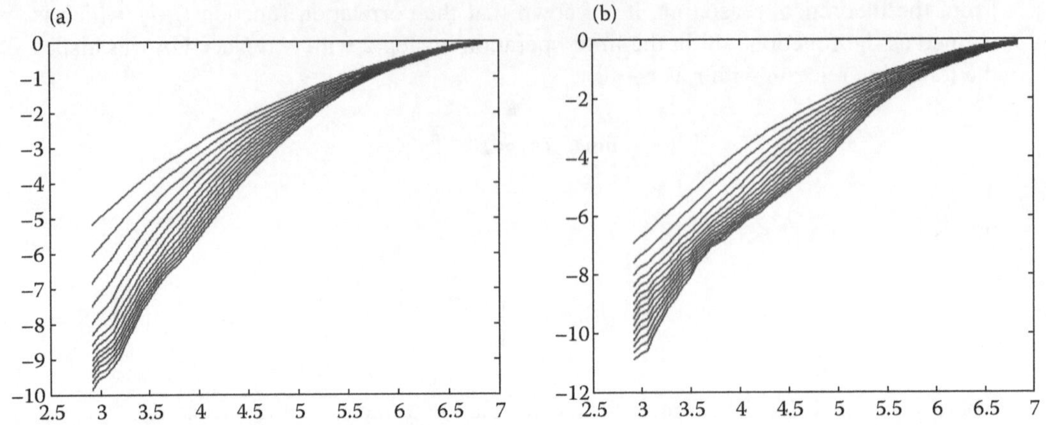

FIGURE 6.8 The ln $C_n(r)$/ln r plot of the data with the correlation dimension m taking values in the interval [4,18], wherein (a) time delay $\tau = 2$, and (b) $\tau = 5$.

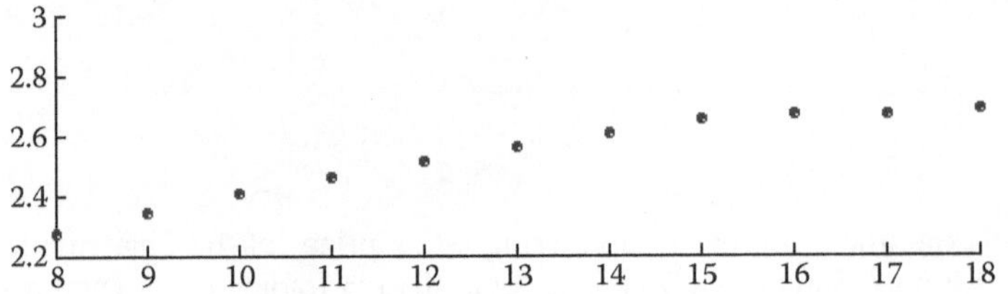

FIGURE 6.9 Plot of $d(m)/m$ of the data with time delay $\tau = 5$.

6.1.5 Applications of Fractals

1. *Applications in geo-sciences.* This part of applications includes the fractals of coastal lines and rivers, earthquake fractals, mineral deposit fractals, and precipitation fractals.

   Within celestial movements, there are many chaos phenomena, while fractals stand for an important mathematical tool from studying chaos.

   General studies of celestial movements most often need to consider perturbations, that is, employing methods of perturbation on the basis of the two-body problem, because in reality, there is no isolated two-body problem. For example, when a satellite travels around the earth, it is affected not only by the geo-gravitation but also by such perturbation forces as the nonspherical shape of the earth, atmospheric resistance, etc. Based on the currently available mathematics, one still cannot directly solve the general three-body problem. Poincare (Barrow-Green 1997) once investigated the three-body problem and found that any analytic solution could not be obtained because the celestial motion trajectories of the three-body problem experience states of chaos. For some recent studies on the three-body problem using the method of general systems, please consult Lin (2008). In his works, Poincare designed some new geometric methods to illustrate the complicated motions. However, as mentioned earlier, because of the level of sophistication involved, his works did not catch much attention at all. As of this writing, the scientific world has gained a good degree of understanding of chaos, leading to the discovery of large amounts of chaos phenomena. For instance, let us imagine that if a man-made space detector is not controlled, it will sooner or later fly chaotically after a period of stable flight in a predetermined manner. Even so, after a long time period of difficult struggle, not much progress has been made in terms of developing the appropriate mathematical methods to deal with problems of chaos. To this end, the fractal theory might be a research field with promising expectations that progress in the fractal theory might shed new light on the study of chaos, because in terms of geometry, strange attractors of chaos are fractals.

2. *Applications in biology, physics, and chemistry.* These applications have led to the study of fractal structures in living bodies, effects of fractal structures on superconducting properties, fractal characteristics of surfaces, and the fractal structures of polymers.
3. *Applications in material sciences.* The fractal structures of cracks in materials and the multirange fractals of materials are investigated.
4. *Applications in computer graphics and image processing.* In this area, the fractal structures of plants and trees are researched, and such techniques of fractal image processing and compression are developed.
5. *Applications in economics and finance.* The fractal principle of economic systems is established, and the fractal dimensions of the distribution of economic incomes and financial market prices are calculated.
6. *Applications in linguistics and intelligence science.* The fractals of word frequency distribution of the linguistics and the statistical fractals of the negative power law of the intelligence science are developed.

6.1.6 Some Philosophical Thoughts on Fractals

6.1.6.1 Universality of Fractal Structures

Fractal is a concept with a wide-ranging significance. There are fractal phenomena in nature, societies, and human thoughts. It can be categorized into four main classes: natural fractals, time fractals, social fractals, and thought fractals. Natural fractals include geometric fractals, functional fractals, information fractals, energy fractals, etc. Time fractals means such systems that have the property of self-similarity along the time axis. Social fractals include those self-similar phenomena seen in human activities and societal phenomena. Thought fractals stand for those self-similarities existing in human knowledge and consciousness.

It is commonly believed (Lin 1998) that nonlinearity, randomness, and dissipation are some of the necessary conditions for fractal structures to appear. Nonlinearity means the existence of nonlinear terms in motion equations, where the evolution of the systems' state experiences bifurcation is a root cause for chaos to appear. Randomness stands for noisy thermal movements along with the internal randomness of chaos motions. Dissipativity destroys the invariance of the time reversal of the laws of macroscopic motions. Regardless of whether it is the fractal structures that study the coordinate spaces or those that concern with phase spaces, the dissipativity of the systems has to be analyzed.

For a nonlinear dissipative system, its solution might be irregular and confined within a bounded region in the phase space. The irregular movement of the dissipative system will eventually converge to its attractor, which, known as a strange attractor, stands for a fractal structure of the phase space. Therefore, the conditions of instability or those of being distant away from the equilibria of dissipative systems might become sufficient conditions for causing strange attractors and the appearance of fractal structures.

6.1.6.2 Fractal Structures and Self-Organization

It is believed in synergetics (Haken 2004) that one common characteristic of self-organizing systems is that the appearance of a new structure is determined by a small number of order parameters. For a high-dimensional or infinite-dimensional nonlinear complex self-organizing system, there are always a small number of unstable and a large amount of stable modules in the region near the critical points. The latter are completely dominated and determined by the former.

The coordination and transformation between regular sets and fractal sets, and between whole number and fractal dimensions, can be materialized through iterations. The difference and coordination between whole number and fractal dimensions reflect the difference and unification between systems' simplicity and complexity, gradual and sudden changes, and quantitative and qualitative changes. Additionally, homogeneity and heterogeneity, and isotropy and anisotropy, also represent dialectical unifications of difference and coordination. In the corresponding fractal systems, they are expressed as the difference and coordination between single-scale and multiscale fractals, between the continuous spectra of single dimensions and multiple dimensions.

6.1.6.3 Dialectical Relationship between Scales and Fractal Dimensions

The problem existing between the structural infiniteness of fractal figures and the scales' finiteness of human understanding in essence represents that about the dialectical relationship between scales used in research and fractal dimensions.

The definition of fractal dimensions requires the existence of limit when the scale approaches zero. However, within naturally existent fractals, there is no such infinitely nesting structure. Instead, there are only finite many-nesting layers. Thus, the length unit of the scale selected in research has to agree with the scale unit of the layers within which the fractals exist.

If formulas appropriate for fractals of infinite layers are employed on objective fractals of finite layers, uncertainties about fractal dimensions might be created. Hence, when studying realistic fractals, one has to first carefully analyze the structural layers and the existent layers. Then and only then can he or she select the appropriate scales and determine the critical points. It is recognized that each self-similar nesting structure that objectively exists in nature contains a finite number of layers and represents an approximate self-similarity. Therefore, when different scales of length are employed to measure and understand "fractal structures" of different domains, the appearance of additional complex scenarios is expected.

6.1.6.4 Philosophical Significance of Fractal Theory

The fractal theory points out the dialectical relationship between wholes and parts of the objective world. It destroys the membrane between wholes and parts and locates the media and bridges that connect parts and wholes. That is, there are self-similarities between wholes and parts.

The fractal theory makes the human understanding of the relationship between wholes and parts evolve from linearity to nonlinearity. Together with systems theory, it reveals the multifaceted, multidimensional, and unidirectional forms of connection between wholes and parts.

The fractal theory provides a new methodology for people to comprehend the whole from parts and a different basis for people to fathom infiniteness from finiteness. Additionally, the fractal theory further deepens and enriches the universal connectedness of the world and the unitarity principle of the cosmos.

6.2 L-SYSTEMS—ANALYSIS AND DESIGN

Although the figures produced out of fractals are generally complicated, the description of fractals can be quite straightforward. Among the commonly utilized methods are the L-system and the iterated function system (IFS). In terms of the fractals they respectively describe, the L-system is simpler than the IF system, where the former contains simple iterations of character strings, while the latter is much more complicated in this regard.

Aristid Lindenmayer (1968), a Hungarian biologist, introduced the Lindenmayer system, or L-system for short, as the mathematics theory for describing the growth of a plant. It is a kind of subsequent string replacement system; its theory focuses on the topology of plants and attempts to describe the adjacency relations between cells or between larger plant modules. While Lindenmayer (1968) and others proposed the initial solution, Prusinkiewicz (1996) used turtle graphics to implement a lot of fractal shapes and herb models based on turtle shapes. His works made turtle the most commonly used form and explanatory schemes of L-systems. In order to avoid the models constructed using L-systems being inflexible, Eichhorst and Savitch (1980) proposed the concept of stochastic L-systems to enhance the flexibility of the earlier L-models. Herman and Rozenberg (1975) generalized the concept of L-systems to a context-sensitive model in order to establish associations between different modules of the plant model. Then, Lindenmayer introduced parameters to make L-systems even more forthright and efficient. The most typical applications of L-systems are done by scholars at Calgary University, Canada. They (Prusinkiewicz 1996) were involved in the parametricalization of L-systems, the establishment of differential L-system and open L-system, and other relevant theoretical research. Additionally, they developed the plant simulation software L-Studio.

Because there has not been any rigorous mathematical definition established for L-systems, all published studies on the subject have stayed only at the level of innovative designs without much theoretical support. On the other hand, as shown in the work of Lin (1999), set theory is a good tool for the theoretical framework of systemics. In this paper, starting from the basics of set theory, we will develop a rigorous mathematical definition for simple L-systems. On the basis of this definition, we establish some of the general properties simple L-systems satisfy so that our constructed theoretical framework is expected to provide the needed fundamental ground for further investigation of L-systems.

6.2.1 Definition of Simple L-Systems

The fact is that L-systems represent a formal language; it can be divided into three classes: 0L-system, 1L-system, and 2L-system. A 0L-system stands for an L-system that is context free. That is, the behavior of each element is solely determined by the rewriting rule; the current state of the system has something to do only with the state of the immediate previous time moment and has nothing to do with any surrounding element. Among L-systems, 0L systems are the simplest; that is why each 0L-system is also referred to as a simple L-system.

Each 1L-system stands for a context-sensitive L-system that considers only one single-sided grammatical relationship. The current state of the system has something to do with not only the state of the immediate previous time moment but also the state of the elements on one side, either left associated or right associated.

Each 2L-system is also a context-sensitive L-system that, different from 1L-systems, considers grammatical relationships from both sides. That is, the current state of the system is related to not only the state of the immediate previous time moment but also the states of the surrounding elements. It represents a method that is most sensitive to the context.

These three classes of L-systems are further divided into deterministic and random L-systems depending on whether or not the rewrite rules are deterministic.

Let $S = \{s_1, s_2, \ldots, s_n\}$ be a finite set of characters of a language, say, English, and S^* the set of all strings of characters from S. Because $S \subset S^*$, S^* is a nonempty set. $\forall \alpha, \beta \in S^*$, define that $\alpha = \beta \Leftrightarrow$ both α and β are identical, meaning that they have the same length and order of the same characters.

Now, define the addition operation $\oplus$ on S^* as follows: $\alpha \oplus \beta = \alpha\beta$ = the string of characters of those in α followed by those in β, $\forall \alpha, \beta \in S^*$. The scalar multiplication on S^* is defined as follows: $\forall \alpha \in S^*$, $k \in Z^+$ = the set of all whole numbers, $k\alpha = \underbrace{\alpha\alpha\ldots\alpha}_{k\ times}$. Then, the following properties can be shown:

1. Both addition and scalar multiplication defined on S^* are closed.
2. $(\alpha \oplus \beta) \oplus \gamma = \alpha \oplus (\beta \oplus \gamma)$, $\forall \alpha, \beta, \gamma \in S^*$.
3. $k(l\alpha) = (kl)\alpha(\alpha \oplus \beta) \oplus \gamma = \alpha \oplus (\beta \oplus \gamma)$, $\forall \alpha, \beta, \gamma \in S^*$.
4. $(k + l)\alpha = k\alpha \oplus l\alpha$, $\forall \alpha \in S^*$, $k,l \in Z^+$.

Before we develop a rigorous definition of simple L-systems, let us first look at a specific mapping $S^* \to S^*$, known as L-mapping.

6.2.1.1 L-Mapping

Definition 6.1

(Production rules). For any $s_i \in S$, $1 \le i \le n$, if an ordered pair $(s_i,\alpha) \in S \times S^*$ can be defined, then this pair defines a relation p_i from s_i to α, known as a production rule from S to S^*. ■

Definition 6.2

(Same class production rules). For a given $s_i \in S$, if there are $r \ge 1$ production rules $p_i^{(1)}, p_i^{(2)}, \ldots, p_i^{(r)}$ defined for s_i such that $\forall j \ne k$ $(1 \le j, k \le r)$, $p_i^{(j)}(s_i) \ne p_i^{(k)}(s_i)$, then $p_i^{(1)}, p_i^{(2)}, \ldots, p_i^{(r)}$ are referred to as r same class production rules of the character s_i and $P_i = \{p_i^{(1)}, p_i^{(2)}, \ldots, p_i^{(r)}\}$ the set of same class production rules of the character s_i.

Definition 6.3

(L-mapping). Assume that m ($\ge n$) production rules $P = \bigcup_{i=1}^{n} P_i$ from S to S^* are given, where each P_i is nonempty and stands for the same class production rules of the character $s_i \in S$, $1 \le i \le n$. For any group of production rules $p_1, p_2, \ldots, p_n \in P$, satisfying $p_i \in P_i$, $i = 1,2,\ldots,n$, the set $\varphi = \{p_1, p_2, \ldots, p_n\}$ defines a mapping $S^* \to S^*$, still denoted φ, by

$$: s_{k_1} s_{k_2} \ldots s_{k_r} \to p_{k_1}(s_{k_1}) p_{k_2}(s_{k_2}) \ldots p_{k_r}(s_{k_r}) \tag{6.1}$$

$\forall s_{k_1} s_{k_2} \ldots s_{k_r} \in S^*$, $k_i \in \{1, 2, \ldots, n\}$, $1 \le i \le r$, and r stands for the length of the character string. Then, this mapping φ: $S^* \to S^*$ is referred to as an L-mapping.

Note: From the definition of L-mappings, it follows that the set of m ($\ge n$) production rules from S to S^*

$$P = \{p_1^{(1)}, \ldots, p_1^{(r_1)}, p_2^{(1)}, \ldots, p_2^{(r_2)}, \ldots, p_n^{(1)}, \ldots, p_n^{(r_n)}\} \tag{6.2}$$

can define $(r_1 r_2 \ldots r_n)$ many L-mappings from S to S^*, where $\sum_{i=1}^{n} r_i = m$. The set of all L-mappings determined by the set P is denoted by $\Phi = \{ \ _1, \ _2, \ldots, \ _{r_1 r_2 \cdots r_n}\}$. ■

Proposition 6.2

(Properties of L-mappings). For any $\alpha,\beta \in S^*$ and $m,n \in Z^+$, each L-mapping φ satisfies the following properties:

(i) $\varphi(\alpha) \in S^*$ is uniquely defined.
(ii) If $\varphi|_S$ is surjective $S \to S$, then $\varphi|_S$ must be bijective.
(iii) If $\alpha = s_{k_1} s_{k_2} \ldots s_{k_p} \in S^*$, then $(\alpha) = (s_{k_1}) (s_{k_2}) \ldots (s_{k_p})$.
(iv) $\varphi(\alpha \oplus \beta) = \varphi(\alpha) \oplus \varphi(\beta)$.
(v) $\varphi(m\alpha) = m\varphi(\alpha)$.
(vi) $\forall \varphi,\phi \in \varphi$, $\varphi\phi(m\alpha \oplus n\beta) = m\varphi\phi(\alpha) \oplus n\varphi\phi(\beta)$.

Proof

Because both (i) and (ii) are evident, it suffices to show (iii)–(vi).

(iii) For any $\alpha = s_{k_1}s_{k_2}\ldots s_{k_p} \in S^*$, the definition of the L-mappings implies that

$$(s_{k_1}s_{k_2}\ldots s_{k_p}) = p_{k_1}(s_{k_1})p_{k_2}(s_{k_2})\ldots p_{k_p}(s_{k_p}). \tag{6.3}$$

Also, $\forall s_{k_i} \in S^* (1 \le i \le p)$, we have

$$(s_{k_i}) = p(s_{k_i}). \tag{6.4}$$

Hence, $(\alpha) = \ (s_{k_1}s_{k_2}\ldots s_{k_p}) = p_{k_1}(s_{k_1})p_{k_2}(s_{k_2})\ldots p_{k_r}(s_{k_r}) = \ (s_{k_1})\ (s_{k_2})\ldots\ (s_{k_r}).$

(iv) For any $\alpha,\beta \in S^*$, from the addition operation on S^* and property (iii), it follows that

$$\varphi(\alpha \oplus \beta) = \varphi(\alpha\beta) = \varphi(\alpha)\varphi(\beta) = \varphi(\alpha) \oplus \varphi(\beta). \tag{6.5}$$

(v) For any $\alpha \in S^*$ and $m \in Z^+$, from the definition of scalar multiplication on S^* and property (ii), it follows that

$$(m\alpha) = \ (\underbrace{\alpha\alpha\ldots\alpha}_{m\ times}) = \underbrace{\ (\alpha)\ (\alpha)\ldots\ (\alpha)}_{m\ times} = m\ (\alpha). \tag{6.6}$$

(vi) For any $\varphi,\phi \in \varphi$, by employing properties (iv) and (v), we obtain

$$\varphi\phi(m\alpha \oplus n\beta) = \varphi(m\phi(\alpha) \oplus n\phi(\beta)) = m\varphi\phi(\alpha) \oplus n\varphi\phi(\beta). \tag{6.7}$$

■

6.2.1.2 Simple L-Systems

With the concept of L-mappings in place, let us now look at how to define simple L-systems. According to the classification of simple L-systems, deterministic and random simple L-systems, we now establish the relevant definitions by using the concept of L-mappings.

Definition 6.4

(Deterministic simple L-systems). Let $S = \{s_1, s_2, \ldots, s_n\}$ be a finite set of characters and S^* the set of all strings of characters from S. Assume that $\varphi: S^* \to S^*$ is a given L-mapping. For any given initial string $\omega \in S^*$, the system that is made up of the nth iterations $\varphi^n(\omega)$, $n \ge 1$, is referred to as a deterministic simple L (D0L) system (of order n), denoted by the ordered triplet $<S,\omega,\phi>$. ■

Definition 6.5

(Random simple L-systems). Let S and S^* be the same as in Definition 6.4, $\Phi = \{\varphi_1, \varphi_2, \ldots, \varphi_k\}$ a set of k L-mappings from S^* to S^*, and ξ the L-mapping randomly drawn from Φ such that the probability $P(\xi = \varphi_i)$ for ξ to be φ_i is π_i, where $\sum_{i=1}^{k} \pi_i = 1$. For a given initial string of characters $\omega \in S^*$, the system that is made up of the nth generalized iterations $\xi_n\xi_{n-1}\ldots\xi_1(\omega)$, the composite of the mappings $\xi_1,\ldots,\xi_{n-1},\xi_n$, $n \geq 1$, is referred to as a random simple L (R0L) system (of order n), where $\xi_1,\xi_2,\ldots,\xi_n$ are independent and identical distributions. This system is written as the following ordered quadruple: $<S,\omega,\Phi,\pi>$. ■

6.2.2 Properties of Simple L-Systems

From the definitions, it follows that each simple L-system is produced by iterations or generalized iterations of L-mappings. Hence, the properties of simple L-systems are determined by those of L-mappings.

Theorem 6.1

(Property of linearity). Each simple L-system is a linear system.

Proof

Assume that the deterministic nth-order simple L-system is given as $L_1 = <S, \omega, \phi>$, and the random nth-order simple L-system $L_2 = <S, \omega, \Phi, \pi>$, where $n \geq 1$. To show that both L_1 and L_2 are linear systems, it suffices to show that both L_1 and L_2 systems, respectively, satisfy the superposition principle.

(I) We use mathematical induction to prove that φ^n satisfies the superposition principle. When $n = 1$, $\forall\alpha,\beta \in S^*$, $k_1,k_2 \in Z^+$, properties (iv) and (v) of L-mappings imply that

$$\varphi(k_1\alpha \oplus k_2\beta) = k_1\varphi(\alpha) \oplus k_1\varphi(\beta). \tag{6.8}$$

Thus, φ satisfies the superposition principle.

Assume that when $n = k$, $k \geq 1$, φ^k satisfies the superposition principle. That is, when $\forall\alpha,\beta \in S^*$, $k_1,k_2 \in Z^+$, the following holds true:

$$\varphi^k(k_1\alpha \oplus k_2\beta) = k_1\varphi^k(\alpha) \oplus k_2\varphi^k(\beta). \tag{6.9}$$

Then when $n = k + 1$, we have

$$\begin{aligned} {}^{k+1}(k_1\alpha \quad k_2\beta) &= (\ {}^{k}(k_1\alpha \quad k_2\beta)) \\ &= (k_1\ {}^{k}(\alpha) \quad k_2\ {}^{k}(\beta)) \\ &= k_1\ {}^{k+1}(\alpha) \quad k_2\ {}^{k+1}(\beta). \end{aligned} \tag{6.10}$$

Hence, $\forall n \geq 1$, φ^n satisfies the superposition principle.

(II) We show that $\xi_n\xi_{n-1}\ldots\xi_1$ satisfies the superposition principle.

From the definition of random simple L-systems, it follows that ξ_1, $1 \leq i \leq n$, stands for the L-mapping φ_j, $1 \leq j \leq N$, where N is the total number of elements in the set Φ, randomly selected from Φ at step i.

Because each $\varphi_i \in \Phi$, $1 \le i \le N$, satisfies the superposition principle, from property (vi) of L-mappings, it follows that $\xi_n\xi_{n-1}\ldots\xi_1$ satisfies the superposition principle. ■

From Theorem 6.1, it follows that in terms of a simple L-system, when its order n is relatively large, to shorten the computational time, one can decompose relatively long strings of characters into sums of several shorter strings, which can be handled using parallel treatments to obtain the state of the next moment.

Theorem 6.2

(Property of fixed points). The restriction of the L-mapping φ in a D0L system on S is bijective from S to S, if and only if for any $\alpha \in S^*$, there is a natural number k such that $\varphi^k(\alpha) = \alpha$. ■

Before proving this result, let us first look at the following lemma.

Lemma 6.1

If the restriction of the L-mapping φ on S is a bijection from S to S, then $\forall s_i \in S$, $\exists k_i \in Z^+$, such that ${}^{k_i}(s_i) = s_i$, $1 \le i \le n$.

Proof

By contradiction, assume that $\exists s \in S$, $\forall k \in Z^+$, $\varphi^k(s) \ne s$. Then let

$$s_{i_1} = (s), s_{i_2} = {}^2(s), \ldots, s_{i_n} = {}^n(s).$$

Thus, we have

$$s \ne s_{i_1}, s \ne s_{i_2}, \ldots, s \ne s_{i_n}. \tag{6.11}$$

Step 1: From the first $(n - 1)$ inequalities in Equation 6.11 and that fact that $\varphi|_S$: $S \to S$ is bijective, it follows that $(s) \ne (s_{i_1})$, $(s) \ne (s_{i_2}), \ldots, (s) \ne (s_{i_{n-1}})$. That is,

$$s_{i_1} \ne s_{i_2}, s_{i_1} \ne s_{i_3}, \ldots, s_{i_1} \ne s_{i_n}. \tag{6.12}$$

Step 2: From the first $(n - 2)$ inequalities in Equation 6.12 and the fact that $\varphi|_S$: $S \to S$ is bijective, it follows that $(s_{i_1}) \ne (s_{i_2})$, $(s_{i_1}) \ne (s_{i_3}), \ldots, (s_{i_1}) \ne (s_{i_{n-1}})$. That is,

$$s_{i_2} \ne s_{i_3}, s_{i_2} \ne s_{i_4}, \ldots, s_{i_2} \ne s_{i_n}. \tag{6.13}$$

By continuing this procedure, we obtain at step $(n - 1)$ that

$$s_{i_{n-1}} \ne s_{i_n}. \tag{6.14}$$

Therefore, $s \neq s_{i_1} \neq s_{i_2} \neq \ldots \neq s_{i_n}$. It means that there are $n + 1$ different elements in the set S. However, S has only n elements: a contradiction. Thus, the assumption that $\exists s \in S, \forall k \in Z^+, \varphi^k(s) \neq s$ does not hold true. In other words, $\forall s_i \in S, \exists k_i \in Z^+$, such that $^{k_i}(s_i) = s_i$, $1 \leq i \leq n$. ■

In the following, let us prove Theorem 6.2.

(⇒) For $\alpha = s_{m_1}s_{m_2}\ldots s_{m_p} \in S^*$, from Lemma 6.1, it follows that $\forall s_{m_i} \in S, \exists k_i \in Z^+$ $^{k_i}(s_{m_i}) = s_{m_i}$, $1 \leq i \leq p$. Let k = $<k_1, k_2, \ldots, k_p>$. Thus,

$$\begin{aligned} ^{k}(\alpha) &= {}^{k}(s_{m_1}s_{m_2}\ldots s_{m_p}) \\ &= {}^{k}(s_{m_1})\ {}^{k}(s_{m_2})\ldots\ {}^{k}(s_{m_p}) \\ &= {}^{k_1}(s_{m_1})\ {}^{k_2}(s_{m_2})\ldots\ {}^{k_p}(s_{m_p}) = \alpha. \end{aligned}$$

Thus, there is a natural number k such that $\varphi^k(\alpha) = \alpha$.

(⇐) If $\forall \alpha \in S^*$, there is natural number k such that $\varphi^k(\alpha) = \alpha$. Let us take $\alpha = s_1, s_2, \ldots, s_n$. Then $\exists k_1, k_2, \ldots, k_n$ such that

$$^{k_1}(s_1) = s_1,\quad {}^{k_2}(s_2) = s_2, \ldots,\quad {}^{k_n}(s_n) = s_n. \tag{6.15}$$

According to property (ii) of L-mappings, to show that the restriction of φ on S is a bijection from S to S, it suffices to prove that $\varphi|_S: S \to S$ is surjective. Again, we prove this end by contradiction. Assume that $\exists s \in S, \forall s_i \in S, 1 \leq i \leq n, \varphi(s_i) \neq s$. Then we have $\forall k \geq 1, \varphi^k(s) \neq s$, which contradicts with Equation 6.15. Thus, $\varphi|_S: S \to S$ is surjective. Hence, $\varphi|_S: S \to S$ is bijective. ■

From Theorem 6.2, it follows that in terms of D0L systems, if the restriction $\varphi|_S: S \to S$ of L-mapping φ is a bijection, then there are at most $k = <k_1, k_2, \ldots, k_p>$ different states. Therefore, to produce complicated fractal figures using D0L systems, one should avoid using any such L-mapping whose restriction on S is a bijection from S to S.

Theorem 6.3

(Property of fixed points). The restriction of the L-mapping φ in a R0L system on S is a bijection from S to S, if and only if for any $\alpha \in S^*$, there is a natural number k such that $\varphi^k(\alpha) = \alpha$.

Proof

Although as a R0L system, the probability of different mappings at each step is different, the maximum number n of total states of each step is fixed. Since the length p of the mapping is finite, the number of overall states is $k = n^p$ at most. Thus, there exists an integer $k = n^p$ to satisfy $\varphi^k(\alpha) = \alpha$. Other parts of the proof are similar to those of Theorem 6.2 and omitted. ■

Remark: In terms of numbers, the finite states of R0L are more than those of D0L. However, from Theorem 6.3, it follows that in order to produce complicated fractal figures using R0L systems, one should still avoid using any such L-mapping that its restriction on S is a bijection from S to S.

6.2.3 Explanation of Holistic Emergence of Systems Using DOL

Although the mechanism for a simple L-system appears quite straightforward, it clearly shows the attribute of holistic emergence of systems. Thus, simple L-systems can be employed as an effective tool to illustrate the emergence of systems' wholeness. In the following, we will design a simple binary system to explain the emergence of systems' wholeness.

In the Cartesian coordinate system $\boldsymbol{R}^2$, assume that the initial location of particle A is (0,0) and its initial angle is 0. Let F stand for moving forward one unit step 1, and + turning an angle of $\pi/3$ counterclockwise. Now, let us construct the following DOL system $L^2 = \langle S, \omega, \phi \rangle$, where $S = \{F,+\}$, $\omega = F$, and

$$= \begin{cases} F \rightarrow F{+}{+}F{+}{+}F \\ + \rightarrow + \end{cases}.$$

The first iteration $\varphi(\omega)$ produces F++F++F; the second iteration $\varphi^2(\omega)$ leads to F++F++F++ F++F++F ++ F++F++F; …. When the nth order, $n \geq 1$, L-system L^2 is applied on the particle A, one can obtain the output state as shown in Figure 6.10.

As shown in Figure 6.10, such an nth-order L-system, a binary system, is very simple. Its function can be comprehended as follows: Drive particle A from the starting point at (0,0) to make 3^{n-1} counterclockwise turns along the triangular trajectory.

Now, let us employ this simple binary system to illustrate the emergence of systems' wholeness:

(i) The whole is greater than the sum of its parts.
(ii) The function of the whole is more than the sum of the functions of the parts.

For our purpose, the word "sum" means the collected pile of parts without any interactions between the parts. As for the word "function," it is understood as follows: As long as a system is identified, its functionality is a physical existence; however, any function of the system has to be manifested through the system's act on a specific object that is external to the system. Thus, to investigate the overall function of a system and the sum of the functions of its parts, one needs to consider the respective effects of the system as a whole and each of its parts on an external object. Through analyzing their effects on this object, one can compare the overall effect of the system and the aggregated effect of the parts.

Before we illustrate the emergence of the system's wholeness, let us first establish the following assumptions:

(a) The whole can always be divided into several distinguishable parts in terms of components, attributes, or functionalities.

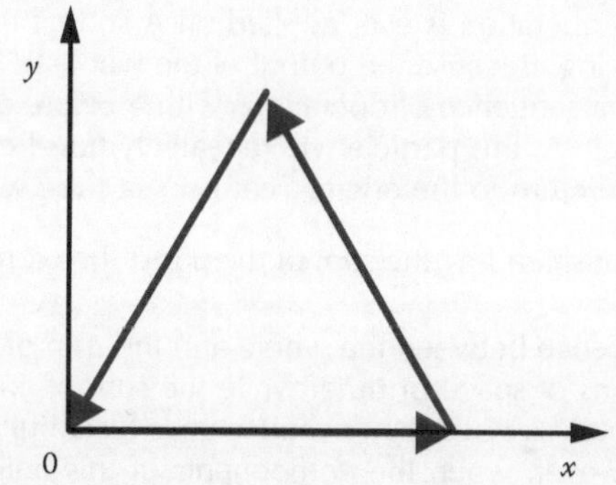

FIGURE 6.10 Output state of the L^2 system.

(b) Similar parts of the system (or parts with similar attributes) satisfy the additive property, while different parts (or parts with different attributes) do not comply with this property. Here, the additivity stands for the algebraic additivity and is different from the meaning of the "sum" in "the sum of parts."

(c) When studying the sum of parts' functionalities, if an identified part does not have any function, then the functionality of this part is seen as "0."

Illustration 1: The whole is greater than the sum of its parts.

Let us symbolically denote the constructed nth-order simple L-system $L^2 = <S, \omega, \phi>$, $n \geq 1$, as $\varphi^n(F)$, and sum of parts as $\sum_i S_i$, where S_i, $i = 1,2,...$, stands for a division of the whole $\varphi^n(F)$, including all characters. In the following, we will discuss from two different angles according to the following different divisions of the whole.

1. The whole is greater than the sum of all the elements (the whole is divided using components).

   If we treat the system L^2 as one containing only two components (operations) F and +, then in the nth-order L^2 system, there are 3^n components F and $2(3^{n-1})$ components +.

   From the basic assumptions, it follows that F and + respectively satisfy the additive property. Their algebraic sums are respectively written as $S_1 = 3^n(F)$ and $S_2 = 2(3^n - 1)(+)$. Then, the sum of the elements of the L^2 system can be expressed as $\sum_{i=1}^{2} S_i = S_1S_2$ or $\sum_{i=1}^{2} S_i = S_2S_1$.

   Applying the output of the elements' sum $\sum_{i=1}^{2} S_i$ on the particle A produces a line segment of length 3^n, while applying the output of the whole $\varphi^n(F)$ on the particle A creates a regular triangle with each side's length being 1. The whole constitutes a figure of the two-dimensional space, possessing a special structure, while the sum of the elements represents a line segment of the one-dimensional space without any qualitative mutation. That is to say, the whole has a spatial structural effect that is not shared by the sum of the parts; therefore, $^n(F) > \sum_{i=1}^{2} S_i$.

2. The whole is greater than the sum of its parts, where the whole is divided using attributes.

   Let us treat the L^2 system as being composed of 3^n unit vectors of the plane: $\vec{i}_1, \vec{i}_2, ..., \vec{i}_{3^n}$. That is, we see the vectors $\vec{i}_1, \vec{i}_2, ..., \vec{i}_{3^n}$ as having the same attributes. Then this L^2 system is made up only of these 3^n components of the same attributes.

   Now, we desire to show $^n(F) > \sum_{k=1}^{3^n} \vec{i}_k$.

   Evidently, $\sum_{k=1}^{3^n} \vec{i}_k = 0$. So, when the output of the sum $\sum_{k=1}^{3^n} \vec{i}_k$ of the parts is applied on the particle A, it is like no operation is ever applied on A so that the particle A is fastened at the origin without moving. Because the output of the whole $\varphi^n(F)$ is a string of characters, which is equivalent to a sequence of operations with a before and after order, when $\varphi^n(F)$ is applied on the particle A, this particle will repeatedly travel counterclockwise along the triangle and eventually return to the origin. That is to say, the whole possesses a structural effect of time that is not shared by the sum of the parts; therefore, we have $^n(F) > \sum_{k=1}^{3^n} \vec{i}_k$.

In fact, the essential difference between the whole and the sum of parts is that the whole has some structural effect in terms of space or time, while the sum of parts does not. For instance, when N bricks are used to build a house, the whole stands for a building along with the spatial structure of rooms, etc. However, when the components of this building are divided, because there is only one kind of component, the bricks, the sum of the parts satisfies the algebraic additivity and is equal to the N bricks, which do not have the spatial structure of the house.

Illustration 2: The functionality of the whole is greater than the sum of parts' functionalities.

Let us first introduce a new kind of set [*S*], where each element is allowed to appear more than once. To avoid creating any conflict with the conventional set theory, we will apply the operation of drawing elements out of the set [*S*] only with the following convention: If a nonempty set [*S*] contains element *s* at least twice, then drawing one *s* from [*S*] means that we take any of the elements *s*'s. For example, [*S*] = {*a*,*a*,*b*,*a*,*b*}. Then, drawing an *a* from [*S*] means that we take any one of the elements *a*'s.

For the *n*th-order simple L^2 system, $n \geq 1$, the individual characters (operations) F and + stand for the smallest units of functionalities. Divide this system into $(3^{n+1} - 2)$ parts, which include 3^n functional units F and $2(3^n - 1)$ functional units +. The set of these $(3^{n+1} - 2)$ functional units is written as set [*S*].

We first consider the effect of the whole on the particle A. The function of this L^2 system is to order the elements of [*S*] according to some specific rules. Then, the effect of the system on A is to force the particle A to counterclockwise travel along the triangle 3^{n-1} times.

Now, let us look at the effect on A of each part. When the set [*S*] is given, the sum of the parts' functions can be understood as follows: There are a total of *m* operations, each of which takes m_i arbitrary elements from [*S*] without replacement to act on A, until all the elements in [*S*] are exhausted. Evidently, as long as the order of the elements taken out of the set [*S*] is different from that of $\varphi^n(F)$, the total effect of these elements that are individually taken out of [*S*] will not reach that of the L^2 system. Therefore, the whole possesses a functionality that the sum of the parts does not share. That is, the function of the whole is greater than the sum of the parts' functions.

In fact, the essential difference between the function of the whole and the sum of parts' functions is that the whole has an organizational effect, while the sum of the parts does not.

6.2.4 Design of Simple L-Systems

6.2.4.1 Basic Graph Generation Principle of Simple L-Systems

In essence, each L-system is a system that rewrites strings of characters. Its working principle is quite simple. If each character is seen as an operation and different characters are seen as distinct operations, then strings of characters can be employed to generate various fractal figures. That is, as long as strings of characters can be generated, one is able to produce figures.

The character strings of L-systems that are used to generate figures can be made up of any recognizable symbols. For example, in the design of programs, the symbols F, –, and + can be used respectively so that "F" means move one unit length forward from the current location and draw a segment, "–" stands for turning clockwise from the current direction a predetermined angle, and "+" turning counterclockwise from the current direction another predetermined angle. When generating character strings, start from an initial string and replace the characters of this string by substrings of characters according to the predetermined rules. That completes the first iteration. Then, treat the resultant character string from the first iteration as the mother string and replace each character in this string by strings determined by the rules. By continuing this procedure, one can finish the required iterations of an L-system, where the length of the resultant string is controlled by the number of iterations.

6.2.4.2 Fractal Structure Design Based on DOL

A tree stands for a fractal structure. In particular, each trunk carries a large amount of branches, and each branch has an end point, representing a figure with one staring point and many ending points. This fact implies that when one draws a branch to its end, he has to return his or her drawing pen to draw other structures. Let us take the following conventions: "F" stands for moving forward a unit length 1, "+" turning an angle of $\pi/8$ clockwise, "–" turning an angle of $\pi/8$ counterclockwise, and the characters within "[]" represent a branch; when the characters within a pair of [] are implemented, return to the position right before "[" and maintain the original direction, and then carry out the characters after "]."

Assume that the starting point is at (0,0) on the complex plane and the initial direction at $\pi/2$. Now, we design a D0L system $G_1 = <S_1, \omega_1, \phi_1>$ as follows:

$$S_1 = \{F,+,-,[,]\};$$

$$\omega_1 = \mathrm{F}; \text{ and}$$

$$_1 = \begin{cases} \mathrm{F} \rightarrow \mathrm{FF{+}[{+}F-F-F]-[-F{+}F{+}F]} \\ + \rightarrow + \\ - \rightarrow - \\ [\rightarrow [\\ ] \rightarrow]. \end{cases}$$

Then, we can produce the fractal structures as shown in Figure 6.11.

We can obtain other illustrations by just changing the design forms as shown in Figures 6.12 through 6.14.

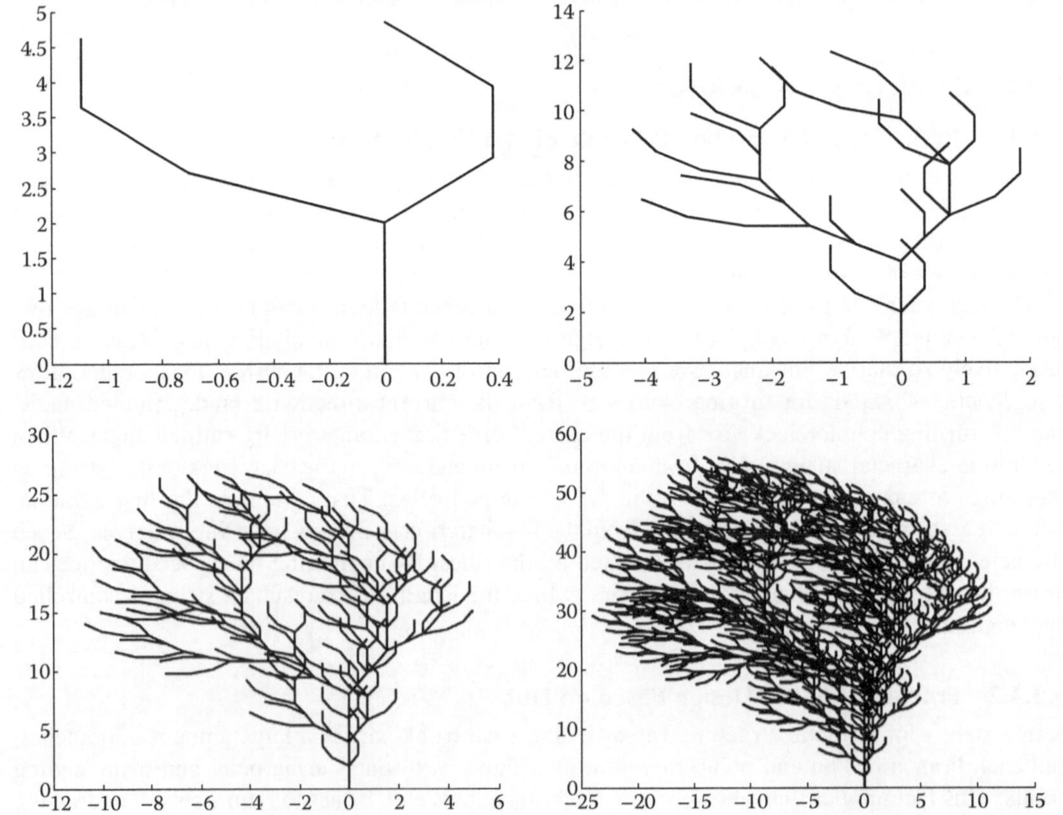

FIGURE 6.11 Fractal structures: a tree dancing in the breeze (of different iteration steps). (a) n = 2; (b) n = 3; (c) n = 4; (d) n = 5.

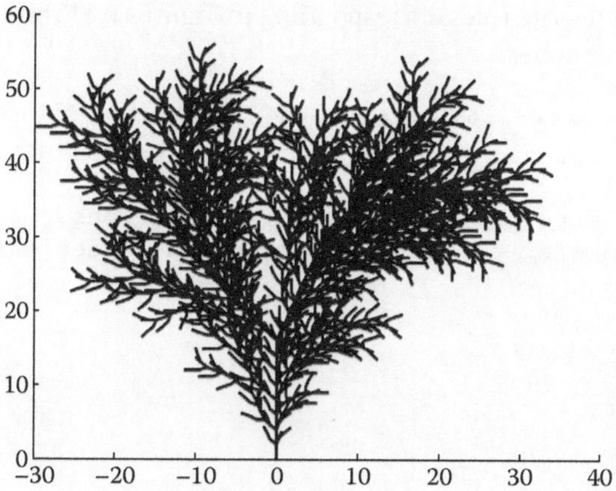

FIGURE 6.12 Fractal structures: a standing tree ($n = 5$, $a = \pi/8$).

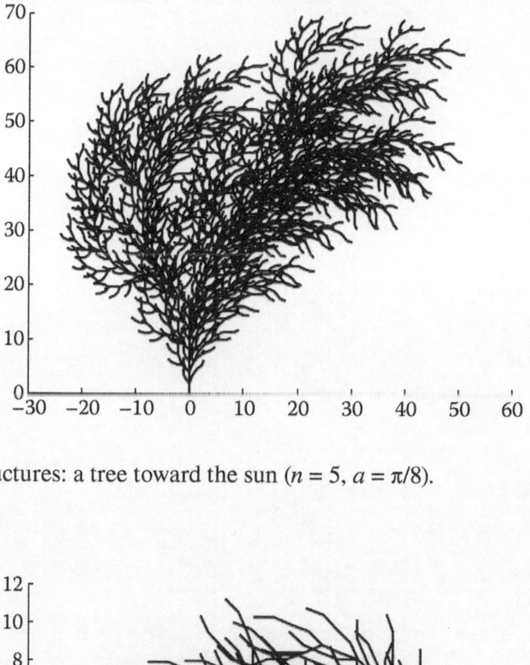

FIGURE 6.13 Fractal structures: a tree toward the sun ($n = 5$, $a = \pi/8$).

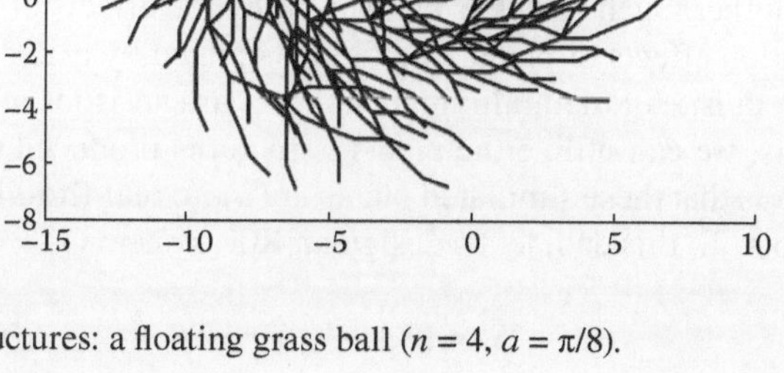

FIGURE 6.14 Fractal structures: a floating grass ball ($n = 4$, $a = \pi/8$).

The different L-system rewrite rules corresponding to Figures 6.11 through 6.13 are shown in the following MATLAB source codes:

```
*********************************************
%% Matlab code for a fractal structure design based on D0L
%%%%%%%%%%%%%%%%%%%%
%%— 1, the input parameter n: the number of iterations, <6 in general
%%— 2, the inner parameter p for L-system, which is designed for requirement
%%%%%%%%%%%%%%%%%%%%
function Ltree1(n);
%%— initialization
S = 'F';
z = 0;
A = pi/2;
a = pi/8;
zA = [0,pi/2];
%%— L mapping
p = 'FF+[+F-F-F]-[-F+F+F]'; %%— a tree dancing in breeze (n = 5)
% p = 'FF+[+F[-FF-]-F]-[-F[-FF+]F]'; %%— a standing tree (n = 4)
% p = 'FF+[+F[-F-F]-F]-[-F[F-FF+F]F]'; %%— a tree towards sun (n = 5)
% p = 'FF+[[+F-FF-F]-[-F+FF+F]]'; %%— a float grass (n = 4)
%%— iteration
for k = 2:n;
S = strrep(S,'F',p);
end
%% drawing the figure
figure;
hold on
for k = 1:length(S);
switch S(k)
       case'F'
       plot([z,z+exp(i*A)],'k','linewidth',2);
       z = z+exp(i*A);
       case'+'
       A = A+a;
       case'-'
       A = A-a;
       case'['
       zA = [zA;[z,A]];
       case']'
       z = zA(end,1);
       A = zA(end,2);
       zA(end,:) = [];
end
end
********************************************
```

6.2.4.3 Fractal Structure Design Based on R0L Systems

In nature, the forms plants take are not invariant. Even for the same kinds of plants, their shapes can vary from one plant to another. Such varieties are caused by the effects of the environment.

In terms of the simulation effects of plants, the figures created by using D0L systems seem to be quite stiff. Under the prerequisite of maintaining the main characteristics of plants, in order to generate varieties on the details, we can utilize the plants' structures produced out of R0L systems. The advantage of these figures is that these simulated plants are more real-life-like and much closer to the true forms of natural plants. To this end, let us design an R0L system $G_2 = <S_1, \omega_1, \Phi, \pi>$ as follows:

$$S_1 = \{F,+,-,[,]\};$$

$$\omega_1 = F;$$

$$\Phi = \{\varphi_1, \varphi_2, \varphi_3\},$$

where

$$_1 = \begin{cases} F \to F[+F]F[-F]F \\ + \to + \\ - \to - \\ [\to [\\ ] \to] \end{cases}, \quad _2 = \begin{cases} F \to F[+F]F[-F[+F]] \\ + \to + \\ - \to - \\ [\to [\\ ] \to] \end{cases},$$

$$_2 = \begin{cases} F \to FF[-F+F+F]+[+F-F-F] \\ + \to + \\ - \to - \\ [\to [\\ ] \to] \end{cases}$$

and $\pi = P(\xi = \varphi_i) = 1/3$, $i = 1,2,3$. Then the fractal structures can be produced. For instance, Figure 6.15 shows the different fractal structures of an R0L system with each iteration unchanged.

The following are the MATLAB source codes for Figure 6.15.

```
*********************************************
%% Matlab code for a fractal structure design based on R0L
%%%%%%%%%%%%%%%%%%%%%
%%- 1, the input parameter n: the number of iterations, <6 in general
%%- 2, the inner parameter p for L-system, which is designed for requirement
%%- 3, Each iteration is chosen randomly, but the rewrite rule in each iteration
%% keep unchangeable
%%%%%%%%%%%%%%%%%%%%%
function Ltree2(n);
%%- initialization
S = 'F';
z = 0;
A = pi/2;
a = pi/16;
zA = [0,pi/2];
%%- L mapping sets
p1 = 'F[+F]F[-F]F';
p2 = 'F[+F]F[-F[+F]]';
p3 = 'FF[-F+F+F]+[+F-F-F]';
p4 = 'FF+[+F-F-F]-[-F+F+F]';
p5 = 'FF+[+F[-FF-]-F]-[-F[-FF+]F]';
%%- produce random numbers
tag = rand(1,n);
for k = 1:n
if(tag(k)<1/5)
        tag(k) = 1;
```

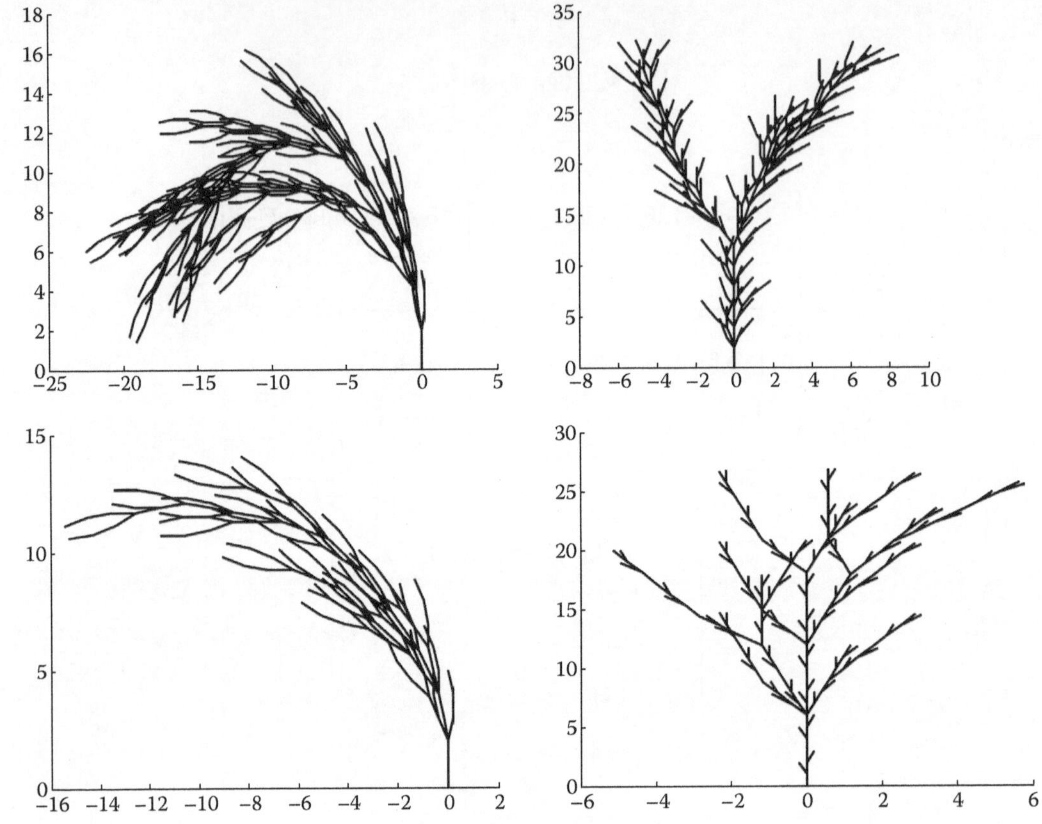

FIGURE 6.15 Different fractal structures for an R0L system with each iteration unchanged ($n = 4$, $a = \pi/16$).

```
elseif(tag(k)<2/5)
        tag(k) = 2;
elseif(tag(k)<3/5)
        tag(k) = 3;
elseif(tag(k)<4/5)
        tag(k) = 4;
else
        tag(k) = 5;
end
end
%%- iteration
for m = 2:n;
if(tag(m) = =1)
        S = strrep(S,'F',p1);
elseif(tag(m) = =2)
        S = strrep(S,'F',p2);
elseif(tag(m) = =3)
        S = strrep(S,'F',p3);
elseif(tag(m) = =4)
        S = strrep(S,'F',p4);
elseif(tag(m) = =5)
        S = strrep(S,'F',p5);
end
end
%%- drawing the figures
figure;
hold on
```

```
for k = 1:length(S);
switch S(k)
       case'F'
       plot([z,z+exp(i*A)],'k','linewidth',2);
       z = z+exp(i*A);
       case'+'
       A = A+a;
       case'-'
       A = A-a;
       case'['
       zA = [zA;[z,A]];
       case']'
       z = zA(end,1);
       A = zA(end,2);
       zA(end,:) = [];
end
end
*********************************************
```

Figure 6.16 shows the different fractal structures of an R0L system where each iteration is different.

The following are the MATLAB source codes of Figure 6.16.

```
********************************************
%% Matlab code for a fractal structure design based on R0L
%%%%%%%%%%%%%%%%%%%%%
%%– 1, the input parameter n: the number of iterations, <6 in general
%%– 2, the inner parameter p for L-system, which is designed for requirement
%%– 3, Each iteration is chosen randomly, and the rewrite rule in each iteration
%% keeps different
%%%%%%%%%%%%%%%%%%%%%
function Ltree3(n);
%%– initialization
S = 'F';
z = 0;
A = pi/2;
a = pi/16;
zA = [0,pi/2];
%%– L mapping set
p1 = ' F[+F]F[-F]F'; %'FF+[[+F-FF-F]-[-F+FF+F]]';
p2 = 'F[+F]F[-F[+F]]';
p3 = 'FF[-F+F]+[+F-F]';
```

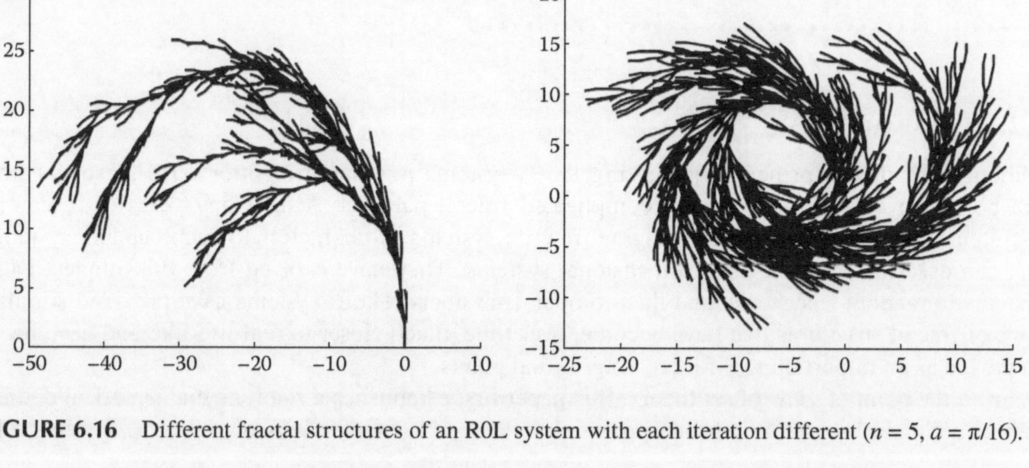

FIGURE 6.16 Different fractal structures of an R0L system with each iteration different ($n = 5$, $a = \pi/16$).

```
%%- iteration
for m = 2:n;
index = find(S = ='F');
N = length(index);
%%- produce the random numbers
tag = rand(1,N);
for k = 1:N
        if(tag(k)<1/3)
        tag(k) = 1;
        elseif(tag(k)<2/3)
        tag(k) = 2;
        else
        tag(k) = 3;
        end
%%- replacing randomly
        if(tag(k) = =1)
        S(index(k)) = 'a';
        elseif(tag(k) = =2)
        S(index(k)) = 'b';
        else
        S(index(k)) = 'c';
        end%-end if
end%-end for
S = strrep(S,'a',p1);
S = strrep(S,'b',p2);
S = strrep(S,'c',p3);
end%end for
%%- drawing
figure;
hold on
for k = 1:length(S);
switch S(k)
        case'F'
        plot([z,z+exp(i*A)],'k','linewidth',2);
        z = z+exp(i*A);
        case'+'
        A = A+a;
        case'-'
        A = A-a;
        case'['
        zA = [zA;[z,A]];
        case']'
        z = zA(end,1);
        A = zA(end,2);
        zA(end,:) = [];
end
end
*********************************************
```

6.2.5 Summary

Although the design principle underlying the L-systems is quite straightforward, these systems can be employed to produce many complicated fractal patterns. After many years of research, L-systems have evolved from the original rewrite systems of characters to such capable systems that can describe complex three-dimensional systems. They have evolved from the simplest D0L systems to random L-systems, and then to open L-systems. The L-systems have provided simulations of fractal structures that have become, over time, much closer to real-life-like and have been employed as an important tool for creating virtual plants.

From the point of view of set theory, this paper first establishes a rigorous mathematical definition of a simple L-system and then proves the common characteristics of simple L-systems. By doing so, this paper develops the badly needed fundamental ground for further investigation of L-systems. It is expected to be applicable to form the theoretical framework of other L-systems.

6.3 ITERATIVE FUNCTION SYSTEMS

Iterative function systems were initially introduced by Michael F. Barnsley (1989, 1993). They form fractal images by using a group of contraction affine transformations. That is, a self-resembling fractal image is produced by employing contractions, rotations, and translations of an original element with respect to the fixed origin and fixed coordinate system. Each set of affine transformations is referred to as an IFS or IF system. It has an internal connection with the iteration fractal of $f(z) = z^2 + c$ of the complex plane, where z and c are complex numbers, except that $f(z)$ is a nonlinear transformation, while an IF system represents a linear transformation. The theory and methods of the IF systems constitute the theoretical basis for simulating natural scenes by using fractal images and the compression of fractal images. Its basic thought is that local parts resemble the whole under affine transformations, on which the well-known collage theorem is established. The merit of IF systems methods lies in that they represent "inverse problems" of fractals generated through iterations. According to the collage theorem, for any given image, such as a photo, by obtaining several rules of generation, the information of the image can be greatly compressed.

6.3.1 Some Theorems of Iterative Function Systems

IFS is a relatively complex method of generating fractal images. It needs a good amount of mathematical knowledge. The following introduces some of the relevant definitions and theorems related to IF systems.

Let us start with the concept of metric spaces. A nonempty set S is known as a metric space, provided that a two-variable real-valued function $\rho(x,y)$ is defined on S satisfying the following conditions:

1. $\rho(x,y) \geq 0$ and $\rho(x,y) \geq 0 \Leftrightarrow x = y$.
2. $\rho(x,y) = \rho(y,x)$.
3. $\rho(x,z) \leq \rho(x,y) + \rho(y,z)$, $\forall x,y,z \in S$.

This function ρ is referred to as a distance defined on S, and the metric space S with ρ as its distance function is written as the ordered pair (S,ρ).

A mapping f:$(S,\rho) \rightarrow (S,\rho)$ defined on the metric space (S,ρ) is named as a compression mapping, if there is an r satisfying $0 < r < 1$ that makes the following hold true:

$$\rho(f(x),f(y)) < r\rho(x,y),\ \forall x,y \in S.$$

Assume that (S,ρ) is a complete metric space and f a compression mapping from (S,ρ) to (S,ρ). Then f has one and only fixed point in the space (S,ρ).

Now, let us look at the definition of fractal space and its characteristics. Assume that (S,ρ) is a metric space, and A and B are subsets of the space S. Then the open r-neighborhood of A is defined as $N_r(A) = \{y \in S\text{: } \rho(x,y) < r,\ \exists x \in A\}$, and the Hausdorff metric between the subsets A and B is defined by

$$D(A,B) = \inf\{r > 0\text{: } A \subseteq N_r(B) \text{ and } B \subseteq N_r(A)\}.$$

However, for a general family of subsets, D may not define a distance. For example, for the family of all intervals of real numbers, we have $D((0,1),(0,1]) = 0$, which makes D not satisfy the conditions of a distance function.

In order to make D become a distance, some additional limitations on subsets A and B have to be added. To this end, the requirement of compact subsets is a good choice. Each compact subset in a metric space is always bounded and closed.

Assume that S is a metric space. Let $H(S)$ denote the set of all nonempty compact subsets of S. Then, $H(S)$ is named as a fractal space over S, and Hausdorff metric D actually becomes a distance between elements of $H(S)$.

For the completeness of our presentation, we will provide the proofs for the following propositions.

Proposition 6.3

$(H(S),D)$ is a metric space. That is, D is a distance on $H(S)$.

Proof

By definition, for any A and $B \in H(S)$, we have

$$D(A,B) = \inf\{r > 0: A \subseteq N_r(B) \text{ and } B \subseteq N_r(A)\}.$$

Because A and B are compact subsets, they are also bounded and closed. Thus, D is well defined.

1. The property of symmetry holds obviously.
2. It is easy to see that $D(A,B) \geq 0$. If $D(A,B) = 0$, then for $\forall \varepsilon > 0$, from the definition of $D(A,B) = 0$, it follows that $A \subseteq N_\varepsilon(B)$. That is, every element in A is a limit point of B. Because B is closed, it means that $A \subseteq B$. Similarly, we can prove $B \subseteq A$. Thus, $A = B$. The opposite that $A = B$ implies $D(A,B) = 0$ is straightforward so that the details are omitted.
3. Let $u = D(A,B), v = D(B,C)$. Because $N_u(A) \subseteq B$, it follows that $N_{u+v}(A) = N_v(N_u(A)) \subseteq N_v(B) \subseteq C$. Similarly, we have $N_{u+v}(C) \subseteq$ A. Thus, $D(A,C) \leq u + v \leq D(A,B) + D(B, C)$. ■

The existence of fractals can be analyzed by using the following propositions. Let f be a compression mapping defined on the metric space (S,ρ). Then, a compression mapping F defined on $(H(S),D)$ can be induced as follows:

$$F(K) = \{f(x), x \in K\}, \text{ for any } K \in H(S). \tag{6.16}$$

If $f_1, f_2, \ldots,$ and f_n are, respectively, $r_1, r_2, \ldots,$ and r_n compression mappings defined on (S,ρ), then the mapping $F{:}H(S) \to H(S)$, defined by

$$F(A) = \cup_{i=1}^{n} f_i(A), \text{ for any } A \in H(S), \tag{6.17}$$

is an r compression mapping, where $r = \max\{r_1, r_2, \ldots, r_n\}$. That is, for any A and $B \in H(S)$,

$$D(F(A),F(B)) < rD(A,B).$$

Proposition 6.4

For any given compression mapping f defined on (S,ρ), the mapping F defined on $(H(S),D)$ in Equation 6.16 is also a compression mapping.

Proof

$\forall A,B \in H(S)$, let $D(A,B) = u$. Then $x_1 \in A,\ y_1 \in B, s.t.\ d(x_1, y_1) \le u,\ y_2 \in A,\ x_2 \in B, s.t.\ d(x_2, y_2) \le u.$ Since f is a compression mapping on S, it follows that $\forall x, y \in S, \exists 0 < r < 1, s.t.\ d(f(x),f(y)) < r{\cdot}d(x,y)$. Thus,

$$d(f(x_1)), (f(y_1)) \le rd(x_1,y_1) \le ru,\ d(f(x_2)), (f(y_2)) \le rd(x_2,y_2) \le ru.$$

Therefore, $\exists x_1 \in A,\ d(f(x_1),f(B)) < ru$; $\exists y_2 \in B,\ d(f(y_2),f(A)) < ru$. That is, $D(f(A),f(B)) < ru = rD(A,B)$. ■

Proposition 6.5

Assume that $f_1, f_2, \ldots,$ and f_n are, respectively, compression mappings defined on the metric space (S,ρ) with compression factors $r_1, r_2, \ldots,$ and r_n. Then, the mapping F defined in Equation 6.17 on $(H(S), D)$ is also a compression mapping with compression factor $r = \max\{r_1, r_2, \ldots, r_n\}$.

Proof

By using induction, it suffers to show the conclusion for $n = 2$. To this end, it suffers to show that $\forall A,B \in H(S),\ D(f_1(A) \cup f_2(A), f_1(B) \cup f_2(B)) \le rD(A,B)$, where $r = \max\{r_1,r_2\}$.

In fact, $\forall A, B, C, D \in H(S),\ D(A \cup B, C \cup D) \le D(A,C) \vee D(B,D)$, assuming that $u = D(A,C) \ge v = D(B,D)$. Thus,

$$N_u(A \cup B) \supseteq N_u(A) \supseteq C,\ N_u(A \cup B) \supseteq N_v(A \cup B) \supseteq N_v(B) \supseteq D, \Rightarrow N_u(A \cup B) \supseteq C \cup D.$$

Similarly, from $N_u(C \cup D) \supseteq A \cup B$, it follows that

$$D(A \cup B, C \cup D) \le D(A,C) \vee D(B,D).$$

Thus, $\forall\ A, B \in H(S)$,

$$D(f_1(A) \cup f_2(A), f_1(B) \cup f_2(B)) \le D(f_1(A), f_1(B)) \vee D(f_2(A), f_2(B))$$

$$\le r_1D(A,B) \vee r_2D(A,B) = rD(A,B),$$

where $r = \max\{r_1,r_2\}$. ■

Assume that A_0 is a nonempty compact subset of S, and $A_{k+1} = \bigcup_{i=1}^{n} f_i\left(A_k\right)$. Then the sequence set $\{A_k\}$ will have a convergence point in $(H(S),D)$; denote $A = \lim_{k\to\infty} A_k$. The collection $\{(S,\rho); (f_i,r_i), i = 1,2,\ldots,n\}$ is referred to as an iterative function system, IFS or IF system for short, where f_i is a compression mapping defined on the metric space (S,ρ) with r_i as its compression factor; $r = \max\{r_i, i = 1,2,\ldots,n\}$ is the compression factor of the IFS and A the attractor of the IF system.

6.3.2 Fractal Images of Iterative Function Systems

Particular fractal images are determined by the affine transformations employed, where an affine transformation is defined as follows:

Given a mapping $f: R^2 \to R^2$, let F be a 2×2 square matrix such that

$$(f(x,y))^T = F\begin{pmatrix} x \\ y \end{pmatrix} + \begin{pmatrix} e \\ f \end{pmatrix} = \begin{pmatrix} a & b \\ c & d \end{pmatrix}\begin{pmatrix} x \\ y \end{pmatrix} + \begin{pmatrix} e \\ f \end{pmatrix}.$$

When and only when $|F| < 1$, f is a compression mapping. The IF system derived from f can be uniquely determined by the parameters a, b, c, d, e, and f.

The following are some fractal images created by using IFS based on MATLAB. The core IF systems function is given as follows.

```
*********************************************
%%%% Basic IFS core function
function IFS_draw(M,p);
N = 30000; % the number of iteration;
for k = 1:length(p);%to form a mapping matrix a1,a2,...
eval(['a',num2str(k), ' = reshape(M(',num2str(k),',:),2,3);']);
end
xy = zeros(2,N);%set the initial value of iteration matrix;
pp = meshgrid(p);
pp = tril(pp);
pp = sum(pp,2);
for k = 1:N-1;
a = rand-pp;
d = find(a< = 0);
xy(:,k+1) = eval(['a',num2str(d(1)),'(:,(1:2))'])*xy(:,k)+eval(['a',num2str(d(1)),'(:,3)']);
end
P = complex(xy(1,:),xy(2,:));
figure;
plot(P,'k.','markersize',5);
axis equal;
*********************************************
```

Example 6.6

Figure 6.17 shows two different kinds of fractal leaves, and the following are the MATLAB source codes.

```
*********************************************
%%% fractal leafs
M = [0.65 0 0 0.65 0.2 0.4;
0.55 0 0 0.55 0.2 0.135;
0.38 -0.28 0.28 0.38 0.3 0.4;
0.38 0.28 -0.28 0.38 0.3 0.1];
p = ones(1,4)/4;
IFS_draw(M,p);
axis image;
title ('fractal leaf 1');
M = [0 0 0 0.15 0 0.2;
0.83 -0.02 0.04 0.83 0 2;
0.2 0.22 -0.3 0.3 0 2;
-0.14 0.24 0.31 0.27 0 0.5;];
p = [0.08 0.8 0.096 0.096];
IFS_draw(M,p);
axis image;
title ('fractal leaf 2');
*********************************************
```

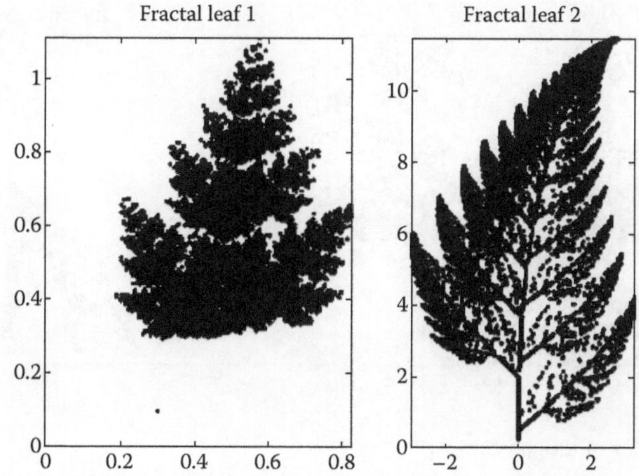

FIGURE 6.17 Two kinds of fractal leaves.

Example 6.18

Figure 6.18 shows two fractal trees. The following are the MATLAB source codes.

```
*******************************************
%%% fractal trees
M = [-0.64 0 0 0.5 0.86 0.25 ;
-0.04 -0.47 0.07 -0.02 0.49 0.51 ;
0.2 0.33 -0.49 0.43 0.44 0.25;
0.46 -0.25 0.41 0.36 0.25 0.57 ;
-0.06 0.45 -0.07 -0.11 0.59 0.1;
];
p = [0.06 0.22 0.23 0.24 0.25];
IFS_draw(M,p);
axis image;
title ('fractal tree 1');
```

FIGURE 6.18 Two kinds of fractal trees.

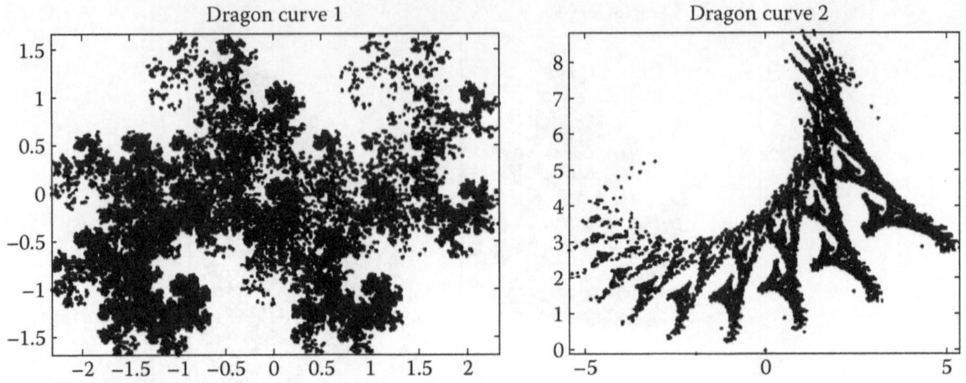

FIGURE 6.19 Two kinds of dragon curves.

```
M = [0.06 0 0 0.6 0 0 ;
0.04 0 0 -0.5 0 1;
0.46 -0.34 0.32 0.38 0 0.6;
0.48 0.17 -0.15 0.42 0 1 ;
0.43 -0.26 0.27 0.48 0 1;
0.42 0.35 -0.36 0.31 0 0.8;]
p = [0.1 0.1 0.1 0.23 0.23 0.24];
IFS_draw(M,p);
axis image;
title ('fractal tree 2');
********************************************
```

Example 6.19

Figure 6.19 shows two dragon curves. The following are the MATLAB source codes.

```
********************************************
%%%% Dragon Curves
M = [0.5 0.5 -0.5 0.5 1 0;
0.5 0.5 -0.5 0.5 -1 0;];
p = [0.3 0.7];
IFS_draw(M,p);
axis image;
title ('dragon curve 1');
M = [0.8 -0.2 0.3 0.9 -1.9 -0.1;
0.1 -0.5 0.5 -0.4 0.8 7;];
p = [0.5 0.5];
IFS_draw(M,p);
axis image;
title ('dragon curve 2');
********************************************
```

6.4 OPEN-ENDED PROBLEMS

1. What is the connection between fractals and chaos?
2. Explore the natural connection between the concept of time systems and that of simple L-systems.

3. By conducting a literature search, study the development history of fractals and find out how Euclidean geometry involves into fractals (how whole-number dimensionality develops into fractional dimensionality).
4. List five examples of fractal dimensions that we find in nature, such as river systems or human lungs, etc.
5. What is the philosophical significance of fractals? What way of thinking has it generalized?
6. Use your own language to describe those invariant characteristics, such as dimension, studied in the theory of fractals. How many different definitions of dimension do you know?
7. Compose a computer procedure that draws a group of fractal graphs. What effects could the procedure have?
8. Given a set of stock data, how could you compute the fractal dimension of the data? Please describe in detail the computational procedure.

7 Complex Systems and Complexity

What is the nature of a complex system? A dictionary definition of the word "complex" is "consisting of interconnected or interwoven parts." Both of the terms "interconnected" and "interwoven" are the essence. Qualitatively, to understand the behavior of a complex system, one must understand not only the behaviors of the parts but also how they act together to form the behavior of the whole.

Famous American physical scientist M. Gell-Mann (1995), one of the three founders of Santa Fe Institute, once pointed out in his book *The Quark and the Jaguar: Adventures in the Simple and the Complex* that the quark and Jaguar represent the two sides of nature, simple and complex. One side is the basic physical laws regarding matter and the universe, and the other is the numerous and complicated structure of the world, which we observe directly and within which we, human beings, the observers, are also situated.

Simplicity and complexity represent two high-level abstractions of various attributes of things, behaviors, and laws in the world. Simplicity means the basic, unchanging, common attributes that the worldly things, events, and organizations generally possess. That is the commonality. However, complexity stands for those particular, varying, individualized attributes that the worldly things, events, and organizations may have. That is, it represents individuality. Simplicity may have a uniform form, while complexity is shown in many different fashions. Along with the deepening exploration and investigation of the universe and nature, the so-called simplicity and complexity of the worldly things, events, and organization scan be transformed into each other.

It was once pointed out by Nicolis and Prigogine (1989) in their book *Exploring Complexity* that the distance between simplicity and complexity, in-order and out-of order, is much smaller than what has been expected in our minds. Each human process of exploration is such a journey that continuously and repeatedly circles from simplicity to complexity and then to simplicity again. However, each seemingly repetitious circle is not just a trivial retrace of the previous path. Instead it represents a process of deepening understanding of nature at a higher level of cognition.

One important challenge facing modern science is that in all areas of knowledge, from elementary particle physics to cosmology and then to complex systems, the world of learning is forced to investigate such objects or phenomena that jointly embody simplicity and complexity, regularity and randomness, in-order and out-of-order. To meet this challenge, one has to understand how with time, the early simplicity, regularity, and orderliness of the universe have been led to the recent formulation of the conditions that exist in between orderliness and disorderliness. It is such scientific transition that makes it possible for the scientific community to consider such complex adaptive systems as biological organisms and the existence of other complicated issues. One has to consider in view of simplicity and complexity how regularities and stochastic models of the universe and various objective matters and events are initially formulated and constructed.

In the past 300 plus years since Newton's *Mathematical Principle of Natural Philosophy* (Dawsons of Pall Mall 1968) was initially published, modern science has explored the objective world by evolving through a developing process from simplicity to complexity. Accordingly, the research objects have changed from the original particles to masses of particles, from mechanical motions to physical and chemical motions, and then to biological and societal motions, from movements of macroscopic objects to the movements of microscopic particles and celestial bodies. The relevant research methods have also been gradually developed from the relatively elementary idea of analyzing local and specific scenarios to those techniques that are capable of handling multidimensional, comprehensive, holistic investigations. A consequence to the previous studies of various complex systems is developed systems science, which collectively studies the evolutionary characteristics and laws of different systems by focusing on the relationship between parts and whole. Systems science tries to explore the evolvement property and rules of different systems, whose research objects may contain mechanics, physics, chemistry, biology, human society, and even brain and thinking systems. This new science investigates the properties and attributes of such systems that range from the lifeless entities considered in the traditional fields such as mechanics, physics, chemistry, etc., to various forms of life studied in biology, humanities, etc., to the mind. From the viewpoint of systems science, systems are classified into four major classes according to their degrees of complexity and relevant methods of study:

1. *Simple systems:* A system is considered simple if it has a small number of subsystems, whose interaction is elementary, and can be treated with the classical analytical method. Simple systems include mechanical systems, electrical systems, etc.
2. *Simple giant systems:* If a system has a large number of subsystems and the interaction among the subsystems is simple, the system is then referred to as a simple giant system. Thermodynamic systems, fluid systems, chemical reaction systems, and various large-scale complicated man-made systems are some examples of simple giant systems.
3. *Complex giant systems:* If a system has a large number of subsystems and the interaction among the subsystems is complicated so that multilevel interactions appear, then the system is referred to as a complex giant system. Examples of complex giant systems include biological systems, ecosystems, human body systems, etc. The characteristics of such systems are mainly about their multilevels and complicated interactions of parts; although most of the interactions are partially or completely known, the overall pattern of evolution of the systems is still not fully understood.
4. *Specific complex giant systems:* These are particularly human systems, such as social systems, including individual humans. As of this writing, there is still a long way for the world of learning to clearly understand how thoughts are formed. Additionally, it has been extremely difficult to quantitatively describe systems that are made of people. These systems include economic systems, military systems, education systems, and all other systems studied in social sciences.

Summing up different research methods of various systems and discovering the basic regularity not only can help deepen our understanding of various kinds of systems but also can provide us with guidelines on how to investigate particular systems.

7.1 SIMPLE GIANT SYSTEMS

7.1.1 Simple Systems

The classical mechanics has systematically analyzed the simple movements of a single particle. At this junction, let us use it as a typical example of a simple system. Strictly speaking, a single particle is not a system. For our purpose, we can regard it as a limited case of a simple system. Assume that a particle has mass m. Its state of three-dimensional movement can be completely described by using three coordinates X, Y, and Z and three momentums p_x, p_y, and p_z. If the mass m is constant, the difference between the particle's momentum and speed is the constant m. Because speed can be computed using derivatives with respect to the coordinates, as long as one knows the three independent coordinate variables, he or she can describe the state of the system perfectly. By using Newton's laws of motion, one can establish the three positional coordinates of the moving particle, also known as the state variables, and the equations that describe how these state variables change with time. If the initial state of the particle is given, including three initial positional coordinates and three initial speeds, one can determine the unique state of movement of the particle at any chosen future time moment.

The basic methods exploring a particle's mechanical movement include determining the system's state variables, finding out the moving rules of the system, setting up the moving equations that the state variables satisfy, using the initial condition, and finally confirming the evolvement process of the system. Those methods are also general research methods of simple systems.

The particular method developed for studying simple mechanical movements of a particle can be summarized as follows: First, determine the state variables of the system (a particle); second, discover the rules of motion of the system so that the equations of motion the state variables satisfy can be constructed; third, by using the given initial values, ultimately establish the pattern of evolution of the system.

All problems regarding the mechanical movement of macroscopic objects, in principle, can be resolved by using Newton's laws of motion. For situations of microscopic particles, because their positional coordinates and momentums cannot be determined at the same time, one generally only employs the three positional coordinates as the state variables of the system. Because the particular values of the three coordinates can be obtained only by using the concept of statistical means, one uses the probability density for the microscopic particle to be at a spatial position (X, Y, Z) as the state variable of the system. This variable no longer satisfies Newton's laws of motion. Instead, it complies with the Schrödinger equation. When considering a particle that travels at a speed near that of light, the concepts of time and space, used to describe the particle, are closely related. Thus, one needs to employ a four-dimensional time–space to describe the state of the system. In this case, the state variables do not satisfy Newton's laws of motion, either. Instead, they satisfy some equations of the relativity theory. For different simple systems of a single particle, due to their varied properties, the variables appropriate for describing the states of the systems are different, and these variables satisfy different sets of laws. In particular, macroscopic particles satisfy Newton's laws of motion; microscopic particles follow the Schrödinger equation; and high-speed particles comply with equations of the relativity theory. That is a very natural phenomenon when seen from the point of view of systems science.

Besides this, in physics, relativistic quantum mechanics investigates the properties of high-speed microscopic particles that travel at as high speed as light, and quantum electrodynamics focuses on the evolution of photons.

Although the mathematical methods employed in the investigation of the motions of a single particle are quite sophisticated, they stand for some particular methods useful for the simple systems.

In conclusion, when we studied the movements of those aforementioned particles, we have used the same method that includes the following: First, select state variables; second, find the relevant laws of evolution; and third, analyze the changes of the state variables. For the evolution of the state variables, we analyzed only the linear changes of the system responsive to the effects of the external environment, considering neither the system's internal structure (particles are approximations of systems without any internal structure) nor nonlinear interactions between the systems and their environments. Thus, we can imagine that if one analyzes the simple particle's movements under some other new conditions, he or she might very well discover new laws of motion. Like Newton's laws, quantum mechanics, and relativity theory, these new laws will specify how a single particle would behave under new sets of conditions. If nonlinear effects are considered, one can expect to produce corresponding laws. Thus, it can be seen that the laws that govern the movements of a single particle may be different, but the basic research methods that are useful for studying the movements of a single particle will be just the same.

Other than the method, as mentioned above, with the basic steps of inductively selecting and confirming variables, discovering the laws of evolution, and calculating the trajectory of evolution of the variables, the method of research of simple systems also possesses the characteristic of "free particle" systems. Speaking more specifically, simple systems can also contain different subsystems (particles). However, when these simple systems are studied, their subsystems must be "free" from each other without any constraints. That is, the subsystems do not have any correlation, and if there are, the correlations can be eliminated by using some kinds of transformations making the subsystem particles "free" from each other. The various kinds of integrable systems considered in Newtonian mechanics are "free particle" systems. Although they are not single-particle systems, they can be dealt with by using the methods of simple systems.

For example, let us look at how to deal with the evolution problem of a two-particle system, a two-body problem. We can establish the interaction of the particles by using Newton's second law of motion; then by introducing a proper transformation, we can eliminate the connection between these two particles so that they become particles "free" from each other. Next, we apply the evolutionary laws each single particle obeys to analyze the two-particle system, producing the desired results. However, for the general system of three particles, known as the three-body problem, because of the complicated interaction among the particles, as of this writing, we still cannot find a proper transformation to eliminate or simplify the interaction so that the system could become one with three free particles under the influence of their respective forces. That is why we cannot completely solve such a three-body problem by just applying Newton's laws of motion. As for problems of interactive multiple bodies, not much has been known in the world of learning.

Newtonian mechanics indeed considers the group movements of n particles. Because every two particles are related, the group of n particles cannot be seen as a simple system. It should belong to the category of simple giant systems. Generally speaking, this kind of system cannot be adequately analyzed by employing the methods developed for simple systems and by studying the movement of each particle. To overcome this difficulty, the concept of center of mass is introduced in Newtonian mechanics. Then a law on how a center of mass moves is established: for a system of n particles, the movement of its center of mass is completely determined by external forces no matter how the particles interact with each other. Here, the movement of the center of mass is equivalent to that of such a super large particle whose mass is equal to the sum of all the n particles under the effect of all external forces acting on the group. In this setup, all internal correlations and interactions between the n particles are totally ignored. By using this method, many practical problems and issues related to the design and production of durable goods have been resolved.

Although in theory, Newtonian mechanics has traditionally treated this method and law as a generalization and an application of the particle dynamics, this method has faced unsolvable (as of this writing) difficulties when it is employed in areas of prediction, such as the forecast of disastrous, small-probability weathers. Additionally, when looking at this law from the angle of systems science, such generalization of the particle mechanics is not quite adequate. It is because when the particles are indeed correlated to and interacting with each other, they constitute a simple giant system. The method applicable to such systems, let us name it as the "center-mass law," should be more than the generalization of the particle mechanics as mentioned above.

Here, let us briefly analyze the characteristics of such a method that is useful for simple giant systems. This analysis will also provide a case study for the following discussion of the general method useful for studying simple giant systems. From the perspective of systems science, when the movement of the center of mass is applied to represent the movement of the system of n particles, the system has been divided into two layers. The position and momentum of each particle stand for two state variables on the "microscopic" level, while the position and momentum of the center of mass are two physical quantities on the "macroscopic" level. They reflect some of the attributes of the whole system.

In general, the position of the center of mass is not that of any single particle of the whole system. It may be either within the inside or outside of the spatial area determined by the n particles. The position of the center of mass, in fact, reflects the mass distribution of the system. Movement of the center of mass does not represent the particular movement of any particle within the system. Instead, it is related to all the particles. The movement of the center of mass reflects the evolutionary tendency of the whole system. If the center of mass does not move, it then means that no movement of the system is observed at the macroscopic level, although every particle within the system can be moving.

The center of mass movement law and single particle movement law represent two methods, respectively, useful for analyzing systems' evolution at two different levels. They are the earliest achievements established when the characteristics of simple giant systems are initially investigated. The center of mass movement law, in fact, provides a general method for dealing with simple giant systems. In the following, when we discuss simple giant systems, we will further elaborate on this general method.

Similarly, according to the viewpoint of systems science, those systems that are composed by multiple microscopic particles or multiple high-speed moving particles should also have their respective descriptions of evolution at their macroscopic levels. However, few such research conclusions have been offered in the textbooks of the classical physics. Quantum mechanics' description of multiple particles merely rests on the level of artificially eliminating the existing connections between the particles so that the system of these particles could be investigated as several independent simple systems. This of course has greatly weakened the capability of quantum mechanics to describe systems of multiple components. Therefore, we can also see that the development of systems science has provided an alternative research direction for physics.

7.1.2 Simple Giant Systems

There are only a few simple giant systems in the real world. Such systems exist widely in various studies when complex problems are simplified for the sake of gaining theoretical convenience. In practice, what exist commonly in abundance are complex systems. As has been pointed out earlier, there are mainly three classes of complex systems: simple giant systems, complex giant systems, and specific complex giant systems (or social systems). Here, included in the class of simple giant systems, are many lifeless systems and some biosystems, which have been, respectively, investigated in various fields with relatively mature research methods. However, for the other two classes of systems, adequate general quantitative methods are still lacking. At the same time, when some of the complex giant systems and specific complex giant systems are appropriately simplified, they can be dealt with as simple giant systems. That is, the research methods of simple giant systems can be utilized to analyze more complex systems in many individual scenarios and can provide clues on how to develop the adequate methods for dealing with complex systems.

The number of subsystems is large; there exists nonlinear relationships among subsystems; the structure of the system is stratified into different levels; and there is a significant distinction between macroscopic and microcosmic states. The system consequently becomes more complicated than ever before. That explains why the research methods employed to study such systems need also to be more varied. It is no longer appropriate to investigate the evolution of systems just from one angle with a fixed set of methods. Instead, the characteristics of a system need to be considered from different angles. When considering a single particle, its mass, position, and speed would be enough to confirm its state. When considering a person, we learn his or her temperament and emotion from the aspect of psychology; we also need to learn his or her social connections, status, etc., from the aspect of physiology. That is, even if only one aspect of a simple giant system is considered, one generally has a set of different methods available. Generally speaking, researchers are interested in the degree of disorderliness of the subsystems' distribution of a simple giant system. This degree is very important in determining the properties of the system and is a key physical quantity. There are different description methods of this degree in different disciplines. For example, in physics, the concept of entropy can be employed to directly describe the degree of disorderliness of subsystems. Then by making use of the equal probability hypothesis, the series of theorems and formulas of the statistical thermodynamics can be derived.

Speaking from the angle of the history of natural sciences, following mechanics, there appeared calorifics. A caloritropic movement means such a "chaotic" movement of a large amount of particles. For a single particle, there are only simple "mechanical movements." That is, there are changes in position without any movement of heat; there is speed with temperature. The analysis of heat movements represents the start of research of simple giant systems. The thermodynamics, developed for the macroscopic analysis of the phenomena of heat, and the statistical physics, advanced for the microscopic investigation of the same phenomena, provide the theoretical foundation and the basic methods of research for the introduction and development of systems science.

Simple giant systems have worked as a bridge connecting the preceding and the following generations of scientific research. The phenomena of heat represent those simple giant systems that have been most well studied. Hence, in the following, when we study simple giant systems, we will often mention the phenomena of heat and analyze ideal gas as examples.

Simple giant systems have the following characteristics:

1. *The number of subsystems is large.* A simple giant system is often composed of over thousands of subsystems. For instance, a laser beam may contain about 10^{18} atoms (subsystems), which interact with each other. Each unit volume (1 cm^3) of liquid contains roughly 10^{18} molecules; each molecule of the Moore gas contains qualities of the magnitude close to 10^{23}. Therefore, it is difficult to analyze the evolution characteristics and laws of simple giant systems by looking at their individual subsystems. Even if every single particle of the simple giant system of concern satisfies the elementary Newton's laws of motion, it is still impossible for one to analyze the particles one by one in order to fully understand the system as a whole.
2. *Interactions of subsystems are relatively complicated.* Due to the large numbers of subsystems, the number of interactive relationships, although they might not be complicated themselves, may become large causing great difficulty for practical purposes. For instance, when considering a nerve network system that is composed of 10^3 neurons, the number of internal interactions among the neurons can reach such a high level of magnitude as $10^7 - 10^8$. Thus, with so many internal interactions, it is impossible to comprehend the whole network system just by analyzing the individual subsystems one by one. At the same time, the task of analyzing the individual subsystems of such a great amount itself is a tedious, if not difficult, task to accomplish.
3. *There is a hierarchical relationship in the system.* Hierarchy is a very important issue in the investigation of systems' evolution. Each simple giant system has at least two different levels: one is the "macroscopic" level, representing the level of the whole system, and the other is the microcosmic level of subsystems. Generally, the macroscopic level can be directly observed; its states and evolutionary characteristics with time can be measured. The microcosmic level cannot be directly observed, and its characteristics and attributes could be understood only through some indirect means, such as logical reasoning.

Ideal gas is a typical example of a simple giant system that is obtained by simplifying the internal interactions. For a given ideal gas, some macroscopic-level variables, such as pressure, volume, temperature, etc., can be observed directly. However, for a particularly chosen molecule, it is impossible for one to derive its specific speed and position. Instead, one has to make use of the methods of statistical physics to derive the characteristics and laws of motion of the molecule.

Each complex giant system can also be divided into layers. However, compared to the layered structures of simple giant systems, complex giant systems possess a much greater number of levels. Generally speaking, the microscopic level of interactive subsystems is further divided into several levels. The magnitudes and mechanics of interaction at a different level may also be different. The existing levels not only affect interactions but also determine the compositional structure of the system.

For example, the complex giant system of the human body is composed of a large number and different kinds of cells, while interactions among the cells are very different. In particular, the cells of the intestines do not have any direct interaction with the cells of the heart. Their connection is established through the organs they are respectively composed of, and materialized through the interactions of these organs. However, the interactions among subsystems of a simple giant system are equally weighted without involving levels. Thus, relatively elementary methods can be employed to analyze the relationship between the holistic level and the subsystems' level. Then on the basis of these well-comprehended relationships, the evolutionary characteristics and laws of multileveled complex giant systems can be investigated.

7.1.3 Traditional Research Method of Simple Giant Systems

One characteristic of simple giant systems is that each system can be divided into two different levels. Connections exist between the macroscopic or the whole level and the microcosmic or local level. Thus, it is very difficult to obtain the evolutionary characteristics and laws of any chosen simple giant system only by analyzing the input and output of the whole system. On the contrary, it is also practically impossible and unnecessary to describe individually the attributes of each subsystem in order to gain an understanding of the simple giant system. Additionally, computing the individual mean values of the large number of subsystems cannot adequately express the holistic characteristics of the system, either. Therefore, the appropriate method is to start with the potential relationship between the two levels, uncover the connection between the state variables of the different levels, and derive the relationship of the state variables of one level from that of the other level. One of the best and most integrated descriptions of this method is materialized by employing statistical physics. In particular, statistical physics investigates various kinds of thermodynamic systems, some specified simple giant systems. The core of the method of statistical physics is the following: First establish a certain quantitative relation for those microscopic variables that are difficult or impossible to measure directly. One of the most basic assumptions is that of identical probability, believing that each state at the microscopic level of the thermodynamic system of interest has an equal probability to appear. Then, statistical averages are computed, and the results are related to the system's macroscopic quantities, leading to the discovery of the laws of change of these macroscopic variables.

7.2 CHARACTERISTIC ANALYSIS OF COMPLEX SYSTEMS

7.2.1 Basic Concepts of Complex Systems

When simple systems (respectively, linear systems) are concerned, there appear complex systems (respectively, nonlinear systems). And in different scientific disciplines, the meanings of complex systems are different.

In 1999, when the journal *Science* published a special issue on "complex systems," the editor in chief pointed out that, if a total and complete understanding of the whole system cannot be achieved by using the knowledge we know about its components (subsystems), the system could be regarded as a complex system. A complex system means such a system that is composed of many parts, and many of their interactive relationships are not easy to comprehend. In each system of this kind, the whole is greater than the sum of parts. In this case, even if one knows the attributes and interactions of the parts, it is still very difficult for him or her to derive the characteristics of the whole system.

Definition 7.1

A system is complex if it is composed of many parts (subsystems) that are complicatedly interwoven and connected, and the behavior, functionalities, and characteristics of the whole cannot be directly acquired from those of the parts. ■

In a complex system, the so-called complicatedly interwoven connections mean countless possible interactions and connections among the constituents, creating for the whole system some overall behaviors and attributes that are not shared by the parts. Highsmith (2000) offered the following equation when he described the complicated behaviors of complex systems:

$$\text{Complicated behavior} = \text{simple rules} + \text{rich interactions}.$$

This equation shows a profound relationship among the three key factors—parts, interactions among the parts, and the behavior and attributes of the overall system—that constitute the complex system of concern. It implies that the complexity is created by the spontaneous interactions among the simple parts and factors.

Contrary to the concept of complex systems, if the behaviors of the whole system can be derived by using superposition principle from the behaviors of the components, then this system is referred to as a simple system. Linear systems are examples of a class of simple systems.

Each component of a complex system, in general, possesses a certain scale, which most likely means biosystems or systems with human participants, such as the human body system, transportation systems, economic systems, ecosystems, etc.

Nonlinear systems generally do not satisfy the superposition principle. However, they may not be complex systems. On the contrary, any complex system must be a nonlinear system. This fact implies that nonlinearity is just a necessary condition, not a sufficient condition, for a system to become complex.

The research objects of complex systems include natural phenomena, physical phenomena, biological phenomena, phenomena of life, eco-environmental phenomena, social phenomena, economic phenomena, and so on. In short, the theory of complex systems investigates systems with the following characteristics:

1. Instead of paying attention to the system's constituents, researchers are interested only in the functionalities, behaviors, and the interactions among the components.
2. Although the rules of the interaction among the components can be relatively simple, through iterations of the rules, complicated behaviors of the system are created, although there might not be any rule governing and sustaining the overall behavior of the system.
3. There are nonlinear, concurrent, and dispersing mutual influences among the components, leading to the creation of peculiar behaviors and functionalities for the whole system. Then these behaviors and functionalities of the whole system are feedback to the various components. This feedback is exactly the characteristic of emergence.
4. The behaviors of the whole system are influenced by the components and their interactions, which, on the other hand, cannot adequately describe and be employed to predict these behaviors of the system.

The concurrency and dispersion of complex systems stand for that all components or subsystems function and mutually influence each other simultaneously, independently, and autonomously. There is not a restriction that can ensure consistent whole system's behaviors, which are created out of the behaviors and interactions of the individual components or subsystems. Thus, even when every single component's behavior and attributes are clear, the behaviors and attributes of the whole system still cannot be satisfactorily explained. That means that complex systems cannot be investigated by using reductionism.

7.2.2 Characteristics of Complex Systems

The key that distinguishes complex systems from simple systems is the different significance of interactions and connections among the subsystems or components. Generally, the components that make up a complex system are not homogeneous and have multileveled structures. There are not only interactions between the components but also very complicated interactions among subsystems and between levels. Especially, some of these interactions are severely nonlinear. Thus, when hoping to understand the behaviors of a complex system, one needs to analyze not only how different components work together to form the behaviors of the whole system, but also the behaviors of the individual parts. Without deep and specific comprehension of the behaviors of the individual parts, there will be no way to capture the behaviors of the complex system. The reason why a system is complex is because some of its parts are difficult to fathom. It is specifically so when the system of concern involves human participants. Thus, to express a complex system, one has to, before anything else, describe adequately the behaviors of the basic elements (individuals) of the system.

Complex systems generally possess the following characteristics (Li 2006).

1. *Nonlinearity.* A certain part or all components of a complex system must have the characteristics of nonlinearity. The essential is the nonlinearity of the interactions between the components or subsystems. This implies that those components or subsystems that interact with each other not only exert unidirectional influences but also affect each other mutually. They place constraints on each other; they rely on each other. It is the nonlinear interactions that make the evolution of complex systems rich and colorful. That is why nonlinearity is seen as the origin of complexity. In other words, nonlinearity causes complexity.
2. *Diversity or multiplicity.* Generally speaking, there are diversities and multiplicities in the holistic behaviors of such complex systems as those existing in nature, in the biological world, in the economic organizations, and in the ecological environments, etc. These diversities and multiplicities are created, on one hand, out of the interactions of the components and factors and, on the other hand, out of the interactions of the parts with the systems' external environments. The diversities and multiplicities of the interactions bring forward the diversities and multiplicities of the system's overall behaviors. Additionally, the diversities and multiplicities in the systems' attributes lead to those of the system's functionalities.
3. *Multileveled hierarchical structure.* Complex systems always have a hierarchical structure of multiple levels and multiple functionalities. Each level constitutes an object of a higher level and helps to materialize a certain functionality of the system. For complex systems, there is not any general superposition principle between their levels. As soon as a new level is formed, some new attributes emerge. Generally speaking, the more complex a system is, the more the levels within the system will be. So, multilevelness and multiscaleness are a fundamental characteristic that should be used for the description of the degree of complexity of complex systems.
4. *Emergence.* Emergence is an essential characteristic of the whole system. However, not every property of the system comes out of the emergence of the whole. Those properties of the whole system that can be obtained by superposition of local properties of the parts do not appear out of emergence. Only those characteristics of the whole system that are dependent on the particular interactions between the parts are produced out of the emergence of the system. They are referred to as constructed characteristics. Emergence stands for a holistic property that appears only in the evolution and evolvement of complex systems.
5. *Irreversibility.* All the fundamental laws of the classical physics are reversible in terms of time. However, the evolutionary processes over time of many complex systems that exist in nature are irreversible. For example, a seed can germinate and grow into a plant. This process of development cannot be reversed back to the original seed. As another example, the constantly varying climate does not reverse its evolution either.
6. *Self-adaptability.* Complex systems possess the characteristics of evolvement, which means that the systems' components, scales, structures, and functionalities can automatically adjust with time and adapt to the changing environment in order to guarantee their viable existence. In the process of constantly adapting to the environment, the systems become more complex. This end reflects the key of creating complex adaptive systems of Holland (1992, 1995).

7. *Criticality of self-organizations.* Criticality of self-organizations means that the complex system of concern is at a critical state that exists far from an equilibrium state; after the critical state, instead of following the past smooth and gradual pattern of development, the system will evolve in an abrupt and chaotic manner like an avalanche that occurs unexpectedly and explosively. For example, earthquakes, tsunamis, societal revolutions, economic crises, etc., are such abrupt, avalanche-like evolutions. Sandpile models are a typical example of systems that are situated at a critical state of self-organization. Theories on the criticality of self-organization have been widely employed in the research of many scientific disciplines, including solar flares, volcanic eruptions, economics, biological evolutions, turbulences, and spread of infectious diseases, etc.
9. *Self-comparability.* There is self-comparability in different levels of complex systems. For example, biological, social, and management systems are all open complex systems; in their evolutions, they are affected by their environments. Because these systems are all life systems and intelligent systems, they have the ability to self-adapt to the changing environment. Thus, along with their evolution, they gradually form, change, and perfect their respective structures. However, within the macroscopic structures of these complex systems, there is self-comparability of a different sort. Self-comparability can mean complex systems' different leveled structures or some self-similarity in any aspects of the systems' morphology, functionality, or information.
10. *Openness.* Openness of a system means that within the inside of the system itself and between the system and its environment, there are exchanges of matter, energy, and information. Openness can make the system's components (subsystems) interact with each other, and the system interacts with the external environment, making the system develop in the direction that will help it better adapt to the environment.
11. *Dynamics.* All complex, self-organizational, and self-adaptive systems have some kind of momentum. This kind of momentum makes them completely different from the other complicated systems, such as integrated circuits. The dynamics of complex systems possesses more spontaneity and disorderliness and is more active. However, such dynamics are very different from chaos. Chaos theory points out the fact (Lorenz 1995; Thompson and Stenmit 1986) that simple dynamic rules may bring forward rather complicated behaviors. However, chaos theory itself still could not explain the structurality and cohesive force in general, and the cohesive self-organizational forces of complex systems in particular. That is, chaos theory could not explain the phenomenon of complexity. However, complex systems possess such a capability to bring order and chaos into equilibrium. The balancing points are often known as the edge of chaos. Of course, systems' factors, in general, can neither stay static in a fixed state nor fluctuate so severely that the systemic wholes would dissolve. The edges of chaos represent where complex systems can spontaneously adjust and survive.

7.2.3 Evolution of Complex Systems

Generally speaking, to understand a complex system, one needs to first describe the system by abstracting all the relevant details and then analyze and simulate the system's behaviors through techniques of mathematics, computer science, and others. Such work includes two aspects. One is the dynamic description of the system; that is, through expressing the relationship of the key factors using mathematical tools, the laws of change of the system are investigated. The other aspect is to reveal the laws of change of the system's movement through reconstructing the state space on the basis of the current state.

Changes in complex systems mostly occur in one of the following three ways: The first is through the system's capability of self-organization, which stands for a key characteristic of complex systems. Self-organization changes the internal structure of the system so that it can better adapt to the surrounding environment. Through learning, the system alters its internal structure. The second way of change is through the system's dissipation, which means that when the system interacts with one or more other organizations, it can enter an organizational state of a higher order through inner disturbances and external forces. For example, in terms of major changes of an economic system and its environment, the economic system can be viewed as a dissipative structure. During the process of an industrial revolution, import of a new technique will make the inner structure of an economy system change very fast. In particular, during the process of an industrial revolution, introducing new technology into a firm can cause rapid structural changes within the economic system.

The third way of change is through a self-organization criticality, which represents the capability of the complex system to maintain itself in balance between random changes and stagnancy. If a system can reach a critical point, the system will face the possibility of breaking down even without the influence of any external force. Self-organization criticality is a form of the system's self-organization. Under this condition, the very rapid adjustment of the internal structure makes the system difficult to adapt to the new environment. However, in order to materialize the eventual survival, the system has to adapt to the rapid change. Most research works of self-organization criticality are limited to the areas of ecological and biological systems, with a few published works in the area of economic systems.

On the background of complex systems, a specifically attractive example in the technology field is about the influence of information technology on social organizations. Helix structures may help one better understand the characteristics of evolution and periodicity of social systems. For example, between the two stable states of centralization and decentralization, there are alternating and exchanging processes. When a simple-leveled grade system, such as a tribe or a small company, continuously grows to such a state that it can no longer be effectively managed with a centralized administrative unit, it will be decomposed into some local cells with a degree of autonomy. After another long period of time, the operational efficiency of the company becomes lower and lower and deteriorates to such a degree that the management needs to be centralized again. During each of these reciprocating transitions, a chaotic period of time might emerge. In other words, the periodical reorganization means that the system sways between centralization and decentralization, bringing the degree of complexity of the system to a higher level. Organizational behaviors similar to what was just described can be observed in many systems of every level and scale from the business world to the universe.

In the research of the evolutionary and self-organizational processes of biological systems, it is found (Coffman 2006) that complex systems develop the best when they operate to benefit themselves and when the structures of their local units are optimal and semiautonomous and do not overlap with each other. For example, in a working group of people or in a biological system that possesses the characteristic of emergence, if the group is stable and efficient, then it tends to contain multiple individuals, and although there are conflicts, the individuals can still compromise and advance together even when no administrator exists. The degree of the best decentralization of power or the optimal scales of component units make the whole organization be situated near the transition point between in-order and chaotic states, that is, the edge of chaos. On top of the present background that information technology is taking dominance in the industrial world, there have appeared such organizations with virtual structures that emphasize cooperation and indigenized globalization. That implies that actualizing equilibrium is extremely important. In the expected forthcoming competition, for an organization to maintain an unprecedented flexibility and inner driving, the organization must install a new networked arrangement.

7.2.4 Classification of Complex Systems

The exploration on complexity initially begins with the research of the complex phenomena of some lifeless natural systems, such as thermodynamic systems, and chemical systems, etc. This class of systems is referred to as complex natural systems. Along with the gradually deepening research, life systems and systems with biological participants, such as biological systems, ecological systems, social systems, and economic systems, etc., become the focus of investigations. It is consequently discovered that complex systems with biological participants are much more difficult to understand than complex natural systems. There are different methods of classification of complex systems depending on which angle and opinion they are based on.

Sunny A. Auyang (1999) points out three classes of complex systems:

1. *Many-body systems*, where each such system is composed of a large amount of components coupled together by only a few relations. This class includes such systems as an evolutionary species of organisms, an organization made up of consumers and manufacturers, etc. The complexity of any chosen many-body system is originated from the diversity and interwoven relationship of the components and the large number of the components.
2. *Organic systems*, where each such system is composed of many highly specialized, closely related, and different kinds of components. It is easier to represent an organic system using function. The functionalities of organic systems can be described with relative ease, where the functionalities of components are defined and expressed by using the roles the components play in maintaining the systems in their expected states. Biological systems are typical organic systems.
3. *Cybernetic systems*, where each such system is composed of the complexities of many-body systems and organic systems, such as neural networks. Each human being represents a very complex cybernetic system. It is inappropriate to describe a human being by using only his or her physical behaviors because he or she also has intentional behaviors. The free market economy is a result of unintentional human behaviors. Economics studies economic components and their relations through abstraction, simplifications, and idealization.

At present, there are three classes of complex systems that have been relatively well investigated.

1. *Nonequilibrium systems:* Each such system is composed of lifeless subsystems, each of which is simple, with relatively elementary interactions. The evolutions of these systems have been studied by using nonequilibrium self-organization theories, such as dissipative structures, synergetics, etc.
2. *Complex adaptive systems:* Each such system consists of subsystems of certain intelligence. The relevant works have brought significance in practice. For instance, the models of complex self-adaptive systems developed by the Santa Fe Institute are based on bio-organisms through sorting and normalizing the key conditions of biological evolution, such as adaptive environments, growth and reproduction, and genetic mutations, etc. The evolutionary characteristics of complex self-adaptive systems are discussed in order to address such complex behaviors as population development, stable existence, and eventual disappearance, etc.
3. *Open complex giant systems:* They represent a class of model systems developed to investigate the most complex evolutions of systems composed of human beings. Because they deal with systems with human participants, although the joint qualitative and quantitative method of metasynthesis, as promoted by Xuesen Qian (for more details, see Chapter 9), provides an epistemological way for solving problems, the theoretical studies still face many difficult challenges.

There exist many phenomena of complexity in the natural world. In the struggle against nature for survival, man has created many complex engineering systems. Also, various social systems have been known to be extremely complex. Thus, complex systems are categorized into three classes.

1. *Complex natural systems:* The study of complexity initially started with natural lifeless systems, such as Rayleigh–Bénard convection, the laser phenomenon, surges of chemistry reactions, climate changes, ecological environments, etc.
2. *Complex engineering systems:* They are mainly manufactured using various systems designs or created out of deterministic engineering systems that become dynamically complex due to the influence of some uncertain factors.
3. *Complex social systems:* They are social organizations, each of which is formed on the basis of some shared physical conveniences, economic conditions, and information resources.

7.2.5 Research Methods of Complex Systems

The research of a complex system (Xu 2000; Li 2006) should start with the system's characteristics under the guidance of systemic philosophical thinking, followed by establishing the interaction and connection between the parts and the whole, the system and its environment, through holistic, dialectic, qualitative, and quantitative analyses, in order to uncover the conditions, patterns, and mechanism of the system's evolution and the holistic emergence.

7.2.5.1 Combined Qualitative Analysis and Quantitative Calculation

Because of the characteristics of multiple levels and nonlinearity, the evolution of complex systems appears to be diverse and variant. Compared to linear systems, it is difficult to develop precise mathematical models for complex systems to make quantitative analysis. Thus, qualitative judgments and quantitative computations should be combined. From qualitative analysis, the motive, direction, development tendency, and other aspects of a system's evolution can be adequately determined. On the basis of the qualitative judgment, quantitative representations are employed to provide detailed and precise descriptions. To this end, the qualitative mathematics, pioneered by Henri Poincare (Nemytskii and Stepanov 1989), is a powerful tool for conducting the desired qualitative analysis of the system.

7.2.5.2 Combined Microcosmic Analysis and Macroscopic Synthesis

The whole of a system is composed of parts, and the behaviors of the parts are constrained and dominated by the whole. Thus, any description of the system should consist of two aspects: the description of the whole and the descriptions of the parts. Under the guidance of systems' wholeness, combining microcosmic descriptions of the parts with the macroscopic synthesis at the level of the whole is a basic method for investigating complex systems.

In a simple system, the difference between the scales of the constituents and the whole also leads to differences between the microscopic and macroscopic views. When a simple system is treated at both the microcosmic and macroscopic levels, statistical methods could be employed to transit microscopic descriptions to the macroscopic representation. For complex systems, especially complex giant systems, effective statistical methods of description are still lacking. Even so, one can still employ the principle of microscopic analysis and macroscopic synthesis to deal with complex systems, where microscopic analysis is applied to understand the hierarchical structures and macroscopic synthesis to fathom the functionality and emergence of the whole.

7.2.5.3 Combined Reductionism and Holism

By using reductionism, one studies the whole by looking at the parts and explains the holistic characteristics through microscopic structures and regional mechanisms. In essence, reductionism is a method of learning the whole through analyses and reconstruction. As science reaches the microscopic level of tiny scales, man has gained much clearer understanding of material systems than ever before. However, his recognition of the whole is becoming more and more blurred. Many complex systems and complicated phenomena come from the emergence of wholeness, and the emergence of wholeness no longer exists when the whole is decomposed into parts. Therefore, reductionism cannot help in resolving the mystery of complex phenomena. On the other hand, the holism recognizes that the macroscopic behavior of systems is determined by the interactions and correlations of the parts. In other words, if the system of concern is not decomposed into parts and the precise structures of the parts are not understood, then any recognition of the whole does not have the needed scientificality. Without the point of view of holism, the holistic behaviors of the system will be impossible to comprehend.

7.2.5.4 Combined Certainty and Uncertainty

Complex systems, in general, contain some uncertain variables, such as random, fuzzy, and gray variables, and some incomplete information. Thus, methods of uncertainty, such as probability, statistics, fuzzy sets, rough sets, etc., should be employed to describe the system of interest.

7.2.5.5 Combined Scientific Reasoning and Philosophical Thinking

Scientific theories are born out of the conceptual system established on the basis of daily activities, production practice experience, and scientific exploration and experiments. When scientists express a scientific theory, they generally formalize and symbolize the basic concepts and employ formal logic to form the necessary systems of axioms. Sometimes scientific theories can also be imperfect and contain inconsistencies. When one such situation occurs, one should apply philosophical reasoning. Isolated phenomena, contingencies, and necessities, which appear in the evolution of the system, should be dialectically analyzed by using such laws as unity of opposites, negation of negation, etc. Only by doing so can one possibly be on the right path of understanding complex systems.

7.2.5.6 Combined Computer Stimulations and Human Intelligence

It is generally hard to establish a precise mathematical model for a complex system. The traditional methods have met with great difficulties when employed to solve problems of complex systems. Through computer simulations, several intelligent optimization methods have been developed, such as genetic arithmetic, immunity arithmetic, particle swarm optimization, colony algorithm, and so on. From using computer simulations, the evolution of complex systems has been investigated. That is, computer intelligent systems are designed to approximate the behaviors of complex systems so that problems involving complex systems can be resolved or optimized.

7.3 COMPLEXITY

The science of complexity focuses on the investigation of the complexity of complex systems. It considers the mechanism behind the emergence of wholeness created by systems' evolution under the interactions between subsystems and the system and its environment.

7.3.1 Concept of Complexity

Both nature and human society continually become more complex in their respective processes of change and development. At the same time, the units or individuals may regularly decline, degenerate, disorganize, and even disappear. With time, on the one hand, through interactions with each other and with the surrounding environment, the units or particles in those simple systems assemble and alter their structures so that they gradually evolve into advanced complex systems. That is the so-called evolution. On the other hand, complex and advanced systems continually degenerate into low-grade simple systems and even particles. That is the so-called decline.

Essentially, the birth, growth, and decline of complexities stand for the processes of evolution and decline of nature and human society. Adaptability begets complexity; that is, the process of evolution creates complexity.

Currently, no unified definition of complexity exists. In terms of the objects of concern, complexity is not an intrinsic property. Instead, complexity is dependent on how and within which range the researcher attempts to describe his or her objects of interest.

The word "complexity" is mainly connected with the meaning of combo. There must be two or more parts and some connection between the parts to constitute a combo. Thus, there appear the binary characteristics between the parts: connection and distinction. Connection (dependency) represents the mutual reliance and constraint, leading to order, of the parts in the combo. Distinction (diversity) stands for the difference, multiplicity, and asymmetry, leading to disorder and chaos, between the parts. Thus, it is readily seen that complexity exists only in between the state of not-total disorder and that of perfect order. A true complexity appears at the edge of chaos.

The concepts of order (or in order) and disorder (or out of order) can be described by using the concept of symmetry. Symmetry means disorder or out of order, while asymmetry stands for order or in order. If symmetry is destroyed, then, on the knowledge of one particular part of the combo, it will be impossible to derive the attributes or behaviors of the rest of the combo. Hence, the concept of symmetry can be employed to describe complexity and the process of growing more complex. If the dependence and diversity of the combo's parts grow, then the quality of coupling and distinction is strengthened, leading to increased complexity of the combo. If one of the characteristics of dependence, diversity, connection, and distinction is strengthened, the complexity of the combo will increase accordingly.

When an open complex system has many different components (subsystems), many structural levels, and a good number of factors with uncertainty, the system will be caused to interact with its environment and show complicated dynamic behaviors and its wholeness characteristics. These behaviors and characteristics change without any pattern and are unpredictable. Thus, they cannot be described and dealt with by using the methods developed on the traditional reductionism.

The complexity of a system depends on how one describes the system. At the same time, the complexity also possesses a degree of objectivity. When speaking from this angle, it is extremely difficult to introduce a complete and accurate definition for the concept of complexity. In the following, let us try to define complexity in the language of set theory.

Definition 7.2

A system $S = (M,R)$ has complexity if it satisfies the following conditions:

1. The system S has a large number of subsystems $S_i = (M_i, R_i)$ and a large number of levels.
2. There is at least one nonlinear relation between the levels.
3. The system S possesses some wholeness property P, which cannot be derived by using any analytical model of the subsystems or components from any level. ■

7.3.2 Classification of Complexities

Complexities can be classified into different types from different angles.

- Organized simplicity, disorganized complexity, and organized complexity (Weaver 1948; Flood and Carson 1993; Klir 1985).
  1. Organized simplicity occurs when a small number of significant factors and a large number of insignificant factors appear. Initially, a situation might seem to be complex, but when insignificant factors are taken out of the picture, a hidden simplicity is found.
  2. Disorganized complexity occurs when many variables exhibit high levels of random behaviors. For example, the gas molecules in a lab container stand for a disorganized complexity.

3. Systems with characteristics of organized complexity appear in abundance. They exist particularly in the life, behavioral, social, and environmental sciences, as well as in applied fields, such as modern technology or medicine.

- Structural complexity, functional complexity, and organizational complexity (Yang 2004)
  1. Structural complexity can increase with the dependence, diversity, connection, and distinction of the combo. In the space dimension, structural complexity is created by integration and distinction, where integration means the process by which the number and strength of connections increase. In other words, integration represents an increasing tendency of dependence.
  2. Function complexity is caused by integration and distinction in the time dimension. When the object of concern increases in size, selecting an adequately appropriate control system becomes increasingly more difficult. An often-employed method of dealing with the complexity is to decompose the decision-making problem into finer parts. That is to divide the original problem into relatively independent subproblems, each of which has its own subobject. However, the establishment of the subobjects should be closely coordinated with the relevant functionalities of the system. The connection of the subproblems is controlled by the object of a higher level so that the object of the overall problem can be materialized. That is, activities of a lower level are connected and integrated by the predetermined objects of a higher level, and this higher level itself is, in turn, integrated by another level that is higher in order to serve the need of this level. In this process of integration from the bottom level to the top level, the higher an object level, the greater the environmental disturbances, the greater the range that needs to be controlled, the greater the strength of control needed, the longer the time required, and the higher the complexity involved.
  3. Organization complexity: Along with the development of productivity and finer social division of labor, the current form of organization resulted from a long process of evolution. Presently, alternative changes in organizational forms still naturally occur. Along with the increase in the complexity in organizational forms, the corresponding organizational structure also shows its diversity and sophistication.

     Organizational complexity means the unevenness existing in the organization of concern. It contains three such closely related directions as the latitudinal, longitudinal, and spatial distribution unevenness. Latitudinal unevenness means the professional difference between components and the difference between the degrees of compartmentality within the organization. Longitudinal difference stands for the number of the vertical administrative layers within the organization and the difference between layers. Spatial distribution difference implies the difference of the organization in the regional distribution of the administrative structure, production arrangements, and locations of workers. It is not hard to see that the greater the latitudinal, longitudinal, and spatial distribution unevenness, the more the complexity the organization possesses.
- Algorithmic complexity, certainty complexity, and integration complexity (Manson 2001)
  1. Each algorithmic complexity is expressed by using the complexity theory of mathematics and information theory. It believes that the complexity of a system is exactly the difficulty of describing the system's characteristics. Usually, algorithmic complexity defines a system's complexity using either the amount of calculation involved or the amount of information needed.
  2. Certainty complexity mainly involves chaos theory and catastrophe theory. It is concerned with the phenomenon of sudden appearance of discontinuity in large stable systems causing the interactions of a few key variables. Certainty complexity has the following features: it is described by deterministic equations and attractors; it contains feedback processes; it is sensitive to initial values and has bifurcation behaviors; and it has strange attractors.

3. Integration complexity mainly analyzes the influence of the system's components on its emerging behaviors and their interactions. Integration complexity consists mainly of the complexity of interactions, complexity of the internal structure, complexity of the external environment, learning to adapt to the environment, harmonizing the capability to form relations, and such characteristics of emergence, evolution, development, etc.

   In terms of the investigation of complex systems, it is very effective to treat systems as structures that constantly evolve and interactions with environments through self-organization, dissipative behaviors, and criticality of self-organizations.

- Physical complexity, biological complexity, and economic and social complexity (Cheng and Feng 1999)
  1. Physical complexity stands for the complexity existing in a physical (natural) system, such as turbulence, laser phenomenon, mud–rock flow, etc. The physical complexity of a system is mainly caused by the irregular thermodynamic movements of a large amount of molecules (particles) of the natural system. Due to the system's openness, under certain environmental conditions, a large amount of microscopic particles interact with each other, making the system far away from any equilibrium state. Through self-organization, a new dissipative structure is formed; or a stable state is broken and the system evolves into another stable state; or the system evolves into chaos through period-doubling bifurcation; and so on. In its process of evolution, this kind of physical (natural) system would show multiple kinds of complexities at the physical level.
  2. Biological complexity: Due to the structural and organizational complexities of biological organisms, these organisms possess the characteristics of growth, adaptability, intelligence, individuality, population multiplicity, strong nonlinearity, etc.
  3. Economic and social complexity: Both economic and social systems are open complex systems with human participants. They have many levels, complicated structures, multiple correlations, many uncertain factors, etc. Thus, the degree of complexity of these systems is much greater than that of physical systems. Studies show (Hommes 1991; Rosser 2000; Serletis and Gogas 2000) that nonlinearity and the phenomenon of chaos exist in economic systems. In their evolutions, under the effect of the external environment, social systems can become instable and experience sudden changes in their structures.

7.3.3 Science and Theories of Complexity

After the 1970s, nonlinear science, as represented by dissipative structures, synergetics, catastrophe theory, chaos theory, fractal theory, hypercycle theory, etc., made great advances and has shaken the determinism that has ruled physics, geometry, and mathematics for over 300 years. Determinism enabled physics to express various behaviors in predictable forms. However, natural, biological, ecological, social, and economic systems are fully filled with nonlinearity and complexity. Thus, investigations of such systems cannot be placed on reductionism. That is, the matter of concern cannot be fully understood by decomposing it into parts or lower-level objects. Instead, the thought of wholeness has to be employed. Complex adaptive systems theory represents an important theoretical achievement. Since entering the 1990s, complex systems and complexity have been a hot topic of scientific activities, producing some important results. That implies that science of complexity, as a new interdisciplinary frontier of science, has been born.

Complexity science studies the complexity (appearance and origin) of complex systems and explores the general mechanism and laws (properties, attributes, behaviors, functionalities) of the emergence that appear in the whole system's evolution under the interactions between the subsystems and between the system and its external environment. Also, because there are complex systems in such a wide range of disciplines such as physics, biology, sociology, economics, environment, ecology, engineering, etc., it determines that complexity science is a new, growing subject matter of interdisciplinary research. This new science has gone beyond the traditional mode of thinking, such as linearity, equilibrium, and reductionism, of modern science and establishes its own new way of reasoning, such as nonlinearity, nonequilibrium, and complex holism.

Some of the most important areas of the complexity theory include

1. The self-organization theory of nonlinear dynamical systems that are far away from any equilibrium state
2. The self-organization criticality theory of complicated natural systems
3. The theory of complex adaptive systems
4. The theoretical framework of the synthesized integration seminar hall of open complex giant systems

For more detailed discussions of these areas, please see the following chapters.

7.4 CASES ON ASSOCIATION OF COMPLEXITY AND SIMPLICITY

Experiences might have told us that simple rules associated with the rich associations could produce complexity. In this section, starting with two particular cases, we demonstrate the association between complexity and simplicity.

7.4.1 The Phenomenon of Three and Complexity

"Three" is a very common and extensively used numeral. The collection of phrases involving "three" contains a large number of elements. In the *Chinese Dictionary*, published in 1986 by Shanghai Dictionary Press, the phrases starting with "three" account for more than 1500 items (Wu and Huang 2005). As a numeral, "three" possesses the intension and characteristics of "maturity," "being complete and satisfactory," "approximation," "stability," etc. In daily life, there are many interesting phenomena involving three. Lao Tzu's claim (time unknown) that "the three begets all things of the world" has been a mystery for thousands of years. Additionally, the frontier research on complexity of modern science is also inextricably linked with the number "three." "Period three implies chaos"; see Chapter 5 for more details. That is, the concept of complexity began from the number "three." Some scholars have even suggested (Li 1997) that "three" should be treated as the third boundary constant of Newton's mechanics and referred to as the third constant of the universe. In the following, we will explore the phenomena of "three" and complexity from several different angles.

7.4.1.1 The Phenomenon of Three in Chinese Idioms

In the Chinese language, there are many idioms, proverbs, and old sayings that are related to "three." For the purpose of demonstration, listed in the following are some commonly used examples: "With three or five guys a group is formed." "Three people form a public." "The combined strength of three people is as strong as a tiger."

Speaking of the numerical characteristics of three, the number "three" stands for the minimum count of many; it represents the minimum number and the minimum agreed-upon factor that constitutes a "group" and a "public." On the other hand, the idiom that "the combined strength of three people is as strong as a tiger" explains in the opposite direction that one person's words may not be trustworthy; however, when more people, such as at least three people, speak of the same matter, even if what they say is incorrect, the strength of mass makes people believe the truthfulness of what they say. This idiom is equivalent to the saying that when a lie is repeated thousands of times, it becomes a truth.

"Standing with three pillars," "visiting the thatched cottage three times," and "separating the land into three" are another set of commonly used Chinese idioms. Here, "three pillars" means either standing extremely stable or a three-way confrontation. During the historical time of three kingdoms in China nearly 2000 years ago, Zhuge Liang, who later became a successful and famous military strategist, announced, when he was still a teenager, his statement that only if the land were divided into three and Liu Bei occupied one of the three pieces, it would become possible for Liu Bei and his power to develop and grow; only after that, Liu Bei could contest the control of central China and eventually unify China. The following development of the history proved the correctness of this prediction. Although Liu Bei never unified China successfully, as expected, the coexistence of three states, Wei, Shu, and Wu, lasted about half a century with Liu Bei being the emperor of Shu for quite a few years. Later, Liu Chan, the only child of Liu Bei, inherited the throne and stayed in power for nearly 40 years, thanks to the stable confrontational balance of the three states. The reason why these three states could have lasted for so long is their mutual containments and restraints, in which each of the states was directly related to each other. It is like the edges of a triangle; when one edge is missing, the triangle is no longer in existence. As soon as state of Shu was annihilated, Wu was also precarious.

What is implied by the idiom that "the combined mind power of three cobblers is better than that of Zhuge Liang" is that the strength of a group is generally greater than that of an individual, no matter who the individual is. It also means the importance of collecting the wisdom of the mass.

The commonly used Chinese proverb that "with three women a drama is made" has utilized the number "three" perfectly, because if only two women are talking, it would be possible for them to run out of topics of common interest. On the other hand, if four women are involved, the number "four" seems to be redundant, because there is always such a chance that one of them could not get a chance to talk.

The proverb "san si er xing" means to think it over three times before taking action; in other words, "think it over three times" means that after coming up with an idea, which is the thinking of the first time, reevaluate the thought carefully, which is the thinking of the second time, and then reconfirm the thought with relevant details involved, which is the thinking of the third time. Here, "three" is a conceptual number, meaning that before taking a leap, one should think through every aspect of the act thoroughly. Could one change the proverb to think two times before taking the action? From the translation above, it can be seen that thinking twice is not enough. Then could one change it to think it over multiple times before taking the action? From the context, this modified version seems to suggest that the person is in doubt and hesitant about what he or she is going to embark on.

In the *Analects* of Confucius (Waley 1989), "three" is used widely. For instance, "I do self-evaluation three times a day." The word "three" in the statement that "among three people, one of them can be my teacher" can be understood readily, because when two people are together, claiming that "the only other person can be my teacher" does seem too strained.

7.4.1.2 The Phenomenon of Three in Fairy Tales and Novels

Many well-known fairy tales have the storyline of "three brothers"; even the very familiar "Cinderella" also has two vicious sisters. At this junction, one might wonder why in these make believe fairy tales, "three" appears more commonly than "two" or "four." In his work, Arnheim (2004) points that in fairy tales, if it is about how the youngest brother defeats his opponent eventually and successfully, the total number of brothers must be three, because the number of two elder brothers is the most appropriate figure to show the little brother's heroic acts. If there were four brothers, it would seem redundant; if there were only two brothers, a closed and symmetric combination would result, while such a combination would present a comparison between two extreme behaviors and characters, such as good and evil, stupid and clever, etc. To this end, the need for trinity (three people or objects) is mainly for the purpose of representing an intertwining relationship instead of a lackluster comparison. When writing in this way, the resultant works will seem neither too simple nor unnecessarily complicated.

The novel *Romance of Three Kingdoms* witnesses the storyline of "three brothers." It has been recognized as one of the four Chinese classics. This novel focuses on the description of Chinese politics and an evolutionary period of fragmented military controls. It is the most successful novel among all those concerned with the topics of politics and the military. It is the trisection of Wei, Shu, and Wu of the land and political control that constitutes a complex yet clear situation. In comparison, the *States in the Eastern Zhou Dynasty*, a novel by Menglong Feng (1574–1646), is way too messy; it involves a great number of extremely complicated relationships between too many states. *The Early and Late Hand Dynasty* (author unknown and time unknown) is too straightforward and consequently has not been that well known. When three parties A, B, and C are involved, their mutual interactions will be intertwined together to form a complicated network. For instance, A and B can join hands to fight against C; or A and C collectively go against B; or C gets his or her personal benefits from A and B fighting against each other; or A and B collaborate openly, while each of them tries to connect with C from behind the closed door; or while in public, A acts justly and plays fair to both B and C, but he or she in fact attempts to hurt either B or C by using all means to sow discords between B and C and by secretly associating with one of them in order to destroy the other. In such a scenario, each party is situated in a difficult position where dangers may appear in any perceivable direction so that no one dares to make a move without carefully considering all possibilities or without full strength. At the same time, with a minor change in strategy, former adversaries become instant "unbreakable" allies or irreconcilable enemies.

7.4.1.3 "Three Begets All Things of the World," as Claimed in Tao De Ching

The origin of systems thinking in China can be traced back to Lao Tzu of the time of Spring and Autumn over more than 2000 years ago. Lao Tzu is one of the earliest philosophers, who considered such questions as: Where and how is the world from? Where is it going to? In chapter 42 of Tao De Ching, Lao Tzu says that, "Tao breeds one, one breeds two, two breeds three, and three begets all things of the world." After stating that one breeds two, two breeds three, why did Lao Tzu jump directly from "three" to "all things?" Of course, one explanation is that Lao Tzu tried to avoid redundancy so that he skipped "four, five, and six." However, if it is so, why did he not simply say that "two begets all things of the world?" It is because throughout Chinese history, yin and yang have represented complementary opposites; through these opposites, the colorful world has been formed. Another explanation is that Lao Tzu has indeed recognized the particular significance of "three" so that only after reaching "three" can all things of the world be created. By using the point of view of the present systems science, the significance of "three" can be quite clearly illustrated.

7.4.1.4 Three and Li–Yorke Theorem

The paper entitled "Period Three Implies Chaos" (Li and Yorke 1975) can be seen as a cornerstone of chaos theory. It reveals the essence of why dynamic systems might experience chaotic phenomena: period 3. As a matter of fact, the Li–Yorke theorem is a special case of the Sarkovskii theorem. However, it is the Li–Yorke theorem that makes the Sarkovskii theorem widely known. If one asks why, he or she will find that the Sarkovskii theorem is very mathematical. When Sarkovskii initially proposes and establishes his result, he develops a very complicated proof. However, he fails to discover the commonly existing natural phenomenon of "chaos" on the basis of his theorem.

7.4.1.5 Summary

Other than what have been analyzed above, there are also the following short stories of "three":

- Fermat's theorem points out (Azcel and Aczel 2007) that "three" stands for the limit of human numerical capability.
- Although Newton and Leibniz invented calculus, nobody seems to know what to do with the three-body gravitation problem.

- Nature has its three primary colors, and the man also has his own "three primary colors." In fact, man's primary colors can be and have been employed to confuse the colors of nature in his reconstructed virtual world.

From all these particular examples, the association between the numeral three and complexity seems to become clearer.

7.4.2 Cellular Automata

7.4.2.1 The Concept

Cellular automata (CA) have also been named as grid automaton points or molecular automata. They are discrete dynamical systems in both time and space. Each of the cells that are scattered in the regular grids, known as lattice grids, takes a finite discrete state, follows the same rules of operation, and gets renewed synchronously according to the predetermined local rules. Through their interactions, a large number of cells constitute the evolution of a simple dynamic system. Unlike the general dynamic model, a CA is not determined by any rigorously defined physics equation or function. Instead, it is composed of a series of rules that constitute the model construction. Those models that satisfy the rules can all be regarded as a CA model. Therefore, each CA model represents a whole class of models or stands for a methodological framework. Its characteristics are that it is discrete in terms of time, space, and state; each variable takes only a finite number of state values; and the rules for state changes are all local in terms of time and space.

Definition 7.3

A CA is a dynamic system that evolves along a discrete time dimension according to some local rules over a cellular space made up of discrete cells of finite states.

A cell's state at an arbitrarily chosen time moment depends on its former state and the states of all its neighboring cells. The cells in the cellular space update their states synchronously in accordance with their individual local rules so that the whole cellular space manifests its changes in the discrete time dimension. ■

Mathematician L.P. Hurd and colleagues (Hurd et al. 1992; Schiff 2008) studied CA by using the language of set theory. In particular, assume that d is the dimension of the cellular space, k is the state of a cell and takes value from within a finite set S, and r is the radius of the neighborhood of the cell. Let Z be the set of all integers, seen as a one-dimensional space, and let t denote time.

For simplicity, let us consider CA on a one-dimensional space. That is, we assume $d = 1$. Then, the entire cellular space is a distribution of the set S of the finite many states over the one-dimensional space Z of integers, denoted by S^Z; the dynamic evolution of a CA represents changes in the combination of states over time. Symbolically, it can be written as follows:

$$F : S_t^Z \rightarrow S_{t+1}^Z$$

which is further determined by the local evolutionary rule f of each cell. This function f is often known as a local rule. For a one-dimensional cellular space, if a cell and its neighbor are written as S^{2r+1}, then the cell's local function f can be expressed as

$$f : S_t^{2r+1} \rightarrow S_{t+1}^{2r+1}.$$

In terms of the local rule f, the input and output sets of the function are finite. As for the cellular space, independent applications of the individual cells on the local function can provide information on the overall evolution:

$$F\left(c_{t+1}^{i}\right)=f\left(c_{t}^{i-r},\cdots,c_{t}^{i},\cdots,c_{t}^{i+r}\right)$$

where c_t^i stands for the cell located at i. As of this step, a CA model is completely established.

7.4.2.2 Classic CA Model

In the development of CA, scholars have constructed a variety of different kinds of CA models (Karp 2001; Langton 1990; Wolfram 2002). The following typical CA models have played significant roles in the promotion of the study of the relevant theoretical methods. Hence, they can be considered as milestones in the historical development of CA.

7.4.2.2.1 J. Conway and the "Game of Life"

The Game of Life was designed by J.H. Conway in the late 1960s as a single-player computer game (Adamatzky 2010). It is somehow similar to the modern game of the chess of go in some aspects: there are black and white pieces. The cells in the Game of Life can have two states: alive or dead, denoted {0, 1}. The board of the game the chess of go is a regularly divided network, where it is determined whether each player is alive or dead by the spatial distribution of the black and white pieces. Similarly, the Game of Life is also a regularly divided network, where each cell is placed within a network grid just like how the chess pieces are placed. That is different from the game the chess of go, where pieces are positioned at the cross points of the network lines. The state of whether a particular cell is alive or dead is similarly determined by its local spatial structure, with easier rules to follow. In the following, let us look at the rules and how to play the Game of Life.

Step 1: Each cell has two states 0 and 1, where 0 stands for "dead" and 1 "alive."
Step 2: Each cell takes the adjacent eight cells as its neighbors; that is the form of Moore neighbors.
Step 3: The state of a cell, either alive or dead, is determined by its own state and the states of its surrounding eight neighbors. In particular, if the state of the cell is "alive" and two or three of its eight neighbors are also "alive," then the state of this cell at the next time moment stays at "alive." Otherwise, this cell would be "dead" at the next time moment. Also, if the current state of a cell is "dead," and out of its eight neighbors, exactly three of them are "alive," then the state of this cell at the next time moment will be "alive"; otherwise, it would stay "dead" for the next time moment.

Although the rules of the Game of Life seem very simple, this game stands for a model of CA that can produce dynamic patterns and dynamic structures, leading to rich and interesting graphs. The optimization of the Game of Life is dependent on the distribution of the initial values of the cells. For a given distribution of the initial cell states, after several steps of operation, some patterns will disappear, while some other patterns will stay invariant. Some patterns would reappear one after another periodically. Some others meander along. Some keep on their directional movements in a fashion similar to the blocks in a military parade that float ahead uninterrupted.

This model of the Game of Life has been practically applied in many areas. Its rules of evolution roughly describe the laws of survival and reproduction of biological masses. When the density of life is too small (the number of adjacent cells is less than 2), due to the problem of loneliness, lack of breeding opportunities, and lack of mutual assistance, there will appear a crisis of life, where the values of cellular states change from 1 to 0 inevitably. When life density is too large (the number of adjacent cells is greater than 3), due to such problems as environmental degradation, resource shortages, and competition for survival, there will also appear a crisis of life, where state values of cells change from 1 to 0. Only when the density of life is moderate (the number of adjacent cells is either 2 or 3) can individual cells possibly survive by maintaining their state values at 1 and reproduce newer generations by switching state values from 0 to 1. It is because of its capability of simulating the survival, extinction, competition, and other complex phenomena of life that it is named the Game of Life. Conway further demonstrates (Conway and Sloane 1988; Adamatzky 2010) that this CA has a computing power to go through the universal Turing machine and is equivalent to a Turing machine. That is to say, when given an appropriate initial condition, the Game of Life model could simulate any kind of computer.

7.4.2.2.2 Lattice Gas Automata

Lattice gas automata (LGA) are a kind of CA particularized in fluid mechanics and statistical physics. They represent a success of application of CA in scientific research. In terms of the Game of Life, LGA place more emphasis on the practicality of the model. They employ the dynamic characteristics of CA to simulate the movements of fluid particles.

LGA stand for a specific model of CA. In other words, they represent a generalized CA model, known as extended CA. As an example, let us look at the lattice gas model of the early version to describe its characteristics as follows.

Step 1: Because fluid particles do not easily disappear from the model space, this characteristic requires that each LGA be a reversible CA model.
Step 2: Neighbors in the LGA model usually take the Margulos type. That is, the neighbors' rules are based on a 2 × 2 grid space and can be described as follows:

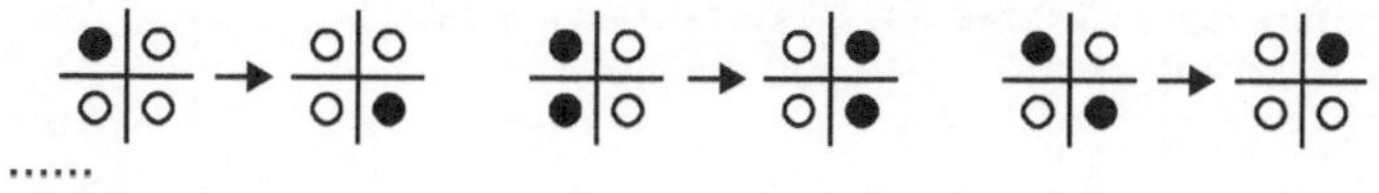

where each black ball represents a fluid particle, while each white ball represents an empty cell.

It can be seen that LGA are different from the general CA model. They do not take individual cells, known as central cells, as their objects of research by considering their state conversions. Instead, they are concerned with square blocks, each of which contains four cells.
Step 3: In accordance with the previous rules and model of neighbors, after the calculation is completed once, the 2 × 2 templates need to slide along the diagonal direction and be counted once again. Hence, the movement of a fluid particle requires a two-step $t \to t+1 \to t+2$ to complete.

In terms of time and space, LGA possess particular characteristics when compared to other models of CA. The research of LGA has become an important area of fluid dynamics, making it almost entirely independent of the studies of CA.

7.4.2.2.3 Odd–Even Rule Model

The corresponding rules are given in the following.

Step 1: Each cell has only two states: either 0 or 1.
Step 2: For each cell, compute the sum of its four neighbors that are located on top, bottom, left, and right, denoted by $M(i, j)$.
Step 3: The state of the cell at the next time moment is determined by the parity of $M(i, j)$. In particular, when $M(i, j)$ is even, the state of the cell at the next time moment is 0; when $M(i, j)$ is odd, the state of the cell at the next time moment is 1.

7.4.2.2.4 Forest Fire Model

The rules of the model are listed as follows.

Step 1: Each cell has three different possible states. State 0 stands for space, state 1 the burning of tree(s), and state 2 tree(s).
Step 2: If the states of one or more of the four neighbors are 1 (burning) and the state of cell itself is 2 (tree), then the state of the cell at the next time moment is 1 (burning).
Step 3: A cell of a forest is in state 2 (tree). The probability for this cell to catch on fire, say, caused by lightning, is very low, such as 0.00005.
Step 4: A burning cell (in state 1) would change into a space (state 0) in the next time moment.
Step 5: Each empty cell would grow with a low probability, say, 0.01, back into the forest in simulation.

7.4.2.3 Case Demonstrations of CA

7.4.2.3.1 Case 1: Even–Odd Rules

Figure 7.1 shows the resultant figures under different numbers of iteration following the even–odd rules. The following are the relevant MATLAB® source codes.

```
************************************
%%% Matlab Code for even-odd rule
N = 512;
t1 = 4;t2 = 6;
S = zeros(N);
S(fix(29*N/59):fix(30*N/59),fix(29*N/59):fix(30*N/59)) = 1;
  %Initialize the state matrix
chushi = imshow(1-S,[]);
axis square;
Sm = zeros(N+2);
k_step = 88; % the steps of iteration, it could be changed for different image
for k = 1: k_step
Sm(2:end-1,2:end-1) = S;
M = Sm(1:end-2,2:end-1)+Sm(3:end,2:end-1)+Sm(2:end-1,1:end-2)+Sm(2:end-1,3:end);%the summation
S = mod(M,2);
set(chushi,'cdata',1-S);
pause(0.01);
drawnow
end
************************************
```

7.4.2.3.2 Case 2: Excitable Media

Figure 7.2 shows the iterative nature of an excited media. In the middle of the figure are the words "System Science." For the sake of convenience for the reader to recreate the animation, the MATLAB source codes are provided below.

```
************************************
%%% Matlab Code for excitable media model
clf;
clear all
n = 128;
z = zeros(n,n);
cells = (rand(n,n))<.1; %relation operation, true for1, false for 0
sum = z;
imh = image(cat(3,cells,z,z)); %Construct the initial RGB image
set(imh, 'erasemode','none') %set the erase mode
axis equal
axis tight
x = [2:n-1]; y = [2:n-1];
t = 6;  %the boundary between the active state and very active state
t1 = 3; % if > = 3 the state is changed to 1
```

FIGURE 7.1 Different iteration figures of the iterative step by even–odd rule. (a) Step = 12; (b) step = 46; (c) step = 88; (d) step = 111; (e) step = 233; (f) step = 250; (g) step = 442; (h) step = 510; (i) step = 980; (j) step = 1007.

FIGURE 7.2 Iterative nature of the excited state media.

```
for i = 1:1200 % make action shot
sum(x,y) = ((cells(x,y-1)>0)&(cells(x,y-1)<t))+((cells(x,y+1)>0)&(cells(x,y+1)<t)) +...
    ((cells(x-1, y)>0)&(cells(x-1, y)<t))+((cells(x+1,y)>0)&(cells(x+1,y)<t)) +...
    ((cells(x-1,y-1)>0)&(cells(x-1,y-1)<t))+((cells(x-1,y+1)>0)&(cells(x-1,y+1)<t)) +...
    ((cells(x+1,y-1)>0)&(cells(x+1,y-1)<t))+((cells(x+1,y+1)>0)&(cells(x+1,y+1)<t));
cells = ((cells = =0)&(sum> = t1))+...
        2*(cells = =1) +...
        3*(cells = =2) +...
        4*(cells = =3) +...
        5*(cells = =4) +...
        6*(cells = =5) +...
        7*(cells = =6) +...
        8*(cells = =7) +...
        9*(cells = =8) +...
        0*(cells = =9);
set(imh, 'cdata', cat(3,z,cells./10,z))
drawnow
end
***********************************
```

7.4.2.3.3 Case 3: Forest Fire

Figure 7.3 shows the evolutionary figures for the start and spread of a forest fire. Again, the MATLAB source codes are provided below.

```
***********************************
%%% Matlab Code for forest fire demonstration
clf
clear all
n = 100;
Plightning =.000005; %the probability of fire caused by natural factors
Pgrowth =.01; %the grow probability of trees
z = zeros(n,n);
o = ones(n,n);
veg = z;
sum = z;
imh = image(cat(3,z,veg*.02,z));
set(imh, 'erasemode', 'none')
axis equal
axis tight
for i = 1:3000
sum = (veg(1:n,[n 1:n-1]) = =1) + (veg(1:n,[2:n 1]) = =1) +...
(veg([n 1:n-1], 1:n) = =1) + (veg([2:n 1],1:n) = =1) ;
veg = 2*(veg = =2)-((veg = =2)&(sum>0 | (rand(n,n)<Plightning))) +...
```

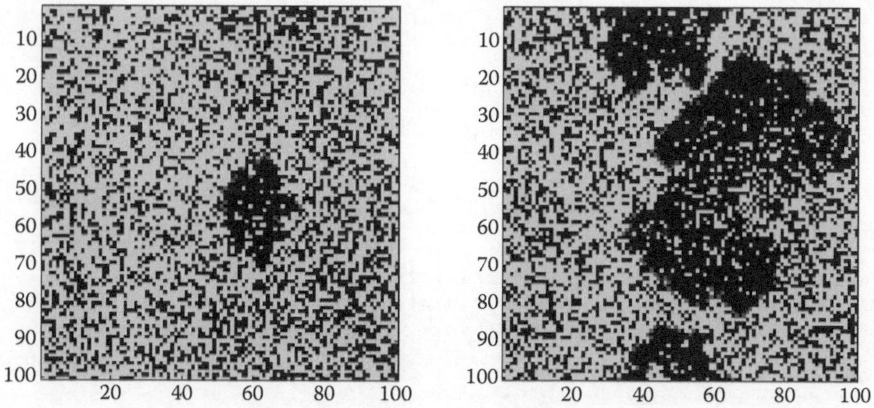

FIGURE 7.3 Start and spread of a forest fire.

```
2*((veg = =0)&rand(n,n)<Pgrowth);
set(imh, 'cdata', cat(3,(veg = =1),(veg = =2),z))
drawnow
end
***********************************
```

7.4.2.3.4 Case 4: Movement of Gas Molecules

Figure 7.4 simulates the movements of gas molecules. There is gas in the airtight container on the left and a vacuum in the airtight container on the right. These two containers are connected through a small hole in the middle. What are shown are the start and the end moments of the experiment. Also, when the experiment is finished, the molecules in both the containers are evenly distributed. The relevant MATLAB source codes are given below.

```
***********************************
%%% Matlab Code for the movement of gas molecules
%HPP-gas
clear all
clf
nx = 52;%must be divisible by 4
ny = 100;
z = zeros(nx,ny);
o = ones(nx,ny);
sand = z ;
sandNew = z;
gnd = z ;
diag1 = z;
diag2 = z;
and12 = z;
```

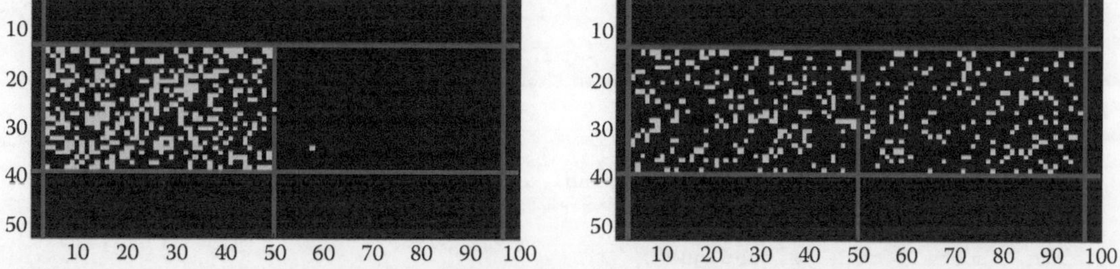

FIGURE 7.4 Simulating the movement of gas molecules. What are shown are for iteration steps 10 and 1000, respectively.

```
or12 = z;
sums = z;
orsum = z;
gnd(1:nx,ny-3) = 1 ;% right ground line
gnd(1:nx,3) = 1 ;% left ground line
gnd(nx/4:nx/2-2,ny/2) = 1;%the hole line
gnd(nx/2+2:nx,ny/2) = 1;%the hole line
gnd(nx/4, 1:ny) = 1;%top line
gnd(3*nx/4, 1:ny) = 1 ;%bottom line
%fill the left side
r = rand(nx,ny);
sand(nx/4+1:3*nx/4-1, 4:ny/2-1) = r(nx/4+1:3*nx/4-1, 4:ny/2-1)<0.3;
%sand(nx/4+1:3*nx/4-1, ny*.75:ny-4) = r(nx/4+1:3*nx/4-1, ny*.75:ny-4)<0.75;
%sand(nx/2,ny/2) = 1;
%sand(nx/2+1,ny/2+1) = 1;
imh = image(cat(3,z,sand,gnd));
set(imh, 'erasemode', 'none')
axis equal
axis tight
for i = 1:1000
p = mod(i,2);%margolis neighborhood
%upper left cell update
xind = [1+p:2:nx-2+p];
yind = [1+p:2:ny-2+p];
%See if exactly one diagonal is ones
%only (at most) one of the following can be true!
diag1(xind,yind) = (sand(xind,yind) = =1) & (sand(xind+1,yind+1) = =1) &...
(sand(xind+1,yind) = =0) & (sand(xind,yind+1) = =0);
diag2(xind,yind) = (sand(xind+1,yind) = =1) & (sand(xind,yind+1) = =1) &...
(sand(xind,yind) = =0) & (sand(xind+1,yind+1) = =0);
%The diagonals both not occupied by two particles
and12(xind,yind) = (diag1(xind,yind) = =0) & (diag2(xind,yind) = =0);
%One diagonal is occupied by two particles
or12(xind,yind) = diag1(xind,yind) | diag2(xind,yind);
%for every gas particle see if it near the boundary
sums(xind,yind) = gnd(xind,yind) | gnd(xind+1,yind) |...
               gnd(xind,yind+1) | gnd(xind+1,yind+1) ;
%cell layout:
%x,yx+1,y
%x,y+1x+1,y+1
%If (no walls) and (diagonals are both not occupied)
%then there is no collision, so move opposite cell to current cell
%If (no walls) and (only one diagonal is occupied)
%then there is a collision so move ccw cell to the current cell
%If (a wall)
%then don't change the cell (causes a reflection)
sandNew(xind,yind) =...
(and12(xind,yind) & ~sums(xind,yind) & sand(xind+1,yind+1)) +...
(or12(xind,yind) & ~sums(xind,yind) & sand(xind,yind+1)) +...
(sums(xind,yind) & sand(xind,yind));
sandNew(xind+1,yind) =...
(and12(xind,yind) & ~sums(xind,yind) & sand(xind,yind+1)) +...
(or12(xind,yind) & ~sums(xind,yind) & sand(xind,yind))+...
(sums(xind,yind) & sand(xind+1,yind));
sandNew(xind,yind+1) =...
(and12(xind,yind) & ~sums(xind,yind) & sand(xind+1,yind)) +...
(or12(xind,yind) & ~sums(xind,yind) & sand(xind+1,yind+1))+...
(sums(xind,yind) & sand(xind,yind+1));
sandNew(xind+1,yind+1) =...
        (and12(xind,yind) & ~sums(xind,yind) & sand(xind,yind)) +...
        (or12(xind,yind) & ~sums(xind,yind) & sand(xind+1,yind))+...
        (sums(xind,yind) & sand(xind+1,yind+1));
sand = sandNew;
set(imh, 'cdata', cat(3,z,sand,gnd))
drawnow
end
***********************************
```

7.5 CONSERVATION LAW OF INFORMATIONAL INFRASTRUCTURE

One of the key reasons why complexity is difficult to deal with is that when one faces a complex system, he or she tends to think within the framework of modern science and tries to attack the problem using the available tools developed on the basis of continuity and differentiability.

In this section, we show that when faced with a complex system, one really should first think about how systems thinking and the systemic logic of reasoning might help. When the tool is right, the seemingly complex matters will immediately become straightforward with much less complication.

7.5.1 Intuitive Understanding of Systems

From a practical point of view, a system is what is distinguished as a system (Klir 1985). From a mathematical point of view, a system is defined as follows (Lin 1987): S is a (general) system, provided that S is an ordered pair (M, R) of sets, where M is the set of objects of the system S and R is a set of some relations on the set M. The sets M and R are called the object set and the relation set of the system S, respectively. The idea of using an ordered pair of sets to define the general system is to create the convenience of comparing systems. In particular, when two systems S_1 and S_2 are given, by writing each of these systems as an ordered pair (M_i, R_i), $i = 1, 2$, we can make use of the known facts of mathematics to show that $S_1 = S_2$, if and only if $M_1 = M_2$ and $R_1 = R_2$. When two systems S_1 and S_2 are not equal, then with their ordered pair structures, we can readily investigate their comparisons, such as how and when they are similar or congruent or one is structurally less than the other, and other properties between systems. For more details about this end, please see Chapter 2 or the work of Lin (1999).

By combining these two understandings of general systems, we can intuitively see the following: Each thing that can be imagined in human minds is a system according to Klir's definition so that this thing would look the same as that of an ordered pair (M, R) according to Lin's definition. Furthermore, relations in R can be about some information of the system, its spatial structure, its theoretical structure, etc. That is, there should be a law of conservation that reflects the uniformity of all tangible and imaginable things with respect to

1. The content of information
2. Spatial structures
3. Various forms of movements, etc.

In the rest of this section, to support this intuition of (general) systems, we will look at examples from several different scientific disciplines. The presentation in this section is mainly from Ren et al. (1998).

7.5.2 Physical Essence of Dirac's Large Number Hypothesis

Dirac (1937), the founder of quantum mechanics, proposed the well-known large number hypothesis. It says that the ratio of the static electrical force and the universal gravitation in an H_2 atom is given as follows:

$$\frac{e^2}{Gm_pm_e} = 2\times 10^{39}, \tag{7.1}$$

where e^2 is the static electric force in an H_2 atom, G is the gravitational constant, m_p is the mass of the proton, and m_e is the mass of the electron. In terms of the atomic unit, the age of the universe is 2×10^{39}. Dirac considered these two nondimensional large numbers as being very close and believed that it could not be a coincidence and that it must stand for something fundamental.

Here, it can be seen that one important contribution Dirac made is that he formally established a connection between the universe and the microscopic world. However, what is the most essential physical meaning of such a connection of the two worlds of different scales? To this end, Dirac concluded that the large number hypothesis implies that the gravitational "constant" G decreases as time advances and causes various matters to be created. Now, let us explore the meaning of the large number hypothesis from a different angle. According to Allen (1976), in an H_2 atom, the static electrical force is

$$e^2 = 2.307113 \times 10^{-19} \text{ (static electricity unit)},$$

the gravitational constant is

$$G = 6.672 \times 10^{-8} \text{ dyne cm}^2/\text{g},$$

the mass of the proton is

$$m_p = 1.6726485 \times 10^{-24} \text{ g},$$

the mass of the electron is

$$m_e = 9.109534 \times 10^{-28} \text{ g},$$

the speed of light is

$$c = 2.99792458 \times 10^{10} \text{ cm s}^{-1},$$

and the radius of a classical electron is

$$r_e = e^2/(m_e c^2) = 2.817938 \times 10^{-13} \text{ cm}.$$

Based on the work of Pan (1980), let us take the Hubble constant

$$H = 55 \text{ km s}^{-1} \text{ Mpc} = 1.782428367 \times 10^{-18} \text{ s}^{-1}.$$

Now, from all these data values, we can compute the following:

1. In an H_2 atom, the ratio of the static electrical force and the universal gravitation is

$$\frac{e^2}{Gm_p m_e} = 2.269 \times 10^{39}. \tag{7.2}$$

2. The ratio of the age of the universe (= $1/H$, approximately a Hubble age) and the time $\left(\frac{e^2/(m_e c^2)}{c}\right)$ needed for a light beam to travel through the distance equal to the radius of an electron is given as follows:

$$\frac{1/H}{e^2/(m_e c^2)/c} = 5.96866 \times 10^{40}, \tag{7.3}$$

which can be rewritten as the ratio of the "radius of the universe" (= c/H, the Hubble distance) and the radius of an electron $\frac{e^2}{m_e c^2}$ as follows:

$$\frac{c/H}{e^2/(m_e c^2)} = 5.96866 \times 10^{40}. \tag{7.4}$$

From Equations 7.2 and 7.3 or Equations 7.2 and 7.4, it can be seen that these two large numbers differ by only ten. Therefore, for numbers of such a magnitude, they can be seen as approximately equal. At this junction, we can tell that some undiscovered important physical essence might be very well implied by such a uniformity in the structural information of the microcosm and the macrocosm. If these two quantities are seen as roughly equal, that is,

$$\frac{e^2}{Gm_p m_e} \approx \frac{1/H}{e^2/(m_e c^2)/c} \approx \frac{c/H}{e^2/(m_e c^2)}. \tag{7.5}$$

Cross-multiplying Equation 7.5 gives

$$Gm_p m_e \times \frac{1}{H} \approx e^2 \times \frac{e^2/(m_e c^2)}{c} \tag{7.6}$$

or

$$Gm_p m_e \times \frac{c}{H} \approx e^2 \times \frac{e^2}{m_e c^2}. \tag{7.7}$$

The meaning of Equations 7.6 and 7.7 is that the product of the universal gravitation and the age of the universe (or the "radius of the universe") approximately equals the product of the static electrical force (in an H_2 atom) and the time for a light beam to travel the distance of the radius of an electron (or "the radius of an electron").

Since the measures for the universe are connected by the universal gravitation, and the measures for atoms in the microcosm are connected by electromagnetic force, the physical meaning of Equations 7.6 and 7.7 becomes clear: the product of physical quantities of the universal scale approximately equals the product of relevant physical quantities of the microcosm. That is to say that there is uniformity between the universe and the atomic world. Does this physical meaning of the large number hypothesis imply the following—besides the unification of the four basic physical forces in the physical world, electromagnetic interaction, weak interaction, strong interaction, and gravitational interaction, there might be a grand unification in which the products of some relevant physical quantities in the universal scale and in the microscale approximately equal a fixed constant?

To this end, we have the following facts: according to Xian and Wang (1987), the radius of an H_2 atom (Bohr radius) is given by

$$r_y = 0.529 \times 10^{-8}\ \text{cm},$$

the course velocity (reaction rate) of electromagnetic interactions is

$$V_{ce} = 10^{16}\text{–}10^{19}\ \text{s}^{-1},$$

the forcing distance of strong interactions satisfies

$$r_q \lesssim 10^{-13}\,\text{cm},$$

and the course velocity of strong interactions is given by

$$V_{cq} = 10^{21}\text{–}10^{23}\,\text{s}^{-1}.$$

Thus, the following are obtained:

$$r_y \times V_{ce} = 0.529 \times 10^{8}\text{–}0.529 \times 10^{11}\,\text{cm s}^{-1} \tag{7.8}$$

and

$$r_q \times V_{cq} \lesssim 10^{8}\text{–}10^{10}\,\text{cm s}^{-1}. \tag{7.9}$$

That is, we have obtained

$$r_y \times V_{ce} \approx r_q \times V_{cq}. \tag{7.10}$$

The meaning of Equation 7.10 is that the product of some physical quantities of electromagnetic interactions approximately equals that of relevant physical quantities of strong interactions. This end implies that there exists uniformity between electromagnetic interactions and strong interactions.

7.5.3 Mystery of Solar System's Angular Momentum

The mystery of our solar system's angular momentum can be stated as follows: In the solar system, the total mass of all planets accounts for 0.135% of the total mass of the solar system, while the total angular momentum of all planets accounts for 99.421% of the total angular momentum of the solar system. In other words, the mass of the sun amounts to 99.865% of the mass of the solar system, while the angular momentum of the sun amounts to only 0.579% of the total angular momentum of the solar system. Two natural questions here are the following: How did this mystery of the solar system's angular momentum form? Are the masses and the relevant angular momentums related?

Here, in this subsection, the masses and the relevant angular momentums will be treated in the same fashion as above: multiply them together; see Ren and Hu (1989) for more details. In the work of Dai (1979), the masses and relevant angular momentums with respect to the center of mass of the solar system of the sun and the planets in the solar system are given. Thus, the sum of all the products of the masses (m_p) and the angular momentums (J_p) of all the planets can be obtained as follows:

$$\sum_p m_p J_p = 4.144 \times 10^{88}\,\text{g}^2\,\text{cm}^2\,\text{s}^{-1}, \tag{7.11}$$

and the product of the mass $m\oplus$ and the angular momentum $J\oplus$ with respect to the center of mass of the solar system of the sun can be obtained as follows:

$$m\oplus \times J\oplus = 4.143 \times 10^{88}\,\text{g}^2\,\text{cm}^2\,\text{s}^{-1}. \tag{7.12}$$

The numerical values in Equations 7.11 and 7.12 are very close and can be seen as the same. That is, we have seen the following law of conservation: the product of the mass and the angular momentum of the sun equals the sum of the products of the masses and the relevant angular momentums of all the planets. In this way, the mystery of the solar system's angular momentum can be resolved as follows: under the conditions of time and space of the solar system, there exists the conservation of phenomenon of equal mass–angular momentum products between the sun and the planets. Hence, it is very possible that the stability of the solar system is realized through the law of conservation: equal mass–angular momentum products between the sun and the planets.

In terms of celestial mechanics, it can be shown (Ren and Hu 1989) that in a system consisting of two celestial bodies (a two-body problem), the motion of the smaller-mass body can be obtained by solving the dynamic equation that describes how the smaller-mass body (m) circles around the larger-mass body (M). From

$$J = mr^2 \frac{d\sigma}{dt} = m\sqrt{G(M+m)a\left(1-e^2\right)}, \tag{7.13}$$

where J stands for the orbital angular momentum, r is the orbital vector, $\frac{d\sigma}{dt}$ is the angular velocity, a is the orbital radius, and e is the orbital eccentric rate, by establishing a rectangular coordinate system with the origin located at the center of mass of the two-body system, we can derive the angular momentum J_1 of the center of the larger-mass celestial body and the angular momentum J_2 of the smaller-mass celestial body as follows:

$$J_1 = \frac{Mm}{(M+m)^2} J \tag{7.14}$$

and

$$J_2 = \frac{M^2}{(M+m)^2} J. \tag{7.15}$$

Since the center of mass of the two-body system is always located on the line segment connecting the two celestial bodies, it now follows obviously from Equations 7.14 and 7.15 that

$$MJ_1 = mJ_2. \tag{7.16}$$

That is, the predicted law of conservation has been obtained: the product of the mass and the angular momentum with respect to the center of mass of the two-body system of the larger-mass celestial body (center body) is the same as that of the smaller-mass celestial body (circling body).

As an example, let us look at the two-body system of the earth and the moon. Based on the numerical values of the masses and the angular momentums with respect to the center of mass of the earth–moon system (Dai 1979), we can calculate and obtain that

$$J_2 = 2.786 \times 10^{41} \text{ g cm}^2 \text{ s}^{-1}$$

$$J_1 = 3.4266 \times 10^{38} \text{ g cm}^2 \text{ s}^{-1}$$

and

$$MJ_1 = 2.0477 \times 10^{67} \text{ g}^2 \text{ cm}^2 \text{ s}^{-1} = mJ_2. \tag{7.17}$$

Equation 7.17 implies that in our earth–moon system, the law of conservation of identical mass–angular momentum products holds true.

For a system of a large-mass celestial body and many small-mass circling celestial bodies, it can be shown (Xian and Wang 1987) that the related multiplicative relationships of mass–angular-momentums approximately hold true in the form of additive components. Thus, it can be seen that in a celestial body system of simple mechanics, there is a new law of conservation of mass–angular momentum products between the central celestial body and the circling celestial bodies.

7.5.4 Measurement Analysis of Movements of Earth's Atmosphere

In meteorology, weather systems, representing movements of the earth's atmosphere, are classified as large (lar), medium (mid), small (li), and micro (mic) scale systems. These systems of different classifications have their scales of spatial level measurements (L) set at approximately 10^8, 10^7, 10^6, and 10^5 cm; their measures of vertical velocity (W) set at approximately 10^0, 10^1, 10^2, and 10^3 cm s^{-1}; and the measures of their life spans (τ) at approximately 10^6, 10^5, 10^4, and 10^3 s, respectively.

Similar to what has been done previously, let us now again multiply relevant quantities for weather systems of the same classification and obtain the following.

1. For the products of spatial level measures and relevant vertical velocities, we have

$$\begin{aligned} L_{lar} \times W_{lar} &\approx 10^8 \text{ cm}^2 \text{ s}^{-1} \\ &\approx L_{mid} \times W_{mid} \\ &\approx L_{li} \times W_{li} \\ &\approx L_{mic} \times W_{mic}. \end{aligned} \tag{7.18}$$

2. For the products of life spans and relevant vertical velocities, we have

$$\begin{aligned} \tau_{lar} \times W_{lar} &\approx 10^6 \text{cm}^2 \\ &\approx \tau_{mid} \times W_{mid} \\ &\approx \tau_{li} \times W_{li} \\ &\approx \tau_{mic} \times W_{mic}. \end{aligned} \tag{7.19}$$

Equations 7.18 and 7.19 obviously mean that no matter which classification of a weather system is in, the product of its spatial level measure and its vertical velocity is always approximately equal to a fixed constant, and the same holds true for the product of the system's life span and its vertical velocity. That is to say, in the earth atmosphere, there also exists a law of conservation of products between different spatial measurements and relevant vertical velocities or between different life spans and relevant velocities.

7.5.5 Conservation of Informational Infrastructure

As is well known, laws of conservations are the most important laws of nature. For example, in classical physics, there are many established laws of conservation dealing with many important concepts, such as mass, energy, momentum, moment of momentum, electric charge, etc. The research of modern physics indicates that each moving object in an even and isotropic space and time, no matter whether the space is microscopic, or mesoscopic, or macroscopic, a particle or a field, must follow the laws of conservation of energy, momentum, and moment of momentum. That is, a unification of space has been achieved, which is called a Minkowski space. In Einstein's relativity theory, the concepts of mass, time, and space have been closely connected, and the mass–energy relation realizes the unification of the concepts of mass and energy.

The discussion above has shown that between the macrocosm and the microcosm, between the electromagnetic interactions of atomic scale and the strong interactions of quark's scale, between the central celestial body and the circling celestial bodies of celestial systems, and between and among the large-, medium-, small-, and microscales of the earth atmosphere, there also exist laws of conservation of products of spatial physical quantities. Based on this fact, we can further conclude, after considering many other cases, that there might exist a more general law of conservation in terms of structure, in which the informational infrastructure, including time, space, mass, energy, etc., is approximately equal to a constant. In symbols, this conjecture can be written as follows:

$$AT \times BS \times CM \times DE = a \tag{7.20}$$

or more generally,

$$AT^{\alpha} \times BS^{\beta} \times CM^{\gamma} \times DE^{\varepsilon} = a, \tag{7.21}$$

where α, β, γ, ε, and a are constants, and T, S, M, E, and A, B, C, D are, respectively, time, space, mass, energy, and their coefficients. These two formulas can be applied to various conservative systems of the universal, macroscopic, and microscopic levels. The constants α, β, γ, ε, and a are determined by the initial state and properties of the natural system of interest.

In Equation 7.20, when two (or one) terms of choice are fixed, the other two (or three) terms will vary inversely. For example, under the conditions of low speed and the macrocosm, all the coefficients A, B, C, and D are equal to 1. In this case, when two terms are fixed, the other two terms will be inversely proportional. This end satisfies all principles and various laws of conservation in the classical mechanics, including the laws of conservation of mass, momentum, energy, moment of momentum, etc. Thus, the varieties of mass and energy in this case are reflected mainly in changes in mass density and energy density. In the classical mechanics, when time and mass are fixed, the effect of a force of a fixed magnitude becomes the effect of an awl when the cross section of the force is getting smaller. When the space and mass are kept unchanged, the same force of a fixed magnitude can have an impulsive effect, since the shorter the time the force acts, the greater the density of the energy release will be. When time and energy are kept the same, the size of the working space and the mass density are inversely proportional. When the mass is kept fixed, shrinking acting time and working space at the same time can cause the released energy density to reach a very high level.

Under the conditions of relativity theory, that is, under the conditions of high speeds and great amounts of masses, the coefficients in Equation 7.20 are no longer equal to 1, Equation 7.21 becomes more appropriate, and the constants A, B, C, D, and a and the exponents α, β, γ, and ε satisfy relevant equations in relativity theory. When time and space are fixed, the mass and energy can be transformed back and forth according to the well-known mass–energy relation:

$$E = mc^2.$$

When traveling at a speed close to that of light, the length of a pole will shrink when the pole is traveling in the direction of the pole, and any clock in motion will become slower. When the mass is sufficiently great, light and gravitational deflection can be caused. When a celestial system evolves to its old age, gravitational collapse will appear, and a black hole will be formed. We can imagine based on Equation 7.20 that when our earth has evolved for a sufficiently long time, say in a billion or trillion years, the relativity effects would also appear. More specifically speaking, in such a great time measurement, the creep deformation of rocks could increase, and solids and fluids would have almost no difference so that solids could be treated as fluids. When a universe shrinks to a single point with the mass density infinitely high, a universe explosion of extremely high energy density could appear in a very short time period. Thus, a new universe is created.

7.5.6 Impacts of Conservation Law of Informational Infrastructure

If the perceived law of conservation of informational infrastructure holds true (all the empirical data, as presented earlier, seem to suggest so), its theoretical and practical significance is obvious.

The hypothesis of the law of conservation of informational infrastructure contains the following facts:

1. Multiplications of relevant physical quantities on either the universal scale or the microscopic scale approximately equal a fixed constant.
2. Multiplications of either electromagnetic interactions or strong interactions approximately equal a fixed constant.

In the widest domain of human knowledge of our modern time, this law of conservation deeply reveals the structural unification of different layers of the universe so that it might provide a clue for the unification of the four basic forces in physics. This law of conservation can be new evidence for the big bang theory. It supports the model for the infinite universe with border and the oscillation model for the evolution of the universe, where the universe evolves as follows:

$$\ldots \rightarrow \text{explosion} \rightarrow \text{shrinking} \rightarrow \text{explosion} \rightarrow \text{shrinking} \rightarrow \ldots$$

It also supports the hypothesis that there exists universes "outside of our universe." The truthfulness of this proposed law of conservation is limited to the range of "our universe," with its conservation constant being determined by the structural states of the initial moment of "our universe." (What is commented on here provides another empirical evidence for the systemic yoyo mode to be learned in Chapter 9).

All examples employed earlier show that to a certain degree, the proposed law of conservation holds true. That is, there indeed exists some kind of uniformity in terms of time, space, mass, and energy among different natural systems of various scales under either macroscopic or microscopic conditions or relativity conditions. Therefore, there might be a need to reconsider some classical theoretical systems so that our understanding about nature can be deepened. For example, under the time and space conditions of the earth's atmosphere, the traditional view in atmospheric dynamics is that since the vertical velocity of each atmospheric huge-scale system is much smaller than its horizontal velocity, the vertical velocity is ignored. As a matter of fact (Ren and Nio 1994), since the atmospheric density difference in the vertical direction is far greater than that in the horizontal direction, and since the gradient force of atmospheric pressure to move the atmospheric system 10 m vertically is equivalent to that of moving the system 200 km horizontally, the vertical velocity should not be ignored. The law of conservation of informational infrastructure, which holds true for all scales used in the earth's atmosphere, might provide conditions for a unified atmospheric dynamics applicable to all atmospheric systems of various scales. As a second example, in the situation of our earth where time and mass do not change, in terms of geological time measurements a (a sufficiently long time), can we imagine the force, which causes the earth's crust movements? Does it have to be as great as what is believed currently?

As for applications of science and technology, tremendous successes have been made in the macroscopic and microscopic levels, such as shrinking working spatial sectors, shortening the time length for energy releasing, and sacrificing partial masses (say, the usage of nuclear energy). However, the law of conservation of informational infrastructure might very well further the width and depth of applications of science and technology. For example, this law of conservation can provide a theory and thinking logic for us to study the movement evolution of the earth's structure and the source of forces or structural information that leads to successful predictions of major earthquakes, and to find the mechanisms for the formation of torrential rains and for the arrival of earthquakes (Ren 1996).

Philosophically speaking, the law of conservation of informational infrastructure indicates that in the same way as mass, energy is also a characteristic of physical entities. Time and space can be seen as the forms of existence of physical entities with motion as their basic characteristics. This law of conservation connects time, space, mass, and motion closely into an inseparable whole. Thus, time, space, mass, and energy can be seen as attributes of physical entities. With this understanding, the concept of mass is generalized, the wholeness of the world is further proved, and the thoughts of never-diminishing mass and never-diminishing universes are evidenced.

7.6 OPEN-ENDED PROBLEMS

1. How do you understand "dealing with the complexity as complexity"? Does it mean that it denies the simplification principle of scientific descriptions?
2. Please describe in your own words a system's complexity from the following two aspects: the structural evolution of the system and the functionality evolution of the system.
3. Based on one of the evolutionary characteristics of complex systems, use examples to illustrate the development tendency of the research on complex systems.
4. Construct an example to demonstrate the relationship between number 3 and complexity.
5. Realize one instance of cellular automata.

8 Complex Adaptive Systems

8.1 BASICS OF COMPLEX ADAPTIVE SYSTEMS

Complex systems represent one of the current main research directions in systems science, while complex adaptive systems (CASs) stand for a class of very representative complex systems. The theory of CASs is proposed by Holland (1996) based on his years of study on complex systems. The basic idea of this theory is that the complexity of CASs originates from the adaptation of individual elements or agents of the systems; it is the interactions between these individual agents and the environment and between these agents themselves that these agents evolve themselves and reshape the environment. The most important characteristic of CASs is their adaptation. That is, the systems' individual agents can directly communicate with the environment and with themselves, and through such communications, these agents "learn" and "accumulate experience" so that they constantly evolve and improve. Based on the learned lessons, these agents alter their internal structures and modify their patterns of behavior. The agents at the very bottom level, through mutual interactions and communications, can produce new structures, phenomena, and more complex behaviors for the higher level and/or the whole level of the system. The interactions and communications of the agents at the very bottom level could create additional levels, dissolve an existing level, or produce multiplicity in the functionality or appearance of the system. They could also lead to the formation of new aggregates, larger agents, etc.

Within a CAS, all individual agents situate in a greater common environment, while each one of them has to deal with its own respective local environment. Through their simultaneous and independent learning and evolution in order to adapt to their individual local environments, these agents manifest their intelligence. Therefore, these elements are also seen as intelligent agents. To meet the need of basic survival and better adapt to their individual local environments, the agents constantly adjust their behaviors and modify their own rules of operation. The various behaviors of adaptive agents reciprocatingly influence and change their respective environments. Combined with the laws of change of the environments, the dynamically evolving environments impose their rules of order and influence on the individual agents in the form of "constraints." With repetition of such reciprocating processes, individual agents and the environments are situated in a process of never-ending interaction, mutual influence, and coordinated evolution.

CASs represent a class of very commonly seen and very important complex systems. For such a class of systems, Holland, on the basis of genetic algorithms (GAs), which he first proposed, established the ECHO model to simulate and investigate the general behaviors of CASs. Scholars at the Santa Fe Institute then developed the corresponding SWARM simulation platform, a modeling tool, on the basis of Holland's model (1996).

8.1.1 SANTA FE INSTITUTE AND BIRTH OF CAS THEORY

Since the mid-twentieth century, with the improvement of productivity and the expansion of human intellectual horizons, the contents and perspectives of almost all disciplines have undergone tremendous changes. That provides the scientific community with inspirations and a great number of examples for furthering the understanding of complex systems. The pioneering works of Josiah Willard Gibbs (Hastings 1909) and Henri Poincare (Belliver 1956) predated the arrival of a brand new way of scientific thinking; the creative theories of Einstein (1987) and Max Planck (Heilbron

2000) broke the illusion that the "scientific building has been completed" and greatly expanded human vision. It was at this occasion that Bertalanffy (1968) once again in history restated that the "total is greater than the sum of its parts." That marked the official rise of a new way of scientific thinking, as represented by the systems thinking.

From the early works of Bertalanffy to the present, these 80 plus years of human history have witnessed how the modern man struggles constantly to foray into new models of scientific reasoning. From the angles of research ideas and focuses, this historical period can be roughly divided into three phases.

Phase I: In the 1940s and 1950s, the attention of the world of learning focused mainly on such areas as automation and control, information feedback, etc. At that time, a so-called "complex system" was in fact established on the background of machines with both Wiener (1948) and Shannon (1948) being the representatives. In this stage, the concepts and methodology of systems brought forward great successes in the areas of engineering and technology, ranging from various kinds of automatic machines to the systems engineering projects of large industrial organizations, leading to well-recognized scientific progress and contributions. However, for complex systems considered in biology, social sciences, etc., some attempts of application have experienced great difficulties. Hence, it indicates that the laws that govern the operation of complex systems still need to be further investigated.

Phase II: From the 1960s to the 1970s, as represented by Prigogine (1961) and Haken (1977), the scientific world turned its attention to such systems that contain very large numbers of elements that move irregularly and randomly. Thermodynamic systems are representative of such systems. Through the introduction of new concepts, such as self-organization, fluctuation, cooperative movement, etc., the scientific exploration of this stage revealed the rich and colorful contents of systems' evolution and development. Meanwhile, in the areas of chaos, fractal, nonlinear science, and others, more and more practical scenarios and examples provide strong supports for formulation of a new way of scientific thinking. The systems thinking and methods developed in this period provide important inspirations for the study of laser, biology, and other disciplines. However, applications attempted in areas of the social spheres experienced limited success. The reason for this end is that the elements of societies—individual human beings and business enterprises or firms—are not such particles that are unconscious and move randomly and aimlessly. Instead, these elements actively pursue after their respective targets; they have the capability of adaptation and learning; they adjust their behavioral patterns and organizational structures according to the information feedback from the environments (Holland 1996).

Phase III: On the tenth anniversary of the Santa Fe Institute in 1994, Holland made a presentation entitled "hidden order" on the Ulam lecture series. In this report, Holland (1994) proposed a relatively more complete theory on CASs on the basis of his years of study on complex systems (Holland 1994). The theory of CASs includes both macroscopic and microscopic levels. At the microscopic level, the basic concept of CAS theory is some adaptable and proactive individuals, referred to as agents. These agents comply with the general stimulus–response model in their interaction with the environment. Their so-called adaptability is shown in their capability to modify their behavioral patterns based on the effects of earlier patterns so that these agents could better survive in the environment. At the macroscopic level, systems, composed of such kinds of agents, evolve and grow along with the interactions between the agents and between the agents and their respective environments. These systems manifest various complex evolutionary processes of decomposition and emergence. This speech was later published as a book entitled Hidden Order: *How Adaptation Builds Complexity* (Holland 1996). The appearance of the CAS theory signals that the modern systems thinking has entered into the third stage, where biological and social systems were the main objects of investigation.

8.1.2 Central Thought of CAS Theory—Adaptation Creates Complexity

In the CAS theory, the elements of a system are known as adaptive agents, or agents for short. The so-called adaptability means that each agent can interact with the environment and with other

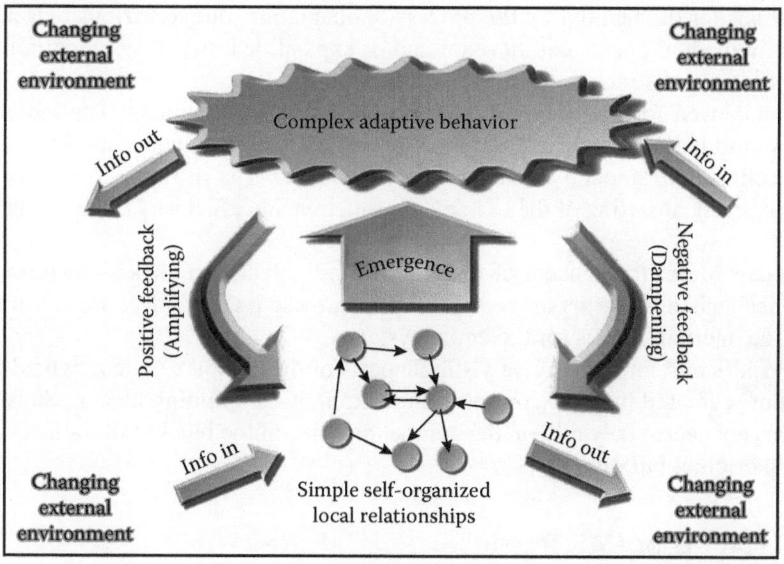

FIGURE 8.1 CAS. (http://en.wikipedia.org/wiki/File:Complex-adaptive-system.jpg, accessed on August 10, 2011.)

agents. The agents constantly "learn" or "accumulate experience" in their ongoing processes of interaction and change their internal structures and behavioral patterns according to the learned lessons. Therefore, the overall evolution and growth of the whole system, including the appearance and dissolution of layers, diversities, greater aggregates, larger agents, etc., gradually emerge.

The CAS theory maintains that it is the agents' proactiveness and repeated mutual interactions with the environment and themselves that the system is motivated to grow and evolve. The root causes for all macroscopic systems' changes and microscopic agents' aggregations and decompositions can be traced to the behavioral patterns of the agents. Holland summed up these proactive, repeated interactions between the individual agents and between the agents and the environment in a single word: "adaptation." That is the central idea underlying the CAS theory: adaptation creates complexity. The network representation of a CAS is shown in Figure 8.1.

8.1.3 Basic Concepts of CAS Theory

Based on his previous experience of studying GAs and systems simulation, Holland proposed the following seven factors that are particularly important in the adaptation and evolution of individual agents: aggregation, nonlinearity, flow, diversity, tag, internal model, and building block.

Aggregation means that through bonding, some individual agents are bonded together to form a multiagent aggregate. Because of such attributes that agents have, under certain conditions, and when a mutual acceptance exists, individual agents are bonded together to form a new individual—an aggregate. This new and larger individual acts within the system as a separate, independent agent.

Nonlinearity refers to the fact that changes in individual agents and their attributes do not follow any simple linear relationship. Especially in the repeated interactions between the system and the environment and between agents and their respective environments, this point becomes abundantly apparent.

Flow means that there exists material, energy, and information flows between individual agents and the system's environment. Whether the channels of these flows are widely open or not and at what rates these flows flow directly affect the evolution of the system.

Diversity stands for the fact that in the process of adaptation, due to various reasons, the differences between individual agents can develop and/or expand, leading to some definite differentiations. That represents a distinctive feature of CASs.

Tagging is employed for the agents to identify and to make selections. The individual agents' tags are very important in the study of the interactions between the agents and between the agents and their respective environments. Thus, no matter whether it is in the modeling of a particular system or in the practical setting of the system, the function and efficiency of tags must be seriously considered.

Internal models reveal the concept of levels. Each individual agent has its own particular complex internal mechanism. In terms of the whole system, these internal mechanisms are collectively referred to as the internal models for the entire system.

Building blocks stand for the relatively simple parts of the complex system, where the complexity of the system is created by the changing combination of the building blocks. Thus, the realistic complexity does not necessarily rely on the number and size of the blocks but, rather, on the reorganizations of the original building blocks.

8.1.4 Main Features of CAS Theory

The core assumption of the CAS theory, "adaptation creates complexity," is very epistemologically significant. It stands for an epistemological leap in the study of movements and laws of evolution of systems. This can be illustrated in four different aspects.

First of all, each adaptive agent is a proactive and living body. That is the key difference between the CAS theory and other modeling methods. This characteristic makes the CAS theory effectively applicable to the studies of the complex systems considered in economics, sociology, ecology, and other humanity areas, where other methods have experienced great difficulties.

From elements to adaptive agents, it is not a simple name change. For the composing parts of a system, they are generally known as elements, units, components, or subsystems. Contrary to those concepts of systems, overall situations, and wholes, the concepts of elements, units, and parts are introduced passively to describe localities. In this sense, the concept of adaptive agents places the proactiveness of individual elements at the height of the fundamental motivation underlying systems' evolution. Thus, agents become the starting point from which one investigates and observes the systems' macroscopic phenomena of evolution. This reversed logic of thinking clearly stands for a breakthrough. It is the simultaneous, proactive interactions and communications between the agents and between the agents and their respective environments that the systems' complexity is formed and developed. In this process of logical thinking, there are neither individual elements that are detached from the whole and the environment nor the abstract whole that is hanging way above the individual parts. The key here is the agents' proactiveness. The degree of agents' proactiveness determines the degree of the behavioral complexity of the whole system.

The proactiveness or adaptability as aforementioned is a very general, abstract concept. It does not necessarily mean being alive in the biological sense. As long as an individual entity is able to adjust its own structure and behavioral pattern according to the varying information received in its interactions with other entities, this entity can be seen as having proactiveness and adaptability. The purpose of adaptability is to survive and to grow. Thus, the question of "purpose" can also be comprehended and explained reasonably well in this context without any need to refer to theology.

Additionally, interactions and mutual influences between agents and between agents and their environments are the main motivation for systems to evolve and to develop. The traditional methods of modeling tend to place the internal attributes of agents on the dominating position without paying enough attention to the interactions between agents and between the agents and their environments. This advantage of the theory of CASs makes the CAS method applicable to the investigation of different scientific fields, where although the problems addressed might have objects of very different attributes, the interactions between the objects share many common properties.

This point of view of interactions is very instructive. When agents are said to be the basis of the whole, it does not mean that the isolated and separated agents form the foundation of the whole. If it were so, one would have returned to the point of view of reductionism. It is the interactions between the agents and between the agents and their environments that constitute the foundation of the whole. When one says that "the whole is greater than the sum of its parts," he or she talks exactly about the "added value" created by these interactions. The rich and colorful behavioral patterns of CASs come exactly from these increased values. The stronger the interactions are, the more complex and more varying the systems' evolution will be.

In addition, the interactions mentioned here mainly refer to those related to the individual agents. There are two special significances for emphasizing this very point. First, there is not any "representative" of the whole that is positioned above each and every agent. In terms of each individual agent, the effect of the whole is embodied in all other individual agents and the relevant interactions. Similarly, each agent also plays the role of the environment for the other agents. Speaking less rigorously, each agent "represents" the effect of the whole, while no individual agent can single-handedly represent the whole. This end well explains the dialectical unity between the whole and the individual agents. On the other hand, because of these interactions, there is a process of development from "equality" to "differentiation" in the relationship of the individual agents. That is to say, in the early stage of the system's evolution, potentials or potential capacities of the individual agents are roughly the same. In principle, each agent has the capability to evolve in various directions. Under the influence of the interactions, various factors and acting forces, including the effect of chance, make the agents evolve in their particular directions, creating a structure for the system while breaking the initial symmetry. In this way, the whole system becomes more complicated than before. This is a simplified description of systems' evolution from simplicity to complexity. In other words, interactions are "memorable"; they are embodied in the evolutionary changes of the structures and behavioral patterns of the individual agents so that they are "remembered" in particular ways in the insides of the individual agents.

Therefore, the study of CASs theoretically developed the thought of interactions that have been historically emphasized in systems science, making the improved concept more specific and practically implemented. Here, the thought of adaptability is introduced into systems research from biology. Evidently, that represents a practical enrichment to and expansion of the thinking logic of systems science.

Third, the thought of adaptability organically links together the macrocosm and microcosm. Through the interactions between agents and between agents and their respective environments, changes in individual agents become the foundation of changes in the whole system so that these changes of different levels are considered at the same time.

The extreme reductionist simply seeks the causes of macroscopic phenomena in the microscopic world, while denying the existence of any qualitative increase from the microscopic world to the macroscopic realm. Another commonly accepted practice is that statistical methods are treated as the only bridge crossing between the microcosm and the macrocosm. It should be recognized that the statistical methods, developed on the basis of probability theory, are indeed one of the important connections between the microcosm and the macrocosm. Some of the properties of macroscopic systems have, as a matter of fact, been understood through the attributes of the microscopic parts by using statistical methods. For example, the temperature of a gas is a reflection of the kinetic energy of the molecules; the overall nation's level of education is determined by the level of education of each individual citizen. That is clearly important and has indeed correctly reflected one aspect of the relationship between the microcosm and the macrocosm. However, the problem is this: Is this the only way to express the relationship between the macrocosm and the microcosm? In fact, if each of the organisms on earth is accidently joined together and formulated purely based on the laws of statistics, then since the time earth was initially born, a single protein molecule as of the present day would not have been produced. Hence, other than the laws of statistics, there must be other mechanisms or methods through which the microcosm and the macrocosm are connected. To this end, the theory of CASs has provided a new way of thinking.

If individual agents do not have their respective proactiveness, such as those molecules in a gas, as assumed so in thermodynamics, then to determine their individual or mass movements and the relevant relationships, one has to, without any choice, rely on the methods of statistics. The main laws that govern the operations of such systems have to be indeed those of statistics. However, if the individual agents are "alive," have their own proactiveness and adaptability, and "memorize" internally the previous experiences, then their movements and changes can no longer be adequately described by using only statistical methods. For example, the previously described process of differentiation cannot be readily specified using only statistical methods.

Therefore, in terms of the relationship between the microcosm and the macrocosm, the theory of CASs provides a new logic of reasoning that is beyond pure statistical methods. If this logic of reasoning is generalized, and if the microcosm and the macrocosm are seen as two relative layers, then the CAS theory provides a beneficial way of thinking for one to recognize, to comprehend, and to go across and between these layers.

Fourth, because of the introduction of the role of stochastic factors, the theory of CASs possesses additional description and expression capabilities.

Although stochastic factors are not unique characteristics of the CAS theory, this theory provides a particular way to deal with stochastic factors. Simply put, the CAS theory absorbs a lot of beneficial clues from the greatly varied phenomena of the biological world. The most representative of the clues is the so-called GA.

Generally, the common approach of considering stochastic factors is to introduce random variables. That is, at a certain stage of evolution, some alien stochastic factors are introduced, assuming that these factors follow some given forms of distribution and affect the process of the system's evolution. In such an approach, the effect of the stochastic factors is "temporary" and shown only at one specific time step. The factors only quantitatively influence some of the state parameters of the system. After the one step effect, only some of the state parameters are altered, while the laws of operation and the internal mechanism stay unchanged. Speaking figuratively, the system does not truly "evolve" drastically from the previous states. Evidently, that is exactly what has been said earlier; it is an expression of locality when the elements of the system are seen as being "dead."

The basic idea of GA lies in that stochastic factors affect not only the system's state but also the organizational structure and the pattern of behavior. Each "live," proactive agent can learn lessons from the past experiences and somehow "memorize" the experiences so that the future patterns of behavior are altered accordingly. Because of this, the CAS theory provides great potential to simulate the complex systems considered in biology, ecology, economics, sociology, etc., with evident advantage over the general statistical methods.

Specifically, GA is a global optimization computational scheme developed in recent years. It improves the adaptability of each individual by borrowing the view point of biological genetics through such functional mechanism as natural selections, inheritances, mutations, etc. This aspect reflects the evolutionary process of "natural selection where the fittest survives" in nature. Since the time when this method was initially developed in the 1960s, it has attracted a large number of researchers and has been quickly extended to such areas of research as optimization, search, machine learning, etc.

A sequential GA processes as follows:

1. Encode the problem to be solved.
2. Randomly initialize the population $X(0) = (x_1, x_2,\ldots,x_n)$.
3. For each individual x_i, $i = 1, 2, \ldots, n$, of the current population $X(t)$, calculate its fitness $F(x_i)$, $i = 1, 2, \ldots, n$, which represents the level of performance of the individual.
4. Apply the selection operator to produce the intermediate generation $X_r(t)$.

5. Apply other operators on $X_r(t)$ to produce a new generation $X(t + 1)$ of the population. The purpose of applying these operators is to expand the coverage of the limited few individuals so that the idea of a global search can be reflected.
6. Let $t = t + 1$. If the condition of termination is not met, continue the procedure by starting over from step 3.

The most commonly used GA operators can be summarized as follows.

1. *Selection operators:* Each selection operator selects a pair of individuals a time from the population according to some predetermined probability, where the probability p_i that the individual x_i is selected is directly proportional to its fitness value $F(x_i)$. The most common implementation is done by using the roulette wheel model.
2. *Crossover operators:* Each crossover operator crosses the gene chains of the pair of selected individuals according to another predetermined probability p_c to generate two new individuals with random locations of crossovers. Here, the probability p_c is a system parameter.
3. *Mutation operators:* Each mutation operator mutates each position of the gene chain of a newly obtained individual according to some predetermined probability p_m. As for a binary gene chain, which has been coded with 0 and 1, the mutation is about taking the opposite value.

Example 8.1

(A linearly constrained population). This example shows how to create a well-dispersed population that satisfies linear constraints and bounds. It also contains an example of a custom plot function.

The problem uses the following quadratic function as the objective:

$$f(x) = \frac{x_1^2}{2} + x_2^2 - x_1x_2 - 2x_1 - 6x_2.$$

This function can be found in lincontest6.m, which is a built-in procedure of MATLAB®. To see the codes of the function, simply enter "type lincontest6" in the command line.

The constraints are the following three linear inequalities:

$$\begin{cases} x_1 + x_2 \le 2 \\ -x_1 + 2x_2 \le 2. \\ 2x_1 + x_2 \le 3 \end{cases}$$

Also, the variables x_i are restricted to be positive. Next, we list the specific steps and their corresponding MATLAB source codes for this example.

1. Create a plot function file containing the following code:

```
**************************
function state = gaplotshowpopulation2(unused,state,flag,fcn)
% This plot function works in 2-d only
if size(state.Population,2) > 2
```

```
return;
end
if nargin < 4% check to see if fitness function exists
fcn = [];
end
% Dimensions to plot
dimensionsToPlot = [1 2];

switch flag
% Plot initialization
case 'init'
     pop = state.Population(:,dimensionsToPlot);
     plotHandle = plot(pop(:,1),pop(:,2),'*');
     set(plotHandle,'Tag','gaplotshowpopulation2')
     title('Population plot in two dimension','interp','none')
     xlabelStr = sprintf('%s%s','Variable ',...
         num2str(dimensionsToPlot(1)));
     ylabelStr = sprintf('%s%s','Variable ',...
         num2str(dimensionsToPlot(2)));
     xlabel(xlabelStr,'interp','none');
     ylabel(ylabelStr,'interp','none');
     hold on;

% plot the inequalities
     plot([0 1.5],[2 0.5],'m-.')% x1 + x2 < = 2
     plot([0 1.5],[1 3.5/2],'m-.');% -x1 + 2*x2 < = 2
     plot([0 1.5],[3 0],'m-.');% 2*x1 + x2 < = 3
% plot lower bounds
     plot([0 0], [0 2],'m-.');% lb = [0 0];
     plot([0 1.5], [0 0],'m-.');% lb = [0 0];
     set(gca,'xlim',[-0.7,2.2])
     set(gca,'ylim',[-0.7,2.7])

% Contour plot the objective function
     if ~isempty(fcn)% if there is a fitness function
     range = [-0.5,2;-0.5,2];
     pts = 100;
     span = diff(range')/(pts - 1);
     x = range(1,1): span(1) : range(1,2);
     y = range(2,1): span(2) : range(2,2);

     pop = zeros(pts * pts,2);
     values = zeros(pts,1);
     k = 1;
     for i = 1:pts
         for j = 1:pts
         pop(k,:) = [x(i),y(j)];
         values(k) = fcn(pop(k,:));
         k = k + 1;
         end
     end
     values = reshape(values,pts,pts);
     contour(x,y,values);
     colorbar
     end
% Pause for three seconds to view the initial plot
```

```
        pause(3);
case 'iter'
        pop = state.Population(:,dimensionsToPlot);
        plotHandle = findobj(get(gca,'Children'),'Tag',...
            'gaplotshowpopulation2');
        set(plotHandle,'Xdata',pop(:,1),'Ydata',pop(:,2));
end
***************************
```

This function plots the lines representing the linear inequalities and bound constraints, the contour curves of the fitness function, and the population as it evolves. It expects to have not only the usual inputs (options, state, flag) but also a function handle to the fitness function, @lincontest6, in this example.

2. Enter the constraints as a matrix and vectors at the command line:
   ```
   A = [1,1;-1,2;2,1]; b = [2;2;3]; lb = zeros(2,1);
   ```
3. Set options to use gaplotshowpopulation2, and pass in @lincontest6 as the fitness function handle:
   ```
   options = gaoptimset('PlotFcn',...
     {{@gaplotshowpopulation2,@lincontest6}});
   ```
4. Run the optimization using options:
   ```
   [x,fval] = ga(@lincontest6,2,A,b,[],[],lb,[],[],options).
   ```

Figure 8.2 shows the result of previous source codes, the linear constraints, bounds, contour curves of the objective function, and initial distribution of the population, where the population is biased to lie on the constraints.

At the end state (Figure 8.3), the population eventually converges to the minimum point.

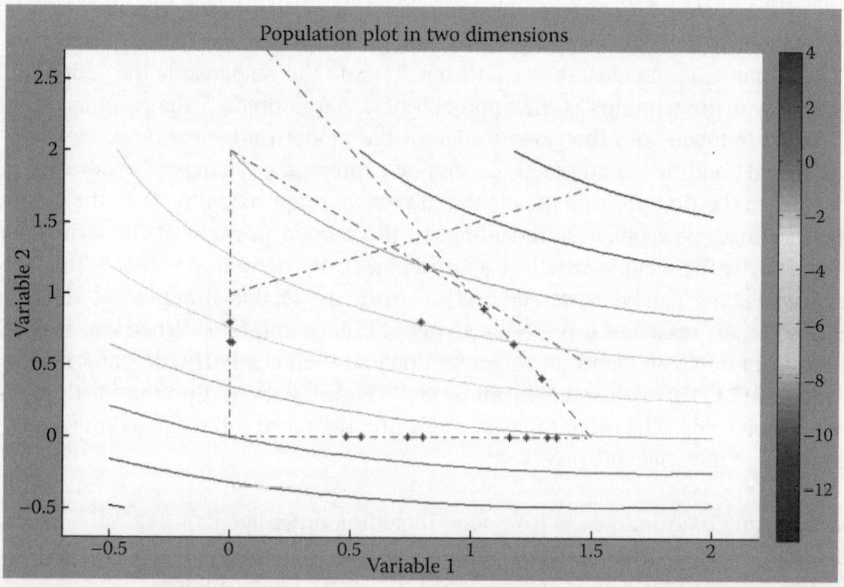

FIGURE 8.2 Initial state of the population.

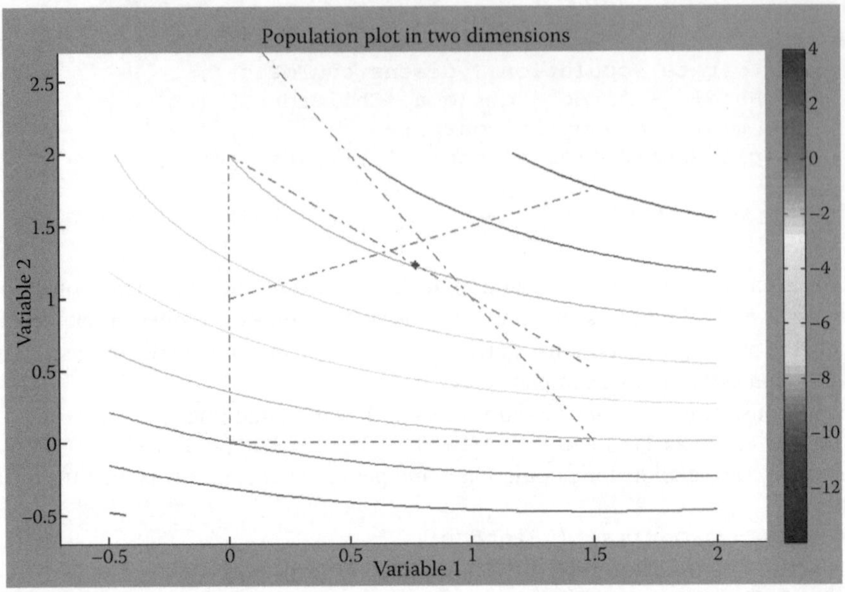

FIGURE 8.3 End state of the population.

8.2 HOW AN AGENT ADAPTS AND LEARNS

To further describe how an agent adapts and learns, Holland (1996) proposed the basic behavior model of agents in three steps: (1) establish the model of the performance system; (2) develop the credit assignment mechanism; and (3) provide a means to discover new rules.

8.2.1 Stimulus–Response Model

The goal of this step is to use a unified approach to represent the most basic behavioral patterns of the agents of the system of concern. The starting point is the basic stimulus–response model. For example, when a frog sees a bug flying across, it then sticks out its tongue to prey upon the bug. Here, the stimulus is the close-by "small bug," while the response is the "sticking out of the tongue." Similarly, if the stimulus is the approach of a "large object," the response might well be to "escape." In accordance with the general idea of the modern information processing, the rules, including the initial conditions and reactions, can be expressed as strings of symbols. For instance, if the symbol "0" in the first position means that no object is approaching, then the symbol "1" will stand for the fact that some object is intruding. In the second position of the string, the symbols "0" and "1", respectively, mean a small or a large object. By assigning symbols this way, the first stimulus discussed above can be expressed as "10," while the second stimulus as "11." Similarly, the reaction can also be expressed as a symbolic string of binary numbers. When two symbolic strings are linked, the front half will stand for the condition and the back half will stand for the response. In the language of GA, that is just a "chromosome" (Figure 8.4). Within the inside of each agent, there are many such rules. The more finer the rules are, the more sophisticated the agent will act.

The following are the concepts involved:

- Input: the stimulus from the environment, including other agents
- Output: the response of the individual agent, usually an action
- Rules: how a particular response is assigned to a particular stimulus
- Detector: the organ or device used to detect and receive the stimulus
- Reactor: the organ or device employed to respond to the stimulus

	Stimulus	Response
Rule:	IF (a small object closes in)	THEN (launch an attack)
	IF (a large object nears)	THEN (escape)
String:	10	01 ....
	11	10 ....

FIGURE 8.4 Representation of chromosomes.

The relationship between these concepts and the message flow among them are shown in Figure 8.5.

As generally expected, the reservoir of rules should contain rules consistent with each other without duplication so that for a given stimulus, there is only one response and without omission (to each stimulus, there must be a response). Otherwise, the reservoir would contain contradictory rules and become inconsistent, and the system would be seen as in an incorrect state. In such a case, there would be no essential difference from the commonly used general statement of "if…then…."

However, at this junction, the theory of CASs introduces an important idea. It maintains that such general expectation does not apply to the modeling of complex systems; on the contrary, the rules in the reservoir should be treated as hypotheses that need to be measured and tested. The evolution of a system represents exactly such a process within which multiple choices are made available. Therefore, contradictions, inconsistencies, and conflicts are indeed desirable and should not be avoided or eliminated. Thus, there should be enough choices for the rules (or chromosomes as known in GA) to take their different forms, among which there can be and there should be contradictions and inconsistencies. Of course, in order to practically operate such a reservoir of rules, a mechanism of comparison and selection needs to be established within the reservoir so that appropriate eliminations can take place. Also, that will be the task of the next step: credit confirmation.

The behavioral system illustrates the ability of the agent at a certain moment of time. It consists of three main parts: a detector, a set of IF/THEN rules, and a reactor. The detector stands for the agent's capability of extracting information from the external environment; the IF/THEN rules represent the agent's ability to handle and to process the collected information; and the reactor shows the agent's ability to react on the environment based on the collected information. These three parts are all abstract concepts with the details of any particular agent removed. Thus, they can be employed to study different kinds of agents.

By carefully analyzing the concept of detectors, one can further understand what is lost and what is gained in such a modeling process. For example, the detectors used in antibiotics are dependent on the local arrangements of chemical bonds, while the detectors of organisms are clearly corresponding to their sensory organs; the detecting function of commercial companies is completed

Message Sequence	Reservoir of Rules
Matches	
00	IF (00 ......) THEN (00001)
01	IF (01 ......) THEN (00010)
10	IF (10 ......) THEN (00011)
11	IF (11 ......) THEN (00100)
Rematches	
Detector	Reactor
Stimulus	Response

FIGURE 8.5 Stimulus–response system.

jointly by its various departments. In each scenario, particularly interesting questions exist regarding how the specific system extracts information from the environment. However, for our purpose here, let us just put these particular questions aside and pay attention to the produced information—the characteristics of the environment to which the agents are sensitive. By recognizing that any such information can be expressed by using a string of binary codes, which is referred to as a message, a natural and unified method of describing an agent's capability of extracting environmental information is obtained.

As for the agent's internal capability to process information, the same idea can be considered. Although the specific mechanisms can be various and take different forms, let us focus on the treatment of information. Combining the IF/THEN rules and collected messages leads to the expression shown in Figure 8.5: IF (there is a particular type of message), THEN (respond accordingly). As a result, already discarded are the details of specific mechanisms used by particular agents to process information. For example, when studying the genetic open–close process in the development of an embryo, the specified details of the mechanisms of inhibition and retro-inhibition are ignored. However, what are kept are a description of the relevant development and each stage's feedback of information. In short, a possibility is created; it enables computers to describe the system's capability of information processing. Because many rules may work at the same time, a natural method of describing the paralleled activities of the CAS of interest is established.

8.2.2 Identifying and Modifying Fitness

In order to compare rules and select the appropriate rules, the assumed credit of every rule has to be first quantified. Thus, each rule is assigned a particular value, known as the strength of or fitness (in the language of GA) of the rule. Each time when rules are needed, the system makes selections according to a predetermined method. The basic idea for selecting rules is to make selections according to the chosen probability so that rules with greater strength or fitness have a better chance to be chosen. On top of this basic algorithm, the thoughts of parallel algorithm, default levels, etc., can also be added in order to make the choice of rules more flexible and more suitable to the realistic behavior of the system.

The essence of credit confirmation is to provide the system with a mechanism to evaluate and to compare rules. Each time when rules are applied, each individual agent will modify its strength or fitness according to the output of the rule application. That is in fact "learning" and "experience accumulation." In the quantitative analysis and research of genetics, economics, and psychology, in order to solve this problem of credit confirmation, it is often the case that according to the requirements, each interested object is assigned a certain value. For example, values of fitness are assigned to pairs of chromosomes; utility values to the consumable goods; and rewards to desirable behaviors. However, the problem is very subtle. Let us consider the behavior of an organism. In general, the mechanism of evolution is established in a certain internal analyzer. Specialized detectors keep timely records of the state and quantity of various types of supplies, such as food, water, etc. The purposeful behavior of the organism is to keep these "reservoirs" not "empty" and, if necessary, keep them constantly refilled. The strength of a rule either increases or decreases depending on the changing state of the reservoirs. Clearly, the phenomena of competition, existing in the realistic economic market places and in the biological world, are the background on which such credit confirmation technology is developed to describe competitive behaviors of adaptive agents.

The significance of the credit confirmation mechanism is that it provides a way to combine quantitative and qualitative research. Originally, the so-called "good" and "bad," "success" and "failure," and "superior" and "inferior" are all qualitative concepts. Although these concepts are often used in processes of adaptation, they have been employed quite randomly with subjectivity. Therefore, it becomes difficult to further discuss "learning," "adaptability," and "accumulation of experience" on the basis of these subjectively assigned values. On the other hand, a credit confirmation mechanism provides a practical method of measurement, and such a mechanism is fundamentally quantitative because it produces exact numerical figures. However, it is not a simple quantitative method of the ordinary sense that attempts to quantitatively measure complex matters using a single or a group of numerical variables. Instead, it contains a set of different, even contradictory, rules, where the differences between the rules are clearly essential and qualitative. The integration point here is "practice," the consequence of the reaction to the changing environment, or the process of interaction with the environment, which is the process of adaptation. It's a new idea to achieve the goal of qualitatively "selecting rules" through the process of quantitative "accumulation of experience." Also, from what will be discussed in the following on the mechanism of creating and discovering new rules, the advantage of credit confirmation will be more clearly seen.

8.2.3 Creation and Discovery of New Rules

Through dialogue and communication with the environment, the existing rules obtain different credit indices. On these indices, the key of the next step is how to discover and formulate new rules so that the agents' capability can better adapt to the changing environment.

The basic idea here is that on the basis of the more successfully tested rules, new rules are created by using such methods as cross-combinations, mutations, etc. (see Figure 8.6). What we need to be careful about here is that because it is based on the prior experience to establish new rules, it would be much faster and more efficient than using solely probability to seek and test all possibilities. In the relevant examples that will follow, this fact will be shown vividly.

Through cross-combination and mutations, new rules can be further established. To understand the effects of these new rules, methods of operations research can be employed to make searches in the state and solution spaces. To search and find either the optimal solution or a feasible solution, various algorithms provide a wide range of ideas. However, because the scale of the spaces is too large, many algorithms become practically impossible or difficult to implement. Although these algorithms are theoretically correct, in many practical scenarios, completely relying on brute forces to make searches and tests is tantamount to finding a needle in a large haystack. That is why the method presented here, mainly the GA, has opened up a different way of thinking: based on the accumulated experience, start with the existing rules, then create new rules from those existing ones that have a better chance to succeed, and then screen out the relatively effective rules and building blocks using the realistic processes of communication with the environment. By going over such a process of creation and selection, more effective rules are established. Depending on the existing achievements and considering the results of practical applications are the characteristics of this new logic of reasoning. That is also the reason why this new logic of reasoning possesses much greater potential than the older ones.

In neurophysiology, each set of cells consists of thousands of neurons that intercross with each other and maintain self-sustained reverberation. The operation of such a set is like a small collection of rules clustered together by using ordinary marks. Multiple sets of cells move in parallel; they disseminate messages widely through a large number of synapses, where the word "synapse" means the places where neurons connect with each other, and a neuron may have tens of thousands

Cross-combinations	Parents	Cross	Offspring
	1110###		1110#1
		Crossover	
	000###1		000####
	Intersection		
Mutation:	1110###	Mutation	1010###
	Mutation		

FIGURE 8.6 Cross-combination and mutation.

of synapses. Each set of cells competes with each other to acquire neurons through recruiting by absorbing parts of other cell collections and through disintegrating by splitting into fragments as offspring. It is easy to treat such a process as regrouping of building blocks through testing. In addition, several sets of cells or cell collections can be integrated into a larger structure, known as phase sequences. In fact, in various scenarios of different types, it is not difficult to find similar phenomena.

Although flags play such an important role in the combination of rules and in providing follow-up activities, what is critical to notice is that they also contain building blocks. In fact, flags are just the modules for the appearance and operation of the rules. Thus, the operations of flags are the same as those of the other parts of the rules. The flags that have been confirmed, that is, those found in the strong rules, will bring forward related flags and provide new combinations, new collections, and new interactions. Flags always try to supply fresh blood (related matters) to the framework of the internal model through default levels in order to enrich the internal model.

With these definitions and the corresponding processes, one has a unified way to depict the behaviors of the adaptive agents in CASs. The possibility of a unified description of adaptive agents brings such hope that all CASs can be described in a common framework. Because of the existence of a common language, cross comparisons of CASs are bestowed a new meaning. One can convert a dominating and visible mechanism of one CAS to a different mechanism, which may be fuzzy but important to another CAS. In searches, metaphors and guiding effects of the general principles become more colorful. Also, each search becomes more straightforward and hopeful.

To understand the result of so doing, making comparisons is useful and beneficial. For example, if an embryo is compared to a city, they possess very similar attributes.

If one looks at the start of New York City four centuries ago and does some appropriate time adjustments, then the development of the city is quite similar to the growth of an embryo. Both grow and evolve and develop the boundaries that separate the interior from the exterior and substructures. They establish the constantly improving infrastructure of communication and resource transportation. Both of them adapt to internal and external changes and maintain critical regional functions so that they individually maintain their necessary cohesions. Also, by continuously strengthening their infrastructures, both of them have a large number of adaptive agents: one contains a wide variety of commercial companies and productive individuals, and the other contains all kinds of biological cells.

8.3 FROM PARTS TO WHOLE—THE ECHO MODEL

8.3.1 Resources and Sites

On the basis of the agent model as discussed earlier, let us now, for the whole system, establish the macroscopic model, which is named as the ECHO model (Holland 1996).

First of all, to provide an external environment in which an agent dwells, let us define the concepts of resources and sites and then propose a basic model. When more complex attributes are further added, we will eventually formulate the ultimate model.

The significance of the concept of resources can be seen in a wide range of scientific areas. For example, wells or spring water in real life can continuously provide some of the required substance and/or energy for agents. Similarly in the economic systems, banks play a role in providing funds to individual citizens or corporations.

To understand the concept of sites, let us imagine it as a "city." It is a "container" within which some agents dwell, and it possesses certain environmental or resource conditions, such as "temperature," "levels of services," etc. Individual agents can move and make choices between different locations. Between locations, there are also the concepts of "distance" and others, which are modeled in computer simulations by using environmental parameters.

It is not difficult to understand the backgrounds of these two concepts—biology and economics.

In the biosphere, land, water, air, food, nutrients, etc., are all resources. In economic management, capitals, raw materials, energy, manpower, technology, information, etc., are also resources. However, in the theory of CASs, the meaning of resource is much broader and more abstract than what is listed. In an arbitrarily chosen CAS, to maintain the survival and development of those "live," proactive, and adaptive agents, some resources will have to be consumed and/or used. In fact, they are not only conditions of livelihood but also signs of quality living. If resources are scarce or fall below a certain extent, the agents will "die of starvation." On the contrary, if resources are more than the minimum adequacy, then the agents will "reproduce," creating additional agents. In addition, the agents will also be capable of processing resources. They can combine several different resources to produce new resources. For example, a plant utilizes different raw materials, coupled with the necessary human power and energy, to manufacture its specific products. All of these properties together describe the concept of resources introduced in the CAS theory. Obviously, the content of this concept is very rich.

Similarly, the concept of sites is also abstracted from many practical situations. It can be a particular market place or a city (in the economic sphere); it can also be a forest or a lake (in the field of environmental and biological sciences). For the individual agents in an adaptive system, sites are just the "place" where they live and operate. For the survival and operations of the agents, different sites provide a number of basic conditions: level of adequacy of resources, quantities and situations of adjacent agents, space to develop, etc. In addition, the proactiveness of the agents can also be shown by the movements of the agents from one location to another. That is, they choose places more suitable for their survival. It is like the migration of animals, movements of population, etc.

8.3.2 Framework of ECHO Model

In this basic model, the agents have the most elementary functions of looking for other agents and resources, storing and processing, and trading resources with other agents.

Therefore, each agent must have three basic parts:

- Offensive flag: it is used to actively communicate with other agents.
- Defensive flag: it is used to decide whether or not to answer the call of other agents.
- Reservoir: it is used to store the processed resources.

Hence, the basic structure of an agent can be shown in Figure 8.7.

Their functions include the following: proactively contact other agents and answer other agents' contacts, exchange resources if a match is made successfully, store and process resources within the interior, and produce new agents if resources are sufficient.

On this basis, the ECHO model becomes the following conditions:

1. The whole of the system includes a number of sites.
2. Within each site, there are several agents.
3. There are contacts and exchange of resources and information between agents.

This is the most basic ECHO model.

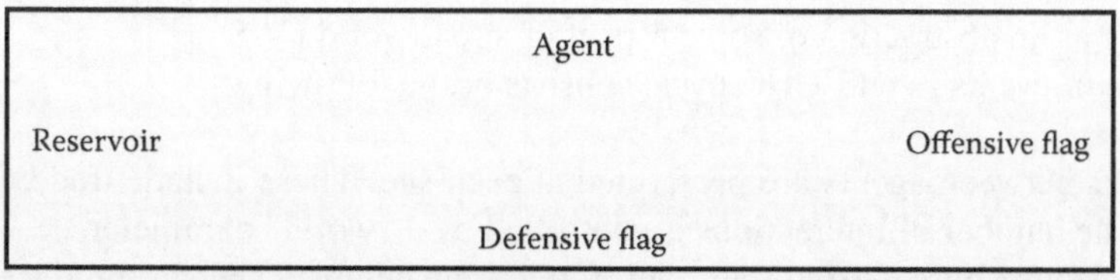

FIGURE 8.7 Basic structure of an agent in the ECHO model.

8.3.3 Extension of ECHO Model

Due to its simplicity, the basic ECHO model is not sophisticated enough to describe complex behaviors of systems. Thus, Holland (1996) gradually added various functions to extend the ECHO model. In his book *Hidden Order: How Adaptation Builds Complexity*, he made a preliminary extension by adding the following:

- *The concept of conditional exchanges:* That is, under the condition that the offensive flag is consistent with the defensive flag, there should be additional conditions for an exchange to take place. For example, when procuring raw materials, other than the existence of the desired raw materials, it is also necessary to consider the purchase price and quantity.
- *The concept of resource transformation:* That is, the agents are capable of processing, utilizing, and assembling resources. The addition of this feature lays the foundation for the division and specialization of the agents.
- *The concept of adhesion:* That is, through the establishment of a fixed link, some agents form an aggregate, which act together as an entity in the operation of the system. Obviously, this concept is introduced on the basis of the symbiosis of living organisms in the biological world and the enterprise groups existing in economic activities.
- *The concept of selective mating:* That is, agents selectively match and combine with other agents. Through cross-combination, new and stronger agents are formed.
- *The function of conditional duplicate:* That is, when resources are sufficient and the conditions are ideal, agents increase their functionalities through duplication.

Through this step-by-step extension, the describing power of the ECHO model is increased so that it becomes adequate to express and investigate various complex systems.

8.4 MODELING CASs WITH ECHO

Forrest and Jones (1994) initially studied the species abundance using the ECHO model. In comparison, GAs focus on the evolutionary components of CASs. They are reasonably well understood and have been developed maturely. However, they ignore several important factors, such as resource allocation, heterogeneity, and endogenous fitness. On the other hand, ECHO extends GAs to an ecological setting by adding the concepts of geography (location), competition for resources, and interactions among individual agents (coevolutions). ECHO is intended to capture important generic properties of ecological systems while not necessarily to model any particular ecology in detail. The original idea of ECHO, including motivation, design decisions, and overall structure, was introduced by Holland (1992, 1994).

8.4.1 ECHO Structure

Each ECHO involves a world that contains a fixed number of sites. Each site may contain an arbitrary number of agents, including zero. Each world specifies certain system-wide parameters, including the number of sites, the number of resource types, the taxation rate, parameters' controlling replication, and the probability of random death. Each of these components is designed by the user of the system, typically as an abstraction of some aspect of a real-world CAS.

The sequence of events in an ECHO cycle consists of the following:

1. Interactions between agents are performed at each site. These include trade, mating, and combat. The number of interactions is controlled by a "world" parameter.
2. Agents collect resources from their site if any are available. The site produces resources according to its "site" parameters.

3. Each agent at each site is taxed (probabilistically). Each site exacts a resource tax from each agent with a given (worldwide) probability. Tax in ECHO can be thought of as economic taxation or as the cost required to live at the site.
4. Agents are killed at random with some small probability. This can be interpreted as bad luck or as a mechanism that prevents agents from living forever.
5. The sites produce resources. Different sites may produce different amounts of each resource. When an agent at a site dies, its occupied resources are returned to its site and become immediately available to other agents at that site.
6. Agents that have not received any resource in this cycle migrate. In particular, if an agent does not acquire any resources through picking them up or through combat or trade during an ECHO cycle either, it will migrate to a neighboring site.
7. Agents that can replicate do so through asexual reproduction. An agent may make a copy of its genome using the resources it has stored in its reservoir. The replication process is noisy. Random mutations may result in genetic differences between the parent and its children.

This cycle is iterated many times during the course of a "run."

8.4.2 Agents

Figure 8.8 illustrates an example ECHO agent. Each agent has a genome that is roughly analogous to a single chromosome in a haploid species. The chromosome has $r + 7$ genes, where r is the number of resources in the world. Six of the genes, the tags, and conditions are composed of variable-length strings of resources (i.e., of the lowercase letters that represent resources), and the last gene trades resources. The mutation operator can alter the allele value at any locus and can also cause a tag or condition to grow or shrink in length.

Tags are genes that produce some easily observable feature of the phenotype. Conditions are genes that do not produce observable phenotypic effects, and their result cannot be detected by other agents. Thus, an agent will interact with another on the basis of its own conditions and the other's tags.

The six tag and condition genes possessed by every agent are the offense tag, defense tag, mating tag, combat condition, trade condition, and mating condition. These genes are used to determine what sort of interaction will take place between a pair of agents and what the outcome will be.

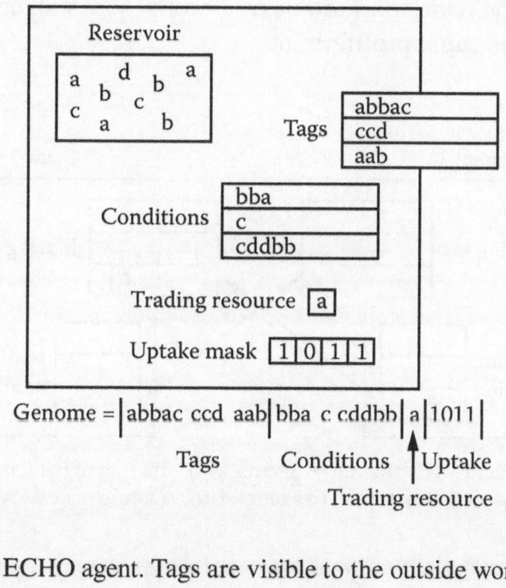

FIGURE 8.8 Structure of an ECHO agent. Tags are visible to the outside world. Conditions and other properties are not.

The r genes correspond to the agent's uptake mask, which determines its ability to collect the resource of each type directly from the environment. If an agent does not have a "1" allele for the uptake gene corresponding to a certain resource, it will not be able to collect the resource of that type if it encounters some amount of the resource at a site. Consequently, if the agent requires this resource, for example, because the site at which it is located charges a tax that includes it, or because the agent needs it to replicate, it will have to either fight or trade for it. The final gene is the trading resource, which is the resource type that the agent will provide to another agent if trading takes place. Each agent also has a reservoir in which it keeps some amount of resources of each type. Resources from the reservoir are used to pay taxes, to produce offspring, and to trade.

Agents at a site are arranged in a one-dimensional array. The probability that a pair of agents will be chosen to interact falls off exponentially with increasing distance between agents in this array. The user must decide which agents initially reside at each site and in what order they should appear in the array.

8.4.3 Agent–Agent Interactions

There are three main forms of agent–agent interaction: combat, trading, and reproduction. All of these forms of interaction take place between agents that are located at the same site, and all involve the transfer of resources between agents

Combat: If two agents are competitive in a real-world system, they would be modeled to engage in combat in ECHO. When two agents encounter each other, the system first checks to see if one agent would attack the other. An agent *A* attacks an agent *B* if *A*'s combat condition is a prefix of *B's* offensive tag. As a result of the combat, the resources that comprise the loser (both its genome and the contents of its reservoir) are given to the winner, and the loser is removed from the population.

Trade. If two agents are chosen to interact, but not engage in combat, they are given the opportunity to trade and mate. Unlike combat, trading and mating must be by mutual agreement. Agents *A* and *B* will trade if *A's* trading condition is a prefix of *B's* offensive tag and vice versa. The offensive tag will also be used here to determine whether combat will occur.

Sexual reproduction. Agents that interact and do not engage in combat may produce one offspring through recombination. In some GAs, the offspring replaces both of the parents in the population. However, in general, this has to be the case. A sexual reproduction occurs between two agents *A* and *B* if *A* finds *B* acceptable and vice versa. Agent *A* finds *B* acceptable if either (1) *A's* mating condition is a nonzero prefix of *B's* mating tag or (2) both *A's* mating condition and *B's* mating tag are of zero length. The restriction to nonzero prefixes is designed to stop agents with zero-length mating conditions from rapid proliferation.

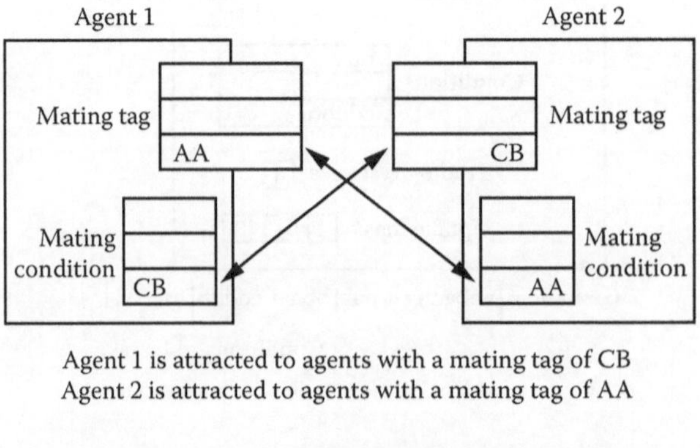

FIGURE 8.9 Simplified view of the two-way tag and condition matching that is used by agents to determine whether mating will occur.

Figure 8.9 shows a simplified view of the two-way matching process used to determine whether a mating will occur.

8.4.4 Species Abundance and ECHO

Suppose one took the catch from a laden fishing boat returning to harbor and sorted the fish according to species. What would the distribution of fish into species look like? The answer, of course, depends on many factors, such as weather, bait used (if any), the depth at which the fish were caught, the water temperature at that depth, the size of the catch, and myriad others. A general perspective on such experiments is to consider the ways in which the n individual entities that are sampled can be partitioned to represent a (typically unknown) number m of species. From a biological perspective, the interesting questions are: Does the distribution into species follow a pattern that can be characterized mathematically? And if it is so, are there biological theories that can account for this pattern?

In examples where this general pattern is seen, Preston's canonical lognormal distribution has often proved the most accurate model. Preston (1948) took the counts for the various species in observed data and grouped them into a series of "octaves." This was simply a (base 2) logarithmic grouping of the species counts. Preston plotted these "species curves" for a number of experiments and found that their general shape was well approximated by a Gaussian (normal) distribution of the form

$$y = y_0 e^{-(aR)^2}$$

where y is the number of species falling into the Rth octave left or right of the modal octave, y_0 is the value of the mode of the distribution, and a is a constant related to the logarithmic standard deviation, to be determined from the data (Preston 1948).

In this subsection, we consider different groupings of ECHO agents. When a grouping is fixed, agents in one group can be considered as being the same species in ECHO.

In all of the figures of this subsection, the populations are taken from ECHO worlds that were stopped after 1000 generations. The parameter settings that have been held constant throughout the experiments reported in this subsection are summarized in Table 8.1. Details on the precise meaning of these parameters are provided in the work of Jones and Forrest (1993).

TABLE 8.1
World and Site Parameters That Were Held Constant throughout This Section

Parameter	Value
Number of resources	4
Trading fraction	0.5
Interaction fraction	0.02
Self-replication fraction	0.5
Self-replication threshold	2
Taxation probability	0.1
Number of sites	1
Mutation probability	0.02
Crossover probability	0.7
Random death probability	0.0001

Note: Those above the line are the worldwide parameters.

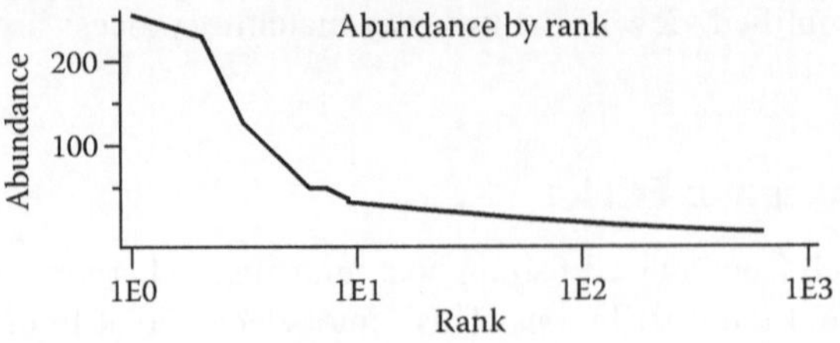

FIGURE 8.10 Example of the abundance of ECHO genomes in a population after 1000 cycles. Abundances are ranked from commonest (left) to rarest (right) with the actual abundance given on the *y*-axis. The final population contained 603 different genomes.

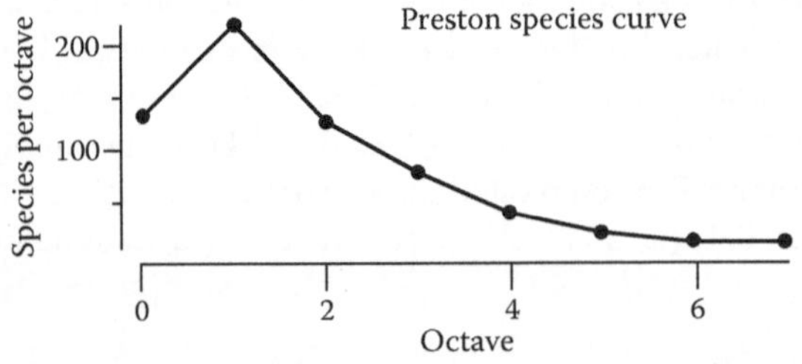

FIGURE 8.11 Population data from Figure 8.10 organized into octaves according to the method of Preston (1948).

The simplest way to study relative abundance in ECHO is to sort the genomes by their abundance and to plot these by rank on the *x*-axis and by number of individuals on the *y*-axis. Figure 8.10 was produced by simply examining the number of copies of individual genomes in the population after 1000 generations of an ECHO run.

Taking the population data from the same ECHO run and organizing them into octaves using the method described by Preston (1948) results in Figure 8.11. This figure bears a strong resemblance to those produced by Preston in his study, especially those in which the veil line is close to the mode of the distribution. That is, the character of genome abundances in ECHO populations tends to follow the general patterns found in some biological systems.

Our ultimate goal is to confirm or disconfirm the hypothesis that ECHO exhibits many of the same broad classes of behaviors as natural ecological systems. Because ECHO emphasizes evolution, a natural starting point in the confirmation process is to ask whether or not evolutions in ECHO produce distributions of agents that are similar to or different from those observed in natural systems. Although this investigation is still in its early stages, the results to date are encouraging.

8.5 SWARM SIMULATION PLATFORM

Swarm is a set of standardized tools of computer simulation and modeling. It was developed by the Swarm Development Group at the Santa Fe Institute for multiagent-based modeling and simulation. Its purpose is to build a shared computer simulation platform. With such a platform, researchers can focus their attention on building the needed models themselves. The agents that are talked about here stand for the intelligent individuals with the capabilities of autonomy and self-adaption in the simulation system of concern. They are the actual workers of the simulation activities. In terms of the general mathematical modeling, multiagent-based simulation and modeling have their significant advantage in the area of modeling and simulating discrete, nonlinear systems. It is a basic method suited for the study of the global and adaptive behaviors of such complex systems.

Swarm was originated in the Santa Fe Institute's research on artificial life. Later, it was evolved into a general system for universal use. As of this writing, it has been applied widely to many scientific areas, including physics, ecology, economics, etc. Swarm is a free open-source software package. It was originally designed for the UNIX system but now supports Windows, with Java programming faculty available. Due to its superior performance in the field of computer simulation, the Swarm platform has been gradually accepted by more people.

8.6 OPEN-ENDED PROBLEMS

1. How do you understand that "adaptation creates complexity"? Give an example to support your point of view.
2. Please describe in your own words the seven basic factors in CAS—aggregation, nonlinearity, flow, diversity, tag, internal model, and building block—and explain the relationship between them.
3. How do you understand the role of genetic algorithm in the development of CAS theory?
4. From the perspective of CAS theory, explain how systemic emergence is generated.
5. Describe how an agent adapts and learns. Do you have some new ideas to model this process?
6. Explain how the ECHO model realizes "from parts to whole."
7. Select an instance of complex adaptive systems, and model it with the ECHO model, including the ECHO structure and how agents interact with each other.
8. Realize the instance mentioned in question 7 by using the swarm platform.

9 Open Complex Giant Systems

Open complex systems that are of giant scales exist everywhere, in nature, in man, and in social cultures or organizations of man. What should be noted is that only until recently, these systems have not been investigated from such an epistemological angle as using relevant particular methodologies. In this chapter, we will focus on the discussion of this class of systems and relevant methods of research.

9.1 CLASSIFICATION OF GIANT SYSTEMS

9.1.1 GIANT SYSTEMS AND THEIR CLASSIFICATION

As we have seen throughout the previous chapters, systems science studies systems and systemhood as its objects, while systems commonly exist in nature and human societies. In terms of a comparison with the conventional science, it can be seen that when the entity of study is seen as an object without size and volume, known traditionally as a particle, one can conveniently employ the concept of numbers, quantitative variables, and relevant theories based on calculus. However, when the entity does have its internal structure, as such internal structure becomes important and not ignorable in the research, one has to deal with systems instead of sizeless and volumeless particles.

For instance, the solar system can be seen as an abstract system; the human body is an (organic) system; each family constitutes a fundamental system of human organizations; a factory is a (business) system; and a nation is also a system that occupies certain geological territory and controls a certain human pollution. There are various kinds of systems in the objective world. For the convenience of investigation, these systems can be classified into different classes using various principles. For example, based on the criteria on formations, functions, and whether or not there are human beings involved, systems can be classified into such classes as natural systems, artificial systems, man–machine systems, etc. For instance, the solar system is natural, each factory is an artificial system, and each university is a man–machine system, in which other than people, there are various kinds of equipments. If one considers whether a system exchanges material, energy, information, and others with the environment, the system can be classified as either a closed or open system. Of course, in nature, a truly closed system does not really exist. However, for the convenience of investigation, one often treats a realistic particular system as an approximately closed system. If one considers whether or not the state of a system varies with time, then he or she can classify systems as being static or dynamic. Just as the concept of closed systems, truly static systems do not actually exist in the objective world. They are employed only to be approximate descriptions of complex systems. If one categorizes systems according to their different physical attributes, then he or she is able to define such classes of systems as physical systems, biological systems, ecological systems, etc. If one separates systems by considering whether or not the systems of concern involve factors of life, then he or she can categorize systems into life systems and nonlife systems, etc. For more details on classification of systems, please consult Section 2.4, entitled "Classification of Systems," in this book.

9.1.2 INTRODUCTION TO THE CONCEPT

Although the different methods of classification of systems as described above are relatively intuitive, the focus of these classifications is placed too much on the particular intensions of the systems of concern, while losing the essence of the systems. However, the essence of the systems, the systemhood, is the very important or the central aspect of the research of systems science. Based on this realization, Xueshen Qian (1989) established the following well-known method of classification of systems.

According to the subsystems that make up the whole, the number of different kinds of subsystems involved, and the degree of complexity of the correlation between the subsystems, the overall system is categorized as either a simple system or a giant system. Here, a system is simple if it is meaningfully composed of a few subsystems, between which only elementary relationships exist. For instance, some nonlife systems, such as a measurement instrument, are simple and small systems. A system is considered large if the number of subsystems is relatively large, such as several tens or over 100. For example, a factory can be seen as a large system. No matter whether a system is small or large, investigations of such a simple system can be accomplished by first starting with the interactions between the subsystems and then obtaining the functionalities of the system's behaviors by directly synthesizing the individual interactions of the subsystems. That can be seen as a straightforward method and will not suffer from any unexpected complication. In the worst scenarios, when dealing with large systems, one might need to employ powerful computers in his or her works.

If a system is made up of a huge number, such as several thousands and/or over several millions, of subsystems, then the system is referred to as a giant system. If the number of different kinds of subsystems of a giant system is relatively small, such as a few or several tens, and their relationships are also relatively simple, then the system is referred to as a simple giant system. For instance, each laser system is a simple giant system. When dealing with such systems, one of course has to employ methods that are different from those used for studying simple small and simple large systems. In such studies, even the most powerful computers currently available in the world will not be enough. Also, in the future, there will not be any such computer that will be powerful enough to deal with the relevant research on simple giant systems. Because the method of direct synthesization is not adequate enough for studying simple giant systems, one naturally and associatively thinks about the major advances of statistical mechanics initiated at the start of the twentieth century. By using this new theory, one can capture the main functionalities of such a giant system that is made up of trillions of molecules by using statistical methods and by ignoring the details of these functionalities. This idea has been quite successful and was developed by Prigogine (1961) and Haken (1982) in their individual theories of dissipative structures and synergetics, respectively.

If a system has a great many subsystems and layers and the relationships between the subsystems and layers are very complicated, then the system is referred to as a complex giant system. If this system is open, then it is referred to as an open complex giant system.

9.2 OPEN COMPLEX GIANT SYSTEMS

9.2.1 EXAMPLES

There are many examples of open complex giant systems, such as the bodies of bio-organisms, human brains, human bodies, human psychological beings, geological systems along with their natural ecological components, social systems, celestial systems, etc. For these systems, no matter whether one looks at their structures, or functionalities, or behaviors, or evolutions, they all look extremely complex so that as of the present day, there are still a great number of phenomena and problems regarding these systems that we do not understand, and we do not have any clue as to how to address these problems.

In particular, for the human brain system, due to its memory, its thinking ability, its reasoning effects, and the role of mind, its characteristics of input–output reactions are extremely complicated. Based on the information (memory) of the past, the expectations (reasoning) of the future, the current input, and the effect of the environment, the brain can produce various sophisticated reactions. In terms of time, these reactions can take place in real time, in delayed time, or even before their times. In terms of types, these reactions can be true reactions or false reactions, or there can even be no reaction. Therefore, human behaviors are definitely not simple conditional reflexes. Their input–output characteristics change over time. As a matter of fact, the human brain contains 1012 neurons and the same amount of glial cells; their interactions are much more complicated than electronic switches. That is why Clemend (1988), a researcher of the IBM Corporation, once said that the human brain is like a large computing network that consists of 1012 giant computers, each of which can perform over 1 billion operations per second.

9.2.2 Layer Structures of Systems' Hierarchies

When one goes one level up from that of the brain, he or she faces such systems each of which contains subsystems of people and various artificially designed intelligent machines. For these systems, the words "open" and "complex" are given new and more general meanings. Speaking generally, like before, "open" implies that the system exchanges with its environment energy, information, or materials. However, more specifically, it means that

1. Both the overall system and its subsystems, respectively, exchange various kinds of information with the environment.
2. Each of the subsystems acquires knowledge through learning.

Because of the effect of human consciousness, the relationship between the subsystems can be not only complicated but also easily altered. Each individual person itself stands for a complex giant system. Now, another level of giant systems, organizations of people, is created, where each of these giant systems consists of such complex giant systems as individual people.

When mankind attempts to comprehend the objective world, other than practice, man also needs to make use of the intellectual wealth acquired in the past. Along with the current exponential growth of information, how to master and apply knowledge has become a prominent problem. If man does not use any known knowledge, then mankind will return to the antiquity of over 1 million years ago. Man has manufactured highly intelligent, powerful computers and is pursuing making robots that move and behave intellectually. The organizations that consist of subsystems of people, those intellectual machines as well as their harmonic working relationships, represent some of the most complex systems the man has ever seen and experienced. Here, not only is the number of different kinds of subsystems involved used to represent the complexity of the system, but it also plays an important role with regard the available knowledge.

The complexity of the systems in this class includes the following:

1. There might be various kinds of communications between subsystems of the system of concern.
2. Other than there being a great number of different kinds of subsystems, these subsystems also have their respective forms.
3. Within these subsystems, their knowledge is expressed in different languages and forms and is, respectively, acquired using different methods.
4. The structures of the subsystems evolve along with the development of the overall system. Therefore, the structure of the whole changes constantly.

These particular systems are referred to as open, specific complex giant systems. They are the commonly known social systems.

9.2.3 General Properties

In principle, the fundamental properties of open complex giant systems are the same as those of general systems, including (1) the wholeness principle; (2) the principle of correlation, the systemhood;

(3) the principle of orderliness; and (4) the principle of dynamics. Beyond these properties, the main characteristics of open complex giant systems can be summarized as follows:

1. *Openness.* The system's objects, subsystems, and the system itself exchange materials, energy, and information with the environment.
2. *Complexity.* There are a great number of different kinds of subsystems and many kinds of mutual and multilayered interactions between the subsystems.
3. *Evolution and emergence.* Subsystems and elements interact with each other and contribute to the formation of some new and particular properties of the overall system in terms of the whole evolution and development.
4. *Hierarchy.* There are many layers between the subsystems that have been quite well understood and the whole system that can be macroscopically measured. As for the specific number of layers that exist in between these levels, it may not be clear.
5. *Giant scale.* The total number of elements and subsystems within the system is very large, reaching the magnitude of several tens of thousands and even over millions.

After more than 10 years of research since the concept of open complex giant systems was initially proposed, through conference presentations, discussions, and relevant published works of scholars from many different scientific fields, further and deeper understandings of the concept and relevant methods have been achieved. The treatments of such open complex giant systems as the human brain system, human body system, and geological system have gone beyond the total scientific achievement of the past 200 plus years developed on the basis of reductionism. It can be concluded that there is a need to organically combine the wholeness theory of the Chinese culture with reductionism in order to guide the investigations of open complex giant systems. Just as Xueshen Qian said on several different occasions, "People generally understand problems through analyzing specific scenarios. To uncover general properties of open complex giant systems, one has to start with resolving problems of these systems one at a time. The path of establishing the theory of open complex giant systems can reference back to that of engineering cybernetics. As for how to determine which open complex giant systems to investigate, one should first consider those that are close to people's ordinary lives, that have close relationships to people, and whose structures, behaviors, functions, and evolutions are more likely to be accepted. Studies of such systems and consequent results will more likely be understood and accepted. Investigations of such systems will play the role of models and will motivate for advanced researches of open complex giant systems in terms of the theory and relevant methodologies."

9.3 INTERNET—AN OPEN COMPLEX GIANT SYSTEM

The Internet system represents an extremely complicated system that involves a great many problems and aspects. Thus, only by using systems science and related terminologies can one possibly analyze and comprehend its essence.

9.3.1 Systemic Components of the Internet

In order to accurately and straightforwardly describe a problem using systems science, one generally does not need to consider how technically one could practically implement a specific detail. Here, he or she needs only to collect the main factors to study his or her Internet-related problems. Currently, information exchanges between users and the Internet are carried out through the World Wide Web (WWW) servers, which represent the most important form of Internet services. These servers are closely related to such other main applications as file transfer protocol (FTP), Telnet (a network protocol used on the internet or local area networks to provide a bidirectional interactive text-oriented communication facility using virtual terminal connection), e-mail, etc. When a user visits the Internet, he or she needs only to get into the target website on the WWW, which is the overall application form of service connecting all the Internet servers. By clicking the mouse on the hyperlinks of hypertexts or hypermedia documents, he or she can immediately and effectively navigate over the Internet between computers; various kinds of applications, such as FTP and telnet; and different forms of information.

Thus, from analyzing the attributes of the WWW, one can potentially comprehend the essence of the Internet system. In this section, without spending too much time on the background technology and the physics of realization, we will show from the point of view of WWW sites that Internet users directly and most often utilize that the Internet system is essentially an open complex giant system.

In particular, the WWW is a component of the Internet system. It reaches almost all corners of the world and represents the organizational form of the Internet information. It simplifies and extracts two key factors: one represents the mainframe computer systems that are realistically connected, and the other represents the relationship of hyperlinks, where the connected computer systems constitute the basic elements of the WWW and are made up of software and hardware. These computers can be either users' laptop or desktop computers or servers. Among servers, their communications are carried out through using the transmission control protocol/internet protocol (TCP/IP). Each hyperlink is located within the hypertext documents, that is, the commonly known web pages, stored in a server system. Communications between mainframe systems are materialized through web pages, while web pages are connected through hyperlinks. Hence, the WWW can be seen as an entity that joins the Internet's mainframe systems and their hypertext documents through hyperlinks (and the basic TCP/IP and communication media, which are omitted for our simplified abstract descriptions). Please note that this simplification is employed on the basis of the fact that mainframe systems, their hypertext documents, and hyperlinks are the key items Internet users directly face and operate. Also, such simplification is helpful for us to recognize the essence of the Internet system without being distracted by other irrelevant aspects. The environment within which the Internet system exists is the worldwide societal system. When seeing from the microscopic angle of social systems, the environment consists of individual users, organizations, states, or regions. When seeing from the macroscopic angle of societal systems, the environment is made up of economic systems, political systems, and human consciousness systems. The WWW and societal systems interact and influence each other and rely on each other.

9.3.2 Attributes of Internet System

From the elements and environment of the Internet system defined above using systems science, it can be seen that the Internet integrates the knowledge of different areas of all peoples acquired through the ages and data and information collected by using various means. It collectively makes use of the technologies and information from almost all scientific disciplines, such as computers, electronics, communications, signals, artificial intelligence, pattern recognition, multimedia, knowledge management, intelligence analysis, etc. It draws all peoples strongly together from different corners of the world, from various fields of activities, and from multiple levels of explorations. It plays the role of a platform for people to make dynamic, random, and uncertain interactions over time. It provides a common ground for people to sufficiently bring their live intuitions and practical experiences into play. Indeed, it is an extremely large, prominently complex system that synthesizes people and machines. This system satisfies all the properties of open complex giant systems as proposed by Xueshen Qian and others.

9.3.2.1 Openness

The mainframe systems, the basic composing elements of the Internet system, can be and have been connected through certain topics, through mutual searches, or through announcements of information. Such blocks of connected networks become shared resources on the Internet; it exchanges information and transfers data. Through hypertexts and hypermedia, the WWW converts all the computers connected together through the Internet into a dynamic superhighway, through which information, knowledge, and data are exchanged and transported. The key people and their mainframe systems in the societal environment of the Internet system are closely related and cannot be separated; without the participation of people, none of the mainframe systems, no matter how technologically advanced they are, will be connected to form an organic whole so that the WWW and Internet will not emerge. When people use the Internet, they acquire large amounts of knowledge and information and different forms of service. Inspirations are created; experiences and wisdom of others are passed on. Thus, people's levels of understanding, capability, thinking logic, and behaviors are altered; their intelligence is lifted up to a higher level. Also, the Internet is and has been changing the politics, economies, cultures, and ideologies of the societies from around the world. The Internet has become one of the sources of motivation for human development and progress. However, it is people who create and construct the networks of various scales that constitute the Internet, the infrastructures of these networks, communication protocols, the large amounts of information of the networks, and different forms of services. The Internet is also constantly improved and perfected through usage. The creation of new tools, new contents, and new forms for the Internet furthermore becomes the social norm and intelligence of the Internet. The constantly strengthening livelihood of the Internet is a good representation of the magnificent achievement of mankind.

9.3.2.2 Complexity

The Internet consists of several hundreds of thousands of networks distributed all over the world. These networks can be either small or large. For example, IBM's wide area network (WAN) is so large that it covers the entire globe, and some networks can be so small that each of them contains only two or three computers within a college dormitory room. The operating system of a network can be a member of the Unix, Linux, or Windows families. The database management system, depending on the type of the operating system, can be Oracle, Sybase, DB2, Informix, or others. The topological structure of the network can be star shaped, ring shaped, of bus architecture, tree shaped, or a combination of these types. The input media can be hard wired, wireless, fiber optic, etc. The Internet input protocol could be TCP/IP, asynchronous transfer mode (ATM), dynamic synchronous transfer mode (DTM), etc. The specific method of connection into the Internet can be by dialing through dedicated lines or wireless, etc. There might be great differences between transmission speeds, operation loads, contents, and servers from one network to another. What is worth noticing is that due to subjective or other people's reasons, or the effect of external forces, all the aforementioned aspects of a network can change with time. Between the mainframe systems that constitute some of the basic elements of the Internet system and between networks, it is within these complicated, diversified, multilayered, and dynamic changes that information, data, and services are exchanged and transferred.

The societal system that is dominated by people and constitutes the environment of the Internet can, at least from the macroscopic angle, be divided into economic systems, political systems, social consciousness systems, etc. These systems can be greatly different from each other depending on the specifics of the civilizations of their particular locations. The Internet helps merge the various types of societies of specific characteristics into a whole, within which these societies impact and affect each other. Through impacts, societies absorb elements from others and evolve into their new forms with respective origins and fresh characteristics, showing the tendency of convergence. When seen from the microscopic angle, the differences of the background to which societies' individuals or organizations belong lead to varied understandings of and consciousness about nature and courses of action. Within the Internet systems, people are not only the users of the Internet and objects to which services are provided but also the designers of the Internet technology, standards, and contents and service providers. It is these varied forms of interaction between people and the Internet that the Internet changes and evolves constantly. It becomes extremely difficult for anyone to accurately grasp and master the Internet information, contents, forms, services, technologies, thoughts, and interactions. In particular, due to the engagements of commercial operations, it is very difficult to predict what will happen to the Internet users, services, concepts, speeds, forms, etc. History is experiencing a fundamental change while facing many new problems and challenges. These problems and challenges have touched on many areas and levels of science, technology, engineering implementations, and social sciences, and are involved with politics, economics, cultures, ideologies, ethics, talents, etc.

9.3.2.3 Self-Organization

The Internet system is unorganized and in a chaotic state of affairs. For instance, there are great differences between the basic elements and networks. The connections and interactions between networks take a great variety of forms. The contents of the Internet are extremely messy and unorganized. The web pages, applications, technical schemes, thoughts, and users are indiscriminately and randomly scattered all over the Internet with their respective characteristics. However, despite the messy appearance, all the mainframe systems and networks are organized seamlessly through TCP/IP and hyperlinks into a whole, which has been evolving with time. Along with updates, the parts of the Internet mutually affect each other, rely on each other, and interact on each other. The whole constantly adapts and adjusts itself. Through competition, some technological achievements, such as Java, Linux, hyperlinks, etc., that can satisfy the needs of users emerge successfully. Under the effects of people and joint efforts of mainframe systems and individual networks, the Internet, a man-made miracle, has been created and maintained. Although the interacting surfaces and contents between the Internet and societal systems are also extremely complicated, it is exactly the effect of the two-way long-lasting exchanges, adjustments, and self-organizations of these individual entities and aspects that the overall Internet system experiences through different periods of development and evolves onto a much higher level of order. That helps the emergence of the present day's wholeness behaviors and intelligence of the Internet.

When a problem attracts widespread attention, a large number of Internet web pages, which are interested in the problem and indiscriminately scattered over the entire Internet, will spontaneously and repeatedly interact with each other by using hyperlinks. Over time, some authoritative web pages and most often-cited central web pages will emerge through these disorderly interactions (see Figure 9.1), which depicts the organizational module of these authoritative and most cited web pages.

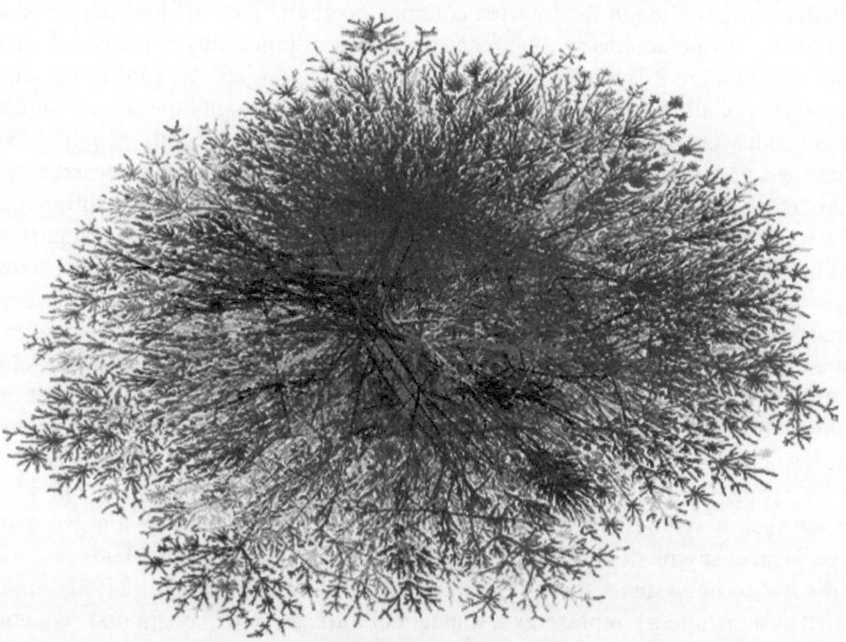

FIGURE 9.1 Organizational module of emerging authoritative and central web pages.

On the WWW, there are many aggregations that center on such authoritative and central web pages. Such modules emerge as a consequence of the self-organization mechanism of the WWW.

Similarly, the operation of the current new economies, the tendency of globalization, the appearance of the Internet culture, the enhancement of American cultural awareness, and the strengthening of American global strategies through the Internet, and other problems, can be seen as an emergence of the Internet system's complexity at the height of its wholeness. They can also be seen as some of the behaviors of the Internet system, created by the interactions between the system's environment and elements, between the elements, and between different types and layers of the environment. Such an emerging wholeness characteristic of the Internet also vividly reveals another aspect of the complex effect of the environment on the system. In particular, in its process of evolution, the system also profoundly affects and alters the environment so that the form of the societal system experiences major alterations, eventually causing the Internet system to evolve constantly.

9.3.2.4 Magnificent Scale

Based on its current popularity and its usefulness, it is not difficult to estimate that the Internet connects millions of different networks and attracts billions of users. More than 1 million new web pages appear each day on the WWW. All the chaotic information is loosely connected together through over 1 billion hyperlinks. Currently, the total amount of information available on the Internet has become a difficult number to estimate. What should be pointed out is that the enormous scale is one of the characteristics that distinguish open complex giant systems from complex self-adaptive systems of the Santa Fe Institute. As of the present day, it would be difficult for anyone to foretell how the Internet will look like 5 years from now. Facing the unstoppable development of the Internet, one of the few sure claims a person can make is that the increasing amount of web information poses a computational challenge if one desires to traverse this forever-expanding web storage of information.

9.3.3 Call for Different Methodology to Deal with Giant Systems

As has been discussed, although the Internet contains great amounts of elements, subsystems, and complicated forms of interactions without central control, it represents a metasynthesized system produced out of evolution; it is a huge intelligent engineering project. Its appearance and development have made use of almost all scientific theories, developed through time since antiquity in every area of human endeavors, data, and knowledge, combined with empirically acquired experiences, judgment abilities, and wisdom. Its development has either consciously or unconsciously employed and materialized the following cognitive and methodological thinking logic through qualitative and quantitative comprehensive syntheses: aim at the need; formulate the goal; establish audacious hypotheses; integrate qualitatively; build models; experiment and test; discover problems; modify the need, goals, and hypotheses; perfect the models; and repeat the experiments and tests. After many layered and repeated supplementations, modifications, and practices, the present state and technology are formed gradually, although the current state is still evolving and developing. The implementation of the aforementioned process starting from qualitative and ending at quantitative synthesization and integration represents exactly the process that evolves gradually from first-hand real-life experiences, hypotheses, and immature theories to the understandings and technologies of quantitative results. Thus, the methods of investigation of the Internet system and relevant problems stand for such a synthesized system of discussion that combines qualitative and quantitative analyses. Hence, any method that is effective in terms of resolving problems of open complex giant systems has to be systems theory based. That is why Xueshen Qian (1981) pointed out in the 1980s that systems theory represents a dialectical unification of holism and reductionism. He described the methodology of open complex giant systems as a synthesized system of study halls, which in essence means an organic combination of relevant experts, data, various information and computers, and information technology, such as the Internet (Qian et al. 1990). It is a system that is made up of scientific theories, knowledge, and people. Such a system has to be network based.

Each synthesized system of study halls consists of discussion terminals, central meeting halls, main communication networks (such as the Internet or WAN), management service systems, and information storages, surrounded by discussion and supporting groups and technologies scattered at different geological locations. It represents a multilayered, bottom-up progressive, man–machine interactive research and decision-making system, which is of large scale and widely distributed, that contains distant participants as well as experts, who make the final decisions at the central meeting halls.

9.4 SIGNIFICANCE OF STUDYING OPEN COMPLEX GIANT SYSTEMS

9.4.1 Characteristics of Metasynthesis

Based on the discussions above, it can be seen that the method of metasynthesis that combines qualitative and quantitative analyses has the following characteristics:

1. Because of the complex mechanism and the characteristic of a great number of variables involved in open complex giant systems, one should organically combine qualitative and quantitative research so that qualitative understandings of multiaspects can be lifted up to quantitative comprehensions.
2. Due to the complexity of the systems, one should combine scientific theories and empirical knowledge and solve problems by collecting bits and pieces of understandings of various objective matters.
3. Based on the thinking logic of systems science, one should investigate open complex giant systems by combining various scientific disciplines.
4. Based on the layered structure of complex giant systems, one should unify macroscopic and microscopic studies.

It is exactly because of these characteristics why the method of metasynthesis has the capability to resolve problems of open complex giant systems. In the following, we will discuss the significance of this method.

9.4.2 Bridges That Connect Physical and Life Systems

One of the objects modern science and technology consider is the entire objective world. However, when investigating problems of the objective world from different angles, along with various opinions, and using varied methods, there have appeared different departments of science and technology. For example, natural sciences study the objective world from the angle of motions of materials, layers of these motions, and the relationships between these layers. Social sciences consider the objective world from the angle of the development and advances of human societies, the effects of the physical world on the human race. The mathematical science investigates the objective world from the angle of quantities and quantitative variables and their transformations. Systems science researches the objective world from the angle of systems viewpoints by applying systems methodology. As the second dimension of science, as argued by Klir (1985), from its theoretical works to practical applications, systems science considers systems and systemhood as its objects of investigation. In the macroscopic world, there is an earth, on which different forms of life appear, including man and the organization of people so that open complex giant systems emerge. However, in the macroscopic cosmos, there are also such systems, such as the galaxies, where the Milky Way is also an open complex giant system. That is, the research of open complex giant systems has a wide range of implications and possesses a general significance. However, as mentioned earlier, none of the conventional scientific theories are adequate to solve problems of open complex giant systems. The cause of such inadequacy can be found in the scientific history.

As is well known, since a long time ago, scholars of various disciplines have realized the clear-cut difference between life systems and physical systems. These two classes of systems seem to follow completely different laws. In particular, nonlife systems tend to satisfy the second law of thermodynamics and always evolve toward equilibrium and disorder, and their entropies increase toward their maximum values. Although these systems automatically evolve from order to disorder, the disorder does not automatically return to order. That is the irreversibility and equilibrium stability of the systems. On the other hand, life systems evolve in opposite directions. Organisms evolve and human societies develop always from simple to complex, from low grade to high grade, from being less ordered to highly organized. These systems can spontaneously form orders and stable structures.

For a long period of time, such contradictory observations of these two classes of systems could not be satisfactorily explained so that some scholars believe that these classes of systems are different and are governed by different sets of laws. At the same time, some other scholars have wondered: are there endogenous connections between these seemingly contradictory observations? The appearance of the theory of dissipative structures and synergetics in the 1960s started to shed new light on potential answers to this question. These theories believe that what the second law of thermodynamics reveals is the law that isolated systems follow when they are at equilibrium or near equilibrium. However, life systems, in general, are open and far from being equilibrium; indeed, they are nonlinear and nonequilibrium systems. Under these conditions, these systems import negative entropy flows from the environments by exchanging materials and energies with the environments. Although the interior of a life system produces positive entropies, its overall level of entropy decreases. When the level of the overall entropy reaches a certain threshold, the original disorderly state of the system will automatically transform into an orderly state in terms of time, space, and/or functions. Such a newly achieved stable orderly structure is referred to as a dissipative structure. Thus, under the condition of not violating the second law of thermodynamics, the theory of dissipative structures establishes the endogenous connection between these two classes of systems. This fact illustrates that there is no clear division between these two classes of systems. Although on the surface, there seems to be a deep gulf, they are governed by a same set of systems laws. Therefore, Nicolis and Prigogine (1989) point out that complexity is no longer a specific characteristic of biology; it has also entered the field of physics. It seems to have been rooted in the laws of nature. Haken (1982) further points out that the key for a system to evolve from a disorderly state to an orderly state is not whether it is an equilibrium system or how far is it from an equilibrium state, either; instead, it is the self-organizing capability of the system that under certain conditions, component subsystems synergize and collaborate with each other through nonlinear interactions, and that creates a stable and orderly structure.

9.4.3 Unification of Natural and Social Sciences

The previous result of modern science of recent decades is very important because it clearly resolves a mystery that has puzzled the world of learning for a long period of time. The successes of the theory of dissipative structures and synergetics at the same time also made a good number of scholars overly joyful, believing that such quantitative methods, developed on the basis of the modern scientific reductionism, can also be employed to attack problems of open complex giant systems. However, as it turns out, they cannot be further from the truth.

In the history of scientific development, sciences that are mainly based on quantitative methods have been known as "exact sciences," while those that are developed on the bases of philosophical reasoning and qualitative descriptions are seen as "descriptive sciences." Hence, natural sciences are considered "exact," while social sciences are considered "descriptive." Social sciences study social phenomena as their objects; the complexity of social phenomena makes it difficult to apply quantitative means. That is one of the reasons why social sciences have not become "exact." Although scholars have made enormous efforts to develop social sciences from "descriptive" into "exact" with some tangible achievements, such as those works in economic sciences, the entire system of social sciences still has a long way to go to become "exact." From our discussions in this chapter, it can be readily seen that the concept of open complex giant systems and related research methods in essence attempt to assemble scattered qualitative experiences, bits and pieces of knowledge, and opinions of the mass into a wholeness structure in order to reach a quantitative understanding. They represent a jump from incomplete qualitativeness to relatively complete quantitativeness, from qualitative experiences to quantitative understandings. Of course, when problems in one area are investigated in this way, there will appear a large amount of intellectual deposits, based on which a new set of higher-level qualitative understandings about the entire area will be derived, materializing another epistemological leap.

Max Planck, a famous German physicist, believes (Marx 1957) that science is an endogenous whole; it is divided into separated wholes not because of the objects of concern themselves but because of the limitations of human ability to understand. As a matter of fact, there are links that connect physics to chemistry, and physiology and anthropology to sociology. These links cannot be artificially broken at any arbitrarily chosen places. The studies of natural and social sciences have covered these links. This process of combining natural and social sciences into one science is referred to as the unification of these sciences. It can be said that the research of open complex giant systems and the establishment of the relevant methods have specifically pinpointed to a realistically implementable way to actualize this unification.

9.5 RESEARCH METHODS OF OPEN COMPLEX GIANT SYSTEMS

9.5.1 VARIOUS ATTEMPTS

As of the present day, there is still an urgent need to formulate a mature theory that bridges the microscopic and the macroscopic worlds for open complex giant systems. There is also an urgent need to construct a theory of open complex giant systems, which might borrow the form of statistical mechanics, based on the interactions of subsystems. Then, is there any method that is available for the investigation of open complex giant systems?

To this end, some scholars have forcefully and wishfully applied the methods developed for simple or simple giant systems, as mentioned in Section 9.1, on open complex giant systems. They did not recognize the limitations and ranges of applicability of these theories and methods, producing conclusions that are neither credible nor usable. For example, in terms of its theoretical framework, the game theory is a good tool for studying social systems. However, considering its maturity and current level of achievements, this theory is still far from being adequate for dealing with complex problems of social systems. One particular reason is that the game theory has oversimplified the social nature and complexity of man, human psychology, and the uncertainty of human behaviors. Thus, to apply the game theory, one has to drastically simplify problems of open complex giant systems to those of simple giant systems or simple systems. For similar reasons, employing systems dynamics and/or the theory of self-organization to the investigation of open complex giant systems cannot be successful, either. Jay Forrester, the founder of systems dynamics, himself warned (Goodman 1974) that when applying his method, one has to be cautious and pay close attention to the reliability of the model.

On the other hand, some scholars have lifted up problems of complex giant systems to the height of philosophy, while phrase-mongering that systems' movements are determined by subsystems, the macrocosm is controlled by microcosm, etc. One typical example is the so-called universe holographic unitics (Wang and Yan 1988). What this author and his followers missed is that human understandings of subsystems, in general, are not perfect or complete, either, and that within any subsystem, there are more finely defined subsystems. Thus, what is the use when one talks about the unknown using what is not really known? These writers even mistakenly concluded that any part has already contained the complete information of the whole; any part is the whole, and the whole is the same as each arbitrary part; part and whole are absolutely identical. Such results completely disagree with the objective reality.

9.5.2 Metasynthesis and Meta-Analysis—Two Potential Methods

Currently, one of the known methods that might be effectively employed to deal with open complex giant systems, including social systems, is the so-called metasynthesis that combines the strengths of qualitative and quantitative analyses. This method is extracted, summarized, and abstracted out of the studies of the following three complex giant systems:

1. In the study of social systems, there exists the systems' engineering technology that combines several hundreds or even over a thousand variables both qualitatively and quantitatively. This technology has been employed in the research and applications of social and economic systems.
2. In terms of human body systems, there are such studies that combine physiology, psychology, Western medicine, Chinese and traditional medicines, qigong, and human special functions.
3. In the area of geological systems, there are works that comprehensively discuss geological problems using the knowledge of ecological systems, environmental protection, regional planning, etc.

In related publications [see Walsh and Downe (2005) and the references therein], empirical hypotheses (or judgments or conjectures) are proposed through some combined efforts based on scientific theories, empirical experiences, and relevant scholars' expertise. These empirical hypotheses tend to be qualitative recognitions and generally cannot be rigorously proven using well-developed scientific methods. Even so, their validity can be checked by using either empirical data and information or models of several tens, several hundreds, or even over thousands of parameters. These models, on the other hand, are generally established on practical experience and/or practical understandings of the systems of concern through quantitative computations and repeated comparisons. The conclusions produced out of these computations and comparisons represent some of the best derived in the study of open complex giant systems. They stand for some of the quantitative recognitions that arise above the underlying qualitative analyses.

Based on the description above, it can be seen that the so-called metasynthesis that combines qualitative and quantitative analyses in essence fuses the total amount of knowledge of relevant experts and available data and various information together with computer technology; it coalesces conventional scientific theories and empirical experiences. These three sources of information and tools of analysis also constitute a system. Each successful application of this method is about how to take advantage of the wholeness of this system. On the other hand, this end also represents a weakness of this approach, because from one person to another, varied understandings of systems' wholeness will inevitably lead to different levels of success or failures of applying this method.

Correspondingly, Gene V. Glass (Cooper and Hedges 1994) in 1976 established the so-called meta-analysis based on the works of Pearson (1904) and others. The main idea of the meta-analysis is to combine the results of several studies that address related research hypotheses. In its simplest form, this is normally done by identification of a common measure of "effect size" for which a weighted average might be the output of meta-analyses. Here, appropriate weights are assigned to allow differences that naturally exist among the relevant studies. The general aim of a meta-analysis is to more powerfully estimate the true "effect size" as opposed to a smaller "effect size" derived in a single study under a given single set of assumptions and conditions. This approach has often been an important component of systematic review procedures, which combine evidences, while leaving other aspects of synthesis, such as combining information from qualitative studies, for the more general context of systematic reviews. However, in terms of applying this method to study open complex giant systems, it is still too elementary and not quite adequate.

9.6 SYSTEMIC YOYOS: COMBINED QUALITATIVE AND QUANTITATIVE MODEL

By pondering over what method is available for us to effectively investigate open complex giant systems, as addressed above, one can readily see the extreme difficulty. In particular, first, the great many variables involved in the study alone would make any such studies hard to manage. Second, the large-scale complexity involved can make the already tedious and difficult problems even more drastically impossible.

Conventionally, each number is identified with a point on the real number line or a point in a high-dimensional Euclidean space. Each (quantitative) variable is seen as a moving point in the space. Thus, when a great many such moving points and variously combined masses of these points are jointly considered, the problem is likely to be beyond the limited capability of human imagination. Among others, this end can be seen as the core reason for the extreme difficulty faced in dealing with open complex giant systems. Thus, the following is a natural question to ask at this particular junction: can we introduce a systemic intuitive background different from that of Euclidean spaces so that such a massive amount of moving particles and masses of various combinations of these particles can be more easily managed? The intuition underneath this question is that because systems science, as the second dimension of science, studies systemhood, while the classical science studies thinghood [for details, see Klir (1985)], the difficulty one faces when dealing with open complex giant systems is really a difficulty one experiences in the first dimension of science. Thus, there should be a relatively more manageable means in the second dimension for one to carry out such large-scale tasks. To geometrically comprehend this reasoning and why it might work in dealing with open complex giant systems, one can imagine a two-dimensional city surrounded by a circular steady wall. If there is no gap on the wall, then it will be difficult for any army to break into the city from within the two-dimensional space. Now, if one smart engineer designs an air-strike by making use of a third dimension, which was not known before, then the forces of his or her side can be easily parachuted into the city along the third dimension. In other words, the difficulty one faces when dealing with open complex giant systems is partially and mainly due to the reason that he or she maintains himself or herself in the first dimension of science without truly taking advantage of systems science, the newly found second dimension of science.

9.6.1 General Form of Motion of Structures

When we study nature and treat everything we see as a system (Klir 1985), we find easily that many systems in nature evolve in concert. When one thing changes, many other seemingly unrelated matters alter their states of existence accordingly. That means that changes in nature should be seen as a whole, and the whole evolution of the system of concern should be considered in order to understand how systems evolve as wholes and how systems related to each other. In whole evolutions, important is discontinuity with which transitional changes (or blown-ups) occur. These blown-ups represent the changes of old structures being replaced by new structures. If the system is described truthfully by a mathematical model, then the model is generally nonlinear, and the blow-ups of the model reflect the destruction of old structures and establishment of new ones. By borrowing the form of calculus, we can write the concept of blown-ups as follows: for a given (mathematical or symbolic) model that truthfully describes the system of our concern, if its solution $u = u(t;\ t_0,\ u_0)$, where t stands for time and u_0 is the initial state of the system, satisfies

$$\lim_{t \to t_0} |u| = +\infty, \tag{9.1}$$

and at the same time moment when $t \to t_0$, the underlying physical system also goes through a transitional change, then the solution $u = u(t; t_0, u_0)$ is called a blown-up solution, and the relevant physical movement expresses a blown-ups. Thousands of real-life evolutionary systems (Lin 2008) indicate that disastrous events appear at the moments of blown-ups in the whole evolutions of systems.

For nonlinear models in independent variables of time (t) and space (x; x and y; or x, y, and z), the concept of blown-ups is defined similarly, where blow-ups in the model and the underlying physical system can appear in time or in space or in both.

One of the key features of blown-ups is the quantitative bloa mathematical indeterminacy. This mathematical symbol in applications has caused instabilities and calculation spills interrupting each and every working computer. To understand this symbol systemically, let us look at the Riemann ball (Figure 9.2), which is a curvature space. The projection of the point x_i of the ball onto x_i' of the plane connects $-\infty$ and $+\infty$ through a blown-up. In particular, when the dynamic point x_i travels through the North Pole N on the sphere, the corresponding image x_i' on the plane of the point x_i shows up as a reversal change from $-\infty$ to $+\infty$ through a blown-up. Thus, the planar points $\pm\infty$ stand implicitly for direction changes of one dynamic point on the sphere at the polar point N instead of indeterminacies. Speaking differently, the phenomenon of being directionless, as shown by blown-ups of a lower-dimensional space, represents a direction change of movement in a higher-dimensional curvature space. Therefore, blown-ups can specifically represent implicit transformations of spatial dynamics. Through blown-ups, problems of indeterminacy of a distorted space are transformed into determinant situations of a more general system in a curvature space.

Corresponding to the previous implicit transformation between the plane and the Riemann ball, one can also relate quantitative $\pm\infty$ in one-dimensional space to a dynamic movement on a circle, a curvature space of a higher dimension through the modeling of differential equations.

This discussion indicates that nonlinearity mathematically stands (mostly) for singularities in Euclidean spaces, and physically stands for eddy motions, the movements on curvature spaces. Such motions are about structural evolutions, a natural consequence of uneven evolutions of materials.

Next, we look at the general dynamic system and how it is related to eddy motions. Newton's second law of motion is

$$m\frac{d\vec{v}}{dt} = \vec{F}. \tag{9.2}$$

Based on Einstein's concept of uneven time and space of materials' evolutions, we can assume that $\vec{F} = - \ S(t,x,y,z)$, where $S = S(t, x, y, z)$ stands for the time–space distribution of the external acting object. Let $\rho = \rho(t, x, y, z)$ be the density of the object being acted upon. Then, Equation 9.2 for a unit mass of the object being acted upon can be written as

$$\frac{d\vec{u}}{dt} = -\frac{1}{\rho(t,x,y,z)} \quad S(t,x,y,z), \tag{9.3}$$

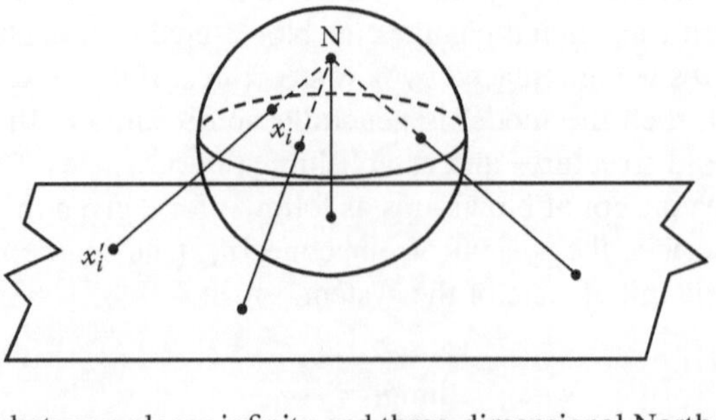

FIGURE 9.2 Relationship between planar infinity and three-dimensional North Pole.

where $\vec{u}$ is used to replace the original $\vec{v}$ in order to represent the meaning that each movement of some materials is a consequence of mutual reactions of materials' structures. Now, if ρ is not a constant, then Equation 9.3 becomes

$$\frac{d(\ x \times \vec{u})}{dt} = -\ x \times \left[\frac{1}{\rho}\ \ S \right] \neq 0, \tag{9.4}$$

which stands for an eddy motion because of the cross products. In other words, a nonlinear mutual reaction between materials' uneven structures and the unevenness of the external forcing object will definitely produce eddy motions.

At this junction, it is important to note that eddy motions are confirmed not only by daily observations of surrounding natural phenomena but also by laboratory studies from as small as atomic structures to as huge as nebular structures of the universe.

What is shown above implies that eddies come from the unevenness of materials' internal structures. Thus, if the world is seen at the height of structural evolutions, then the world is simple with only two forms of motions: clockwise rotation and counterclockwise rotation. Also, the vertical vectority in structures has been very practically implemented in the common form of motion of the universe.

Now, the fundamental problem is why all structures of the universe are in rotational movements. According to Einstein's uneven space and time, we can assume that all materials have uneven structures. Out of these uneven structures, there naturally exists gradients. With gradients, there appears forces. Combined with uneven arms of forces, the carrying materials will have to rotate in the form of moments of forces.

Based on what is discussed above, we can now imagine that the universe is entirely composed of eddy currents, where eddies exist in different sizes and scales and interact with each other. That is, the universe is a huge ocean of eddies, which change and evolve constantly. One of the most important characteristics of spinning fluids, including spinning solids, is the difference between the structural properties of inwardly (converging) and outwardly (diverging) spinning pools and the discontinuity between these pools. Due to the stirs in the form of moments of forces, in the discontinuous zones, there exist sub-eddies and sub-sub-eddies (see Figure 9.3, where sub-eddies are created naturally by the large eddies *M* and *N*). Their twist-ups (the sub-eddies) contain highly condensed amounts of materials and energies. In other words, the traditional frontal lines and surfaces (in meteorology) are not simply expansions of particles without any structure. Instead, they represent twist-up zones concentrated with irregularly structured materials and energies [this is where the so-called small probability events appear and small-probability information is observed and collected, so such information (event) should also be called irregular information (and event)].

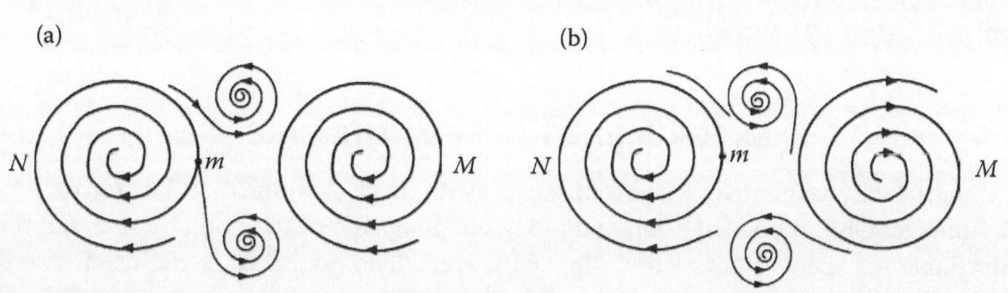

FIGURE 9.3 Appearance of sub-eddies.

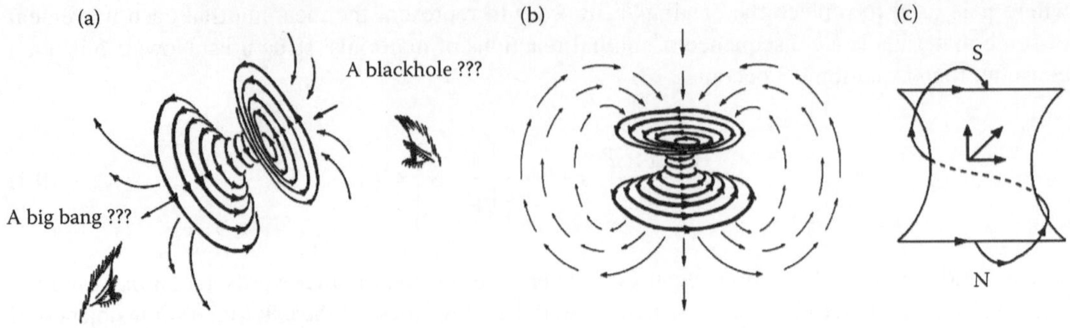

FIGURE 9.4 Eddy motion model of the general system.

9.6.2 Systemic Yoyo Structure of General Systems

Based on what is discussed above, the concepts of black holes, big bangs, and converging and diverging eddy motions are coined together (Wu and Lin 2002) in the model shown in Figure 9.4 for each object and every system imaginable. In particular, each system or object considered in a study is a multidimensional entity that spins about its invisible axis. If we fathom such a spinning entity in our three-dimensional space, we will have a structure as shown in Figure 9.4a. The side of the black hole sucks in all things, such as materials, information, energy, etc. After funneling through the short narrow neck, all things are spit out in the form of a big bang. Some of the materials, spit out from the end of a big bang, never return to the other side, and some will (Figure 9.4b). Such a structure, as shown in Figure 9.4a, is called a yoyo due to its general shape. More specifically, what this model says is that each physical entity in the universe, be it a tangible or intangible, a living being, an organization, a culture, a civilization, etc., can all be seen as a kind of realization of a certain multidimensional spinning yoyo with an invisible spin field around it. It stays in a constant spinning motion as depicted in Figure 9.4a. If it does stop its spinning, it will no longer exist as an identifiable system. What Figure 9.4c shows is that due to the interactions between the eddy field, which spins perpendicularly to the axis of spin, of the model, and the meridian field, which rotates parallel to the axis of spin (Figure 9.4b), all the materials returning to the black hole side travel along a spiral trajectory.

As for why materials in the universe and matters of the world rotate in the first place and continue to do so, according to the concept of uneven space and time of Einstein's theory of relativity (1987), we know that all materials have uneven structures. Out of these uneven structures, there naturally exists gradients. With gradients, there will appear forces. Combined with uneven arms of forces, the carrying materials will have to rotate in the form of moments of forces (Lin 2008, p. 31).

9.6.3 Empirical Justifications

The multidimensional yoyo model in Figure 9.4 is manifested in different areas of life. For example, each human being, as we now see it, is a three-dimensional realization of such a spinning yoyo structure of a higher dimension. To illustrate this end, let us consider two simple and easy-to-repeat experiments.

EXPERIMENT 1.1 FEEL THE VIBE

Let us imagine that we go to a sport event, say, a swim meet. Assume that the area of competition contains an Olympic-sized pool, and along one long side of the pool, there are about 200 seats available for spectators to watch the swim meet. The pool area is enclosed with a roof and walls all around the space.

Now, let us physically enter the pool area. What we find is that as soon as we enter the enclosed area of competition, we immediately fall into a boiling pot of screaming and jumping spectators, cheering for their favorite swimmers competing in the pool. Now, let us pick a seat a distance away from the pool deck anywhere in the seating area. After we settle down in our seat, let us purposelessly pick a voluntary helper standing or walking on the pool deck for whatever reason, either for her beauty or for his strange look or body posture, and stare at him or her intensively. Here is what will happen next: magically enough, before long, our stare will be felt by the person from quite a good distance; he or she will turn around and locate us in no time out of the reasonably sized and boiling audience.

By using the systemic yoyo model, we can provide one explanation for why this happens and how the silent communication takes place. In particular, each side, the person being stared at and us, is a high-dimensional spinning yoyo. Even though we are separated by space and possibly by informational noise, the stare of one party on the other has directed that party's spin field of the yoyo structure into the spin field of the yoyo structure of the other party. Even though the latter party initially did not know the forthcoming stare, when his or her spin field is interrupted by the sudden intrusion of another unexpected spin field, the person surely senses the exact direction and location where the intrusion is from. That is the underlying mechanism for the silent communication to be established.

When this experiment is done in a large auditorium where the person being stared at is on the stage, the aforedescribed phenomenon does not occur. It is because when many spin fields interfere with the field of a person, these interfering fields actually destroy their originally organized flows of materials and energy so that the person who is being stared at can feel only the overwhelming pressure from the entire audience instead of from individual persons.

This easily repeatable experiment in fact has been numerously conducted by some of the local high school students. When these students eat out in a restaurant and after they run out of topics to gossip about, they play the game they call "feel the vibe." What they do is to stare as a group at a randomly chosen guest of the restaurant to see how long it takes the guest to feel their stares. As described in the situation of swim meet earlier, the chosen guest can almost always feel the stares immediately and can locate the intruders in no time.

EXPERIMENT 1.2 SHE DOES NOT LIKE ME!

In this case, let us look at the situation of a human relationship. When an individual A has a good impression about another individual B, magically, individual B also has a similar and almost identical impression about A. When A does not like B and describes B as a dishonest person with various undesirable traits, it has been clinically proven in psychology that what A describes about B is exactly who A is himself (Hendrix 2001).

Once again, the underlying mechanism for such a quiet and unspoken evaluation of each other is based on the fact that each human being stands for a high-dimensional spinning yoyo and its rotational field. Our feelings toward each other are formed through the interactions of our invisible yoyo structures and their spin fields. Thus, when person A feels good about another person B, it generally means that their underlying yoyo structures possess the same or very similar attributes, such as spinning in the same direction, both being either divergent or convergent at the same time, both having similar intensities of spin, etc. When person A does not like B and lists many undesirable traits B possesses, it fundamentally stands for the situation that the underlying yoyo fields of A and B are fighting against each other in terms of either opposite spinning directions or different degrees of convergence, or in terms of other attributes.

Such quiet and unspoken evaluations of one another can be seen in any working environment. For instance, let us consider a work situation where quality is not and cannot be quantitatively measured, such as a teaching institution in the United States. When one teacher does not perform well in his or her line of work, he or she generally uses the concept of quality loudly in day-to-day settings in order to cover up his or her own deficiency in quality. When one does not have honesty, he or she tends to use the term "honesty" all the time. It is exactly what Lao Tzu (exact time unknown, Chapter 1) said over 2000 years ago: "The one who speaks of integrity all the time does not have integrity."

When we tried to repeat this experiment with local high school students, we found the following: when two students A and B, who used to be very good friends, turned away from each other, we ask A why she does not like B anymore. The answer is exactly what we expect: "Because she does not like me anymore!"

9.7 EXAMPLES OF APPLICATIONS

In this section, we look at four preliminary applications of the systemic yoyo model on the study of open complex giant systems. Each of these examples shows the fact that by relying on this systemic intuition, an advantage of the second dimension of science, investigations of open complex giant systems can become manageable, and practically tangible results can be obtained.

9.7.1 Systemic Law underneath Binary Star Systems

For two celestial bodies m_1 and m_2, which are distance r apart from each other, the attraction between them is given by Newton's law of universal gravitation:

$$F_{\text{grav}} = G\frac{m_1 m_2}{r^2} \tag{9.5}$$

where G is the universal gravitation constant.

Because there exist binary star systems, each of which consists of two stars that travel around their common center of mass in elliptical orbits, a natural question is: why are these stars not pulled together to form a single start? According to the law of universal gravitation, when $r \to 0$, the gravitational pull F_{grav} between the masses m_1 and m_2 should be approximately ∞. Thus, no masses should be able to fight against such an infinitely large force of attraction.

Now, let us apply the yoyo model to answer this question.

Assume that the masses m_1 and m_2 are the masses of the yoyo structures N and M, respectively. Then, no matter how these spin fields are situated, objects N and M act and react on each other with the gravitational pull F_{grav} given in Equation 9.5. However, at the same time, their spin fields keep them apart. More specifically, for the scenario in Figure 9.5a, when the eddy flows of N and M run into each other from the downside, they push each other away. In Figure 9.5b, the flow directions of N and M are totally opposite of each other. In Figure 9.5c, because the spin fields of N and M are both converging and their spinning directions are harmonic, the yoyo structures N and M have a tendency to combine and become one bigger eddy motion. Even though there is such a tendency, they still will not become one spin field (for details, see the discussion below). The scenario in Figure 9.5d is similar to that in Figure 9.5a, except that the pushing between N and M occurs at the upper side of the adjacent area.

In Figure 9.5e through h, due to the fact that all the spin fields of N and M do not rotate harmonically against each other, other than their gravitational pulls on each other, as given in Equation 9.5, there are uncompromising pushes between N and M. Unlike Figure 9.5c, none of the scenarios in Figure 9.5e through h has a tendency to have the spin fields of N and M combined and become a unified spin field.

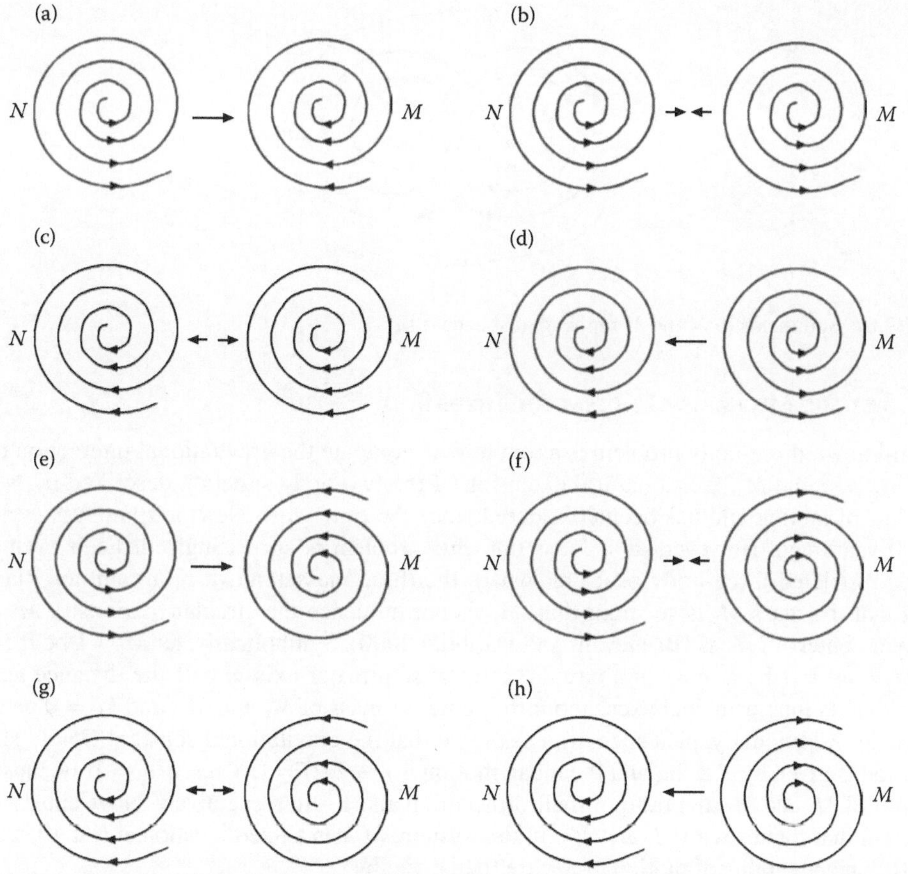

FIGURE 9.5 Interactions between two yoyo fields. (a) *N* diverges and *M* converges. (b) Both *N* and *M* diverge. (c) Both *N* and *M* converge. (d) *N* converges and *M* diverges. (e) *N* diverges and *M* converges. (f) Both *N* and *M* diverge. (g) Both *N* and *M* converge. (h) *N* converges and *M* diverges.

To address the previous question, we can see that each binary star system can only possibly come into existence out of the scenarios in Figure 9.5b, c, e, or h. It is because in all these cases, the field of *N* spins against that of *M*. However, in Figure 9.5(c) because both *N* and *M* are converging fields, even though they might form a binary star system (yes, they will—see discussion below for details), to us, we cannot actually see them because they are black holes. For the scenarios in Figure 9.5e and h, because one of the spin fields is convergent, we do not see a binary star system, either. That is, only the scenario in Figure 9.5b can lead to visible binary star systems.

Now let us study why the binary stars *N* and *M*, as shown in Figure 9.5b, travel individually in their own nearly elliptical orbits. When we see the layout in Figure 9.5b, the other sides of the yoyo structures of *N* and *M* look the same as the fields in Figure 9.5c. These two fields attract each other and have the tendency to combine into one spin field. Thus, the yoyo structures of *N* and *M* are pulled toward each other. However, when they are too close to each other, the diverging fields of *N* and *M*, as shown in Figure 9.5b, start to repel against each other (Figure 9.6). The force of repellence comes from the opposite spinning directions of the fields of *N* and *M*. Under the influence of this force, *N* is pushed away along the (i) direction and *M* along the (ii) direction (Figure 9.6). When *N* and *M* travel away from each other to a certain distance, the attractions of the other sides of *N* and *M* (Figure 9.5c) start to once again pull them together. Such an alternating effect of repellence and attraction keeps the stars *N* and *M* together and on their individual orbits.

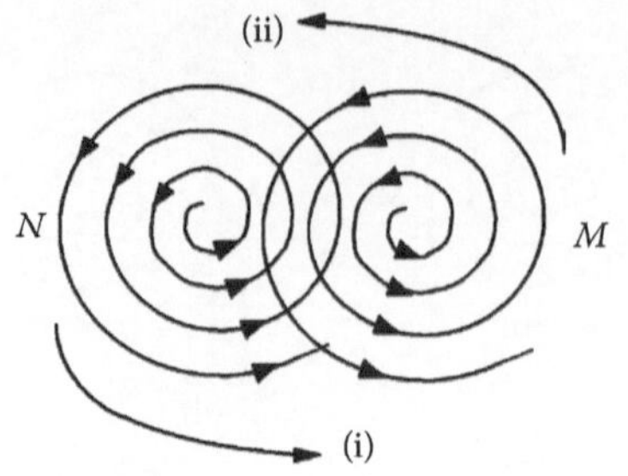

FIGURE 9.6 Spin fields of N and M repel against each other.

9.7.2 Systemic Mechanism underneath Three-Body Problem

The well-known three-body problem is about how to compute the gravitational interaction of three masses M_1, M_2, and M_3. It is a natural extension of the two-body situation described by Newton's second law of motion and has been considered since the time when Newton's law was established over 300 years ago. However, as it turns out, this problem is surprisingly difficult even for the so-called restricted three-body problem, where the three masses move in a common plane, and satisfies that the mass M_3 is so small that it does not influence the circular motions of M_1 and M_2 about their centers of mass (Basdevant and Dalibrad 2000). Symbolically, let $M_1 = 1$ be the largest mass; $M_2 = m \ll 1$ be a mass in a circular orbit of semimajor axis a, half the distance across an ellipse along its long principal axes, about the center of mass of M_1 and M_2; and $M_3 = 0$ be a massless particle. Additionally, pick such dimensions so that the gravitational constant $G = 1$. Then, the orbit period is given by $T \equiv 2\pi$, and the mean motion is $n \equiv 2\pi/T = 1$. Thus, $n^2a^3 = 1$. Because of the definitions of M_1 and M_2, the radii of their orbits are μ and 1 – μ, respectively. Now, enter a coordinate system that rotates with M_1 and M_2. In this system, M_1 has a fixed location at (–μ, 0), and M_2 at (1 – μ, 0). The equations of motion of M_3 are then given by

$$\ddot{x} = 2\dot{y} + x + (1-\mu)\frac{x+\mu}{r_1^3} - \mu\frac{x-1+\mu}{r_2^3} \tag{9.6}$$

$$\ddot{y} = -2\dot{x} + y - (1-\mu)\frac{y}{r_1^3} - \mu\frac{y}{r_2^3} \tag{9.7}$$

where $r_1 = \sqrt{(x+\mu)^2 + y^2}$ and $r_2 = \sqrt{(x-1-\mu)^2 + y^2}$. This is then converted to a Hamiltonian system with two degrees of freedom, with $q_1 = x,\ q_2 = y,\ p_1 = \dot{x} - y,\ p_2 = \dot{y} - x$, and

$$H = \frac{1}{2}\left(p_1^2 + p_2^2\right) + p_1q_2 - p_2q_1 - \frac{1-\mu}{r_1} - \frac{\mu}{r_2}. \tag{9.8}$$

There is one integral of motion known as the Jacobi integral, defined as follows:

$$C = -2H = x^2 + y^2 + \frac{2(1-\mu)}{r_1} + \frac{2\mu}{r_2} - \dot{x}^2 - \dot{y}^2. \tag{9.9}$$

No other integral is known as of this writing. Making a mapping of $\dot{x}$ versus x for $C = 4.5$, there are four elliptical fixed points corresponding to periodic orbits around M_1 and M_2 in either direction.

At $C = 4$, the trajectory is chaotic. Also, primary resonance occurs when M_3 makes $J + 1$ orbits in the same time that M_2 makes J, or $n = (J + 1)/J$, where J is an integer. The period between conjunctions is then $2\pi J$. For $J \geq \mu^{-2/7}$, there is chaos.

In terms of practical use of the work discussed above, if there is any such potential, we have to examine the unrealistic assumptions:

1. M_3 is massless; the mass of M_2 is much smaller than that of M_1.
2. M_2 travels along a circular orbit.

Even with such strict assumptions in place, we either run into the technical difficulty of not being able to get much or are trapped in chaos. In other words, it is necessary to introduce an alternative method to help out the current study of the three-body problem. At the same time, this brief recall of the existing literature also indirectly explains why the current study of open complex giant systems has faced great difficulty, because in these studies, we have to consider hundreds and even thousands of variables and their associations. However, what is presented here indicates that we have great difficulty even with just three bodies, let alone with hundreds or thousands of them.

In the rest of this subsection, let us see what results we can obtain with the figurative analysis method developed along with the yoyo model.

In order not to make this presentation too long, let us consider only three visible bodies M_1, M_2, and M_3. For other possibilities, the interested reader is advised to consult the work of Lin (2007, 2008). This is also the case of the traditional consideration. By visible, in terms of the yoyo model, we mean that the individual yoyo structures (Figure 9.7a) of M_1, M_2, and M_3 have their big bang sides situated side by side against each other. Then there are four possibilities for us to consider (Figure 9.7), where the locations of M_1, M_2, and M_3 are relative. After analyzing and comparing the four possibilities in Figure 9.5, it can be seen that all the scenarios in Figure 9.7a, c, and d are essentially the same. Thus, without loss of generality, let us analyze cases a and b only.

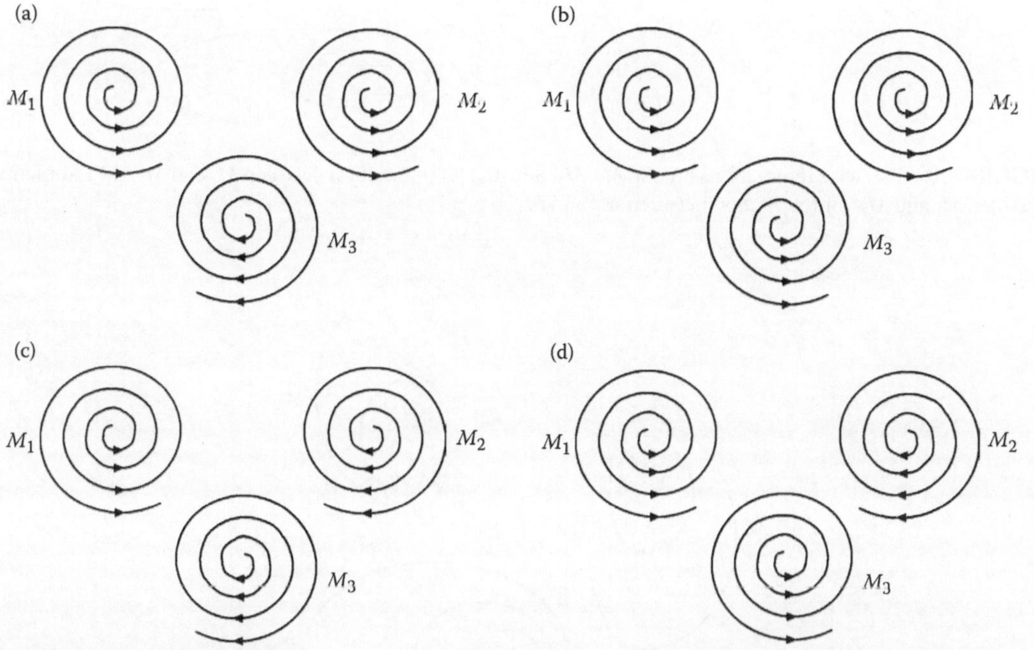

FIGURE 9.7 Three visible bodies interacting with each other. (a) M_3 does not spin harmonically with M_1 and M_2. (b) M_3 spins harmonically with M_1 and M_2. (c) M_3 does not spin harmonically with M_1 and M_2. (d) M_3 spins harmonically with M_1 and M_2.

For the scenario in Figure 9.7a, let us first consider the shape of the spin fields of M_1, M_2, and M_3. When each of the spin fields exists alone without any interference from the others, the spins will be nearly circular. Interference between any two of them will force their circular spins to the shape of an ellipse (Figure 9.8). Thus, when the effect of the third spin field is added, the elliptical spin fields will be further twisted from their original circular shapes. For example, the ellipses of M_1 and M_2 in Figure 9.8a will be pressured further to the right, and the elliptical field of M_3 in Figure 9.8b and c will be further squeezed to the upside direction. Thus, a more accurate representation of the spin fields of M_1, M_2, and M_3, considering them interacting with each other at the same time, is given in Figure 9.9.

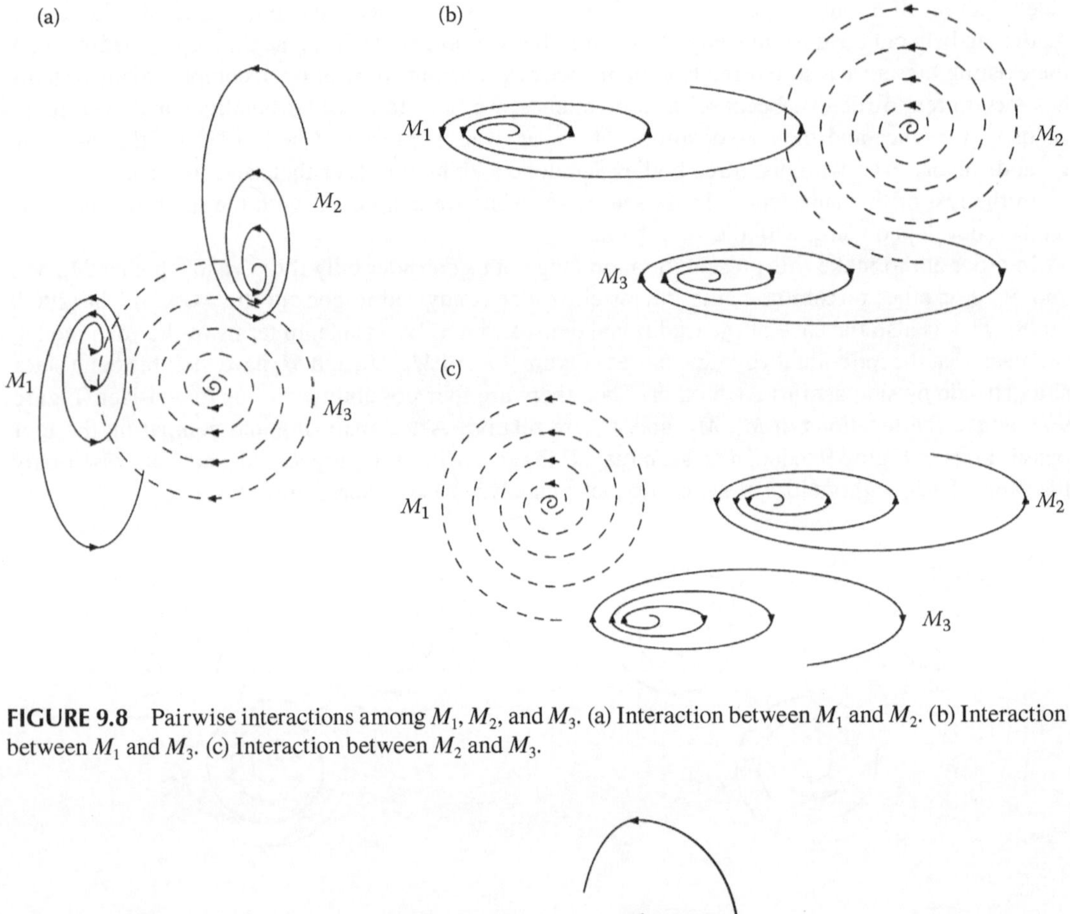

FIGURE 9.8 Pairwise interactions among M_1, M_2, and M_3. (a) Interaction between M_1 and M_2. (b) Interaction between M_1 and M_3. (c) Interaction between M_2 and M_3.

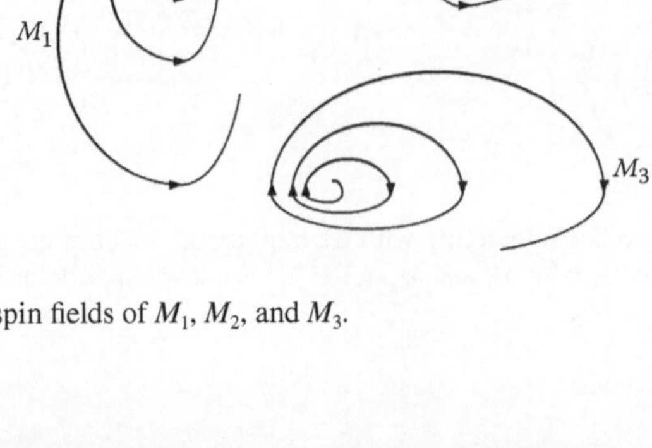

FIGURE 9.9 Interacting spin fields of M_1, M_2, and M_3.

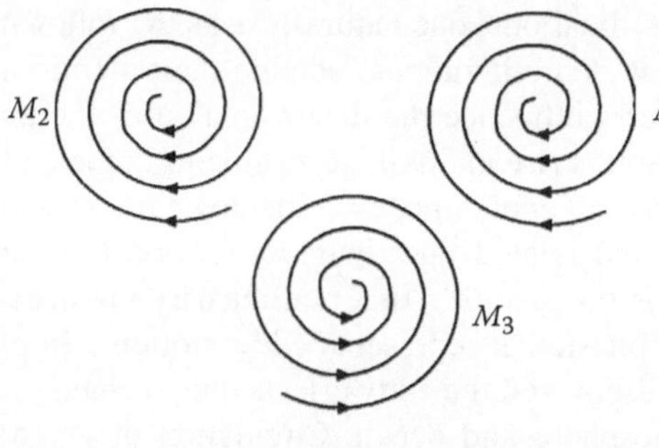

FIGURE 9.10 Other sides of M_1, M_2, and M_3.

If we look at the black hole side of the masses M_1, M_2, and M_3 (Figure 9.10), it can be seen that M_3 is only loosely attached to the binary star system (M_1, M_2). If M_3 stays with M_1 and M_2, M_3 will travel with M_2 (in a similar direction) more closely than with M_1. Chances are, if there appears a yoyo structure M_4 in the nearby region, M_3 might very well leave M_1 and M_2 and join M_4 to form a binary star system.

9.7.3 Formation of Civilizations

In this subsection, we will see how the systemic yoyo model can be employed to investigate such open complex giant systems as civilizations, hoping that we can produce useful and practically beneficial results that can assist policymakers at national and international levels. For more detailed and intensive research along this line, please consult the work of Lin and Forrest (2010a,b,c).

Generally, the international history is composed of stories of civilizations from ancient Sumerian and Egyptian to Classical and Mesoamerican to Christian and Islamic civilizations and through successive manifestations of Sinic and Hindu civilizations. On different perspectives, methodology, focus, and concepts, many distinguished historians, sociologists, and anthropologists have well studied the causes, emergence, rise, interactions, achievements, decline, and fall of civilizations (Kissinger 1994).

Civilization means to be a space, a cultural area, a collection of cultural characteristics and phenomena (Braudel 1980, pp. 177, 202), a particular concatenation of worldview, customs, structures, and culture (both material culture and high culture) that forms some kind of historical whole and that coexists with other varieties of this phenomenon (Wallerstein 1991, p. 215), and a kind of moral milieu encompassing a certain number of nations, each national culture being only a particular form of the whole (Durkheim and Mauss 1971, p. 811). For example, blood, language, religion, and way of life were what the ancient Greeks had in common and what distinguished them from the Persians and other non-Greeks. When people are divided by cultural characteristics, they are grouped into civilizations, which are comprehensive; just as in the concept of general systems (Lin 1999), none of their constituent units (or elements) can be fully understood in isolation without reference to the encompassing whole (civilization). A civilization is a maximal whole or totality in terms of its fundamental values and philosophical assumptions. Civilizations comprehend without being comprehended by others (Toynbee 1937, p. 455), and (they) have a certain degree of integration. Their parts are defined by their relationship to each other and to the whole. If the civilization is composed of states, these states will have more relation to one another than they do to states outside the civilization. They might fight more and engage more frequently in diplomatic relations. They will be more interdependent economically. There will be pervading aesthetic and philosophical currents (Melko 1969, pp. 8–9). For example, European states share cultural features that separate them from others such as Chinese or Hindu communities. Also, Chinese, Hindus, and Westerners are not part of any broader cultural entity so that they are from different civilizations.

Related to the concept of civilizations, one naturally asks the following questions: How do civilizations or cultures form? How do civilizations redefine their identities, composition, and shapes throughout history? What factors influence the determination of the ambiguous but real dividing line between civilizations? Empires rise and fall, governments come and go. However, civilizations remain and survive political, social, economic, even ideological upheavals. Why?

To address some of these and related questions, let us first look at Bjerknes' (1898) circulation theorem (Hess 1959). This result shows that nonlinearity mathematically stands (mostly) for singularities, and in terms of physics, it represents eddy motions. In particular, at the end of the nineteenth century, Bjerknes discovered the eddy effects due to changes in the density of the media in the movements of the atmosphere and ocean. Circulation means a closed contour in a fluid. Symbolically, each circulation Γ is defined as the line integral about the contour of the component of the velocity vector locally tangent to the contour. In symbols, if $\vec{V}$ stands for the speed of a moving fluid, S is an arbitrary closed curve, and $\delta\vec{r}$ is the vector difference of two neighboring points of the curve S (Figure 9.11), then a circulation Γ is defined as follows:

$$\Gamma = \oint_S \vec{V}\delta\vec{r} \,. \tag{9.10}$$

Through some ingenious manipulations (Wu and Lin 2002), the following well-known Bjerknes' circulation theorem is obtained:

$$\frac{d\vec{V}}{dt} = \iint_\sigma \left(\frac{1}{\rho}\right) \times (- \quad p)\delta\sigma - 2\Omega\frac{d\sigma}{dt}, \tag{9.11}$$

where σ is the projection area on the equator plane of the area enclosed by the closed curve S, p is the atmospheric pressure, ρ is the density of the atmosphere, and Ω is the earth's rotational angular speed.

The left-hand side of Equation 9.11 represents the acceleration of the moving fluid, which according to Newton's second law of mechanics is equivalent to the force acting on the fluid. On the right-hand side, the first term is called a solenoid term in meteorology. It originated from the interaction of the p- and ρ-planes due to uneven density ρ so that a twisting force is created. Consequently, materials' movements must be rotations with the rotating direction determined by the equal p- and ρ-plane distributions. The second term in Equation 9.11 comes from the rotation of the earth.

Because uneven densities create twisting forces, fields of spinning currents are naturally created. Such fields do not have uniformity in terms of types of currents. Clockwise and counterclockwise eddies always coexist, leading to destructions of any preexisting smooth fields of currents. What is important is that the concept of uneven eddy evolutions reveals that forces exist in the structures of evolving entities and do not exist independently outside of the entities.

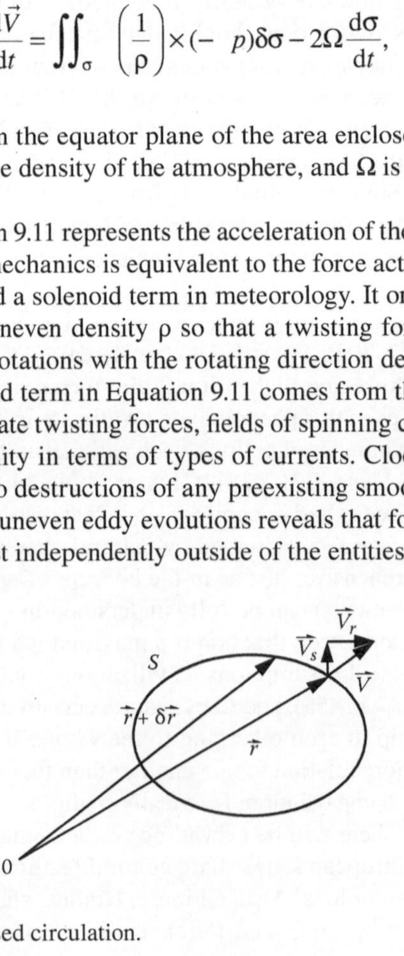

FIGURE 9.11 Definition of a closed circulation.

Now, let us look at the previous and related questions about civilizations.

At the start of time when still living in primitive conditions, due to the existing natural conditions and available resources within the environment, people formed their elementary beliefs, basic values, and fundamental philosophical assumptions. Since the population density was low, and tools available for production, conquering surroundings, etc. were extremely limited and inefficient, minor obstacles of nature in today's standard easily divided people on the same land into small tribes. [For theoretical studies and computer simulations for the formation of communities in living and nonliving beings, please consult the work of Vicsek et al. (1995) and the references therein.] Since these individual and separated tribes in reality lived in the same natural environment, they naturally held an identical value system and an identical set of philosophical assumptions, on which they reasoned in order to explain whatever was inexplicable; developed approaches to overcome hardships; and established methods to administrate members in their individual tribes. As time went on, better tools for production and transportation and better practices of administration were designed and employed in various individual tribes. The natural desire for better living conditions paved the way for the inventions of new tools, discovery of new methods of reasoning, and the introduction of more efficient ways of administration to pass around the land through word of mouth. Thus, a circulation of information, carried around and spread throughout the land by people with special skills, started to form. Along with these talented and mobile people, knowledge and commercial goods were also part of the circulation. As a circulation started to appear, Bjerknes's circulation theorem guarantees the appearance of abstract eddy motions over the land consisting of migration of people, spread of knowledge and information, and transportation of goods. That explains how civilizations were initially formed and why different civilizations have different underlying assumptions and values of philosophy.

At this junction, a natural question that needs to be addressed is to justify the validity for us to employ the Bjerknes' circulation theorem, because in theory, this theorem holds true only for fluids. To this end, first, each human organization, as a whole, is made up of a physical body, its internal structure, and its interactions with the environment. It is the cultural, philosophical, spiritual, psychological, etc., aspects of the organization that guarantee the body to be an identifiable whole. Hence, the organization can be seen as a conceptual fluid of information, knowledge, and energy flows, which circulate within the inside of, going into, and giving off from the organizational body, with individual human beings as local concentrations of these and other abstract material flows. Second, the universe is a huge ocean of eddies, which changes and evolves constantly. That is, the totality of the physically existing world can be studied as fluids with solids as local concentrations of energy of various fluids. Third, as described in the previous paragraph, people in the land helped to circulate beliefs, basic values, fundamental philosophical assumptions, knowledge, commercial goods, etc., all of which are studied conventionally using continuous or differentiable functions in social sciences in general and economics in particular. When these aspects of a civilization are modeled by such functions, they are generally seen in physics and mathematics as flows of fluids and are widely known as flow functions. Specifically, in the formation of a civilization, it is these commonly shared aspects (of fluids) that make the land to have a living culture, where individual persons are simply local "impurities" of the fluids, and each of the "impurities" carries some concentrated amount of "energy," "information," etc.

As for how civilizations redefine their identities, composition, and shapes throughout history, the previous discussion indicates that it was all accomplished through the greater desire of controlling more natural resources that would make one's own yoyo structure more powerful and fighting off different beliefs and value systems that might very well destroy one's own yoyo structure. That is, when each civilization is seen as a spinning yoyo, differences in the natural environments make the abstract but very realistic civilizational yoyos spin in different ways, such as different spinning speeds, angles, and directions. When two civilizations met either in cooperation or in conflict, in order to sustain themselves viably in the contact, each of them absorbed some elements of the other to benefit themselves individually. To accomplish this without destroying the existing set of values and philosophical assumptions, the receiving civilization had to reformulate the needed elements from the other civilization in its own terms before adding these external elements into their basic system of values and philosophical assumptions. Even though such activities have been carried out throughout history by various cultures, nations, and civilizations unconsciously, the underlying theoretical guarantee for success is given in the well-known Godel's theorem below:

Theorem 9.1

[Godel's incompleteness theorem (Hewitt 2008)]. For any consistent formal, recursively enumerable theory that proves basic arithmetic truths, an arithmetical statement that is true, but not provable in the theory, can be constructed. That is, any effectively generated theory capable of expressing elementary arithmetic cannot be both consistent and complete.

Here, the word "theory" refers to an infinite set of statements, some of which are taken as true without proof, which are called axioms, and others (the theorems) that are taken as true because they are implied by the axioms. The phrase "provable in the theory" means derivable from the axioms and primitive notions of the theory using standard first-order logic. A theory is consistent if it never proves a contradiction. The phrase "can be constructed" means that some mechanical procedure exists that can construct the statement, given the axioms, primitive notions, and first-order logic. The elementary arithmetic consists merely of additions and multiplication over the natural numbers. The resulting true but not provable statement is often referred to as the Godel sentence for the theory, although there are infinitely many other statements in the theory that share with the Godel sentence the property of being true but not provable from the theory. ■

9.7.4 Dynamics of Large and Small Business Projects

In this subsection, from the point of view of the systemic yoyo model, we will investigate the roles of small and large projects in the development and evolution of a commercial company and why companies with a history of taking on large projects tend to eventually fail with large projects. In particular, when a spinning yoyo focuses on taking in only smaller and weaker rotational fields, this specific yoyo will have a better chance to be long-lasting and to grow with each acquisition of a smaller and weaker field than when it also takes in powerful rotational fields. On the other hand, when a spinning yoyo engages in conflicts only with same-size or larger and stronger rotational fields, sooner or later, this specific yoyo will have to face off with a much more powerful rotational field than it is. Thus, as soon as it faces with such a field that is so much more powerful than itself, it just simply gets itself destroyed.

With this systemic reasoning in place, assume that in time period 1, the company of concern commits I_s of investment to a small project, and in time period 2, the fundamental value of the project is v_s for the company when the project is fully completed and the benefit of the completed project is materialized in the marketplace. (To make the analysis relatively simple, throughout this subsection, we assume no discount of the future over time.) Thus, the expected profit from embarking on this small project is given as follows:

$$(v_s - I_s R_s) P_s, \tag{9.12}$$

where $R_s > 1$ stands for the gross interest spent on the total investment amount I_s, and $P_s \approx 1$ is the probability for the small project to be successful.

Similarly, assume that in period 1, the company invests as much as $I_L \gg I_s$ for a large project. In time period 2, the project is completed, and its benefit is fully materialized so that the fundamental value the project creates for the company is v_L. Thus, in this period, the expected profit from engaging in this project is

$$(v_L - I_L R_L)P_L, \tag{9.13}$$

where $R_L > R_s > 1$ is the gross interest spent on the total investment I_L, and $P_L \approx 0$ is the probability for the large project to be successful for the reason that large projects tend to involve many unexpected factors that may turn out to be detrimental to the success of the projects. The reason why $R_L > R_s$ is that the large project tends to be much riskier than the small project so that the lender of funds charges a much higher rate of interest.

Now, the large project is a lot riskier with much greater return if successful than the small project. Therefore, if the company decides on investing in the large project, the return has to be proportionally much higher than that of the small project. Thus, we have the following equation:

$$k(v_s - I_s R_s)P_s = (v_L - I_L R_L)P_L \tag{9.14}$$

where k is the constant of proportionality. By rewriting this equation, we obtain

$$kv_s P_s = v_L P_L + kI_s R_s P_s - I_L R_L P_L. \tag{9.15}$$

If $P_s = \ell P_L$, for some large real number $\ell > 1$, Equation 9.15 can be rewritten as follows:

$$k\ell v_s = v_L + k\ell I_s R_s - I_L R_L \text{ or } k\ell(v_s - I_s R_s) = v_L - I_L R_L. \tag{9.16}$$

Equation 9.16 provides an analytic explanation for why some decision-makers like to take on large projects, because the success of just one such project would produce as much financial profit as $k\ell$ small projects completed one by one. Considering the meanings of the constants k and ℓ, the product $k\ell$ can potentially be a really large number. In particular, it means that if one is successful with a large project just once, twice, or a few times, he or she would strike it huge in the business world. On the other hand, if one is risk-averse and takes on small projects only, it may very well take him or her forever to build his or her business to a certain respectable magnitude.

The definition of small and large projects is asset dependent. That is, when companies have different market capitalizations, their definitions of small and large projects change from one company to another. A specific project could be considered small for one company and immensely large for another company. Thus, Equation 9.16 provides an explanation for why most new start-ups fail because of a lack of funds, since due to their limited financial resources, almost all projects that new start-ups take on would be considered large. In this case, if they do not make it with just one project, the companies would be over permanently.

Next, let us consider the market response to large projects. Assume that a company has a choice of either taking on relatively safe small projects or one large relatively risky project in time period 1. If the company chooses the option of small projects, then its stock is traded at p_s ($<$, $=$, or $>v_s$) in time period 1, where v_s is the fundamental share value of the company. In this time period, an investor buys $n(p_s)$ shares of the stock at the market value p_s a share. His or her total cost of investment is $I_s = n(p_s)p_s$. Assume that in time period 2, the trading price equals the fundamental value v_s. Thus, the profit of this investor is given as follows:

$$n(p_s)v_s - I_s R_s = \frac{I_s v_s}{p_s} - I_s R_s = I_s\left(\frac{v_s}{p_s} - R_s\right), \tag{9.17}$$

where, as before, $R_s > 1$ is the gross interest spent on the total investment I_s.

On the other hand, if the company takes on the choice of one large, risky project, the same investor buys in period 1 $n(p_L)$ shares of the company stock at p_L a share, where p_L<, =, or $>v_L$, with v_L being the fundamental share value of the company stock. Therefore, the total cost of investment to the investor is $I_L = n(p_L)p_L$. In time period 2, assume that the project is completed with its consequence known so that the fundamental share value v_L is materialized. Thus, the total profit for the investor is given as follows:

$$n(p_L)v_L - I_L R_L = \frac{I_L v_L}{p_L} - I_L R_L = I_L\left(\frac{v_L}{p_L} - R_L\right), \tag{9.18}$$

where $R_L > R_s > 1$ is the gross interest spent on the total investment I_L.

If the investor wants to produce more return on his or her investment of the large project option, then from Equations 9.17 and 9.18, we have

$$I_s\left(\frac{v_s}{p_s} - R_s\right) < I_L\left(\frac{v_L}{p_L} - R_L\right). \tag{9.19}$$

From the assumption that $I_s = I_L$, meaning that the investor allocated the same amount of funds to each option of investment, we have

$$\frac{v_s}{p_s} < \frac{v_L}{p_L} - (R_L - R_s) < \frac{v_L}{p_L}. \tag{9.20}$$

From the assumption that the choice of small projects contains a huge number of relatively safe small projects completed one by one so that the totality of the small projects does not become a large project, we can take $\frac{v_s}{p_s} \approx 1$. That is, the market read on the potential values of the relatively safe small projects in time period 1 is very close to that of the fundamental value materialized in period 2. Therefore, we have from Equation 9.20 that

$$\frac{v_L}{p_L} > 1. \tag{9.21}$$

This end implies that for the option of taking on one large risky project, the marketplace tends to underestimate the benefit of the potential success that the large project would create for the company. That is, investors value a large number of relatively safe small projects completed one by one more than one single large project.

9.7.5 Systemic Structure of Mind

The totality of each human being is four-dimensional: body, mind, heart, and spirit. It is physically made up of flesh, bone, blood, hair, and brain cells, and systemically of self-awareness, imagination, conscience, and free will (Covey 1989, p. 70). By using self-awareness, humans are able to examine their own thoughts and have the freedom to choose their response to whatever they come across or whatever is imposed on them. With imagination, they are able to create a (fantasy) world in their mind beyond the present reality. With conscience, they are deeply aware of what is right and wrong, of the principles that govern their behavior, and of a sense of the degree to which their thoughts and actions are in harmony with the principles. They have free will to act based on their self-awareness, free of all other influences. In this subsection, we will see how the systemic yoyo model can be employed to provide new insights as for what the human endowments—self-awareness, imagination, conscience, and free will—are and to address some of the very important questions related to the phenomenon of these human endowments. This presentation is based on the works of Lin and Yi (2010) and Lin and Forrest (2011a,b).

9.7.5.1 Phenomenon of Self-Awareness

Self-awareness means the human awareness that one exists as an individual and an entity that is separate from other people and objects with private thoughts and individual rights (Cooke 1974, p. 106). It also includes the understanding that other people are similarly self-aware. Self-awareness is a self-conscious state in which attention focuses on oneself. It makes people more sensitive to their own attitudes and dispositions (Branden 1969, p. 41). At various circumstances, self-consciousness is used synonymously with self-awareness. It is credited only with an individual's development of identity; it is nonpositional and not attached to any particular location (Gerassi 1989).

Self-consciousness includes such elements as self-discipline, carefulness, thoroughness, organization, deliberation (the tendency to think carefully before acting), and need for achievement (Salgado 1997). With different degrees of ability to mobilize their self-consciousness, people are greatly affected by their own self-awareness, as some people scrutinize themselves more than others. Also, the importance of self-consciousness is emphasized differently from one culture to another. Even so, individuals maintain a varying degree of self-motivation and self-determination (Cohen 1994, p. 136).

Now, we use the systemic yoyo model to explain where human self-awareness or self-consciousness is from.

For any person, his or her self-awareness is really a natural consequence of his or her underlying multidimensional systemic yoyo structure. Specifically, in theory, we can think of the totality of all materials that can be physical, tangible, intangible, or epistemological, and that all these matters are contained in the yoyo structure of the person, if he or she is situated in isolation from other yoyo structures. That is, he or she is a whole being of his or her own. However, in reality, no man lives in isolation; systems are of various kinds and scales; and the universe can be seen as an ocean of eddy pools of different sizes, where each pool spins about its center or axis. To this end, one good example in our three-dimensional physical space is the spinning field of air in a tornado. In the solenoidal structure, at the same time when the air within the tornado spins about the eye in the center, the systemic yoyo structure continuously sucks in and spits out air. In the spinning solenoidal field, the tornado takes in air and other materials, such as water or water vapor on the bottom, lifts up everything it took in into the sky, and then continuously spays out the air and water from the top of the spinning field. At the same time, the tornado also breathes in and out with air in all horizontal directions and elevations. If the amounts of air and water taken in by the tornado are greater than those given out, then the tornado will grow larger with increasing effect on everything along its path. That is the initial stage of formation of the tornado. If the opposite holds true, then the tornado is in its process of dying out. If the amounts of air and water that are taken in and given out reach an equilibrium, then the tornado can last for at least a while. In general, each tornado (or a systemic yoyo) experiences a period of stable existence after its initial formation and before its disappearance.

Because each person is a system with his or her yoyo structure, he or she also constantly takes in and spits out materials. Other than breathing in and out materials from the black hole and big bang sides, the yoyo structure also takes in and gives out materials in all horizontal directions and elevations, just as in the case of tornadoes discussed earlier. As the spin field constantly takes in and gives out materials, there does not exist any clear boundary between the yoyo structure and its environment, which is analogous to the circumstance of a tornado that does not have a clear-cut separation between the tornado and its surroundings.

This description of the general yoyo structure of a human being provides a fundamental theoretical explanation for where human self-awareness or self-consciousness is from. In particular, because each system breathes in and out materials throughout each part of its surface area, between different systems, there is a constant battle for the following.

1. Pushing against each other when materials that are emitted outward are thrown against each other. This is how each person feels that he or she is separate from other people and objects.
2. Attracting toward each other when different systems try to absorb a piece of material of common interest. This is how the feeling of individual rights and entitlements is created.

These constant battles between systems collectively make them aware of their own existence, the existence of others, and their private thoughts, be they human beings or objects. Of course, the strength of such self-awareness is determined by the intensity of individual yoyo's spin field.

The reason why people focus on themselves and are sensitive to their own attitudes and dispositions is because their very own viability is at stake, where their viability is determined by how strongly their yoyo field spins and how much and how effectively the field can absorb materials from the environment.

9.7.5.2 Systemic Structure of Imagination

Imagination means the faculty of the human mind to imagine, to form mental images or concepts of what is not actually present to the biological senses, and the action or process of forming such images or concepts. With this faculty of the mind, one derives meaning to experience and understanding to knowledge. It is a fundamental human endowment through which people make sense of the world (Norman 2000, pp. 1–2) and plays a key role in the learning process (Egan 1992, p. 50). Only through imagination do we encounter everything in life. Whatever we touch, see, and hear coalesces into a mental picture via the faculty of imagination. The ability to imagine is the innate ability through which we form our complete personal philosophical value and belief systems within the mind based on elements derived from sense perceptions of the shared physical world (Harris 2000, p. 94).

Now, let us address the following question: Is there any underlying mechanism over which the human imagination works and functions in its action and process to imagine, to form mental images or concepts of what is not actually present to the biological sense organs?

Going along with the systemic yoyo model of self-awareness, imagination is also a natural consequence of the human yoyo structure. Because biological sense organs are simply the body parts of the three-dimensional realizations of the underlying multidimensional spinning fields of human yoyos, a great deal of the field structures of the world, these sense organs cannot really pick up. From the fact that each human yoyo constantly interacts with other spin fields in the forms of pushing against each other or grabbing over materials of common interests, it follows that the experience and knowledge gained from the human field interactions are much richer than what is known based on the information collected by the sense organs. This end provides a systemic model for human imagination. That is, the so-called human imagination is in fact the collection of all the conscious (meaning through the sense organs) and unconscious (meaning not through any of the sense organs) records of the interactions between the underlying yoyo fields. When one needs to establish an abstract image or concept, he or she relies on his or her various trainings to tap into this reservoir of experience and knowledge collected both consciously and unconsciously from his or her field interactions with others. Here, the level of training on his or her self-awareness determines how deeply he can reach into this reservoir. Consequently, it determines how thought-provoking his or her established concepts and images will be.

It is shown in the work of Lin (2008) that the material world and human thoughts share the same structure, the yoyo field structure. Thus, when human imagination is called for action, it simply matches what is given, be it a difficult problem, a challenging situation, or a difficult task, with some of the relevant knowledge or experience from the reservoir. If the match is nearly perfect, the action will be considered successful. If the match is not good or no match is found, then new experience or knowledge will be added into the reservoir. Remember, this reservoir is composed of two parts. One part is the collection of all learned knowledge and experience through the sense organs, and the other part is the collection of all knowledge and information collected not through the sense organs. With adequate training on self-consciousness, one can "magically" apply a great deal of materials out of the reservoir of his or her imagination.

9.7.5.3 Makeup of Conscience

Conscience means the ability with which people distinguish whether their actions are right or wrong and are deeply aware of the principles that govern their behavior and a sense of the degree to which their thoughts and actions are in harmony with the principles. When people do things not in agreement with their moral values, their conscience will make them feel remorse; when their actions conform to their moral values, their conscience brings them feelings of rectitude or integrity. Conscience also represents the attitude that informs a person's moral judgment before performing any act.

Many authors from different angles, including religious views, secular views, and philosophical views, have studied the concept of conscience and its role in ethical decision-making. Religious views of conscience link this concept to an inherent morality, to the universe, and/or to divinity with many nuances. Modern-day scholars in fields of ethology, neuroscience, and evolutionary psychology treat conscience as a function of the brain that developed to facilitate reciprocal altruism within societies (Tinbergen 1951; Pfaff 2007; Buss 2004). It is found that conscience can prompt different people in quite different directions, depending on their beliefs. This observation suggests that while the capacity for conscience is probably genetically determined, its subject matter is probably learned, like language, as part of a culture. For example, numerous case studies of brain damage have shown that damage to specific areas of the brain can result in reduction or elimination of inhibitions with a corresponding radical change in behavior patterns. When the damage occurs to adults, they may still be able to perform moral reasoning; however, when the damage occurs to children, they may never develop that ability (Hare 1970).

Let us now employ the systemic yoyo model to discuss the following questions: Where does our conscience come from? Why can it help distinguish whether an action is right or wrong?

Based on the systemic understanding of self-awareness and imagination, it can be seen that the so-called conscience is simply a partial function with two output values such that the partial function is defined on some of the spin patterns and interactions of these patterns stored in the reservoir of imagination. Here, certain kinds of field flows and interactions of the flows are assigned the value of +, known as being right or moral, and some other kinds of flows and interactions are assigned the value of –, known as being wrong or immoral. The reason why this function is only partial is because other than the flow patterns and their interactions that are assigned either a + value or a – value, there are still plenty of other flow patterns and interactions in the reservoir of imagination that are not assigned with any + or – value. For the sake of convenience of communication, let us call this partial function the ± function. By the domain of this function, we mean the totality of all the flow patterns and the flow interactions, each of which has been assigned either a + or – value.

The assignments of + and or – values of certain flow patterns and interactions of flows start when the person is still an infant and continues throughout the entire life of the person. Initially, the domain of the ± function is the empty set. As the person grows older, the domain starts to expand. Also, if some dramatic event happens during his or her lifetime, the assignments of the + and/or – values may be altered.

As suggested by the formations of sub-eddies in Figure 9.12, some of the very first assignments of + values should be given to the flows heading in the same direction, and – values to currents flowing in the opposite direction. It is because same directional flows feed into the existence of the sub-eddies, while opposite directional currents slow down or even attempt to stop the rotations of

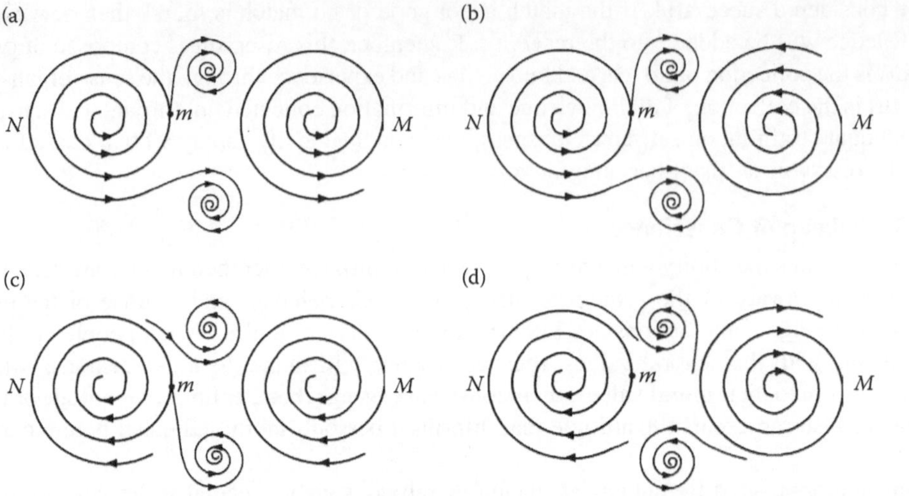

FIGURE 9.12 *N* and *M* jointly produce sub-eddies in areas of their control. (a) Sub-eddies created by two diverging yoyos. (b) Sub-eddies created by one convergent and one divergent yoyo. (c) Sub-eddies created by two converging yoyos. (d) Sub-eddies created by one convergent and one divergent yoyo.

the sub-eddies. In daily language, the initial + values are given to behaviors that help strengthen the wellbeing of all children involved and the initial – values to the behaviors that damage or destroy the wellbeing. For instance, when two children fight over a toy, most likely, the parents would demand the children to share, reinforcing a + value on making everybody happy, even though the parents know very well which child actually owns the toy. When two little children are hitting each other, the parents would simply stop the fight forcefully, indicating the assignment of a – value to violence.

After people reach a certain age or level of maturity, their ± function will be quite well defined and rooted deeply in their head. Thus, whatever action they take or thought they think of, they would unconsciously compare it to the elements in the domain of their ± function. If the action or the thought has a + value, they sense the feeling of rectitude as the consequence of being constantly appraised by adults for such occasions; if the action or thought has a – value, they feel remorse, since similar situations have been cursed regularly by grownups, and they have let their caretakers down one more time. If the action or the thought does not have a well-assigned + or – value, people will feel afraid of the potentially uncertain reactions from others. Afterward, they receive either a + or – value, or they will start their journey to explore the potential value for their specific action or thought by pushing this action further or thinking of the thought deeper until they reach a certain outcome or they are attracted to some other more interesting, urgent, or important topic before reaching any meaningful + or – value.

Therefore, is one's capacity for conscience genetically determined? Our answer is an indirect *yes*, because our discussion on conscience indicates that conscience is completely established on imagination, which in turn is dependent on self-awareness, while self-awareness is an innate ability. On the other hand, the subject matter of conscience, the specific content of the domain of the ± function and the assignments of the + or – values, is learned.

9.7.5.4 How Free Will Works

Free will means the human ability to keep the promises one makes to himself or herself and others. It is the human ability to make decisions and choices and to act in accordance with those choices and decisions. The extent to which personal free will is developed is tested in day-to-day lives in form of personal integrity. It stands for one's ability to give meaning to his or her words and walk the walk. It is an integral part of how much value is placed on oneself. The concept of free will appears when people study the question of whether or not and in what sense rational agents exercise control over their actions and decisions.

Let us now look at how to address the following question: what is the systemic yoyo mechanism for the existence of the human endowment free will?

Based on what has been discussed, the so-called free will is the human ability to predict at least for the short term what one can or cannot accomplish and what choices are better or the best for the situation involved. In particular, with self-awareness, people form their reservoir of imagination and their ± function of moral values with its constantly expanding domain. What matters here is where their self, as identified by their self-awareness, is located, inside the domain of the ± function or outside the domain.

If their identified self is inside the domain of their ± function with a specific value assigned to it, then the assignment of a + value to their self will force them to make as accurate predictions as possible on what they can or cannot accomplish and what choices are better or the best for the situation involved. If they were uncertain with respect to a situation, their promise either to themselves or others would be that they would try their best without any guarantee for success. On the other hand, if a – value is assigned to their self, they will still force themselves to make as honest predictions as possible on what they can or cannot accomplish and what choices are better or the best for the situation involved. However, in this case, most likely, they will make promises opposite to what they foresee based on their imagination and their ± function. Now, when the third scenario occurs, where their identified self is not inside the domain of their ± function, then whatever appears, they would make their random promise, because no matter whether or not they keep their promise, the outcome would not bear any conscientious consequence to them.

That is, in terms of the development of individuals' conscience, by placing them in an appropriate environment during their upbringing, their ability to keep promises can be drastically different from one kind of environment to another.

9.8 OPEN-ENDED PROBLEMS

1. List three examples of open complex giant systems and explain why they are considered open complex giant systems.
2. Construct one example based on your real-life environment by dealing its hierarchical structure and describing how different layers interact with each other to form a whole.
3. In your own words, what is the essence of complexity?
4. If the Internet is seen as a system, then what are the emergent attributes of this system?
5. Describe a situation in which a human organization evolves from simple to complex, from low grade to high grade, from being less ordered to highly organized.
6. Describe a difficulty you experience in life and apply the systemic yoyo model to explain what is underneath the problem and how you might resolve the difficulty for the better outcome for everybody involved.
7. Referencing Figure 9.12, list all possible scenarios when sub-eddies could be potentially created by two large rotational fields and how likely these sub-eddy fields are to be sustained. Assume that all yoyo fields exist and interact with each other on the same plane.
8. According to the systemic yoyo model, describe in your own words the fundamental difference between individualism and collectivism. Is it possible that a balance between the two could be achieved at least theoretically?
9. Based on the systemic structure of the mind, design a procedure for yourself to become a leader in a well-defined circumstance.

10. If the inner drive for a person to accomplish is imagined to be the intensity with which the person's yoyo structure spins, then how would you design a way to make an ordinary person possess an inner drive to accomplish in life?
11. Both complex adaptive systems and open complex giant systems represent two different angles from which complex systems are investigated. Please spell out your understanding to this regard.
12. Are there other angles from which complex systems could be illustrated and investigated? What would be the range of applicability of your imagined method(s)?

10 Complex Networks

10.1 INTRODUCTION

For decades, it has been believed by default that the components, such as cells, societies, and even networks, of network systems are connected together by chance. In the past 10 years, the advance in research, which has appeared as overwhelming as an avalanche in the scientific area of complex networks, indicates that many realistic network systems can be investigated by using and abstracting into similar mathematical structures without specifying the relevant ages, functionalities, scales, and other characteristics of the components. Such theoretical generality allows researchers from different fields to employ a unified framework to investigate their respective networks. The concept of scale-free networks, developed 10 years ago, is one of the several important events that have helped in catalyzing the appearance of the network science, a new field of study that has been full of challenges and advances.

10.1.1 What Are Complex Networks?

Through how many mutual friends will any two chosen people on earth know each other? What is the average number of clicks of the mouse necessary to browse from one given page to another predetermined page on the World Wide Web (WWW)? How does the endless stream of computer viruses spread on the Internet? How do the infectious diseases, such as AIDS, SARS, avian flu, etc., spread among humans and animals? How does the global or a regional financial crisis happen? Why and how can the brain think? Although these problems seem different on the surface, every question involves a very complex network. These networks often have a large number of nodes that are connected through some complicated relationships. For example, each human society is a network made up with people along with a variety of social interactions; the nerve system of the human body is a network formed with a large number of nerve cells connected by nerve fibers; the Internet is a network interconnected by routers and computers; the WWW is a network composed of a large number of web pages connected through hyperlinks; and even world trades, municipal economics, etc., can be described as a complex network. Three examples are graphed in Figure 10.1. The research on complex networks, especially scale-free networks, has become a new hotspot of complex science, information science, and many other disciplines. Since 1999, remarkable results have been established in the research of complex networks, and the researchers come from such diverse areas as graph theory, statistical physics, computers, ecology, sociology, economics, and others. With the discovery of the small-world effect and free scales in complex networks, the in-depth exploration, scientific understandings, and possible applications of the qualitative characteristics and quantitative laws of complex networks have become an extremely important challenge in the scientific research of complexity in the network age of the twenty-first century.

Xueshen Qian (unpublished manuscript) presents a relatively more rigorous definition of complex networks as follows: a complex network is such a network that possesses some or all of the properties of being self-organized, self-similar, an attractor, a small world, and scale free.

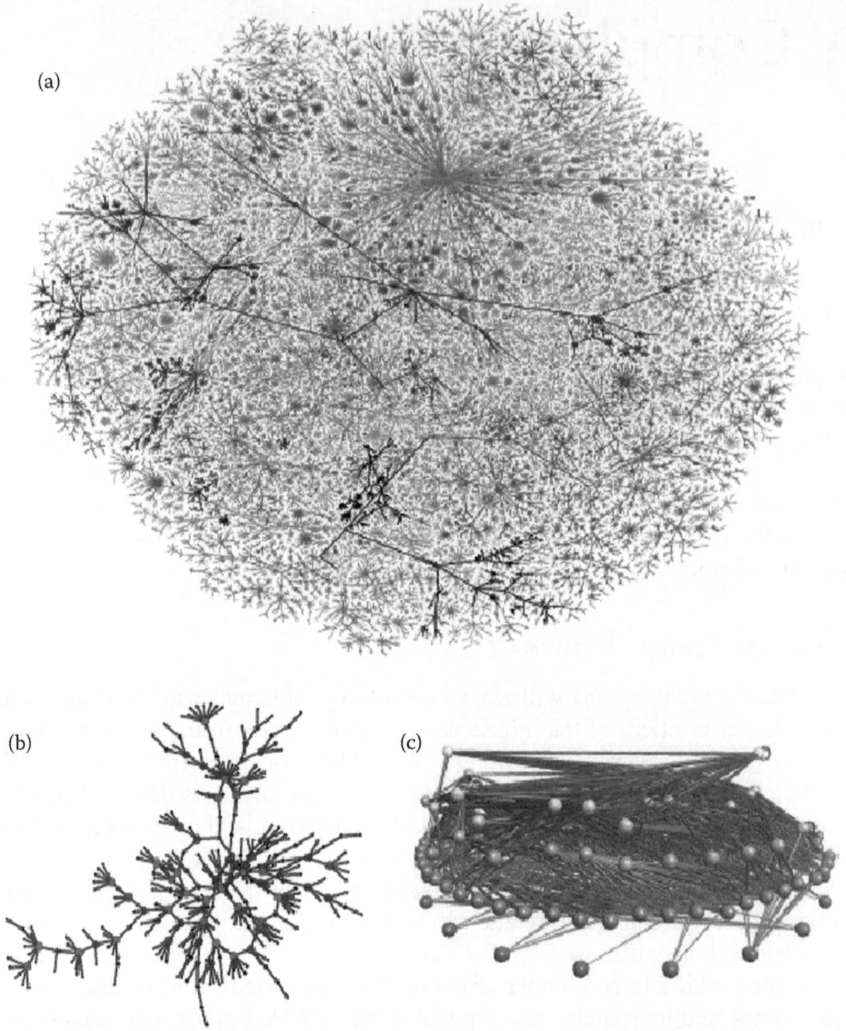

FIGURE 10.1 Three examples of different kinds of networks: (a) Internet; (b) social network of sexual contacts; and (c) food web.

In short, a complex network is such a network that presents a high-level complexity, and this complexity is reflected in the following aspects:

1. *Structural complexity:* it is reflected in the large number of nodes, and the network structure demonstrates the many different characteristics.
2. *Network evolution:* each network reveals the phenomenon of appearance, disappearance, and/or generation of nodes and/or connections between the nodes. For instance, in the WWW, web pages or links may either appear or be disconnected at any time, resulting in changes in the network structure.
3. *Connection diversity:* the weights of the connection between nodes are different, and there may be directions for the connections.
4. *Dynamic complexity:* the set of nodes may belong to a nonlinear dynamic system. For example, the node states might change in a complex way with time.

5. *Node diversity:* the nodes in a complex network can represent anything. For example, a complex network that represents a human relationship has people as its nodes; and different web pages are the nodes in the WWW network.
6. *The integration of multiple complexities:* any interaction of the aforementioned complexities can potentially lead to unpredictable results and consequences. For example, designing a power supply network needs to consider the evolution of this network, and the evolutionary process determines the network topology. When energy transfers frequently between two nodes, the connection between these nodes will increase the weight through continuous learning, and the memory will gradually improve the performance of the network.

The contents of the study of complex networks include geometric properties of networks, the mechanism of network formation, the (statistical) laws of network evolution, the nature of network models, the structural stability of a network, and dynamic mechanisms of network evolution. In natural sciences, the basic measurements of network research include various degrees and their distributions, degree correlation, clustering coefficient and its distribution, the shortest distance and its distribution, betweenness and its distribution, and size distribution of connected clusters.

The key research objectives of the complex network theory are as follows:

1. *To discover:* reveal the networks of real-life systems and describe the statistical properties of the network structures of these systems.
2. *To model:* establish appropriate network models to help understand the significance of these statistics and mechanisms of nature.
3. *To analyze:* analyze and predict the nature of the behaviors of each network based on the characteristics of each single node and the network structure.
4. *To control:* improve the existing network performance and design a new effective method for the investigation of networks.

10.1.2 History of Complex Networks Research

In 1736, Euler published a famous article on the "rainbow bridge problem." Euler's idea of abstraction and arguments on the "seven bridges problem" creates a branch of mathematics—graph theory. However, the study object of traditional graph theory includes just simple small-scale networks, in each of which the number of nodes is small and the connection is already established. With an increased number of nodes in a network, connections between the nodes become more complex, and sometimes, they even become impossible to determine. Thus, it is inevitable for graph theory to enter the area of complexity of networks. For a long period of time after Euler initially solved the "seven bridges problem," graph theory did not catch much attention in the world of learning. Until 1936, no monograph on graph theory had been published, and after then, graph theory entered the fast lane of development and created breakthroughs one after another.

In the 1960s, Erdös and Rény established papers on random graph theory, which have been well recognized as the works that pioneered the systematic study of complex networks in mathematics. Over the next 40 years, random graph theory has been the foundation of complex networks, even though most of the actual structures of complex networks are not completely random. For example, whether two people are friends or not, whether there exists a fiber connection between two routers on the Internet or not, and whether there are hyperlinks between any two web pages or not will not be completely determined by coin flips.

At the end of the twentieth century, important changes were taking place in scientific exploration of complex networks, and complex network theory is no longer confined to the area of mathematics. It received wide-ranging attention in different disciplines and has made gratifying progress. At the turn of the century, the major breakthrough in complex networks, which, following the computerization of data collection and the development of topology information, and due to the rise of complexity science, which possesses the important feature of combining reductionism and holism, encouraged the scientific world to deal with network topologies and mechanisms. So, it becomes possible to study complex networks. The overlaps between various disciplines allow researchers to compare a wide range of different types of network data and further reveal the complex networks of similarities. The empirical research on realistic network systems helps in establishing a more comprehensive understanding of the general properties of complex networks. Based on these general properties, models of complex networks are developed one after another, thus promoting the overall development of the theory of complex networks.

Two groundbreaking articles can be viewed as the beginning of a new era of complex networks research: one is entitled "Collective Dynamics of 'Small-world' Network" (Watts and Strogatz 1998) in the journal *Nature* in June 1998; the other one is entitled "Emergence of Scaling in Random Networks" (Barabási and Albert 1999) in the journal *Science*. These two articles reveal the small-world effect and scale-free property of complex networks, respectively, and construct the corresponding models to explain the generation mechanism of these characteristics.

10.1.3 Basic Definitions and Notations

The fundamental unit, also called a site (physics), a node (computer science), or an actor (sociology), of a network is a *vertex* (pl. vertices). A line segment that connects two vertices is referred to as an edge, or a bond (physics), or a link (computer science), or a tie (sociology). A loop is an edge whose endpoints are equal. Multiple edges are such edges that have the same pair of endpoints.

A simple graph is a graph that contains neither loops nor multiple edges. When u and v are the endpoints of an edge, they are *adjacent* and are referred to as *neighbors*. An edge is directed if it runs in only one direction (such as a one-way road between two points) and undirected if it runs in both directions. Directed edges, which are sometimes called *arcs*, can be thought of as sporting arrows indicating their orientation. A graph is directed if all of its edges are directed. An undirected graph can be represented by a directed one having two edges between each pair of connected vertices, one in each direction.

The *degree* of a vertex v is the number of adjacent edges. Note that the degree is not necessarily equal to the number of vertices adjacent to a vertex, since there may be more than one edge between any two vertices. The *component* to which a vertex v belongs is such set of vertices that can be reached from v by paths running along edges of the graph. In a directed graph, a vertex has both an in-component and an out-component, which are the sets of vertices from which the vertex can be reached and which can be reached from it.

A *path* is a simple graph whose vertices can be ordered so that two vertices are adjacent if and only if they are consecutive in the list. A *cycle* is a graph with an equal number of vertices and edges whose vertices can be placed around a circle so that two vertices are adjacent if and only if they appear consecutively along the circle. A *geodesic path* is the shortest path through the network from one vertex to another. Note that there may be and often is more than one geodesic path between two vertices. The *diameter* of a network is the length (in number of edges) of the longest geodesic path between any two vertices. A *subgraph* of a graph G is a graph H such that $V(H)$ is contained in $V(G)$ and $E(H)$ is contained in $E(G)$. A graph G is *connected* if each pair of vertices in G belongs to a path; otherwise, G is *disconnected*. A *complete graph* is a simple graph whose vertices are pairwise adjacent.

Complex systems are often composed of many elements that interact with each other. Hence, before considering the nature of the elements, the first step is just to describe the set of interactions by means of a network. From this point of view, each element is represented by a site (physics), node (computer science), actor (sociology), or vertex (graph theory), and correspondingly, the interaction between elements corresponds to a bond (physics), link (computer science), tie (sociology), or edge (graph theory). Graphically, nodes are depicted as dots and links as segments connecting two of these dots. Any complex system can be described as a network system with unrelated details ignored. Symbolically, we have the following [for more detailed presentation of relevant concepts and results, please consult the work of Lin (1999)].

Definition 10.1

Given a graph $G = (V,E)$, where V is the vertex set and E the edge set of the graph G, let $S = (V,R)$, where $R \subseteq 2^E$ (the power set of E) is a set of some relations defined on V based on the edges from E. Then $S = (V,R)$ is a network system constructed from the graph $G = (V,E)$. In particular, if every edge in E is a relation in R and vice versa, then $S = (V,R)$ is known as the network system that can be trivially identified or identifiable with $G = (V,E)$. In this case, the network system $S = (V,R)$ will still be denoted by $G = (V,E)$. The sets V and R are, respectively, called the node set and the relation set of S.

Thus, we can derive that the cardinalities $|V|$ and $|R|$ are referred to as the system's size and the relation's size. ■

Definition 10.2

Let $S = (V, R)$ be a network system constructed based on a graph $G = (V,E)$ and $r \in R$ a relation. The support of r, denoted Supp(r), is defined by

$$\text{Supp}(r) = \{v \in V : \exists v' \in V \text{ and } e \in r, v \text{ and } v' \text{ are the endpoints of edge } e\}.$$

Specifically, for $e \in E$,

$$\text{Supp}(e) = \{v_1, v_2 : v_1 \text{ and } v_2 \text{ are the endpoints of edge } e\}.$$ ■

Definition 10.3

Let $S = (V, R)$ be a network system constructed based on a graph. Define the neighborhood of a subset $V^* \subseteq V$, denoted Neighbor(V^*), as follows:

$$\text{Neighbor}(V^*) = \cup\{\text{Supp}(e) : e \in E \text{ and } \text{Supp}(e) \cap V^* \neq \varnothing\}$$

by applying mathematical induction on $n \in N$, where the following are defined:

$$\text{Neighbor}^1(V^*) = \text{Neighbor}(V^*) \text{ and } \text{Neighbor}^{n+1}(V^*) = \text{Neighbor}(\text{Neighbor}^n(V^*)).$$

Specifically, for any $v \in V$, we have that Neighbor(v) = Neighbor($\{v\}$), which is defined to be the node set consisting of v and all nodes that are adjacent to v. ■

Proposition 10.1

Suppose that $S = (V,R)$ is the network system that is trivially identified with a graph $G = (V,E)$. Then for any subset $V^* \subseteq V$, $\text{Neighbor}^\infty(V^*) = \bigcup_{n=1}^{\infty} \text{Neighbor}(V^*)$ is a component (connected subgraph) of G, if V^* is connected. ■

Proof

To show that Neighbor$^\infty$(V^*) is a component, it suffices to prove that for any two nodes $v, v' \in$ Neighbor$^\infty$(V^*), there exists a path between v and v'. To this end, there are three possibilities as follows:

1. Both $v, v' \in V^*$.
   In this case, because V^* is connected, according to the definition of connected graphs, there exists a path between v and v'.
2. $v \in V^*$ but $v' \in \text{Neighbor}^\infty(V^*) - V^*$.
   Because $v' \in \text{Neighbor}^\infty(V^*) - V^*$, there exists a node $v'^* \in V^*$ such that there is a path, denoted by path(v', v'^*), connecting v' and v'^*. Although $v, v'^* \in V^*$ and V^* is connected, there exists a path between v and v'^*, denoted by path(v, v'^*). Therefore, path(v, v'^*) ∪ path(v', v'^*) is a path connecting v and v'.
3. Both $v, v' \in \text{Neighbor}^\infty(V^*) - V^*$.
   In this case, there exist nodes $v^*, v'^* \in V^*$ such that there are two paths, respectively denoted by path(v, v^*) that connects v and v^* and path(v', v'^*) that connects v' and v'^*. Although $v^*, v'^* \in V^*$ and V^* is connected, there exists a path between v^* and v'^*, denoted by path(v^*, v'^*). Therefore, path(v, v^*) ∪ path(v^*, v'^*) ∪ path(v', v'^*) is a path connecting v and v'.

■

Proposition 10.2

Assume the same as in Proposition 10.1. If there exists a connected subset $V^* \subseteq V$ such that $\text{Neighbor}^\infty(V^*) = V$, then $G = (V,E)$ is a connected graph. ■

Proof

Based on the conclusion of Proposition 10.1, $\text{Neighbor}^\infty(V^*) = V$ is a component (connected subgraph). So $V = \text{Neighbor}^\infty(V^*)$ is connected, that is, $G = (V,E)$ is a connected graph. ■

Proposition 10.3

Assume the same as in Proposition 10.1. If for any node $v \in V$ such that Neighbor(v) = V, then $G = (V,E)$ is a complete graph. ■

Proof

It suffices to prove that for any two nodes $v, v' \in \text{Neighbor}^\infty(V^*)$, there exists an edge between v and v'.

Because for any node $v \in V$ such that Neighbor(v) = V, for any other node $v' \in V$, we have $v' \in$ Neighbor(v). According to the definition of neighborhoods, there exists an edge connecting the nodes v and v'. ■

10.2 COMPLEX NETWORK MODELS

10.2.1 Random Networks

More than two centuries have passed since Euler's inspiring work of the seven bridges problem was initially completed. During this period of time, the scientific community has focused its research on topics ranging from the properties of different figurative graphics to the causes of network formations. How does a realistic network form? What is the rule, if any, that controls the appearance and structure of a network? These concerns were not raised until the 1950s, when the first solution for the first problem was provided. During that specific decade, Paul Erdos and Alfred Renyi jointly published eight papers on the subject matter and explored for the first time in history the most basic question of how to understand the interconnectedness of the universe we ourselves live in by addressing, "How is a network formed?" Their answer lays down the foundation for the development of the theory of random networks. Their works have produced a profound impact on how the following generations think of networks.

The ultimate goal of the theory of random networks is to find the simplest explanation to the very complex phenomenon of the interconnectedness of the universe. To this end, Erdos and Renyi developed the complex network diagrams with a single structure. Although different systems follow different rules to form their networks, Erdos and Renyi ignored dissimilarity between the systems and provided the simplest answer that nature could provide: random connections. Erdos and Renyi think that the network graph and the world represented by network graphs are fundamentally random.

In their first classic article on random graphs, Erdos and Renyi (1960) defined a random graph as N labeled nodes connected by n edges that are chosen randomly from the $N(N-1)/2$ possible edges. In total, there are $C^n_{N(N-1)/2}$ graphs with N nodes and n edges, forming a probability space in which every realization is equiprobable. This model is referred to as the ER model.

An alternative and equivalent definition of a random graph is known as the binomial model. To introduce this model, let us start with N nodes and assume that the probability for an arbitrarily chosen pair of nodes to be connected is p (Figure 10.2). Then consequently, the total number of edges is a random variable with the expected value $E(n) = p\frac{N(N-1)}{2}$. If G_0 is a graph with nodes $P_1, P_2, \ldots, P_N$ and n edges, then the probability of obtaining such a graph, based on how this graph is constructed, is $P(G_0) = p^n(1-p)^{\frac{N(N-1)}{2}-n}$.

Since the theory of random graphs was founded in 1959, it has dominated the scientific thinking about networks, and the random graph model has been used as the dominant network model. It is believed that realistic complex networks are fundamentally random, and complexity is equivalent to randomness. As a reference, Table 10.1 shows the basic statistics of a number of published networks.

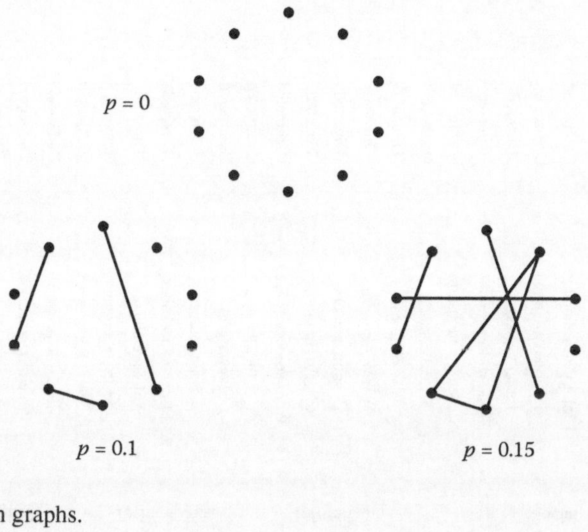

FIGURE 10.2 Random graphs.

TABLE 10.1
Basic Statistics for a Number of Published Networks

Category	Network	Type	n	m	z	ℓ	α	C	r
Social	Film actors	Undirected	449,913	25,516,482	113.43	3.48	2.3	0.20	0.208
	Company directors	Undirected	7673	55,392	14.44	4.60	–	0.59	0.276
	Math coauthorship	Undirected	253,339	496,489	3.92	7.57	–	0.15	0.120
	Physics coauthorship	Undirected	52,909	245,300	9.27	6.19	–	0.45	0.363
	Biology coauthorship	Undirected	1,520,251	11,803,064	15.53	4.92	–	0.088	0.127
	Telephone call graph	Undirected	47,000,000	80,000,000	3.16		2.1		
	Email messages	Directed	59,912	86,300	1.44	4.95	1.5/2.0		
	Email address books	Directed	16,881	57,029	3.38	5.22	–	0.17	0.092
	Student relationships	Undirected	573	477	1.66	16.01	–	0.005	–0.029
	Sexual contacts	Undirected	2810				3.2		
Information	WWW nd.edu	Directed	269,504	1,497,135	5.55	11.27	2.1/2.4	0.11	–0.067
	WWW Altavista	Directed	203,549,046	2,130,000,000	10.46	16.18	2.1/2.7		
	Citation network	Directed	783,339	6,716,198	8.57		3.0/–		
	Roget's Thesaurus	Directed	1022	5103	4.99	4.87	–	0.13	0.157
	Word co-occurrence	Undirected	460,902	17,000,000	70.13		2.7		
Technological	Internet	Undirected	10,697	31,992	5.98	3.31	2.5	0.035	–0.189
	Power grid	Undirected	4941	6594	2.67	18.99	–	0.10	–0.003
	Train routes	Undirected	587	19,603	66.79	2.16	–		–0.033
	Software packages	Directed	1439	1723	1.20	2.42	1.6/1.4	0.070	–0.016
	Software classes	Directed	1377	2213	1.61	1.51	–	0.033	–0.119
	Electronic circuits	Undirected	24,097	53,248	4.34	11.05	3.0	0.010	–0.154
	Peer-to-peer network	Undirected	880	1296	1.47	4.28	2.1	0.012	–0.366
Biological	Metabolic network	Undirected	765	3686	9.64	2.56	2.2	0.090	–0.240
	Protein interactions	Undirected	2115	2240	2.12	6.80	2.4	0.072	–0.156
	Marine food web	Directed	135	598	4.43	2.05	–	0.16	–0.263
	Freshwater food web	Directed	92	997	10.84	1.90	–	0.20	–0.326
	Neural network	Directed	307	2359	7.68	3.97	–	0.18	–0.226

Source: Newman, M. E. J. *SIAM Review*, 45, 167–256, 2003.

Notes: The properties measured are as follows: total number of vertices n; total number of edges m: mean degree z; mean vertex–vertex distance l; type of graph, directed or undirected; exponent a of degree distribution if the distribution follows a power law (or "–" if not, in/out-degree exponents are given for directed graphs); clustering coefficient C; degree correlation coefficient r. Blank entries indicate unavailable data.

10.2.2 Small-World Networks

In the 1960s, Stanley Milgram, a social psychologist from Harvard University, did an interesting experiment of sociology. He first selected two target subjects: one is the wife of a seminary graduate student from Sharon, MA, and another a stock market broker in Boston. Then he recruited a group of volunteers in the distant Kansas and Nebraska and asked each of the volunteers to deliver a letter to one of the two target subjects with as few deliveries as possible through people they know themselves. Milgram (1967) analyzed the letters delivered successfully by using statistics and found that, on the average, the volunteers needed only six deliveries to pass the letter to one of the target subjects. The results of this experiment to some extent reflect the "small world" effect in interpersonal relationships.

Along a different line of work, Mark Granovetter studied a microsociological issue in his doctoral thesis: how do people get to work? Granovetter remembered what he learned in his basic chemistry lessons—how weak hydrogen bonds keep a great number of water molecules to stay together. Inspired by this basic chemistry fact, he wrote the first paper, "The Strength of Weak Ties" (Granovetter 1973), and discussed the importance of weak social relationships for human lives. Today, this paper has been recognized as one of the most influential sociological papers ever published.

In the mid-1990s, Duncan J. Watts and his graduate study mentor Steven Strogatz (1998) considered the question: what is the likelihood of mutual understanding between two of my friends? Watts and Strogatz introduced the concept of clustering coefficients to show the closeness of your circle of friends and showed the fact that there indeed exists the cluster phenomenon in social systems. For a certain attribute of social networks, only if it reflects an essence of many networks, a good number of scholars will be interested in knowing more about the attribute. Therefore, to this end, the most important discovery Watts and Strogatz made is that the cluster phenomenon exists not only in social networks but also in networks of other scientific areas. In their study of the neural network of *Caenorhabditis elegans*, a small, free-living round worm, an ideal organism for the study of gene regulation and function, which contains only 302 nerve cells, Watts and Strogatz found that the network has a high clustering coefficient—the possibility that an arbitrary nerve cell links with its adjacent cell is five times as high as it does in a random network. Cluster is pervasive, such as it is in the WWW, the Internet, the network of a joint venture relationship, the food web, and the molecular networks in cells. That changes the cluster phenomenon, a unique characteristic of social networks in the past, into a common characteristic of complex networks and poses a challenge to the point of view that real networks are all random.

There is a shorter average path in realistic networks and networks constructed by using the ER model. However, in a large number of realistic networks, especially social networks, the individuals as the networks' nodes often have group consciousness, that is, these realistic networks tend to have large clustering coefficients. In order to construct a network model that can simulate the characteristics of realistic networks, Watts and Strogatz (1998) introduced an interesting model, known as the small world model or the WS model. That is a one-parameter model that interpolates between an ordered finite-dimensional lattice and a random graph. The algorithm behind the model is given as follows (Figure 10.3):

1. Start with order: start with a ring lattice with N nodes in which every node is connected to its first K neighbors (in particular, $K = 2$ on either side). In order to have a sparse but connected network at all times, consider $N \gg K \gg \ln(N) \gg 1$.
2. Randomize: randomly rewire each edge of the lattice with probability p such that self-connections and duplicate edges are excluded. This process introduces $pNK/2$ long-range edges, which connect nodes that otherwise would be part of different neighborhoods. By varying the p-value, transitions between order when $p = 0$ and randomness when $p = 1$ can be closely monitored.

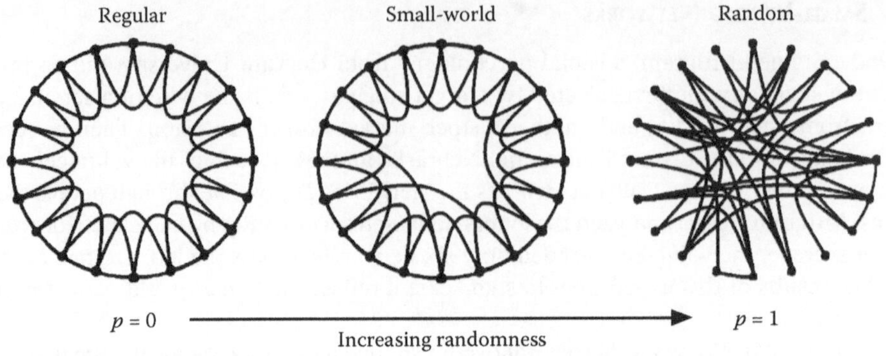

FIGURE 10.3 Random rewiring procedure of the WS model, which interpolates between a regular ring lattice and a random network without altering the number of nodes or edges.

10.2.3 Scale-Free Networks

Malcolm Gladwell published a book entitled *The Tipping Point*, in which he discovered the existence of the connecter, something of mostly "human." A connector or a central node, that is, a node with a very large number of links, exists in many complex systems that range from the economy to the cells. The set of connectors represents some of the essential components of most networks. The existence of connectors has inspired the interest of a number of experts in different scientific fields from biology to computer science to ecology. Their findings overturned all the seemingly quite-intuitive knowledge of networks. The clustering phenomenon is the first crack in Erdos–Renyi's view of a random world, to which Watts and Strogatz's model can be treated as a temporary solution, substantiating six degrees of separation. However, the connector phenomenon stands for a fatal blow to the Eros–Renyi model. To explain the existence of nodes with large numbers of links, one must totally abandon the view of a random world. To this end, Figure 10.4 shows the degree distribution of a typical network.

In any network with central nodes, these central nodes play a key role in the network structure and make the network present small world effect. The fact that the central nodes connect with an unusual number of nodes creates a link shortcut between any two nodes in the system. If the world is observed from the position of the central nodes, the world is indeed very small. As affected by Eros and Renyi's works, it has been believed for decades that networks are random. This point of view has recently been challenged in many different ways. In particular, Watts and Strogatz's model put forward a simple explanation on the cluster and integrated random networks with the clustering phenomenon, but the discovery of central node in networks further changes the state of research of networks. The two models that have been mentioned so far cannot yet explain the existence of central nodes. Therefore, the existence of central nodes forces the world of learning to rethink about the knowledge of networks and raises three fundamental questions: How do central nodes appear? In an established particular network, what is the possible number of central nodes? Why can the models previously mentioned not foresee the existence of central nodes?

The existence of central nodes is not of a rare coincidence in the interconnected universe. Instead, it is consistent with the rigorous mathematical laws developed in systems science, and their wide-range appearance forces the world of learning to consider network problems in different angles with completely new ideas.

Barabási and Albert (1999) proposed a scale-free model (SF model), and the corresponding algorithm for constructing an SF model is the following:

1. Growth: Start with a network with m_0 nodes. At every time step, add a new node with $m(\leq m_0)$ edges that link the new node to m different nodes that are already present in the network.

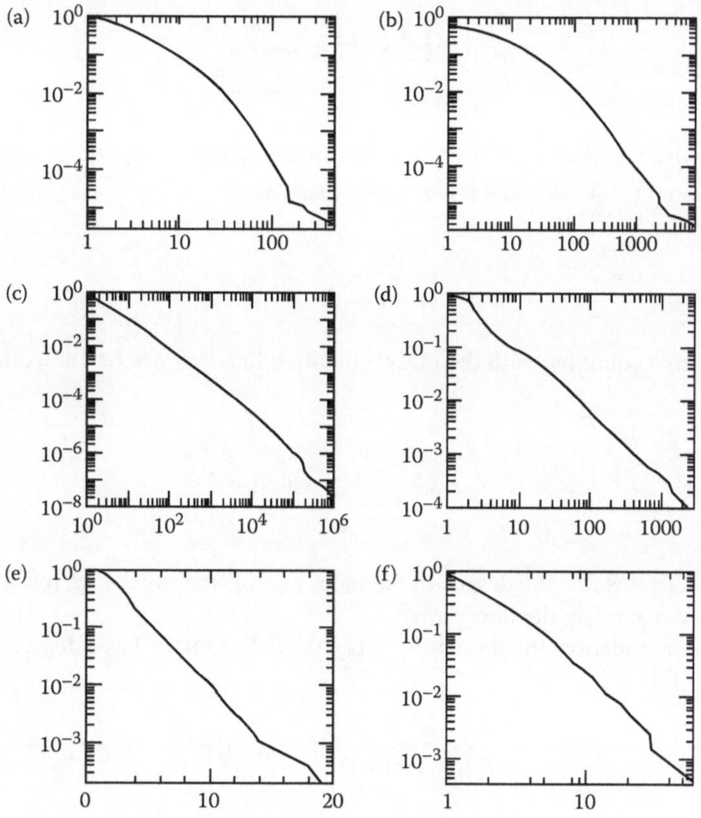

FIGURE 10.4 Degree distribution of a typical network. (a) Collaborations in mathematics. (b) Citations. (c) WWW. (d) Internet. (e) Power grid. (f) Protein interactions.

2. Preferential attachment: The probability for a new node to connect to an existing node i is directly proportional to the degree k_i of node i; symbolically

$$\prod(k_i) = \frac{k_i}{\sum_j k_j}. \tag{10.1}$$

Each scale-free network possesses a good degree of robustness. With the emergence of the Internet and WWW, the requirement for network stability becomes particularly prominent. It is because a minor local error appearing in a network system or a local attack on the system may lead to the collapse of the entire network. However, the evolutionary mechanism of SF networks guarantees that the distribution of node degrees of the entire network system stays in a steady state. Many studies on large natural and social networks indicate that these networks are all scale-free and possess the same characteristics of steady states.

The continuum approach was introduced by Barabási and Albert (1999), and these authors calculated the time dependence of the degree k_i of a given node i. This degree increases every time a new node enters the network and links to node i, where the probability for the new node to connect to node i is $\prod(k_i)$. Assuming that k_i is a continuous real variable, the rate at which k_i changes is expected to be proportional to $\prod(k_i)$. Thus, k_i satisfies the dynamical equation

$$\frac{\partial k_i}{\partial t} = m\prod(k_i) = m\frac{k_i}{\sum_{j=1}^{N-1} k_j}. \tag{10.2}$$

The sum in the denominator goes over all nodes in the system except the newly introduced one. Thus, the sum is equal to $\sum_j k_j = 2mt - m$, which leads to

$$\frac{\partial k_i}{\partial t} = \frac{k_i}{2t}. \tag{10.3}$$

The solution of this equation, with the initial condition that every node i at its introduction satisfies $k_i(t_i) = m$, is

$$k_i(t) = m\left(\frac{t}{t_i}\right)^{\beta} \quad \text{with } \beta = \frac{1}{2}. \tag{10.4}$$

This equation indicates that the degree of all nodes evolves the same way, following a power law with the only difference being the intercept.

Using the precious equation, the probability $P(k_i(t) < k)$ that a node has a degree $k_i(t)$ smaller than k can be written as follows:

$$P(k_i(t) < k) = P\left(t_i > \frac{m^{1/\beta}t}{k^{1/\beta}}\right). \tag{10.5}$$

Assume that nodes are added at equal time intervals to the network. Then the t_i values have a constant probability density

$$P(t_i) = \frac{1}{m_0 + t}. \tag{10.6}$$

Substituting this into $P(k_i(t) < k) = P\left(t_i > \frac{m^{1/\beta}t}{k^{1/\beta}}\right)$ produces

$$P\left(t_i > \frac{m^{1/\beta}t}{k^{1/\beta}}\right) = 1 - \frac{m^{1/\beta}t}{k^{1/\beta}(m_0 + t)}. \tag{10.7}$$

The degree distribution $P(k)$ can be obtained by

$$P(k) = \frac{\partial P(k_i(t) < k)}{\partial k} = \frac{2m^{1/\beta}t}{m_0 + t}\frac{1}{k^{1/\beta+1}}, \tag{10.8}$$

which predicts that asymptotically ($t \to \infty$),

$$P(k) \to 2m^{1/\beta}k^{-\gamma}, \quad \text{with } \gamma = \frac{1}{\beta} + 1 = 3. \tag{10.9}$$

The result of the previous analysis indicates that the exponent γ of the power-law distribution is independent of m; its value is close to the exponents of the majority of the real-life networks (Newman 2003).

10.3 GENERATION MECHANISM

In the past, studies on the structures of social networks and the Internet have assumed that the node degrees have a Poisson distribution. However, the attention of the recent literature has focused on graphs with significantly different node degree distributions. In particular, Newman et al. (2001) developed a detailed theory of random graphs with arbitrary degree distributions.

10.3.1 Generation Functions

A so-called generation function is introduced for generating the probability distribution of vertex degrees k. In particular, suppose that an undirected graph has N vertices, where N is a large natural number. Then a generation function $G_0(x)$ can be defined as follows:

$$G_0(x) = \sum_{k=0}^{\infty} p_k x^k \tag{10.10}$$

where p_k is the probability that a randomly chosen vertex on the graph has degree k. The distribution p_k is assumed to be correctly normalized so that

$$G_0(1) = 1.$$

The normalization of the probability distribution also implies that the function value $G_0(x)$ is finite for all x satisfying $|x| \le 1$. To make this chapter more accessible to as many scholars as possible, all the calculations of this chapter will be confined to the region $|x| \le 1$.

The function $G_0(x)$, and indeed any probability generation function, has a number of properties that will prove useful in subsequent developments of the theory of complex networks.

1. *Derivatives:* The probability p_k for a randomly chosen vertex to have degree k is given by the kth derivative of $G_0(x)$ according to the following:

$$p_k = \frac{1}{k!} \frac{\mathrm{d}^k G_0}{\mathrm{d}x^k}\bigg|_{x=0}. \tag{10.11}$$

   Thus, that one particular function $G_0(x)$ encapsulates all the information contained in the discrete probability distribution p_k. That is why it is said that the function $G_0(x)$ generates the probability distribution p_k.
2. *Moments:* The mean over the probability distribution generated by a generation function, for instance, the average degree $\langle k \rangle$ of a vertex in the case of $G_0(x)$, is given by

$$\langle k \rangle = \sum_k k p_k = G_0'(1). \tag{10.12}$$

   Thus, if one can calculate a generation function, he or she can then also calculate the mean of the probability distribution that the generation function describes. Higher-order moments of the distribution can be calculated from higher-order derivatives:

$$\langle k^n \rangle = \sum_k k^n p_k = \left[\left(x \frac{\mathrm{d}}{\mathrm{d}x} \right) G_0(x) \right]_{x=1}. \tag{10.13}$$

3. *Powers:* If the distribution of a property k of an object is generated by a given generation function, then the distribution of the totality of k summed over m independent realizations of the object is generated by the mth power of that generating function. For example, if m vertices are chosen at random from a large network, then the distribution of the sum of the degrees of those vertices is generated by $[G_0(x)]^m$. To see why this is so, consider the simple case of just two vertices. The square $[G_0(x)]^2$ of the generation function for a single vertex can be expanded as

$$[G_0(x)]^2 = \left[\sum_k p_k x^k \right]^2 = \sum_{jk} p_j p_k x^{j+k}$$

$$= p_0 p_0 x^0 + (p_0 p_1 + p_1 p_0) x^1 + (p_0 p_2 + p_1 p_1 + p_2 p_0) x^2 \tag{10.14}$$

$$+ (p_0 p_3 + p_1 p_2 + p_2 p_1 + p_3 p_0) x^3 + \cdots$$

It is clear that the coefficient of x^n term in this expression is precisely the sum of all products $p_j p_k$ such that $j + k = n$. Hence, it correctly describes the probability that the sum of the degrees of two vertices will be n.

Another quantity that will be important in the study of complex networks is the distribution of the degrees of the vertices, which we arrive at by following a randomly chosen edge. Such an edge arrives at a vertex with a probability proportional to the degree of that vertex, and the vertex therefore has a probability distribution of a degree proportional to kp_k. The correctly normalized distribution is therefore generated by

$$\frac{\sum_k k^n p_k x^k}{\sum_k k^n p_k} = x \frac{G_0'(x)}{G_0'(1)}. \tag{10.15}$$

If one starts at a randomly chosen vertex and follows each of the edges at that vertex to reach the k-nearest neighbors, then the vertices arrived at have the distribution of remaining outgoing edges generated by this function, less one power of x to allow for the edge that he or she arrived along. Thus, the distribution of the outgoing edges is generated by the function

$$G_1(x) = \frac{G_0'(x)}{G_0'(1)} = \frac{1}{z} G_0'(x) \tag{10.16}$$

where z is the mean vertex degree as before. The probability that any of these outgoing edges connects to the original vertex that was started off with initially, or to any of its other immediate neighbors, goes as N^{-1} and hence can be neglected at the limit of a large N. Thus, by making use of the "power" property of the generation function described above, the generation function for the probability distribution of the number of second neighbors of the original vertex can be written as

$$\sum_k p_k [G_1(x)]^k = G_0(G_1(x)). \tag{10.17}$$

Similarly, the distribution of the third-nearest neighbors is generated by $G_0(G_1(G_1(x)))$, and so on. The mean number z_2 of the second neighbors is

$$z_2 = \left[\frac{\mathrm{d}}{\mathrm{d}x} G_0(G_1(x))\right]_{x=1} = G_0'(1)G_1'(1) \tag{10.18}$$

where the fact that $G_1(1) = 1$ has been applied.

10.3.2 Poisson-Distributed Graphs

The simplest example of a graph of this type is one for which the distribution of vertex degrees is binomial, or Poisson at the large N limit. This distribution yields the standard random graph. In this graph, the probability $p = \langle k \rangle / N$ for the existence of an edge between any two vertices is the same for all vertices, and a generation function $G_0(x)$ is given by

$$G_0(x) = \sum_{k=0}^{N} \binom{N}{k} p^k (1-p)^{N-k} x^k = (1-p+px)^N = e^{z(x-1)} (N \to \infty). \tag{10.19}$$

The mean degree of a vertex is given by

$$z = \langle k \rangle = G_0'(1) = Np. \tag{10.20}$$

The probability distribution of vertex degrees is given by

$$p_k = \frac{1}{k!} \frac{\mathrm{d}^k G_0}{\mathrm{d}x^k}\bigg|_{x=0} = \frac{z^k e^{-z}}{k!}, \tag{10.21}$$

which is the standard Poisson distribution.

Notice also that for this special case, one has $G_1(x) = G_0(x)$. Thus, the distribution of outgoing edges at any chosen vertex is the same regardless of whether one arrived there by choosing a vertex at random or by following a randomly chosen edge. This property, which is peculiar to Poisson-distributed random graphs, is the reason why the theory of random graphs of this type is especially simple.

10.3.3 Exponentially Distributed Graphs

Perhaps the next simplest type of graph is such a graph with an exponential distribution of vertex degrees:

$$p_k = (1 - e^{-1/\kappa})e^{-k/\kappa} \tag{10.22}$$

where κ is a constant. The generation function for this distribution is

$$G_0(x) = (1 - e^{-1/\kappa}) \sum_{k=0}^{N} e^{-k/\kappa} x^k = \frac{1 - e^{-1/\kappa}}{1 - xe^{-1/\kappa}} \tag{10.23}$$

and

$$G_0(x) = \left[\frac{1 - e^{-1/\kappa}}{1 - xe^{-1/\kappa}} \right]^2 . \tag{10.24}$$

10.3.4 Power-Law Distributed Graphs

For a power-law distributed graph, its probability can be written as

$$p_k = Ck^{-\tau}e^{-k/\kappa} \text{ for } k \geq 1 \tag{10.25}$$

where C, τ, and κ are constants. The reason for including the exponential cutoff is two-fold: First, many real-world graphs appear to show this cutoff; second, it makes the distribution normalizable for all τ and not just for $\tau \geq 2$.

The constant C is fixed in order to satisfy the requirement of normalization, which in turn gives $C = [\text{Li}_\tau(e^{-1/\kappa})]^{-1}$. Hence, one has

$$p_k = \frac{k^{-\tau}e^{-k/\kappa}}{\text{Li}_\tau(e^{-1/\kappa})} \quad \text{for } k \geq 1, \tag{10.26}$$

where $\text{Li}_n(x)$ is the nth polylogarithm of x, a function familiar to those who have worked with Feynman integrals (Feynman and Hibbs 1965).

Substituting p_k into the generation function $G_0(x)$ leads to the fact that the generation function for graphs with this degree distribution is

$$G_0(x) = \frac{\text{Li}_\tau(xe^{-1/\kappa})}{\text{Li}_\tau(e^{-1/\kappa})} . \tag{10.27}$$

At the limit state $\kappa \to \infty$, this simplifies to

$$G_0(x) = \frac{\text{Li}_\tau(x)}{(\tau)} \tag{10.28}$$

where $\varsigma(x)$ is the Riemann ς-function.

The function $G_1(x)$ is given by

$$G_1(x) = \frac{\text{Li}_{\tau-1}(xe^{-1/\kappa})}{x\text{Li}_{\tau-1}(e^{-1/\kappa})} . \tag{10.29}$$

Thus, for example, the mean number of neighbors of a randomly chosen vertex is

$$z = G_0'(1) = \frac{\text{Li}_{\tau-1}(e^{-1/\kappa})}{\text{Li}_\tau(e^{-1/\kappa})} , \tag{10.30}$$

and the mean number of second neighbors is

$$z_2 = G_0'(1)G_1'(1) = \frac{\text{Li}_{\tau-2}(e^{-1/\kappa}) - \text{Li}_{\tau-1}(e^{-1/\kappa})}{\text{Li}_\tau(e^{-1/\kappa})} . \tag{10.31}$$

10.3.5 Graphs with Arbitrarily Specified Degree Distribution

In some cases of investigation, one might like to model specific real-world graphs that have known degree distributions. The distributions are known because one can measure them directly. A number of the graphs described in the introduction of this chapter fall into this category. For these graphs, one knows the exact number n_k of vertices that have degree k. Hence, he or she can write down the exact generation function for that probability distribution in the form of a finite polynomial:

$$G_0(x) = \frac{\sum_k n_k x^k}{\sum_k n_k} \tag{10.32}$$

where the sum in the denominator ensures that the generation function is properly normalized.

10.4 PROCESSES ON COMPLEX NETWORKS

10.4.1 Network Resilience

Most of the networks widely investigated for their functionality rely on their connectivity, that is, the existence of paths leading from one vertex to another. If some individual vertices are removed from a network, the typical length of these paths will increase; ultimately, some pairs of vertices will become disconnected, and communication between them through the network will become impossible. Networks vary in their levels of resilience to such vertex removal.

There are also a variety of different ways with which vertices can be removed, and different networks show varying degrees of resilience to vertex removals. For example, one could remove vertices at random from a network, or one could target some specific class of vertices, such as those with the highest degrees. Network resilience is of particular importance in epidemiology, where "removal" of vertices in a contact network might correspond, for example, to vaccination of individuals against a particular disease. Because vaccination not only helps prevent the vaccinated individuals from catching the disease but may also destroy paths between other individuals along which the disease might have been spreading, better understanding of network resilience on vertex removal can have a much wider-reaching effect than one might at first think. Also, careful consideration of the efficacy of different vaccination strategies could lead to substantial advantages for public health services and administration.

Albert et al. (2000) studied the effect of vertex deletion in two example networks, a 6000-vertex network representing the topology of the Internet at the level of autonomous systems and a 326,000-page subset of the WWW. In terms of their forms, both the Internet and the web have been observed to have approximate power-law degree distributions. These authors measured the mean vertex-to-vertex distance as a function of the number of vertices removed for both the cases of random removal and progressive removal of the vertices with the highest degrees. What is shown in Figure 10.5 are the relevant results of the Internet obtained by these authors. They found for both of these networks that the distance is almost entirely unaffected by random vertex removal. That is, the networks studied are highly resilient to this type of removal. This fact is intuitively reasonable because the degrees of most of the vertices in these networks are low; therefore, the vertices lie on a few paths between each other. Thus, their removal rarely affects the overall communication substantially. On the other hand, when the removal is targeted at the vertices with the highest degrees, it produces a devastating effect. The mean vertex-to-vertex distance increases very sharply with the fraction of vertices removed, and typically, only a few percentage points of vertices need be removed before essentially all communication throughout the network is destroyed. Albert et al. (2000) expressed their results in terms of failure or sabotage of network nodes. The Internet (and

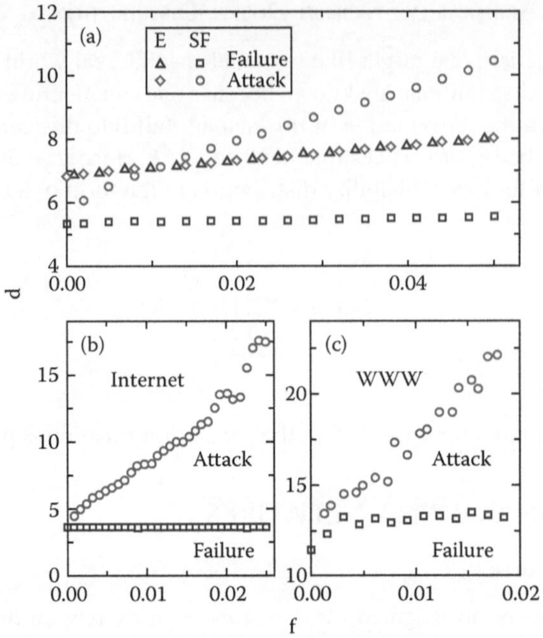

FIGURE 10.5 Changes in the diameter of the network as a function of the fraction of the removed nodes. (From Albert, R. et al., *Nature*, 406, 378–382, 2000. With permission.)

the web), they suggest, is highly resilient against the random failure of vertices in the network but highly vulnerable to deliberate attacks on its highest-degree vertices.

Most systems of high fault tolerance have a common characteristic: their functionality is ensured through the highly interconnected complex network of their objects. For instance, cell stability is supported by a complex regulatory network and a metabolic network; each society's resilience is deeply rooted in the interconnected social network of the members; economic stability is ensured by elaborate financial regulations and networks; the stability of ecological systems is achieved by the well-knitted network interaction of different species. In nature, each high degree of stability is achieved by mutual connectivity.

The "9-11" terrorist attack also showed two points: first, the strong influence of the central node and, second, the network tolerance. This incidence shows that not much of the implicit relationship between robustness and vulnerability is understood.

10.4.2 Epidemiological Processes

Throughout human history, the pandemic of any infectious disease, such as malaria, smallpox, measles, plague, typhoid fever, etc., is caused by the communication process and interactions between and within individual human civilizations. In turn, it can be expected that each future large-scale infectious disease will again have far-reaching impacts on the human race in various ways. The human societal network of growing complexity promotes the constant improvement of the modern public health system in order to reduce the threat of plagues. On the other hand, this network makes the movement of people and goods more frequent and convenient; unfortunately, it also greatly speeds up the rate of diffusion of infectious diseases. For example, thousands of people worldwide have died of AIDS. In recent years, such health problems as mad cow disease, foot and mouth disease, avian influenza, etc., have also caused tremendous economic losses to the world.

Compared with biological viruses, computer viruses have been spreading more easily across borders and intruding into every corner of the world through the convenience of the huge Internet.

To this end, some scholars even exclaimed that if no intervention is made available, the Internet can completely collapse within a period of time between a few seconds and several minutes, because of the spread of computer viruses. The recent harsh health issues have forced scholars to reconsider the following questions: With such a high level of developed medical/biological technology and modern public health system, why and how can a new virus still spread rapidly? With costly control and preventive measures installed, why and how can some computer viruses on the Internet remain active and others hard to detect?

One can define the individuals in either a biological species or a computer network, such as individual organisms or individual computers, as (abstract) nodes and define the correlations between individuals as edges between nodes. The rapid development of the theory of complex networks has effectively enhanced the understanding of the mechanism underneath the outbreak of large-scale biological and computer viruses. In particular, the traditional theory maintains that only when the effective transmission rate exceeds some positive threshold value does a large-scale dissemination become possible; Pastor-Satorras and Vespignani (2001) show that when the network size increases infinitely, the threshold value of a scale-free network tends to be zero. That means that even if the source of infection is small or minor, the infection is still sufficient to spread substantially throughout a large network.

Assume that the spread of an infectious disaster in a network follows the SIS model, ...→ susceptible → infected → susceptible →..., where the nodes of the network stand for people and the disaster runs through this cycle stochastically. Let the probability for a susceptible node to be infected from a susceptible state be v, and the probability for an infected node to recover to a susceptible state be δ. Then an effective spreading rate of the disaster can be defined as $\lambda = v/\delta$. Without loss of generality, $\delta = 1$ can be assumed, which affects only the time scale used in the study of the disaster spreading. By using the mean-field theory, we can analytically study the SIS model.

In particular, let us first consider a homogeneous network, such as the WS model. Assume that $\rho(t)$ is the relative density of infected nodes in the entire network. Then the mean-field reaction equation for the density of infected nodes $\rho(t)$ is

$$\frac{\partial \rho(t)}{\partial t} = -\rho(t) + \lambda \langle k \rangle \rho(t)(1-\rho(t)). \tag{10.33}$$

The first term on the right-hand side in Equation 10.33 represents the recovery of the infected nodes at the unit rate. The second term stands for the mean density of newly infected nodes that are generated by the existing infected nodes. It is directly proportional to the rate λ of effective spreading; the node degree, which is assumed to be equal to the mean degree of the network; and the probability $(1-\rho(t))$ that an infected node is connected to a healthy node. Because we are concerned only with the case of $\rho(t) \ll 1$, the other higher-order terms in Equation 10.33 have been ignored.

By letting Equation 10.33 be equal to zero and solving the resultant equation, we obtain the steady-state density of infected nodes in the network as follows:

$$\rho = \begin{cases} 0, & \lambda < \lambda_c \\ \dfrac{\lambda - \lambda_c}{\lambda}, & \lambda \geq \lambda_c \end{cases} \tag{10.34}$$

whose epidemic threshold is

$$\lambda_c = \frac{1}{\langle k \rangle}. \tag{10.35}$$

For any nonhomogeneous network, such as the scale-free model, let us define $\rho_k(t)$ to be the relative density of infected nodes with degree k in the network. That is, $\rho_k(t)$ is the probability for a node of degree k to be infected. Then the mean-field reaction equation for the density $\rho_k(t)$ of infected nodes of degree k is

$$\frac{\partial \rho_k(t)}{\partial t} = -\rho_k(t) + \lambda k(1-\rho_k(t))\Theta(\lambda). \tag{10.36}$$

The first term on the right-hand side of Equation 10.36 represents the recovery of the infected nodes of degree k at the unit rate. The second term stands for the mean density of newly infected nodes of degree k that are generated by the existing infected nodes. This mean density is directly proportional to the effective spreading rate λ, the node degree k, the probability $(1 - \rho_k(t))$ that a healthy node of degree k is connected to an infected node, and the probability $\Theta(\lambda)$ for an edge to connect to an infected node. Once again, all other higher-order terms in Equation 10.36 have been ignored.

If Equation 10.36 is equal to zero, the steady-state density of infected nodes of degree k in the network can be obtained:

$$\rho_k = \frac{k\lambda\Theta(\lambda)}{1+k\lambda\Theta(\lambda)}. \tag{10.37}$$

From the steady-state density solution of the system, it follows that the higher the degree of a node is, the higher the probability that the node could be infected. In order to find the probability $\Theta(\lambda)$ that any given link points to an infected node, the nonhomogeneity of the network should be considered. The probability that an arbitrary link points to a certain node should be directly proportional to the degree of the node. That is, a randomly chosen link is more likely to connect to a node of high degree. Symbolically, we have

$$\Theta(\lambda) = \sum_k \frac{kP(k)\rho_k}{\sum_s sP(s)}. \tag{10.38}$$

The mean density of infected nodes in the network can be expressed as

$$\rho = \sum_k P(k)\rho_k. \tag{10.39}$$

In the BA model, the degree distribution can be written as $P(k) = 2m^2/k^{-3}$, where m stands for the minimum number of links at the individual nodes. Noticing that $\langle k\rangle = \int_m^\infty sP(s)\,\mathrm{d}s = 2m$, by treating k as a continuous variable, we can write $\Theta(\lambda)$ in the following integral form:

$$\Theta(\lambda) = m\lambda\Theta(\lambda)\int_m^\infty \frac{1}{k^3}\frac{k^2}{1+k\lambda\Theta(\lambda)}, \tag{10.40}$$

which yields the following solution:

$$\Theta(\lambda) = \frac{e^{-1/m\lambda}}{\lambda m}(1-e^{-1/m\lambda})^{-1}. \tag{10.41}$$

In order to calculate the mean density of infected nodes in the network, we solve Equation 10.39 and produce

$$\rho = 2m^2\lambda\Theta(\lambda)\int_m^{\infty}\frac{1}{k^3}\frac{k}{1+k\lambda\Theta(\lambda)}. \tag{10.42}$$

By substituting $\Theta(\lambda)$ and calculating the resultant integral, we can find the following lowest-order expression of ρ in λ:

$$\rho \approx 2e^{-1/m\lambda}, \tag{10.43}$$

which shows that the spreading threshold in a BA scale-free network tends to be zero.

10.4.3 Search on Networks

Milgram's small world experiment, as discussed in Section 10.2.2, revealed two properties of social networks. The first is the small world effect of networks: although the size of a network is huge, the mean distance between any two persons is amazingly small. The second is that the networks are searchable; in particular, although the number of paths connecting two arbitrarily chosen persons may be large and the length difference between diverse paths may be great, it is possible to find a quite-short path that connects these two strangers. Watts and Strogatz's small-world network model shows that as long as a small amount of random long-range connections is introduced in a regular network, the mean distance between any two nodes in the network becomes very small, which to some extent explains the small world effect in social networks. However, in general, the network's small world effect does not necessarily mean that the network can be quickly searched. In a large-scale network, whether a shorter or even the shortest path between two nodes can be found or not depends on the information about the network structure, the searching algorithm used, and the actual structure of the entire network. In Milgram's small world experiment, scholars have used a simple decentralized algorithm to search the target objects. In particular, the current letter holder transmits the letter in such a way to reach the target people that are most likely based on the available local information. In other words, the current letter holder passes on the letter to one of his or her friends, whom he or she thinks is the closest to the target audience within the information available to him or her. This algorithm is also known as the "greedy algorithm." Since the algorithm itself is very simple without anything special and a certain number of letters can reach the targeted eventual receivers successfully in a relatively small number of steps, it shows that the structure of a social network itself must have its special characteristics. Following the initial experiment, Watts et al. (2002) have also made a further study on this issue.

Problems of searching a complex network have many practical applications, including that of finding the shortest path between two persons in a social network, that of locating certain pages in the WWW, and that of pinpointing to a specified file or data in peer-to-peer networks.

Kleinberg (2000a,b) first studied theoretically the search capabilities of complex networks. That is, he looked at the properties of how quickly searches can be made in networks.

In Kleinberg's model, there are n nodes distributed in a two-dimensional lattice. The distance between any two nodes u and v, denoted by $d(u,v)$, is defined as the steps between the two nodes in the lattice. If the coordinates of u and v are, respectively, (i,j) and (k,l), then we have

$$d(u, v) = |k - 1| + |l - j|. \tag{10.44}$$

In the lattice, each node directly connects to its neighboring nodes with directed edges within some constant number $p \geq 1$ of steps. These connections are known as the short-range connections of the node. In addition, there are q directed edges that connect this node to other q nodes in the network. These q connections are known as the long-range connections of the node. See Figure 10.6 for an illustration.

In Kleinberg's model, the probability that there exists a long-range connection between nodes u and v is

$$\frac{[d(u,v)]^{-r}}{\sum_{v}[d(u,v)]^{-r}} \tag{10.45}$$

where r is a parameter. For fixed values of p and q, when $r = 0$, the number of long-range connections follows a uniform distribution. As the value of r increases, the probability that there exists a long-range connection between two grid nodes that are far apart becomes smaller and smaller.

Based on Milgram's small world experiment, let us randomly select two nodes in the network and consider how to pass a piece of information from one of the nodes, known as the source node, to the other node, known as the destination node, with the least number of steps. Assume that a decentralized algorithm is used at each node. That is, the current letter holder has only the local information, with which the holder passes the letter to one of his or her neighbors with either a short-range or a long-range connection. The information that the current holder knows includes

1. The set of all local contacts among all nodes
2. The location on the lattice of the target node
3. The location on the lattice of all the nodes that have passed letters and the long-range connections among them

Kleinberg proved the following three theorems.

Theorem 10.1

There is a constant α_0, depending on p and q but independent of n, so that when $r = 0$, the expected delivery time of any decentralized algorithm is at least $\alpha_0 n^{2/3}$.

What Theorem 10.1 says is that despite the shortest path between any two chosen nodes in a small-world network being small, which means that it is a polynomial in log n, such a path cannot be found by using an arbitrarily selected decentralized algorithm. ■

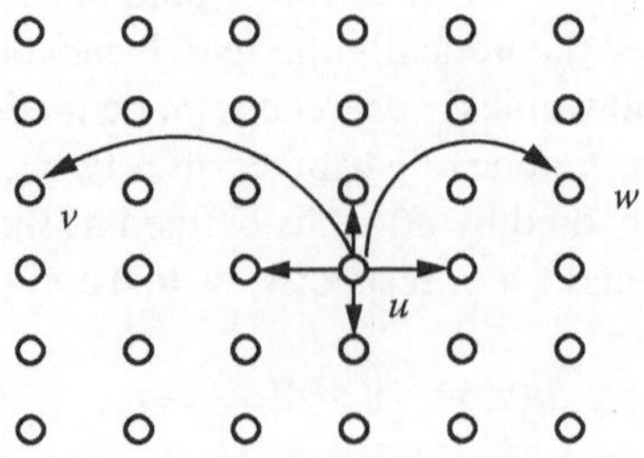

FIGURE 10.6 Two-dimensional grid network with $n = 6$. The contacts of a node u with $p = 1$ and $q = 2$. Nodes v and w are the two long-range contacts.

Theorem 10.2

There is a decentralized algorithm A and a constant α_2, which is independent of n, so that when $r = 2$ and $p = q = 1$, the expected delivery time of A is, at most, $\alpha_2(\log n)^2$.

What Theorem 10.2 says is that when $r = 2$, the required number of steps to pass a letter by using a decentralized algorithm is, at most, a polynomial in log n. ■

Theorem 10.3

(a) Let $0 \le r < 2$. There is a constant α_r, depending on p, q, and r, but independent of n, so that the expected delivery time of any decentralized algorithm is at least $\alpha_r n^{(2-r)/3}$.

(b) Let $r > 2$. There is a constant α_r, depending on p, q, and r, but independent of n, so that the expected delivery time of any decentralized algorithm is at least $\alpha_r n^{(r-2)/(r-1)}$.

What Theorem 10.3 says is that when $r \ne 2$, the required number of steps to pass a letter by using any selected decentralized algorithm is at least a polynomial in log n. ■

10.4.4 Community Structures

With the natural and physical meanings of networks and the relevant mathematical properties discovered through in-depth studies, research has found (Gibson et al. 1998; Adamic and Adar 2003; Holme et al. 2003) that many realistic networks have a common characteristic: the community structure. For instance, Figure 10.7 shows the network of Zachary's karate club, where membership clusters can be clearly seen. In other words, each network is composed of a number of "groups" or "clusters." Also, each group is identified by using dense connections between nodes, while groups are separated from each other with relatively sparse connections.

In general, associations or connections between nodes can mean any concept that is significant to the researcher, such as a module, class, group, groups, or anything else. For example, the WWW can be seen as a large network composed of various organizations, such as a community discussion site where topics of common interest connect all the participants together, and the same individual can at the same time belong to a different discussion group of a completely different set of topics. Similarly, in a biological network or a circuit network, a chosen node can also play different roles depending on how the node is seen from within individually different communities. Revealing the community structure of networks, for the purpose of understanding and analyzing networks' characteristics, is very important. The community structure of networks appears widely in biology, physics, computer graphics, and sociology. Thus, it is expected to have a great potential for practical applications.

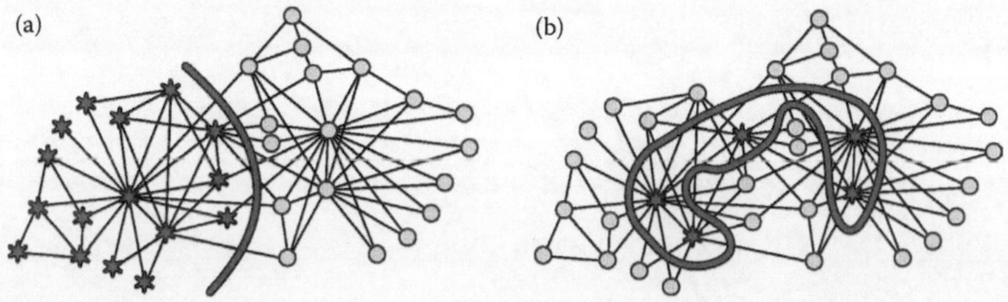

FIGURE 10.7 Zachary's karate club.

10.5 NETWORK AND COMPLEXITY

The founder of the theory of complex networks has made great contributions to systems science. The concept of network not only stands for the structure of many complex systems but can also serve as a model for the investigation of the topological characteristics of systems. Each system contains objects and, more importantly, their interactions. It is the objects that make the system to have a steady-state existence, while it is the interactions between the objects that bring the objects into an organic whole, a system. That explains why physics studies the elementary interactions between sizeless and volumeless objects, chemistry considers molecular interactions, biology looks at the interactions between living beings or organisms, and sociology investigates the interactions between the various humans and human organizations. That is why when a collection of things is seen as a system, the structure of the system can be abstracted into an appropriate network, where the things are abstracted into the nodes and the interactions into connections between the nodes. Thus, one can employ the theory, methods, and tools of graph theory and network analysis to analyze the topological properties of systems and their structures. Each system's structure can be described as a network structure. Most complex systems are open, possess the functionality of self-organization, and evolve dynamically. The research results of complex networks reflect some of the most talked-about characteristics of complex systems and represent some of the substantial breakthroughs achieved in the study of systems.

10.6 OPEN-ENDED PROBLEMS

1. Prove that degree distribution of a random graph follows Poisson distribution.
2. Construct a network with five nodes, each of which represents a friend of yours. Two nodes are connected with an edge if these two persons like the same sport.
3. Show that a graph $G = (V, E)$ is connected if and only if $\text{Neighbor}^{\infty}(v) = V$, for any $v \in V$.
4. From the angle of the development history of network science, explain why the concept of complex networks emerges. Is there some inevitability?
5. Why do we call the networks whose degree distributions follow the power law distribution scale-free networks?
6. What is the essential difference between random, small-world, and scale-free network?
7. Are there further statistical distributions that can provide insights on the structure and classification of complex networks?
8. Are there universal features of network dynamics?
9. How do dynamical processes taking place on a network shape the network topology?
10. How do you understand that "topology determines function" in complex networks?
11. Please describe in your own words the relationship between networks and complexity.

11 Case Studies

The purpose of this chapter is to show the reader how systems thinking and methods have been practically employed in various scientific investigations, leading to magnificent theoretical and practical values. This chapter includes four case studies. The first one is about systems evaluation, which specifically looks at how to make estimates and evaluations for such systems as guided missile systems. The second example applies the holistic thinking to analyze the positioning accuracy of satellite navigation systems, especially when the system is composed of different kinds of satellites. The third case study looks at the temporal analysis of social networks. To show the practical usefulness of this work, two real-life networks are investigated: the world trade and the e-mail communication network of Enron Corporation. The fourth case study investigates the formation of values among Chinese college students. Other than discovering the current intellectual state of this age group, beneficial suggestions are provided.

In short, based on the previous chapters, what is presented in this chapter shows the fact that systems science entails not only a new way of thinking and a logic of reasoning but also methodologies that can actually produce tangible consequences.

11.1 SYSTEMS EVALUATION

11.1.1 Engineering Background

A giant complex system is often a "super system" materially composed of many subsystems. Testing a complex system is a key process of evaluating its performance against a certain criterion. Considering the restriction and effects of many factors relevant, such as cost, periods, feasibility, measurability, etc., it is impossible to design a test to practically cover all possible states that might appear under different external environments. The idea of substitute tests has been effectively employed to resolve this problem. Each substitute test contains mainly computer simulations, small-scaled tests (Neuberger et al. 2007; Obermark 2004), varying parameter tests (Wang et al. 2002; Zeng et al. 2007), indirect metrical tests (Angrisani et al. 2006a, 2006b; Yun et al. 1999), equivalent tests (Qi et al. 2007; Shang and Gao 2006; Duan et al. 2005; Zhao et al. 2006; Huang et al. 2008; Wang et al. 2006; Yang et al. 2008; Li et al. 2004), and others, all of which combined could be employed to indirectly support the results of the realistic, but practically impossible, evaluation of a system's performance. In order to obtain the evaluation results regarding the system's performance indexes for the situation that an overall test is practically impossible, the data fusion process of evaluation needs to be improved by exploiting more efficiently the available small sample information from all relevant data. The entire process of the substitute equivalent test (SET) and the corresponding data analysis should be regarded as a process of verifying the theoretical model.

In terms of testing fully armed operations, the theory of SET, and the relevant comprehensive evaluations of different kinds of complex systems, scholars have developed various methods to convert substitute equivalent data between an overall operational test (OOT) and SET. In particular, the detecting distance of radar in SET can be converted into that in OOT by deriving the radar equation in finer detail (Wang et al. 2002). This conversion method is also extended to estimate the jamming effectiveness of radar networks in electronic countermeasure test (Zeng et al. 2007). Equivalent methods are also used in the tests of infrared detecting systems to estimate the detecting distance based on the traditional equation of infrared detecting distance (Angrisani et al. 2006a, 2006b; Yun et al. 1999). Because it is practically impossible to have a realistic enemy system against which to test the true effectiveness of complex radar electronic equipment in an actual military conflict, a highly economically meaningful method is proposed (Qi et al. 2007) to comprehensively evaluate the performance of the radar system by jointly making use of numerical simulation, fully armed operations, and substitute equivalent manipulations.

In the area of designing and evaluating aircrafts, in order to conduct six-degree-freedom simulations when not enough data are available based on the results of the SET and corresponding data conversion methods, Shang and Gao (2006) developed an equivalent moment-derivative model. Zhao et al. (2006) established a method for evaluating the flying quality of an airplane. In particular, under the same external interference, the method first introduces a simpler system that is equivalent to the original complex system, and then computes the performance parameters of the simpler system by applying optimization schemes so that the quality of the original system can be assessed. As for the analysis and conversion of guided missile precisions, a substitute trajectory is used to identify and evaluate the actual precision of the realistic trajectory without OOT (Duan et al. 2005), where various substitute errors are also converted to the actual scenario. As for the heterogeneous data fusion, Huang et al. (2008) proposed a novel posterior-weighted Bayesian estimation method for considering the credibility of the physical sources of the prior information. Facing the problem of evaluating various modern complex systems, the multihierarchical analytic method and its extensions (Wang et al. 2006) are applied in the evaluation of complex systems of multiple types of attributes. In the area of reliability evaluation, a method of making use of the information of similar products is proposed (Yang et al. 2008) to produce a comprehensive evaluation of reliability of electronic equipment. The method identifies the comparability between the historical sample and the current operational one based on the analysis of similar systems and then fuses all the information for evaluation based on the derived successive factors. There is also a method (Tang et al. 2001) that evaluates the use performance of giant complex systems based on extenics theory and hierarchy analysis. In short, the conversion method based on SETs of similar systems has been widely studied and applied in the evaluation of aircraft design, performance of weaponry, radar electronic jamming, reliability, and use performance of complex systems.

In general, the method of converting the information of SETs first finds the successive or corrective relation between the target quantities of measurement and the substitute quantities based on the analysis of the similarity and difference between OOT and SET. Then, it derives the target measurements of the OOT quantities through the measurements of the substitute quantities by fusing all relevant information obtained in actual tests. At present, in the field of impact point precision evaluation of missiles, the main consideration has been the historical data and prior information of impact points without paying much attention to the observed flight process information (Duan et al. 2005; Huang et al. 2008; Wang et al. 2006; Wang et al. 2009). As intuition suggests, impact points should be highly related to the flight process. However, rarely has any published research ever focused on the analysis of the relationship between flight processes and impact points in homing radar precision evaluation. That is, measured data in flight processes are not adequately utilized in homing radar equivalent tests and evaluations.

11.1.2 Key of System Evaluation

In order to obtain reliable evaluation results of the performance indexes under the condition of small samples due to the limited number of OOTs, the data fusion process should and has to be improved by more efficiently exploiting information from all available data. However, how can all relevant information be exploited? To this end, for example, the observed flight process data can be adequately used to validate the established model and estimate systematic errors and the characteristics of random errors so that due to too much randomness, the current unreliable estimates, produced by employing only the isolated few impact point data, can be improved with increasing robustness. Now, let us address how to implement the process of verifying the model by using all kinds of available information.

11.1.3 Systemic Thought (Duan and Lin 2011)

11.1.3.1 Some Briefings on Laws of Conservation

As is well known, laws of conservation are some of the most important laws of nature. In the classical physics, there is a law of conservation for each important concept, such as mass, energy, momentum, moment of momentum, electric charge, etc. Studies of modern physics indicate that each moving object in an even and isotropic space and time, no matter whether the space is microscopic, mesoscopic, or macroscopic, a particle, or a field, must follow the laws of conservation of energy, momentum, and moment of momentum. That is, a unification of space, called a Minkowski space, has been achieved. In Einstein's relativity theory, a unification of the concepts of mass, time, and space is realized.

Based on what is discussed earlier in this book, it is known (Lin 2008) that between the macrocosm and the microcosm, between the electromagnetic interactions of the atomic scale and the strong interactions of quark's scale, between the central celestial body and its circling celestial bodies of celestial systems, and between the large-, medium-, small-, and microscales of the earth's atmosphere, there are laws of conservation of products of spatial physical quantities. Thus, it is conjectured that there might be a more general law of conservation in terms of structure, in which the informational infrastructure, including time, space, mass, energy, etc., is approximately equal to a constant. In symbols, this conjecture can be written as follows:

$$AT \times BS \times CM \times DE = a \tag{11.1}$$

or more generally

$$AT^{\alpha} \times BS^{\beta} \times CM^{\gamma} \times DE^{\varepsilon} = a \tag{11.2}$$

where α, β, γ, ε, and a are constants, T, S, M, E and A, B, C, D are time, space, mass, and energy, and their relevant coefficients, respectively. These two formulas hold true in both macroscopic and microscopic scales, and the constants α, β, γ, ε, and a are determined by the initial state of the system of concern.

It can be seen from Equation 11.1 that when two of the terms of time, space, mass, and energy are fixed, the other two terms will evolve in opposite directions.

11.1.3.2 Quantitative Measurements of System Evaluation Model, Observational Data, and Prior Knowledge

Each evaluation in reality is a recognition process of the underlying system. Only with adequate supply of input information can one obtain the needed accuracy in the evaluation and prediction of the system's behaviors.

A standard for measuring the usability of the available information is the statistical measures of the information. For a given purpose, how should one measure the increase in the amount of information brought forward by the model, observational data, and/or prior knowledge?

11.1.3.2.1 Information Metric of Models

It is difficult to directly measure the information content of a model. However, it is relatively easier to make the measurement by using the data statistically. One way to do this is to transform the original task of measuring the model's information content to that of the residuals according to the following criterion: the less the information content in the residuals, the more the information content provided by the model.

Suppose that there is some quantitative verification. In this case, if one does not have any prior knowledge on the form of the underlying model, then he or she has to use the available data to simultaneously estimate the model and the relevant parameters. If the form of the underlying model is known according to the prior knowledge, then he or she can directly estimate the parameters. The difference between these two scenarios can be seen as an informational realization of the model. If the model is accurate (i.e., its information content is sufficient), the residual of the model fitting should be Gaussian white noise; if the model is not accurate (i.e., its information content is insufficient), then the residual of the model fitting would not be Gaussian white noise; it may contain some kind of systematic deviation or correlated random error. Hence, one can measure the information content of models using the distributions of the residuals of model fittings.

At this junction, let us introduce an index of information divergence to describe the difference between the residual distributions. What this index describes is the structural difference between two distributions (Lee and Landgrebe 1993; Carreira-Perpinan 2001). We will make use of this concept to measure the difference between two distribution functions. Assume that f is a one-dimensional density function, and g is a one-dimensional standard normal density function. Then, the relative entropy of f to g is given as follows:

$$d(f\|g) = \int_{-\infty}^{+\infty} g(x)\cdot \log\frac{f(x)}{g(x)}\,\mathrm{d}x.$$

The index of information divergence is defined by

$$Q(f, g) = |d(f\|g)| + |d(g\|f)|. \tag{11.3}$$

When $f = g$, $d(f\|g) = 0$, and the further f deviates from g, the greater the value $d(f\|g)$. Thus, the value $d(f\|g)$ indeed depicts the degree of deviation from f to g. When using sampling data to estimate the density functions f and g, one first needs to analyze the accurate expressions of the density functions f and g and then apply methods of optimization to estimate the parameters of the distributions. Therefore, a more convenient method is that under a certain requirement of accuracy, one can directly use the discretized probability distributions p and q to substitute for the continuous density functions f and g.

By using the discrete probability distributions p and q to compute the value of information divergence, the index in Equation 11.3 becomes

$$Q(p, q) = D(p\|q) + D(q\|p) \tag{11.4}$$

where $D(p\|q) = \sum q\cdot \log\left(\frac{p}{q}\right)$. What can be seen here is that the greater the index of information divergence, the greater the difference between the distributions. In general, we take q to be a standard normal distribution and p to be the distribution of the residuals of the model fitting so that $Q(p,q)$ stands for the information content of the model.

11.1.3.2.2 Information Metric of Observational Data

After one obtains the form of his or her model, in terms of data fitting processes involving parameters, an increase in information is realized in the recognition of the parameters of the model. Based on the amount of increase in accuracy of the estimate of the parameters, obtained from an increased number of observations or prior knowledge on the parameters, one can analyze the amount of increase in information. For example, for a sample $(x_1, x_2, \cdots, x_n)$ collected from a population of normal distribution $N(\theta, \sigma^2)$, if the prior knowledge of θ is $N(\mu_0, \tau_0^2)$, then the posterior mean is

$$\mu = \frac{n\sigma^{-2}}{n\sigma^{-2} + \tau_0^{-2}}\bar{x} + \frac{\tau_0^{-2}}{n\sigma^{-2} + \tau_0^{-2}}\mu_0. \quad (11.5)$$

The posterior variance τ satisfies

$$\frac{1}{\tau^2} = \frac{n}{\sigma^2} + \frac{1}{\tau_0^2}. \quad (11.6)$$

Evidently, the estimate accuracy of the posterior mean is relevant to those of the sample mean and the prior mean. Thus, both increasing the sample size and decreasing the variance of the prior distribution can enhance the estimate accuracy of the posterior estimate.

In the following, we look for such a quantitative index that can measure the information content contained in a collected sample satisfying the addition rule. To this end, Fisher statistical information may meet the demand. That is, the increased information content gained from increasing the sample size should be a statistic whose expected value is exactly the Fisher statistical information.

From the well-developed methods, it follows that when the joint distribution density of $x_1,\cdots, x_n$ is $p(x_1,\cdots, x_n;\theta)$, then the Fisher information of the parameter θ is given below (Mao 1999):

$$I(\theta) = \mathrm{E}\left(\frac{\partial \ln p(x_1,\cdots,x_n;\theta)}{\partial\theta}\right)^2, \quad (11.7)$$

where the partial derivative of $\ln p(x_1,\cdots, x_n;\theta)$ with respect to θ is involved. If $x_1,\cdots, x_n$ are identically and independently distributed (i.i.d.) and $x_i \sim f(x_i;\theta)$, $i = 1,2,\cdots,n$, then we have

$$I(\theta) = \mathrm{E}\left(\sum_{i=1}^{n}\frac{\partial \ln f(x_i;\theta)}{\partial\theta}\right)^2 = \sum_{i=1}^{n}\mathrm{E}\left(\frac{\partial \ln f(x_i;\theta)}{\partial\theta}\right)^2 = n\mathrm{E}\left(\frac{\partial \ln f(x_1;\theta)}{\partial\theta}\right)^2. \quad (11.8)$$

This expression indicates that the information content about the parameter θ, provided by n independent samples, is n times that provided by one sample. In terms of normal distributions, the increase in Fisher information gained along with an additional independent sample point is $\frac{1}{\sigma^2}$. Thus, the Fisher information about the parameter θ provided by n independent samples is $\frac{n}{\sigma^2}$. This result agrees with the minimum variance, namely, the Cramer–Rao minimum bound, of n independent samples $\frac{1}{nI(\theta)} = \frac{1}{n\sigma^{-2}} = \frac{\sigma^2}{n}$.

If there is more than one parameter, then the information content corresponding to Equation 11.7 becomes the following information matrix:

$$I(\boldsymbol{\theta}) = \mathrm{E}\left(\frac{\partial \ln p(x_1,\cdots,x_n;\boldsymbol{\theta})}{\partial\boldsymbol{\theta}}\right)\left(\frac{\partial \ln p(x_1,\cdots,x_n;\boldsymbol{\theta})}{\partial\boldsymbol{\theta}}\right)^T = -\mathrm{E}_{\boldsymbol{\theta}}\frac{\partial^2 \ln p(x_1,\cdots,x_n;\boldsymbol{\theta})}{\partial\boldsymbol{\theta}^2}. \quad (11.9)$$

11.1.3.2.3 Information Metric of Prior Knowledge

The information content of prior knowledge can be measured similarly to that of observational data. However, in this case, we must consider the credibility of the prior knowledge. This credibility in general is obtained by analyzing the consistency between the prior knowledge and the data collected through autoptic tests. It stands for the probability for these two classes of data to come from the same population. If the data contains a systematic deviation, after having estimated the credibility, the prior data needs to be weighted when fused with the data collected through autoptic tests. We can determine the weight using the magnitude of the credibility. The precision (i.e., the random error) is used as the measurement of the information content.

The currently most applied method of measuring the credibility of prior knowledge is based on the data consistency test. This method uses rank-sum test or other test methods to check the consistency of the data. If the significance level is α and if the two kinds of data, namely, the prior data and the autoptic test data, are consistent, then the credibility p of the prior data is given as follows (Huang et al. 2008):

$$p = P(H_0 \mid A) = \frac{(1-\alpha)P(H_0)}{(1-\alpha)P(H_0) + (1-P(H_0))\beta} = \frac{1}{1 + \frac{(1-P(H_0))}{P(H_0)} \cdot \frac{\beta}{1-\alpha}}, \tag{11.10}$$

where β is the probability of type II error. Because its computation is quite complicated, one can employ the bootstrap method and Monte Carlo simulation to estimate its value. Let $P(H_0)$ be a prior probability. If there is no other prior information available, then one can take $P(H_0) = 1/2$. Thus, the computation of the credibility can be simplified to

$$p = \frac{1-\alpha}{1-\alpha+\beta}. \tag{11.11}$$

The previous method is about how to compute the credibility of the prior data at the quantitative level. No matter whether we define the credibility using consistency tests or differences in distributions, they are essentially the same.

Assume that the credibility of the prior knowledge is r_S $(0 \le r_S \le 1)$ and the variance of the prior data is τ_R. Then the Fisher information provided by the prior information is

$$r_S \frac{1}{\tau_R^2}. \tag{11.12}$$

11.1.3.3 Conservation Law of Information for System Evaluations

It is well known that the information provided and revealed by the model, autoptic test data, and prior knowledge can help accurately recognize the underlying system. Intuitively, it can be imagined that for any given evaluation accuracy, if the information content of the model is relatively large, then the needed amount of autoptic test data can be relatively decreased. If the information content of the model is very little or there is no model available, then more autoptic test data would be needed for producing reliable system evaluation. When there are not enough autoptic test data, it is possible that an increased amount of prior knowledge can help make up the shortage of these data. According to the analysis in Section 2.2, we can employ Fisher information to measure the effects of different models, autoptic test data, and prior knowledge. Therefore, we can establish the conservation law of information for quantified system evaluations.

Given an evaluation or prediction precision, imagine that the information, provided by the inputting parts of the model description, autoptic test data, and prior knowledge, satisfies a certain condition of conservation as follows:

$$Ae^{I_M} \times Be^{I_D} \times Ce^{I_P} = \mathrm{a} \tag{11.13}$$

where I_M stands for the information content described by the model, I_D is the information content of the autoptic test data, and I_P is the information content of the prior knowledge. The constant a should somehow depict the minimum amount of information required to satisfy the given precision (in the estimate of the model or parameters), where the precision is given in terms of the model accuracy and parameter estimation precision.

According to the analysis of the previous subsection, we have

$$I_M = Q(p,q), \quad I_D = I(\theta) = \mathrm{E}\left(\frac{\partial \ln p(x_1,\cdots,x_n;\theta)}{\partial \theta}\right)^2,$$

$$I_P = \frac{1}{\tau_R^2}, \quad A = B = 1, \quad C = e^{r_s}.$$

Evidently, Equation 11.13 is a special case of Equation 11.2. For the current situation, we consider only changes in information flows instead of changes in time, space, and energy. Due to the fact that the basis for measuring information is the information entropy or Fisher information under the meaning of statistics, we choose the exponential form of base e in Equation 11.13. Considering the credibility of prior information, the coefficient of the measurement of the prior information is no longer an invariant constant like those of the model information and the autoptic test data.

In particular, in order to obtain the accurate estimate for systems attributes or prediction outcome, one has to satisfy that the product of the information contents is equal to a constant a. Of course, there are different ways to meet this requirement.

1. If the form of the model is unknown, then to estimate the form of the model and its parameters, one needs the background knowledge of the relevant systems dynamics and autoptic test data.
2. If the form of the model is correctly determined with some parameters unknown, then one needs a good amount of autoptic test data to make up the shortage of information in order to estimate the relevant parameters.
3. If the form of the model is correctly determined with some parameters unknown, and if the amount of autoptic test data is not sufficient to accurately estimate the parameters, then one needs to find an adequate amount of consistent prior information.

11.1.4 Particular Evaluation Method on Complex Guidance Systems

11.1.4.1 Error Model of Compound Guiding Radar

11.1.4.1.1 Error Model of Radar

The impact point dispersion of navigation missiles equipped jointly with inertial navigation equipment and radar is caused by measurement errors, such as radar navigation system error, instrumental error of inertial navigation systems (INS), model error of terminal guidance, methodical error, random error, and others. Specifically, the final impact point dispersion is mainly caused by the measurement error of radar navigation systems.

Radar measurement error can be modeled (Wu et al. 2005) by a response function of its impact factors, including radar and environmental parameters. Such an error can be expressed as a random process:

$$\xi(t) = \sum_{i=1}^{3} [m_{xi}(\boldsymbol{\theta}, t) + \sigma_{xi}(\boldsymbol{\theta}, t)\eta_i(t)] \tag{11.14}$$

where $\boldsymbol{\theta}$ is the parameter vector of different error types, t is the time, m_{xi} is the mean, while σ_{xi} are the variances of systematic error (when $i = 1$), slow fluctuant error (when $i = 2$), and fast fluctuant error (when $i = 3$), and $\eta_i(t)$ is the random function when the correlative function $k_{xi}(\tau) = 1$, satisfying that the mean of $\eta_i(t)$ is zero, and the variance an identity entity. Under typical conditions, $k_{x1}(\tau) = 1$ and $k_{x2}(\tau)$ are determined by the slow fluctuation error, and $k_{x3}(\tau) = \delta(\tau)$. That is, this part of the model describes the correlative noise error of samples.

11.1.4.1.2 Impact Point Error Model

For a missile with an active homing radar warhead, the measurement error of the warhead–target relative positions x, y, and z is mainly equal to its impact dispersion caused by the error of radar measurement. The measurement of actual radar includes the distance R, the azimuth A, and the elevation angle E. The relationship between x, y, z and R, A, E is shown in Figure 11.1 (Duan and Lin 2011).

According to Gauss law of error propagation, the transfer relation between the errors of x, y, z and those of R, A, E can be described as follows (Duan and Lin 2011):

$$\begin{bmatrix} \Delta_x \\ \Delta_y \\ \Delta_z \end{bmatrix} = \begin{bmatrix} \frac{\partial x}{\partial R} & \frac{\partial x}{\partial A} & \frac{\partial x}{\partial E} \\ \frac{\partial y}{\partial R} & \frac{\partial y}{\partial A} & \frac{\partial y}{\partial E} \\ \frac{\partial z}{\partial R} & \frac{\partial z}{\partial A} & \frac{\partial z}{\partial E} \end{bmatrix} \begin{bmatrix} \Delta_R \\ \Delta_A \\ \Delta_E \end{bmatrix} = \begin{bmatrix} \cos E \cos A & -R\cos E \sin A & -R\sin E \cos A \\ \sin E & 0 & R\cos E \\ \cos E \sin A & R\cos E \cos A & -R\sin E \sin A \end{bmatrix} \begin{bmatrix} \Delta_R \\ \Delta_A \\ \Delta_E \end{bmatrix} \tag{11.15}$$

where Δ_x, Δ_y, and Δ_z are the measurement errors of warhead–target relative positions x, y, and z in the OOT, and Δ_R, Δ_A, and Δ_E are the errors of R, A, and E measured by radar in the OOT, respectively.

11.1.4.1.3 Analysis

In the impact point error test of warhead missiles with active homing radar, it is expected to convert the result of impact dispersion in the SET to that of the OOT. With the classical method, the conversion result of impact dispersion is analyzed based on the final impact point data, which is usually not

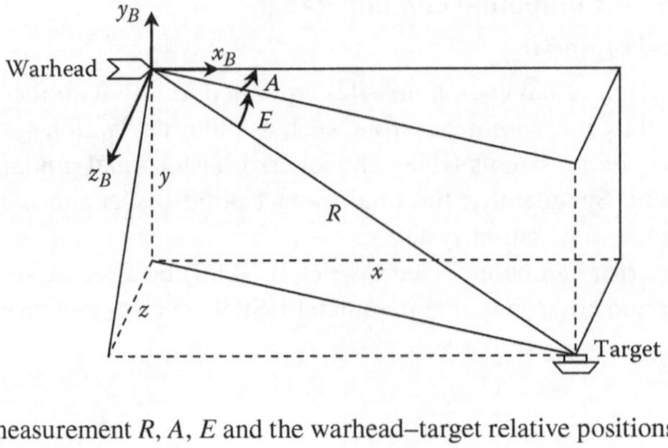

FIGURE 11.1 Radar measurement R, A, E and the warhead–target relative positions x, y, z.

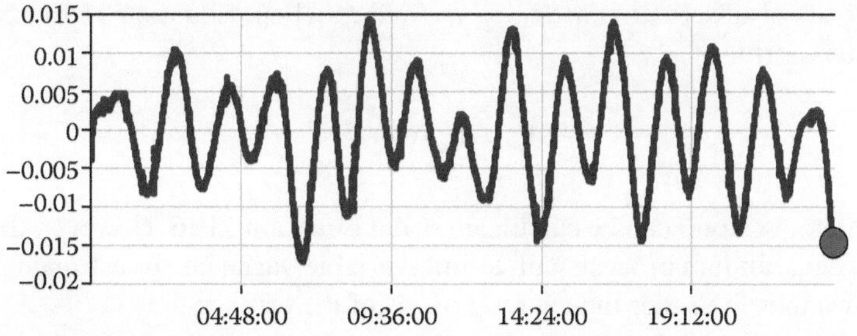

FIGURE 11.2 Observed flight process data are used to construct a time series model to produce more robust results for the final impact point error.

robust because the sample size is very small and the measurement of the single final impact point is greatly influenced by random factors.

To resolve this problem, many observational flight process data can be utilized. In particular, it will be more beneficial to the equivalent substitutes of the performance indexes if the flight process error is first described by using a time series model. After the error time series model is validated by the observed flight process data, it is then fused to the impact dispersion conversion in the OOT.

This method is superior to the previous equivalent conversion of the performance indexes that is based only on the impact point error information, because it exploits additional relevant data to construct a full-scale time series model to support the eventual evaluation (Figure 11.2).

The information of the performance index, such as the impact point dispersion, can be regarded as final when $t = t_n$ and as in process when $t \in [t_0, t_n)$. We try to estimate θ, the characteristic parameter vector of the model, with additional accuracy by using process information, and then obtain the converted results of the final performance index for when $t = t_n$. All particular details will be described in the following.

11.1.4.2 Converting Observed Data/Deducing Performance Index Based on Characteristics of Random Processes

This subsection mainly considers how to convert the measurement errors of different types in Equation 11.14. For systematic error, a mathematic model can usually be established by using the corresponding physical background.

11.1.4.2.1 Conversion of Systematic Error

First, we discuss the systematic error in Equation 11.14. Let us take one kind of systematic error of radar ranging as our example. The dynamic lag error can be expressed as

$$\sigma_R = \frac{\dot{R}}{K_v} + \frac{\ddot{R}}{K_\alpha}$$

where $\dot{R}$ is the radial velocity of the target relative to the moving radar, $\ddot{R}$ is the radial acceleration, K_v is the velocity error constant of the range tracking loop, and K_α is the acceleration error constant. The relevant conversion results could be easily calculated after the relative error constants were obtained with observed data in equivalent tests. Therefore, the key of systematic error conversion is to compute the relative error coefficients based on the error model and observed data. Let us construct the following model for this situation:

$$y_m(t) = f_m(\boldsymbol{X}_m(t); \boldsymbol{\theta}) + e_m(t), \quad \boldsymbol{X}_m(t) \in \boldsymbol{D}_m \tag{11.16}$$

where the time nodes of observation are t_1, t_2, ..., and t_n. Then, Equation 11.16 can be rewritten in the following matrix form:

$$Y_m = F_m(\boldsymbol{\theta}) + \vec{\boldsymbol{\varepsilon}}_m. \tag{11.17}$$

Now the parameter vector $\boldsymbol{\theta}$ can be calculated using Equation 11.16. However, the problem is that the error $\vec{\boldsymbol{\varepsilon}}_m$ may be a random process with a time-variable variance. To estimate parameter vector $\boldsymbol{\theta}$ with accuracy, we must consider the characteristics of the error $\vec{\boldsymbol{\varepsilon}}_m$.

11.1.4.2.2 Conversion of Random Error with Time-Dependent Variance

In this subsection, we will focus on the random error in Equation 11.14. Errors observed by missile-borne radar can also be expressed with models with time-dependent variances. For our purpose, let us take clutter and jamming error of radar as an example. This error can cause ranging error, no matter whether the target's echo in the range tracking system is intensive or not. The estimation formula for the ranging error caused by the clutter and the jamming error is given as

$$\sigma_R = \frac{1}{\beta\sqrt{2(S/C)n_e}}$$

where β is the effective bandwidth of radar signal, S/C is the signal/clutter ratio (SCR), and n_e is the effective accumulated pulse number of the range tracking loop.

Because SCR varies with time, the variance of the ranging error caused by clutter and jamming is also diverse with time. Suppose that the random error in one equivalent test is

$$\varepsilon_s(t) = \sum_{h=1}^{H} e_{sh}(t) \tag{11.18}$$

where $e_{sh}(t) \sim \mathrm{N}(0, \sigma_{sh}^2(t))$ and $\sigma_s^2(t) = \sum_{h=1}^{H} \sigma_{sh}^2(t)$, and the random error in the OOT is

$$\varepsilon(t) = \sum_{h=1}^{H} e_h(t), \text{ where } e_h(t) \sim \mathrm{N}(0, \sigma_h^2(t)).$$

That is,

$$\varepsilon(t) \sim \mathrm{N}(0, \sigma^2(t)), \text{ where } \sigma^2(t) = \sum_{h=1}^{H} \sigma_h^2(t) \tag{11.19}$$

where the expression of $\sigma_h^2(t)$ is known (Duan and Lin 2011). That is, the transfer relation between $\sigma_h^2(t)$ and $\sigma_{sh}^2(t)$ can be explicitly obtained when the relevant physical parameters are given.

11.1.4.2.3 Measurement Data Conversion Based on Time-Dependent Process Information and Parameter Estimation of a Single Test

In this subsection, under the assumption that the random error has time-dependent variance, we estimate the model parameter vector $\boldsymbol{\theta}$ using observed data and then calculate the corresponding indexes of the characteristics of the random error. Take

$$y_m(t) = f_m(\boldsymbol{X}_m(t);\boldsymbol{\theta}) + e_m(t), \quad \boldsymbol{X}_m(t) \in \boldsymbol{D}_m \tag{11.20}$$

as our example, assuming that the time nodes of the sampling point are $(t_1, t_2, \cdots, t_{n_m})$. Then, Equation 11.20 can be rewritten in the following matrix form:

$$\boldsymbol{Y}_m = \boldsymbol{F}_m(\boldsymbol{\theta}) + \vec{\boldsymbol{\varepsilon}}_m. \tag{11.21}$$

By making use of the least-squares method, the estimation of the parameter vector $\boldsymbol{\theta}$ in Equation 11.20 can be written as the following optimization problem:

$$\left\|\boldsymbol{Y}_m - \boldsymbol{F}_m(\hat{\boldsymbol{\theta}})\right\|_2^2 = \min \left\|\boldsymbol{Y}_m - \boldsymbol{F}_m(\boldsymbol{\theta})\right\|_2^2. \tag{11.22}$$

By using Gauss–Newton iteration, the parameter $\boldsymbol{\theta}$ can be estimated. Considering the transfer relation between the random error $\sigma_{mh}^2(t)$ of the substitute test and that $\sigma_h^2(t)$ of the OOT $\sigma_{sh}^2(t)$, the conversion result of the corresponding observed data is obtained as follows:

$$\hat{y}(t) = f(\boldsymbol{X}(t);\hat{\boldsymbol{\theta}}) + \frac{\sigma(t)}{\sigma_m(t)} \varepsilon_m(t) \tag{11.23}$$

where $\varepsilon_m(t)$ is the regression residual $\hat{y}(t) - f(\boldsymbol{X}(t);\hat{\boldsymbol{\theta}})$. After obtaining the conversion result of the observed data in one substitute test, the corresponding statistical characteristics, such as the mean, variance, confidence interval, and so on, can also be calculated.

11.1.4.2.4 Fusion of Converted Observed Data from Multiple SETs

What follows is an extension of what is presented in the previous subsection. We will estimate performance indexes with observed data in three SETs. Similar to what had been done earlier, we define the symbols in the other two equivalent substitute tests as $\boldsymbol{Y}_l$, $\boldsymbol{F}_l(\boldsymbol{\theta})$, $\vec{\boldsymbol{\varepsilon}}_l$, $\boldsymbol{\Sigma}_l$ and $\boldsymbol{Y}_k$, $\boldsymbol{F}_k(\boldsymbol{\theta})$, $\vec{\boldsymbol{\varepsilon}}_k$, $\boldsymbol{\Sigma}_k$, and let

$$\boldsymbol{Y}_T = \begin{bmatrix} \boldsymbol{Y}_m \\ \boldsymbol{Y}_l \\ \boldsymbol{Y}_k \end{bmatrix}, \boldsymbol{F}_T(\boldsymbol{\theta}) = \begin{bmatrix} \boldsymbol{F}_m(\boldsymbol{\theta}) \\ \boldsymbol{F}_l(\boldsymbol{\theta}) \\ \boldsymbol{F}_k(\boldsymbol{\theta}) \end{bmatrix}, \vec{\boldsymbol{\varepsilon}}_T = \begin{bmatrix} \vec{\boldsymbol{\varepsilon}}_m \\ \vec{\boldsymbol{\varepsilon}}_l \\ \vec{\boldsymbol{\varepsilon}}_k \end{bmatrix}, \boldsymbol{\Sigma}_T = \begin{bmatrix} \boldsymbol{\Sigma}_m & & \\ & \boldsymbol{\Sigma}_l & \\ & & \boldsymbol{\Sigma}_k \end{bmatrix}.$$

Then the measurement model of these SETs can be expressed as

$$\boldsymbol{Y}_T = \boldsymbol{F}_T(\boldsymbol{\theta}) + \vec{\boldsymbol{\varepsilon}}_T. \tag{11.24}$$

Thus, the estimation of the parameter vector $\boldsymbol{\theta}$ can be simplified to the following optimization problem:

$$\left\|\boldsymbol{Y}_T - \boldsymbol{F}_T(\hat{\boldsymbol{\theta}})\right\|_2^2 = \min \left\|\boldsymbol{Y}_T - \boldsymbol{F}_T(\boldsymbol{\theta})\right\|_2^2. \tag{11.25}$$

Similar to the previous subsection, the corresponding conversion result of the observed data in all the SETs is obtained as

$$\hat{y}(t) = f(\boldsymbol{X}(t);\hat{\boldsymbol{\theta}}) + \frac{\sigma(t)\sigma_m^{-3}(t)}{\sum\limits_{j=m,k,l}\sigma_j^{-2}(t)}\varepsilon_m(t) + \frac{\sigma(t)\sigma_k^{-3}(t)}{\sum\limits_{j=m,k,l}\sigma_j^{-2}(t)}\varepsilon_k(t) + \frac{\sigma(t)\sigma_l^{-3}(t)}{\sum\limits_{j=m,k,l}\sigma_j^{-2}(t)}\varepsilon_l(t), \tag{11.26}$$

and the corresponding statistical characteristics, such as the mean, variance, confidence interval, and so on, can also be calculated.

11.1.4.3 Method and Conversion Precision of Composite Performance Indexes

In a simple case, the observed data are exactly the same as the performance indexes; the conversion results of the observed data are what are needed. The previous analysis can be applied to obtain the desired conclusion.

In a more complicated case, the desired performance index might be a composite function of the observed data. In this case, the observed data can be converted using the method just discussed above. Then, the computation of the performance index is carried out. As an illustration, let us look at some observed data of two types. The corresponding conversion results of the observed data are given below:

$$y_1(t) = f_1(\boldsymbol{X}(t);\boldsymbol{\theta}_1) + \varepsilon_1(t),$$
$$y_2(t) = f_2(\boldsymbol{X}(t);\boldsymbol{\theta}_2) + \varepsilon_2(t),$$

The desired performance index is not simply the conversion results $y_1(t)$ and $y_2(t)$. Instead, it is the following composite function of the two:

$$z(t) = g(y_1(t), y_2(t)) = g(f_1(\boldsymbol{X}(t);\boldsymbol{\theta}_1) + \varepsilon_1(t), f_2(\boldsymbol{X}(t);\boldsymbol{\theta}_2) + \varepsilon_2(t)) \tag{11.27}$$

from which the final index can be obtained.

We next develop such a method that a compound response function on the performance index can be constructed directly. Continuing the previous equation (Equation 11.27), we have

$$z(t) = s(\boldsymbol{X}(t), \boldsymbol{\theta}) + \xi = g(f_1(\boldsymbol{X}(t);\boldsymbol{\theta}_1) + \varepsilon_1(t), f_2(\boldsymbol{X}(t);\boldsymbol{\theta}_2) + \varepsilon_2(t)).$$

The relationship between the index $z(t)$ and the parameter vector $\boldsymbol{\theta}$ is directly established. We can similarly use the method discussed in the previous subsection to estimate the parameter. Thus, the previous equation can be rewritten in the following matrix form:

$$\boldsymbol{Z}_T = \boldsymbol{G} \circ \boldsymbol{F}_T(\boldsymbol{\theta}) + \vec{\xi}_T. \tag{11.28}$$

Thus, the integrated estimation of the parameter $\boldsymbol{\theta}$ can be simplified to the following optimization problem:

$$\left\|Z_T - G \circ F_T(\tilde{\boldsymbol{\theta}})\right\|_2^2 = \min \left\|Z_T - G \circ F_T(\boldsymbol{\theta})\right\|_2^2. \tag{11.29}$$

By using the Gauss–Newton iteration algorithm, the particular iteration scheme is

$$\begin{cases} \text{Initial value } \boldsymbol{\theta}^{(0)} \\ W_j = \ G \circ F_T\left(\boldsymbol{\theta}^{(j)}\right) \\ \boldsymbol{\theta}^{(j+1)} = \boldsymbol{\theta}^{(j)} + \lambda\left(W_j^{\mathrm{T}} \boldsymbol{\Lambda}_T^{-1} W_j\right)^{-1} W_j^{\mathrm{T}} \boldsymbol{\Lambda}_T^{-1}\left(Y_T - G \circ F_T\left(\boldsymbol{\theta}^{(j)}\right)\right). \end{cases}$$

Thus, using the weighted least-squares method produces the estimated $\tilde{\boldsymbol{\theta}}$ as follows:

$$\tilde{\boldsymbol{\theta}} = \left(W_j^{\mathrm{T}} \boldsymbol{\Lambda}_T^{-1} W_j\right)^{-1} W_j \boldsymbol{\Lambda}_T^{-1}\left(W_j \boldsymbol{\theta} + \boldsymbol{\Delta}\tilde{s}\right) \tag{11.30}$$

where $\boldsymbol{\Delta}\tilde{s}$ is the linearization error of the nonlinear model $g \circ f$. Its expected value is

$$\mathrm{E}[\tilde{\boldsymbol{\theta}}] = \mathrm{E}\left[\left(W_j^{\mathrm{T}} \boldsymbol{\Lambda}_T^{-1} W_j\right)^{-1} W_j \boldsymbol{\Lambda}_T^{-1}\left(W_j \boldsymbol{\theta} + \boldsymbol{\Delta}\tilde{s}\right)\right] = \boldsymbol{\theta} + \left(W_j^{\mathrm{T}} \boldsymbol{\Lambda}_T^{-1} W_j\right)^{-1} W_j \boldsymbol{\Lambda}_T^{-1} \boldsymbol{\Delta}\tilde{s}. \tag{11.31}$$

The corresponding precision $\mathrm{MSE}(\tilde{\boldsymbol{\theta}})$ of the parameter estimation is

$$\mathrm{MSE}(\tilde{\boldsymbol{\theta}}) = \mathrm{tr}\left[\mathrm{E}(\tilde{\boldsymbol{\theta}} - \boldsymbol{\theta})(\tilde{\boldsymbol{\theta}} - \boldsymbol{\theta})^{\mathrm{T}}\right] = \left(\sigma_\xi^2 + \sigma^2\right) \mathrm{tr}\left(W^{\mathrm{T}} W\right)^{-1} \tag{11.32}$$

where σ_ξ^2 is the variance of the measurement error, and σ^2 is the variance of the model error.

At this junction, the relationship between the index conversion precision and parameter estimation precision is

$$\mathrm{E}[y] = \mathrm{E}\left[g \circ f\left(\tilde{\boldsymbol{\theta}}\right)\right]. \tag{11.33}$$

11.1.5 Radar Precision Conversion: A Particular Case

As an illustration, let us look at the impact dispersion caused by the homing radar system of a combined navigation missile (Wang and Cui 2006). We convert the performance index by using the method described above based on the relevant data observed in the flight process and the final impact point data, and then compare the result with that obtained using only the final impact point error information (Wang et al. 2009).

Let ΔL_{Radar} denote the assessed impact dispersion caused by the error of radar measurement in the OOT, and $\Delta \tilde{L}_{\text{Radar}}$ the impact dispersion caused by the error of radar measurement in a SET. Here, the performance index (namely, the warhead–target relative position error) can be represented as a composite response function g of observed data in the flight process in terms of distance R, azimuth A, and elevation angle E, as discussed in the previous subsection.

First, we obtain radar measurement errors Δ_R, Δ_A, and Δ_E in the SETs with corresponding observed data. Second, the conversion process of ranging error and angle measurement error is analyzed. In the following, we choose the monopulse radar system as an illustration to specifically analyze the main impact factors of ranging and angle measurement error (Wang and Cui 2006). The systematic error of

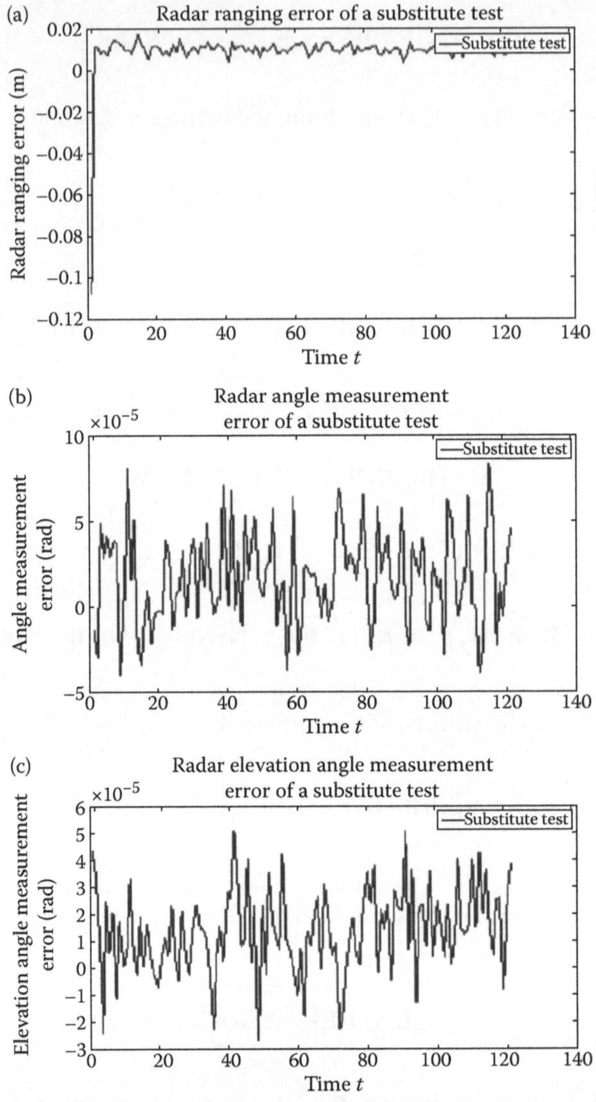

FIGURE 11.3 Simulation results of distance, azimuth, and elevation errors of a substitute test.

radar ranging is mainly the dynamic lag error. Its time-dependent random errors are mostly composed of clutter and jamming error, thermal noise error, and distant glint error of the radar.

Next, we use different conversion methods to produce the impact dispersion for missiles with homing radar in two different tests. Simulated radar measurement elements are the distance R, the azimuth A, and the elevation angle E. We compute the result by using the classical conversion method, with only the information of the final impact points, and the new method proposed in the previous section, in which the observed data are first converted between different states; the conversion results of indexes are derived consequently.

Observed data are simulated separately in the SET and the OOT, which contain radar ranging error, angle measuring error, real-time velocity and acceleration in ranging, and angle measurement. Suppose that the random errors contain independent clutter and jamming error, thermal noise error, and distant glint error. Also, the main part of systematic errors is the dynamic lag error.

For the simulation of radar ranging error and angle measuring error in the SET, we use the works of Wu et al. (2005) and Li and Mao (2006) as reference for the quantity levels of simulated

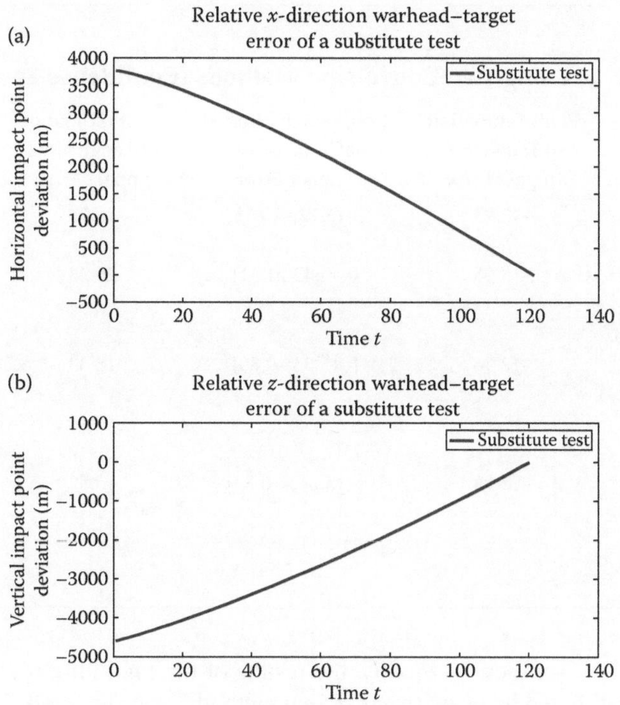

FIGURE 11.4 Simulation results of the warhead–target relative error in a substitute test.

data. The ranging error gradually decreases as the missile gets closer to the target, while the angle measurement error fluctuates within a range of dispersion. The results are shown in Figure 11.3. It can be seen from Figures 11.3 and 11.4 that the simulated warhead–target relative error converges gradually with time.

Next, we use the two aforedescribed methods to convert impact dispersion in the SET into that in the OOT.

1. We use the classical conversion method on the final impact point information to produce the systematic error. The basic idea is as follows: First, factors that influence the impact dispersion are analyzed, and then the impact dispersion difference of two tests is propagated to the measurement devices of the homing radar according to the law of error propagation; then, the systematic error is produced based on analysis of the error sources of radar measurement elements. By analyzing the sources of error at the measurement devices of the homing radar, different radar measurement errors are derived according to the corresponding parameters of error sources in different tests. At the end, these measurement errors are propagated to the impact point precision.
2. We employ the method proposed above. The basic idea is as follows: First, respectively, construct models for the ranging error and angle measurement error of radar; second, the regression coefficients of the systematic error in the models are derived by using the least-squares method according to the Gauss–Newton iteration algorithm; third, the systematic error and random error in these models are converted; and then, derive the results by using the transfer matrix between the impact dispersion and the observed data.

For the conversion method emphasized above, there are two cases: (1) there is only one substitute test and (2) there are multiple substitute tests. Similar to the simulation of one substitute test, three other samples of substitute tests are simulated. That is, there are a total of four classes of SET samples. Then, the observed data are converted, and the results are analyzed with the method

TABLE 11.1
Estimated Impact Error Using Two Conversion Methods (Confidence Level $\alpha = 0.99$)

Unit (m)	Point Estimation of Transverse Impact Error	Confidence Interval of Transverse Impact Error	Point Estimation of Lengthways Impact Error	Confidence Interval of Lengthways Impact Error
Simulated impact error in the OOT	**–12.93**	**[–15.72,–10.13]**	**18.78**	**[18.09,19.47]**
Classic conversion method using only the impact point information (1 SET)	–13.96	[–49.43,21.51]	20.23	[–29.56,70.02]
Previous conversion method using also flight process information (1 SET)	–12.56	[–15.33,–9.80]	18.33	[17.61,19.05]
Previous conversion method using also flight process information (4 SETs)	–12.99	[–15.66,–10.33]	18.83	[18.16,19.51]

described in the previous subsection. Finally, the results of the final impact error caused by radar error in the OOT, as estimated by using two different methods, are compared. The particular results are shown in Table 11.1.

Remark 11.1

The classical conversion method does not use the observed flight process information of homing radar. It computes the impact point error caused by radar error by using only the final impact point information. The obtained result is greatly influenced by the random error of a small sample size, a fatal weakness of this method. On the other hand, the aforementioned method employs the prior knowledge of the physical background and observed flight process information. Confidence estimates can be directly obtained by using the results of the parameter estimates. Therefore, comparing to the single point measurements, the influence of random error is correspondingly decreased. ■

Remark 11.2

Comparing the conversion results of a single SET with those of multiple SETs, it is found that the latter is more robust. The reason is that the influence of random error decreases as the sample size increases. Thus, with multiple SETs, the confidence interval estimate is more appropriate. ■

11.1.6 Some Comments

Basing on the thought of systems science, a novel viewpoint of complex systems evaluation is introduced in this case study by exploiting information from different angles, such as model, prior knowledge, data, etc. Aiming at the evaluation of a particular weapon system, an equivalent conversion and fusion evaluation method for the precision indexes of combined navigation missiles between SET and OOT is established. An illustration on the evaluation of homing radar impact precision demonstrates the reliability of this method. Comparing with the classical method, which uses only the final impact point information, this systemic method also exploits the observed flight process information to verify the validity of the model, producing a more robust and reasonable result. The systemic idea underneath this method can be widely applied to other fields in which the performance index in SETs needs to be converted into the result of another test state in order to test and evaluate a large complex system.

11.2 SYSTEMS SCIENTIFIC ANALYSIS FOR POSITIONING ACCURACY OF SATELLITE NAVIGATION SYSTEM

11.2.1 Satellite Navigation System and Its Accuracy Index

Navigation is a technique or method that guides a desired object to its destination from an initial point of location (Parkinson et al. 1994). The system that can offer the navigated object some timely navigation service, such as positioning, velocity, course, etc., is called a navigation system. The commonly seen navigation methods include navigation marks, position calculation, astronomical navigation, inertial navigation, radio navigation, satellite positioning navigation, and so on. Satellite positioning navigation uses man-made satellites as the basis of navigation, which helps the user to make a precise point-by-point measurement. The positioning signal transmitted by a navigation satellite determines the position and state of motion of the object of concern and provides the guidance for the object to safely reach its predetermined destination. Comparing with the commonly used ground navigation systems, satellite navigation systems possess good characteristics like high degree of accuracy, timely positioning, and low cost for the receivers of the terminal user.

High accuracy is not the only important service that a satellite navigation system provides, but so are the safety, continuity, and reliability of the navigation service. The service performance of a navigation system is determined by its own structure and operation characters, which can be described by using the signal characteristics and performance indexes of the system. Accuracy is a characteristic of central importance for each satellite navigation system; it is also a key impact factor for other performance indexes of the system, such as integrity, continuity, availability, etc. As one important factor of accuracy and one important measure for evaluating the positioning performance of the satellite navigation system, the dilution of precision (DOP) possesses significance in aspects of measurement and mathematics. In terms of measurement, DOP reflects the ratio between pseudorange error and user's time service error, which is caused by the spatial geometric distribution of the available satellites and the receiver. It represents the main content of evaluation of the user's timely positioning precision. Based on the differences in users' performance needs of measurement, DOP is additionally divided into geometric dilution of precision (GDOP), position dilution of precision (PDOP), horizontal dilution of precision (HDOP), vertical dilution of precision (VDOP), and TDOP. For the definitions of these finer divisions of DOP, the reader is advised to consult with the earlier literature on global positioning system (GPS) (Parkinson et al. 1994). In terms of mathematics, DOP is an important part of the diagonal of the weighted inverse matrix based on the least-squares adjustment solution and also the ratio between the diagonal elements of the unknown parameters' covariance matrix and the user's equivalent pseudorange error. It reflects the degree of contribution of the measurement information to the solution of the unknown parameters. Thus, from the perspective of either measurement or mathematics, DOP reflects the quality of the navigation solution and possesses a position of equal importance as the navigation solution. However, due to historical reasons, DOP is initially introduced in the GPS theory to mean the "dilution of precision." Thus, in most traditional circumstances, it is often known as a decaying factor of the precision index. It is because in the general scenario, DOP has indeed been used as an enlargement factor to embody the influence of the spatial geometric distribution of the available satellites and the user on the positioning and time service error. That is the reason why Parkinson et al. (1994) points out that the type value of GDOP is 1–100. However, as a matter of fact, if we take GDOP as an example, its minimum value can be mathematically less than

1. Thus, GDOP is not only a decaying factor of precision but also an enhancing factor of precision. This conclusion also holds true in measurements and possesses some positive significance.

For the positioning of a satellite navigation system, four or more available satellites are commonly used to calculate the proper navigation and to position an object. The covariance matrix of the user's position and the parameter error of the user's time deviation is expressed as $G = (H^T H)^{-1}$, where H is the coefficient matrix of the equation of the navigation and positioning service and is commonly known as the observation matrix. It is written as follows:

$$H = \begin{bmatrix} a_{x1} & a_{y1} & a_{z1} & 1 \\ a_{x2} & a_{y2} & a_{z2} & 1 \\ a_{x3} & a_{y3} & a_{z3} & 1 \\ \cdots & \cdots & \cdots & \cdots \\ a_{xn} & a_{yn} & a_{zn} & 1 \end{bmatrix} \tag{11.34}$$

where a_{xi}, a_{yi}, a_{zi}, $i = 1,2,\cdots,n$, respectively, represent the cosine of the direction between the user receiver and available navigation satellites, and n is the number of available navigation satellites. Thus, the following relation holds true:

$$a_{xi}^2 + a_{yi}^2 + a_{zi}^2 = 1, \ (i = 1, 2, \cdots, n). \tag{11.35}$$

Because GDOP = $(\text{trace}(H^T H)^{-1})^{1/2}$, we need to compute $H^T H$ as follows:

$$H^T H = \begin{bmatrix} \sum_{i=1}^{n} a_{xi}^2 & \sum_{i=1}^{n} a_{xi} a_{yi} & \sum_{i=1}^{n} a_{xi} a_{zi} & \sum_{i=1}^{n} a_{xi} \\ \sum_{i=1}^{n} a_{xi} a_{yi} & \sum_{i=1}^{n} a_{yi}^2 & \sum_{i=1}^{n} a_{yi} a_{zi} & \sum_{i=1}^{n} a_{yi} \\ \sum_{i=1}^{n} a_{xi} a_{zi} & \sum_{i=1}^{n} a_{yi} a_{zi} & \sum_{i=1}^{n} a_{zi}^2 & \sum_{i=1}^{n} a_{zi} \\ \sum_{i=1}^{n} a_{xi} & \sum_{i=1}^{n} a_{yi} & \sum_{i=1}^{n} a_{zi} & n \end{bmatrix}. \tag{11.36}$$

Suppose that the receivers' measurement errors for each satellite are independent with equal variance. Then the positioning accuracy and time service accuracy can be expressed by the following formulas:

$$\sigma_{pos} = \sigma_{\rho\rho} * \text{PDOP}$$

$$\sigma_{tt} = \sigma_{\rho\rho} * \text{TDOP}$$

$$\text{PDOP} = \sqrt{H_{11} + H_{22} + H_{33}}$$

$$\text{TDOP} = \sqrt{H_{44}}$$

where PDOP and TDOP are, respectively, known as the positioning accuracy factor and the time service accuracy factor, and $\sigma_{\rho\rho}$ is the root mean square error of the pseudodistance measurement or known as the user-equivalent range error (UERE).

The previous process of solving the precision is based on the assumption that the UEREs of the individual available satellite are i.i.d. However, as for Beidou-2 satellite navigation system, it consists of geostationary earth orbit (GEO), medium earth orbit (MEO), and inclined geosynchronous orbit (IGSO) satellites, three different types. Thus, there are clear differences between the UERE indexes. That is, the UERE index of each satellite type needs to be individually modeled and estimated.

11.2.2 The Question

Beidou Satellite Navigation System is a heterogeneous satellite constellation, consisting of the classes of different GEO, MEO, and IGSO satellites. How can the precision of such a heterogeneous satellite constellation be analyzed and evaluated from the perspective of systems science, where the constellation can be treated as an open system?

11.2.3 Systemic Model and Analysis of Satellite Navigation System

The basic model of a system S is $S = (M, R)$, where S stands for the whole of the system, M is the set of all factors or elements of the system, and R is a set of relation between the factors.

Considering how Ma (2006) describes a complex giant system, we expand the basic model of systems to construct a systemic model of a satellite navigation system as follows:

$$S = f(\{M_i\}, [R], s, t|E)$$

where S represents the whole of the satellite navigation system; f is the framework of the system; M is the set of all factors or principles of the navigation system, which contains the navigation satellites in a specified space area, ground region of control, and user area of receivers; R is the set of relations among the factors; s represents the space factors; t is the time; and E is the environment of the system, including all the noises produced by the sun, the earth, the geoatmosphere, the electromagnetic waves, etc.

Let us take the GDOP accuracy index of the satellite navigation system as an example to see how the strong characteristic of the whole system's emergence can be analyzed. Suppose that a minimum macroscopic state M* is built up with a ground observation station and four navigation satellites that are available in sight. By treating GDOP as an attribute of the whole system, we can see that for any macroscopic state that is smaller than M*, neither the covariance matrix nor a precise position setting can be calculated. This fact implies that the state cannot have the GDOP, an emergent holistic attribute of the system. Thus, it should be recognized that the satellite navigation system has a strong holistic emerging characteristic with the GDOP accuracy index as an attribute of the system that is not shared by any of the parts.

11.2.4 Detailed Analysis

11.2.4.1 Theoretical Analysis for Minimum Value of GDOP

According to the definition of a matrix's trace, we have trace$(H^T H) = 2n$. Suppose that λ_i, $i = 1,2,3,4$, are the characteristic values of $H^T H$. Then, we have

$$\sum_{i=1}^{4} \lambda_i = 2n \tag{11.37}$$

where GDOP can be expressed as follows:

$$\text{GDOP} = \sqrt{\text{trace}(H^T H)^{-1}} = \sqrt{\text{trace}\left(\text{diag}\left(\frac{1}{\lambda_1}, \frac{1}{\lambda_2}, \frac{1}{\lambda_3}, \frac{1}{\lambda_4}\right)\right)} = \sqrt{\sum_{i=1}^{4} \frac{1}{\lambda_i}}. \tag{11.38}$$

Now, the problem of solving for the minimum value of GDOP is transformed into the problem of evaluating the range of the matrix $H^T H$'s characteristic values. If the difference of the ranges of $H^T H$'s characteristic values is not considered, which means that the four characteristic values are identical to each other, then the minimum of GDOP can be obtained by using the fact that the geometric average of some positive numbers is always smaller than the arithmetic average. However, from Equation 11.36, it follows that the fourth entry on the diagonal is always equal to n, while the other diagonal entries have similar characteristics. In the following, we will use Gerschgorin's disk law of matrix theory (Bu and Luo 2003) to prove that the ranges of the first three characteristic values are the same.

According to Gerschgorin's disk law, for the matrix $H^T H$ defined on the complex number field Z, the corresponding four disks in the Z plane can be expressed, respectively, by the following:

$$\begin{aligned}
R_1 &= \left\{ z \left| \left| z - \sum_{i=1}^{n} a_{xi}^2 \right| \le \left| \sum_{i=1}^{n} a_{xi}a_{yi} + \sum_{i=1}^{n} a_{xi}a_{zi} + \sum_{i=1}^{n} a_{xi} \right| \right. \right\} \\
R_2 &= \left\{ z \left| \left| z - \sum_{i=1}^{n} a_{yi}^2 \right| \le \left| \sum_{i=1}^{n} a_{xi}a_{yi} + \sum_{i=1}^{n} a_{yi}a_{zi} + \sum_{i=1}^{n} a_{yi} \right| \right. \right\} \\
R_3 &= \left\{ z \left| \left| z - \sum_{i=1}^{n} a_{zi}^2 \right| \le \left| \sum_{i=1}^{n} a_{xi}a_{zi} + \sum_{i=1}^{n} a_{yi}a_{zi} + \sum_{i=1}^{n} a_{zi} \right| \right. \right\} \\
R_4 &= \left\{ z \left| \left| z - n \right| \le \left| \sum_{i=1}^{n} a_{xi} + \sum_{i=1}^{n} a_{yi} + \sum_{i=1}^{n} a_{zi} \right| \right. \right\}.
\end{aligned} \tag{11.39}$$

This law points out that there is a certain connection between the characteristic values of a matrix and their corresponding disks. From Equation 11.39, it follows that the first three equations have a rotational symmetry, while the fourth equation is obviously different from the first three equations. Thus, this realization implies that the first three characteristic values of $H^T H$ have the same range, while the range of the fourth characteristic value is different. In fact, the fact that $\lambda_4 \ge n$ can be proved by contradiction. For more details, please consult the work of Sheng et al. (2009).

Therefore, the minimum value of Equation 11.38 can be obtained by using the following inequality:

$$\text{GDOP} = \sqrt{\sum_{i=1}^{4} \frac{1}{\lambda_i}} = \sqrt{\sum_{i=1}^{3} \frac{1}{\lambda_i} + \frac{1}{\lambda_4}} \geq \sqrt{3\left[\prod_{i=1}^{3} \frac{1}{\lambda_i}\right]^{1/3} + \frac{1}{\lambda_4}} \tag{11.40}$$

where the equal sign holds true if and only if λ_i, $i = 1,2,3$, are equal to each other. Considering Equation 11.37, GDOP can be expressed by using λ_4. Because $\lambda_4 \geq n$, it follows that when $\lambda_4 = n$, GDOP reaches the following minimal value:

$$\text{GDOP}_{\min} = \sqrt{\frac{10}{n}}. \tag{11.41}$$

This conclusion is also proved in the work of Li et al. (2011) from the perspective of measurement.

In the case that the user's position is known, there is only one unknown variable, which is the clock error of the receiving device. In this case, the observation equation is $H_4 = (1 \ \ldots \ 1)^T$; thus, $H_4^T H_4 = n$, and GDOP degenerates to the TDOP under special conditions. That is, the following holds true:

$$\text{TDOP} = \sqrt{\text{trace}(H_4^T H_4)^{-1}} = \sqrt{\frac{1}{n}}. \tag{11.42}$$

Thus, the corresponding characteristic value of $H_4^T H_4$ is $\lambda_4 = n$.

Similarly, if the receiver's clock error is not considered or known, then $\text{trace}(H_3^T H_3) = n$, where H_3 stands for the observation matrix with the clock error removed. Thus, $H_3^T H_3$ is the symmetric submatrix of the first three rows and first three columns of $H^T H$ of Equation 11.36. That is, GDOP in this case degenerates into PDOP so that the following holds true:

$$\text{PDOP} = \sqrt{\text{trace}(H_3^T H_3)^{-1}} = \sqrt{\sum_{i=1}^{3} \frac{1}{\lambda_i}} \geq \sqrt{3\left[\prod_{i=1}^{3} \frac{1}{\lambda_i}\right]^{1/3}}. \tag{11.43}$$

The equal sign holds true when $\lambda_i (i = 1, 2, 3)$ are equal to each other. By combining Equations 11.37 and 11.42, we have

$$\text{PDOP} \geq \sqrt{\frac{9}{n}}. \tag{11.44}$$

Combining Equations 11.42 and 11.44 leads to

$$\text{GDOP} = \sqrt{\text{PDOP}^2 + \text{TDOP}^2} \geq \sqrt{\frac{10}{n}}. \tag{11.45}$$

It is also shown (Bu and Luo 2003, pp. 164–174) that only when the following holds, GDOP reaches its minimum:

$$\sum_{i=1}^{n} a_{xi} = 0, \sum_{i=1}^{n} a_{yi} = 0, \sum_{i=1}^{n} a_{zi} = 0. \tag{11.46}$$

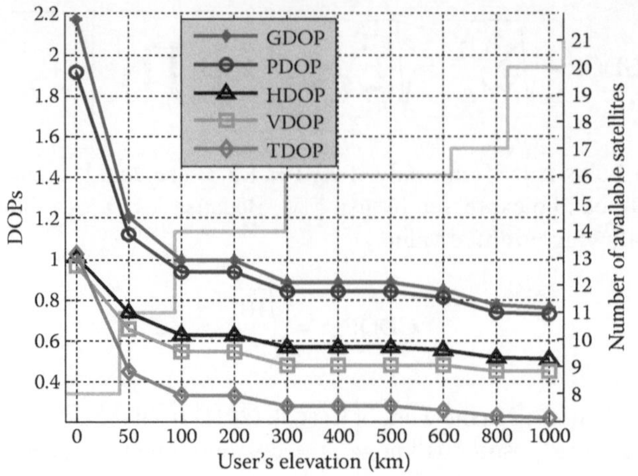

FIGURE 11.5 Comparison of DOPs at different elevations near the ground level.

Thus, when the user is located at the center of the uniform polyhedron formed by multiple available navigation satellites, GDOP reaches the minimum value, while the positioning accuracy reaches the highest level. The previous equation also indicates that at this very moment, the navigation satellites in every orbital plane are uniformly distributed. In addition, Equation 11.41 also implies that the minimum value of GDOP is also a function of the number of the available navigation satellites, so that the greater the number of available satellites is, the smaller the GDOP minimum value reaches.

When the service space of global navigation satellite system (GNSS) is higher than 100 km, the structure of the geometric spatial distribution of the user and available satellites will improve greatly. In such a circumstance, the available satellites will no longer be distributed on one side of the user's plane. Those available satellites of negative elevation angles can also provide effective service so that the user's positioning and time service accuracy in navigation application can be consequently improved. In addition, this kind of enhanced performance will bring forward essential changes to the attributes of the classical GDOP, ranging from the "decaying factor" of the precision index to the "enhancing factor" (less than 1) (Li et al. 2011). The particular comparison is shown in Figure 11.5, where the horizontal axis stands for the user's elevation, and the right-side vertical axis stands for the number of available satellites.

Presently, such enhanced GNSS performance based on the structure of spatial geometric distribution has been practically applied. In particular, these applications basically employ pseudosatellites to enhance the navigation quality of higher-space users.

11.2.4.2 Specific Cases

In deriving the precision of navigation positioning service, each satellite's UERE is generally supposed to be independent and subject to the same distribution. However, in a real satellite navigation system, because of the relevance of such errors as caused by the ionospheric delay and other reasons existing in the navigation signal transmission, the assumption that the precision index = (UERE)*(DOP) may not always hold true. As for the MEO, IGSO, and GEO, different kinds of navigation satellites of three altitudes, their navigation signal transmissions experience rather large differences in environmental conditions. Their respective orbit accuracies vary due to the different orbit determination algorithms employed and the quality of the tracking data. Thus, there is also a large difference among the UEREs of the individual satellite types. When analyzing the DOP performance index of the Beidou satellite navigation system, treating the UEREs as being i.i.d. is not accurate enough. Because of this reason, this subsection applies a generalized least-squares method to estimate the DOP, assuming that the UEREs of these three kinds of satellites are distributed differently.

It is planned to deploy 12 navigation satellites before the end of 2012 to form a regional navigation system, which will eventually be a part of the completed Beidou satellite system consisting of

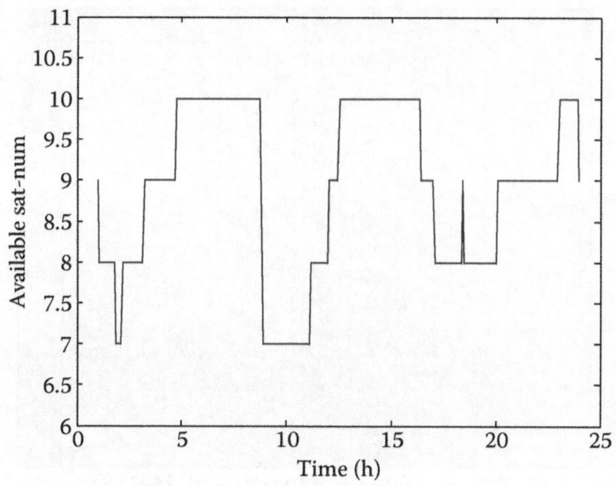

FIGURE 11.6 Change in the number of available satellites in 1 day (30°N, 120°E).

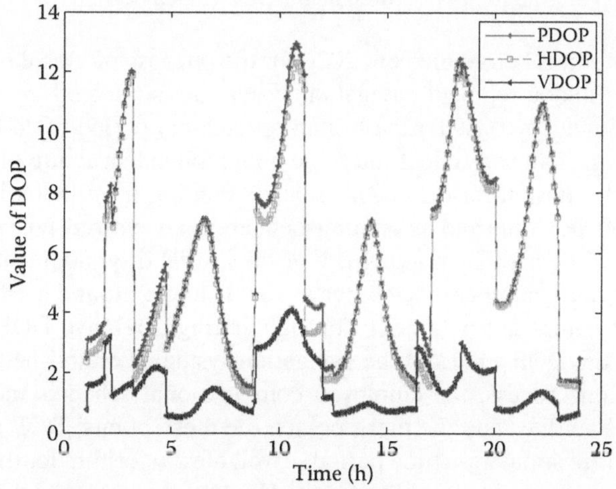

FIGURE 11.7 Change of different DOP indices in 1 day (30°N, 120°E).

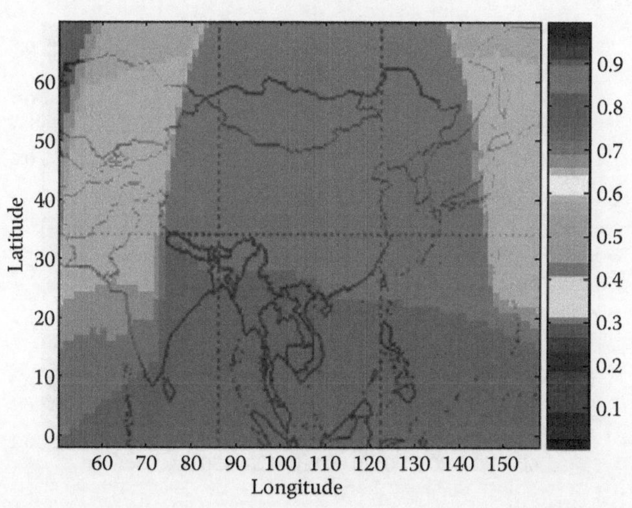

FIGURE 11.8 Cover map of the number of available satellites (limit set at 8).

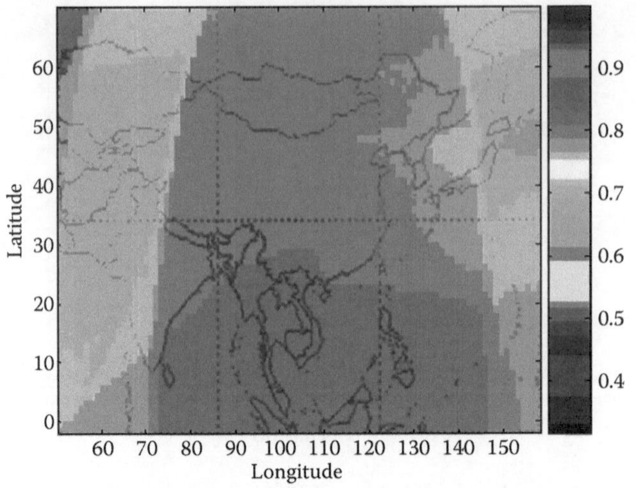

FIGURE 11.9 Cover map of availability of unweighted PDOP (PDOP < 6).

a planned total of 35 satellites before the year 2020 for the purpose of global navigation. Each day's positioning data of the Beidou regional navigation system are simulated by using satellite tool kit (STK). Because the navigation constellation has an operational period of 24 h, 24-h simulation of the DOP of the Beidou system can reflect the basic geographical structure of the whole constellation. By making a mesh division for the earth's surface ranging from latitudinal 0°N to 60°N and longitudinal 60°E to 150°E, China and its circumjacent areas are entirely covered. Then, once every 5 min, sample data are collected. By making use of the availability judgment between the ground mesh and satellites, the daily number of satellites available to the ground mesh and the corresponding mesh points' DOP values are obtained. Through setting a relative DOP threshold, the daily availability analysis of the DOP values of the navigation system is established.

During the simulation process, the employed computational schemes include the availability algorithm between the satellites and the mesh points, the mesh points' DOP algorithms, the availability of DOP in the entire simulation time period, a weighted algorithm for the UEREs of the three satellite types, etc. In the weighted algorithm for the UEREs, it is assumed that the estimated value

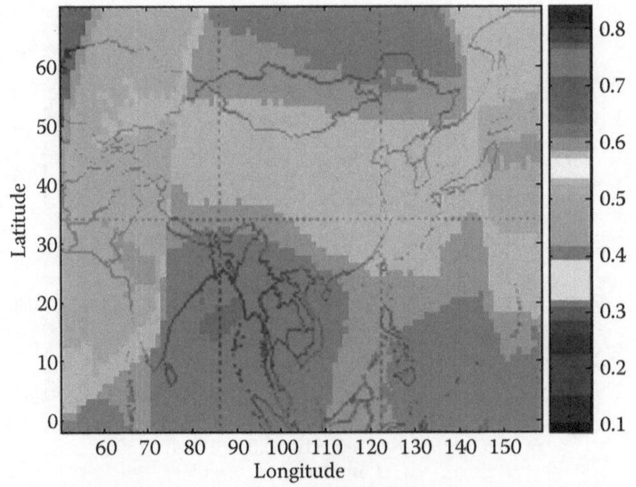

FIGURE 11.10 Cover map of availability of weighted PDOP (PDOP < 6).

of GEO's UERE is two times larger than the estimated UERE values of MEO and IGSO, and that each satellite's UERE is independent of each other.

The simulation results are given in Figures 11.5 through 11.9.

From Figures 11.6 and 11.7, it can be seen that at the mesh grid point (30°N, 120°E), which is the Hangzhou area, the number of available satellites varies from 6 to 10. That meets the demand of continuous coverage of four available signal sources. Considering information redundancy, it also meets the requirement of the integrity receiver autonomous integrity monitoring/fault detection and exclusion (RAIM/FDE) algorithm. From the daily changes of DOP at this particular point, it can also be seen that PDOP, HDOP, and VDOP at this point all follow a similar pattern of change with time. Additionally, PDOP, HDOP, and VDOP often experience dramatic sudden jumps. For instance, at the time moments of 2.5, 8, 12, 16, and 20 h, the DOP values see jump changes, which might be caused by the sudden changes in the number of available satellites. Also, from these figures, we can see that VDOP contributed much less to PDOP than HDOP does to PDOP.

When setting the average number of available satellites to 8, Figure 11.8 reflects the numbers of available satellites throughout one day period of time, where different colors represent the different time percentages of the number of available satellites that is larger than the set limit of 8.

From Figure 11.8, it can be seen that when using 8 as a limit number, only a small area in western China has 60% of availability, while all other parts get more than 90% of availability. That implies that the Beidou navigation system has an obvious dominance in terms of the number of available satellites, which is determined by the operational positioning advantage of both GEO and IGSO. In covering areas of GEO and IGSO, the number of available satellites is much better than other areas. Also, as the degree of latitude decreases, the number of available satellites gradually increases.

Under the condition that PDOP < 6, analyzing the PDOP's availability of the meshed area produces the results of availability shown in Figures 11.9 and 11.10, where colors represent the different time percentages when PDOP is smaller than 6.

Comparing Figures 11.9 and 11.10 shows that with a weight matrix introduced, the resultant availability of PDOP becomes worse. For the most areas of China, the availability of PDOP drops from the original 90% to about 60%. Also, when the limit number of PDOP is set to smaller than 6, the corresponding deterioration in availability will be much worse. The heterogeneous constellation structure of the Beidou satellite navigation system dictates that the UEREs of the individual satellites could not be simply treated as being i.i.d. Instead, a weight matrix needs to be introduced. However, the design of the weight matrix should be based on a quality estimation of the UEREs of the individual satellites.

11.3 TEMPORAL ANALYSIS OF SOCIAL NETWORKS

Network refers to a variety of associations, while social network can be simply posed as the structure of some social relations. Thus, on one hand, each social network represents a structural relationship; on the other hand, it can reflect the social relations between actors. The following are the main elements that constitute social networks.

11.3.1 Terminology

The main elements that constitute a social network include the following. *Actors* refer to not only specific individuals but also groups, companies, or other collective social units. The position of an actor in the network is known as a "node." *Relational ties* are simply the associations between actors. The relationships between people take many different forms, such as kinship, cooperation, exchange relations, antagonism, etc., all of which constitute different relational ties.

Social network analysis stands for a set of norms and methods with which one analyzes the relational structures, attributes, and properties of social networks. It is also known as structural analysis because it mainly analyzes the structures and properties of social relations that bridge different social units, such as individuals, groups, or societies.

In this sense, social network analysis is not only a set of techniques developed for analyzing relationships or structures, but also a theoretical approach consisting of ideas useful for structural analysis. This is because from the point of view of the scholars in the area of social network analysis, the research object of sociology is exactly about social structures, and each of these structures is manifested as a relational pattern of actors' behaviors.

Barry Wellman, a social network analyst, pointed out (Wellman and Berkowitz 1988) that network analysis aims to explore the deep-level structures—some network pattern that is hidden under the surface of complex social systems. For example, network analysts are particularly concerned about how the association patterns, in a particular network, affect the behaviors of people through providing different opportunities and/or constraints.

As a basic method of studying social structures, social network analysis is developed on the premises of the following basic principles:

1. Relational ties often interact with each other in asymmetric ways; they are also different in terms of the contents and intensities.
2. Relational ties connect all the network members either directly or indirectly; they must be analyzed in the context of a much larger network structure.
3. Structures of relational ties produce nonrandom networks, thus forming clusters of networks, network boundaries, and cut–cross correlations.
4. The cut–cross correlations associate clusters of network and individuals together.
5. Asymmetric ties and complex networks lead to unequal and uneven distributions of scarce resources.
6. Networks create cooperation and competition with the purpose of obtaining scarce resources.

The epistemological significance of such structural analysis is that the research objective of social science should be social structures rather than individuals. Through the study of networks, the relationships between individuals, "microscopic" networks, and macroscopic structure of large-scale social systems could potentially be linked together.

As an example, let us consider the following simple problem. To investigate the preference relationship in a group of 20 people, a sociologist studies these 20 people by following them as an outsider. Based on his or her observations, he or she then develops quantitative expressions to describe the preference between each pair of these people, forming a matrix $\mathbf{X}$ of size 20×20. The (i, j) entry x_{ij} of $\mathbf{X}$ represents how much the ith person likes the jth person. Now, the problem becomes how to analyze this relationship matrix. The method of multidimensional scaling does not apply here due to the fact that the matrix $\mathbf{X}$ is not symmetric, and the indicated numerical preferences between two people are often not equal. Because there tends to be a direction in the relationship between two people, this relationship matrix in essence is asymmetric. Intrinsic asymmetry is often a typical characteristic of human relationships.

Multidimensional scaling models the available data based on a spatial method, where the magnitudes of the coefficients in the model indicate the relational intensity of the relevant variables in the original data. Because spatial relationships are inherently symmetric, the final expression is limited only to the description of the symmetric nature of the data. In other analytical techniques developed to describe the relationship matrix of a set of variables, such limitation also exists. For example, factor analysis can also be considered as a method based on spatial modeling, in which angles of vectors represent the correlations of the relevant variables. When the relationship matrix is intrinsically asymmetric, the aforementioned methods become invalid and can no longer be applied. The preference relationship matrix that we just mentioned earlier is one such case; other examples include world trade relations. Therefore, when an intrinsically asymmetric relationship matrix is encountered, the methods that are spatial model–based have their obvious deficiencies.

How to analyze social networks with intrinsically asymmetric relations is one of the first problems that needs to be resolved. At the same time, how to analyze the temporal structural changes of social networks in order to help explain the evolutionary nature of friendships, collaborative relationships, and organizational structures, represents an increasingly serious challenge in social network analysis.

11.3.2 Social Network Systems

In this subsection, we start from a systemic point of view to analyze and to resolve the problems of asymmetric relations and their evolutions widely existing in social networks. According to the study in Chapter 10, each social network can be seen as a network system, denoted by an ordered pair $S = (M, R)$, where M is the collection of all the individuals in the social network, and R is the collection of relationships between the individuals in M. In a social network system $S = (M, R)$, M describes the actors and R the relational ties between the actors. Thus, the system $S = (M, R)$ contains two major elements of the social network of concern.

In order to give an explanation to the asymmetric relationship in social networks, we can specifically define R, the collection of relationships in the social network system $S = (M, R)$, as follows:

$$R = \{r = (m_1, m_2): \forall m_1, m_2 \in M\} \tag{11.47}$$

where $r = (m_1, m_2)$ represents the direct relation between individual m_1 and m_2 in the social network. This relation is defined in the same way as "edge" in the traditional graph model of a social network, where only direct relations between two individuals are considered while all complex relations among more than two individuals are ignored. The relation set R, as defined above, can be expressed by an adjacency matrix $\mathbf{X}$ of size $n \times n$, where $n = | M |$ and the entry x_{ij} is a measure of the directional relation from the ith individual to the jth individual. This social network system is especially denoted by $S = (M, \mathbf{X})$.

For example, the aforementioned group of 20 individuals and their preference relationships can be expressed as a social network system $S = (M, \mathbf{X})$, where the (i, j) entry x_{ij} of matrix $\mathbf{X}$ represents how much the ith person likes the jth person. Based on the previous analysis, the matrix $\mathbf{X}$ in general is asymmetric. Therefore, the problem of analyzing the asymmetric relationships between individuals in a social network is translated into the problem of analyzing the adjacency matrix $\mathbf{X}$ in the social network system $S = (M, \mathbf{X})$.

Each realistic social network is evolutionary over time. Thus, the concept of social network systems, as just defined above, is naturally a temporal system. Therefore, any social network system can be expressed by a temporal system or a time system $S_t = (M_t, R_t)$, where t stands for the numerical time. For each social network system with its relations specifically defined in Equation 11.47, it can be expressed as a time system $S_t = (M_t, \mathbf{X}_t)$, where $\mathbf{X}_t$ represents the adjacency matrix that describes the relationship of the social network at time t. Thus, the problem of analyzing the temporal changes of social networks is translated into the problem of analyzing the temporal social network system $S_t = (M_t, \mathbf{X}_t)$.

Next, we will investigate the issues of asymmetric relationships and their temporal changes in social networks from a systemic point of view, specifically, by using a three-way decomposition into directional components (DEDICOM) model to decompose the relationship matrix. As a result, the

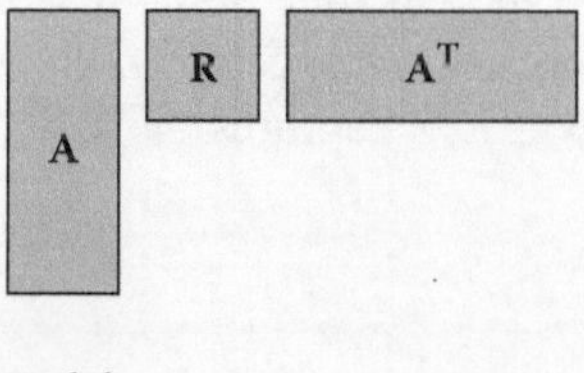

FIGURE 11.11 Two-way DEDICOM model.

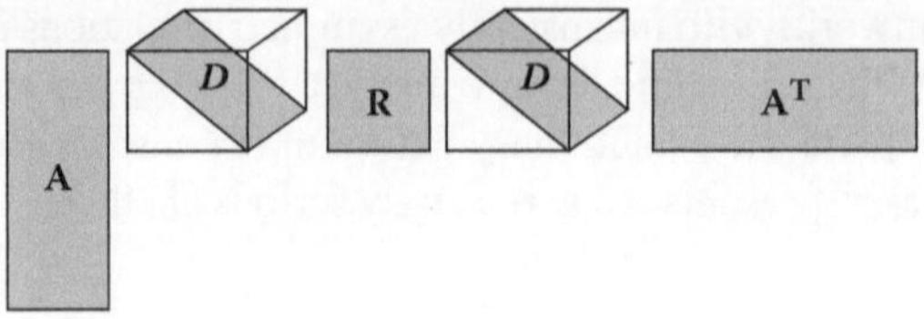

FIGURE 11.12 Three-way DEDICOM model.

key components implicitly existing in social network systems are uncovered, and the evolutionary characteristics of these components are revealed.

For the sake of convenience of communication, let us specially make the following notational conventions. Scalars are denoted by using lowercase letters, such as a, b, ... Vectors are denoted by using bold-faced lowercase letters, such as **a**, **b**, ..., and the ith entry of vector $\boldsymbol{a}$ is written as a_i. Matrices are denoted by bold-faced capital letters, such as **A**, **B**, ...; the jth column of a matrix **A** is denoted by $\mathbf{a}_j$ and the (i, j) entry by a_{ij}. Tensors (multiway arrays) are denoted by bold-faced Euler letters, such as **X**, the (i, j, k) entry of a three-way array **X** is denoted by x_{ijk}, and the kth frontal slice of **X** is denoted by $\mathbf{X}_k$, which is a matrix formed by holding the last index of **X** fixed at k. The symbol $\otimes$ denotes the Kronecher product of matrices. The Frobenius norm of a matrix **Y**, denoted by $\|\mathbf{Y}\|_F$, is the square root of the sum of squares of all the entries.

11.3.3 DEDICOM Model

DEDICOM is an algebraic model (Harshman 1978) that is designed to analyze asymmetric data. This model can help find out the implicit principal components inherently existing in the data as well as the relational patterns between these components so that the system's attributes, as implicitly described by the data, can be expressed.

Given a social network with n actors, we generate the corresponding social network system $S = (M, \mathbf{X})$, where **X** is a matrix of size $n \times n$ and describes the relationship between the n actors. That is, **X** is an adjacency matrix describing the directed graph that characterizes the n actors and their relationships. Usually **X** is an asymmetric matrix. Applying a two-way DEDICOM to the matrix **X** produces

$$\mathbf{X} \approx \mathbf{A}\mathbf{R}\mathbf{A}^{\mathrm{T}} \tag{11.48}$$

where $\mathbf{A} \in R^{n\times p}$, $n < p$, R the set of all real numbers, is an $n \times p$ matrix of real entries (where p is the number of components as described by the model). The matrix **A** represents the weights for the n vertices to respectively belong to the p components. The symbol $\mathbf{R} \in R^{p\times p}$ stands for the matrix that captures the asymmetric relationship between the p components (Figure 11.11).

For a temporal social network system $S_t = (M_t, \mathbf{X}_t)(t = 1, 2,\ldots,m)$, by taking time as the third mode, DEDICOM can be extended to a three-way model:

$$\mathbf{X}_k \approx \mathbf{A}\mathbf{D}_k\mathbf{R}\mathbf{D}_k\mathbf{A}^{\mathrm{T}}\ (k = 1,2\ ,\ldots,m), \tag{11.49}$$

where $\mathbf{X}_k$ is both the kth slice of tensor $\mathbf{X} \in R^{n\times n\times m}$ and the adjacency matrix corresponding to the directed graph at the kth moment; $\mathbf{A} \in R^{n\times p}(p < n)$ is a matrix of weights for the n vertices that belong to the p components; and $\mathbf{D}_k$ is the kth slice of the tensor $\mathrm{D} \in R^{p\times p\times m}$ and a diagonal matrix. Each of the diagonal entries in $\mathbf{D}_k$ represents, respectively, the weight of the corresponding column of the matrix **A** in time dimension; $\mathbf{R} \in R^{p\times p}$ captures the asymmetric relationship matrix between the p components (Figure 11.12).

11.3.4 ASALSAN Algorithm

The problem of solving a three-way DEDICOM can be transformed into the following minimization problem:

$$\min_{\mathbf{A},\mathbf{R},\mathbf{D}} f(\mathbf{A},\mathbf{R},\mathbf{D}) \tag{11.50}$$

where

$$f(\mathbf{A},\mathbf{R},\mathbf{D}) = \sum_{k=1}^{m} \left\| \mathbf{X}_k - \mathbf{A}\mathbf{D}_k\mathbf{R}\mathbf{D}_k\mathbf{A}^T \right\|_F^2 . \tag{11.51}$$

To solve Equation 11.50, the ASALSAN algorithm (Bader et al. 2007) is needed. In particular, this algorithm is carried out as follows: First, randomly initialize the matrices **A** and **R**, let $\mathbf{D}_k = \mathbf{I}$, and then make an update and iterate as follows:

1. Updating **A**.
   We expand the tensor X as follows:

$$\left(\mathbf{X}_1\ \mathbf{X}_1^{\mathrm{T}}\ \cdots\ \mathbf{X}_m\ \mathbf{X}_m^{\mathrm{T}}\right) = \mathbf{A}\left(\mathbf{D}_1\mathbf{R}\mathbf{D}_1\ \ \mathbf{D}_1\mathbf{R}^{\mathrm{T}}\mathbf{D}_1\ \cdots\ \mathbf{D}_m\mathbf{R}\mathbf{D}_m\ \ \mathbf{D}_m\mathbf{R}^{\mathrm{T}}\mathbf{D}_m\right)\left(\mathbf{I}_{2m}\quad \mathbf{A}^{\mathrm{T}}\right) \tag{11.52}$$

   where $\mathbf{I}_{2m}$ is the $2m \times 2m$ identity matrix. We use the least-squares method to solve the nonlinear problem above. Without loss of generality, let us assume that $\mathbf{A}^{\mathrm{T}}$ in Equation 11.52 is a constant matrix. By using normal equations to calculate the least-squares solution of **A**, we have

$$\mathbf{A} = \left[\sum_{k=1}^{m}\left(\mathbf{X}_k\mathbf{A}\mathbf{D}_k\mathbf{R}^{\mathrm{T}}\mathbf{D}_k + \mathbf{X}_k^{\mathrm{T}}\mathbf{A}\mathbf{D}_k\mathbf{R}\mathbf{D}_k\right)\right]\left[\sum_{k=1}^{m}(\mathbf{B}_k + \mathbf{C}_k)\right]^{-1} \tag{11.53}$$

   where

$$\mathbf{B}_k \equiv \mathbf{D}_k\mathbf{R}\mathbf{D}_k(\mathbf{A}^{\mathrm{T}}\mathbf{A})\mathbf{D}_k\mathbf{R}^{\mathrm{T}}\mathbf{D}_k \tag{11.54}$$

$$\mathbf{C}_k \equiv \mathbf{D}_k\mathbf{R}^{\mathrm{T}}\mathbf{D}_k(\mathbf{A}^{\mathrm{T}}\mathbf{A})\mathbf{D}_k\mathbf{R}\mathbf{D}_k. \tag{11.55}$$

2. Updating **R**.
   The objective function in Equation 11.50 can be rewritten as

$$f(\mathbf{R}) = \left\| \begin{pmatrix} \mathrm{Vec}(\mathbf{X}_1) \\ \vdots \\ \mathrm{Vec}(\mathbf{X}_m) \end{pmatrix} - \begin{pmatrix} \mathbf{A}\mathbf{D}_1 & \mathbf{A}\mathbf{D}_1 \\ & \vdots \\ \mathbf{A}\mathbf{D}_m & \mathbf{A}\mathbf{D}_m \end{pmatrix} \mathrm{Vec}(\mathbf{R}) \right\|.$$

   Minimizing $f(\mathbf{R})$ through $\mathrm{Vec}(\mathbf{R})$ is a multiple regression problem. Its solution is

$$\text{Vec}(\mathbf{R}) = \left(\sum_{k=1}^{m} \left(\mathbf{D}_k \mathbf{A}^{\mathrm{T}} \mathbf{A} \mathbf{D}_k\right) \ \left(\mathbf{D}_k \mathbf{A}^{\mathrm{T}} \mathbf{A} \mathbf{D}_k\right) \right)^{-1} \sum_{k=1}^{m} \text{Vec}\left(\mathbf{D}_k \mathbf{A}^{\mathrm{T}} \mathbf{X}_k \mathbf{A} \mathbf{D}_k\right). \tag{11.56}$$

3. Updating **D**.

$$\min_{\mathbf{D}_k} \left\| \mathbf{X}_k - \mathbf{A}\mathbf{D}_k \mathbf{R} \mathbf{D}_k \mathbf{A}^{\mathrm{T}} \right\|_F^2 . \tag{11.57}$$

For details of the ASALSAN algorithm, please consult the work of Bader et al. (2007).

11.3.5 Applications

11.3.5.1 World Trade Analysis

The first example is the analysis of world trade. The data consist of import and export trade data for the time period from 1981 to 1990 from 18 different countries in Europe, North America, and the Pacific coasts. This set of data is a tensor X of size 18 × 18 × 10. Here, the ASALSAN algorithm is used to conduct a decomposition of three principal components.

	#1	#2	#3
#1 North America	4589	187	178
#2 Europe	126	896	89
#3 Japan	60	168	37

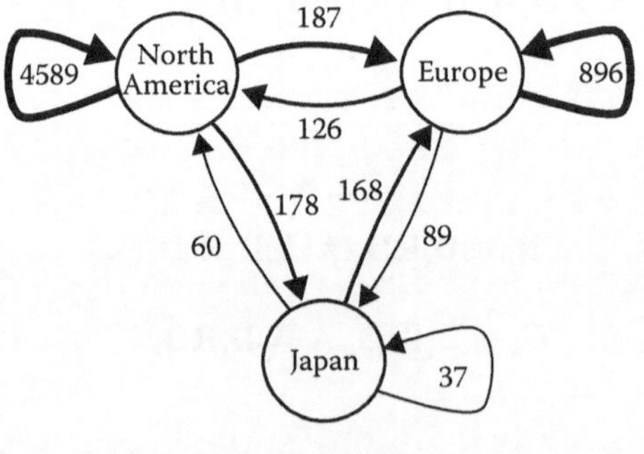

FIGURE 11.13 World trade: **R** matrix and associated directed graph.

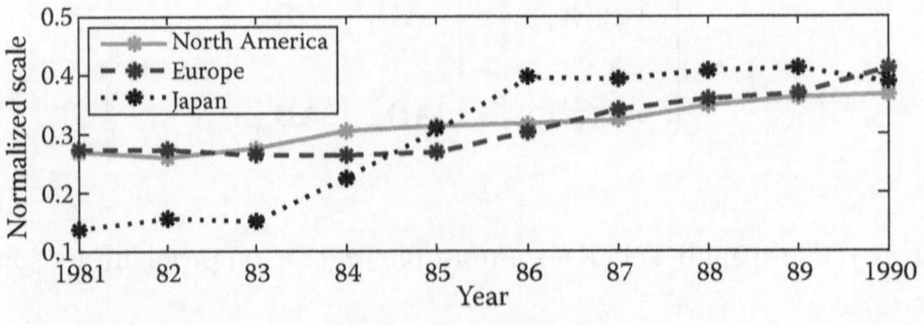

FIGURE 11.14 Scales in D for trade blocks indicate the level of commerce over the said time period.

In this analysis of the world trade, **A** is an 18 × 3 matrix, where the rows represent the 18 countries; the columns, the components set by the algorithm, can be interpreted as three economic groups that have different trading patterns; and any entry a_{ij} in **A** represents the weight for the *i*th country to belong to the *j*th economic group. Therefore, matrix **A** identifies how much these 18 countries belong to the three economic blocks. From the results of the analysis, it is not difficult to find that the three economic groups, as decomposed by the model, have strong correlations with the regions where they are respectively located. The results show that the first principal component is mainly the North American economy led by the United States and Canada; the second principal component is mainly the European economy led by Germany, France, the Netherlands, Italy, Belgium, and the United Kingdom; and the third principal component is mainly dominated by Japan, with minor participation of the United Kingdom and Italy. Regarding the geographical distribution and trade characteristics of these 18 countries, the previous results evidently have their practical significance. For relevant numerical figures, see Figure 11.2.

Matrix **R** is of size 3 × 3 and represents the cumulative pattern of trade relations among the three economic blocks over the 10 years. Any entry r_{ij}, $1 \le i \le 3$, $1 \le j \le 3$, in **R** represents the export from the *i*th economic block to the *j*th block; naturally, it means the trade within the same economic group when $i = j$. Figure 11.13 shows the content of matrix **R** and its corresponding directed graph. From this figure, it can be seen that there are a large number of internal trades in North America and Europe, while Japan's exports to Europe are much larger than European exports to Japan, reflecting the trade imbalance.

Tensor D is of size 3 × 3 × 10 and records the trade weight of each of the three economic blocks over the 10-year period. The *k*th slice $\mathbf{D}_k$ of D is a diagonal matrix of size 3 × 3 and represents the trade weight of each of the three economic blocks in the *k*th year. Figure 11.14 shows the change in weights of the three economic blocks in their trade development over the 10 years. The three curves indicate that the trades in these three economies are steadily growing; in particular, a rapid growth of trade appeared in Japan.

11.3.5.2 E-Mail Network Analysis

The second example is the analysis of communication pattern in the e-mail network of Enron Corporation. The set of e-mail data was made public by the U.S. Federal Energy Regulatory

	#1	#2	#3	#4
#1 Legal	440.2	13.4	−7.9	−5.6
#2 Exec/Government affairs	13.8	286.7	157.8	0.4
#3 Executive	−23.6	93.5	211.6	−4.8
#3 Pipeline	−4.8	−5.9	−6.5	172.4

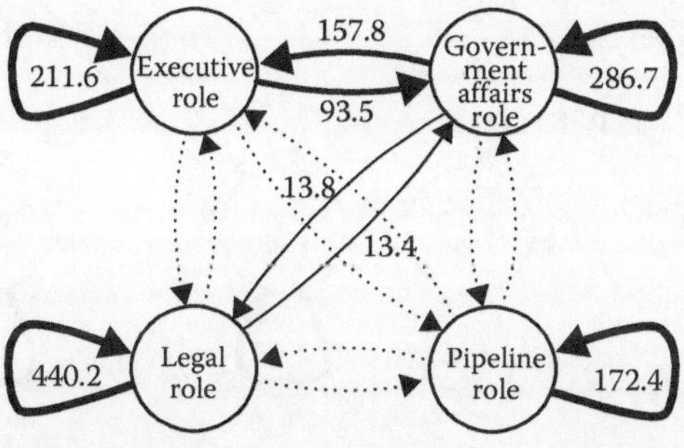

FIGURE 11.15 Enron corpus: **R** matrix and associated directed graph.

Commission (FERC) during its investigation of the company. The set consists of the e-mail communication data of Enron employees during 3 1/2 years. During the investigation of Enron business operations in its energy price manipulation of the Western market, FERC seized more than 500,000 e-mail messages of 158 Enron employees and made this information public. In the following years, researchers had conducted a number of filtering and updating processes for this e-mail database. The data set used here is a simplified version that involves 184 Enron e-mail addresses that existed during the time period from November 13, 1998 to June 21, 2002, including a total of 34,427 messages of over 44 months of e-mail communications. For our purpose, this data set is a tensor X of size 184 × 184 × 44.

This data set also includes the department and position information of the former Enron employees that correspond to the 184 e-mail addresses. Based on this information, these 184 individuals can be broadly grouped into five categories: executives (56), legal department (15), transportation department or pipeline (13), energy trading (29), and other unaffiliated (71). Executives were mainly from the director level or higher. Legal employees were from the law department in Enron. Transportation employees were mainly from a branch of Enron Transportation Services. Energy traders were those who traded gas and electricity in the energy markets. Others were those former Enron employees for whom we had little information. With the information of previous four categories, including the category of "other," it is natural to consider using the ASALSAN algorithm to conduct a decomposition of these four principal components. Matrix **A** is of size 184 × 4, where those 184 rows represent 184 former Enron employees, and those four columns represent four roles that embody different communication patterns. Any entry in **A**, say $a_{ij}(1 \le i \le 184, 1 \le j \le 4)$, represents the weight that the ith employee belongs to the jth type of roles. Therefore, matrix **A** identifies how much these 184 former

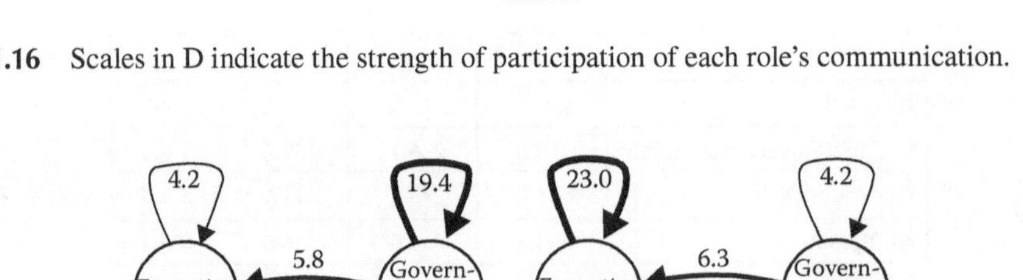

FIGURE 11.16 Scales in D indicate the strength of participation of each role's communication.

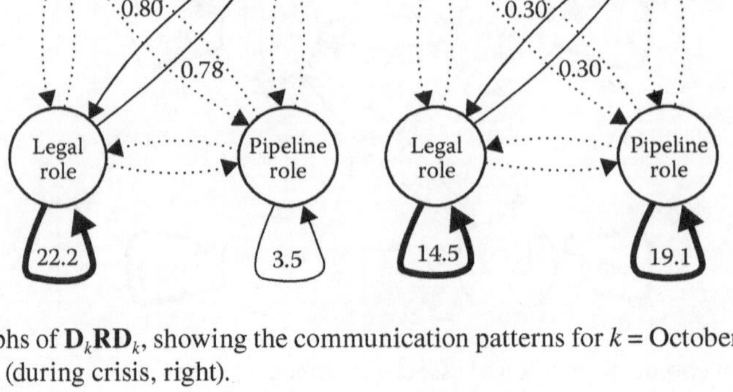

FIGURE 11.17 Graphs of $\mathbf{D}_k\mathbf{R}\mathbf{D}_k$, showing the communication patterns for k = October 2000 (precrisis, left) and k = October 2001 (during crisis, right).

Enron employees belong to the four types of roles. The results show that the four types of roles are legal, general executive that is responsible for government affairs, top executive, and transportation. These results are a little different from the four previously known categories defined by using the department and position information, where the category of energy does not form a separate role, but some belong to the role of transportation, and the category of executive is cut into the role of general executive and the role of top executive. Matrix **R** is of size 4 × 4 and represents the cumulative patterns of communication of the four roles over the specified time period of 44 months. Figure 11.15 shows the matrix **R** and its corresponding directed graph. Because the magnitudes of the entries on the principal diagonal of the **R** matrix are significantly larger than those of the other entries, it can be concluded that most of the e-mail communications were internal within the same role. There were also some communications between the role of general executive and the role of top executive. However, the communication between these two roles is clearly unbalanced; the number of messages sent from the role of general executive to the role of top executive (157.8) is significantly larger than the number of messages sent from the latter to the former (93.5). The off-diagonal entries in the fourth row and the fourth column are generally small, indicating that other than internal communications, the role of transportation had very little communication with other roles. The negative entries that are off of the principal diagonal of matrix **R** could be interpreted as undesirable communications, suggesting that the role of top executive avoided communicating with the role of legal.

Tensor D is of size 4 × 4 × 44 and records the weight of the e-mail communication of each of the four roles over the said time period of 44 months. The *k*th slice $\mathbf{D}_k$ of D is a diagonal 4 × 4 matrix and represents the weight of the e-mail communication of each of the four roles in the *k*th month. Figure 11.16 shows the changes in weights of the four roles in e-mail communications over the 44-month time period. The legal role always had the same amount of e-mail communications. However, the role of general executive had more frequent communication during the time period from October 2000 to October 2001 and then decreased significantly. The role of top executive and the role of transportation had similar e-mail communication patterns. They both had frequent communication after October 2001.

To observe the communication pattern at a particular time, such as the *k*th time moment, we can compute $\mathbf{D}_k\mathbf{R}\mathbf{D}_k$. For example, Figure 11.17 shows, respectively, the communication patterns of the four roles in October 2000 and October 2001. These two time periods correspond to the phases before and during the Enron crisis. Comparing these two phases, it can be read from Figure 11.17 that the communication between the role of general executive and the role of law had reduced, while the communication between the role of top executive and the role of transportation increased.

11.3.6 Conclusions

Social network analysis has become an important technique for the research of a wide range of scientific areas, especially modern sociology. In analyzing social networks, the asymmetric relationships between the individuals of the social networks and temporal changes of social structures have been important research topics. This case study, from a systemic point of view, models social networks using the concept of temporal network systems and introduces the multiway DEDICOM model. Two illustrative examples, one on world trade and the other on Enron e-mail network, show the significance of systems scientific points of view in solving problems of social sciences.

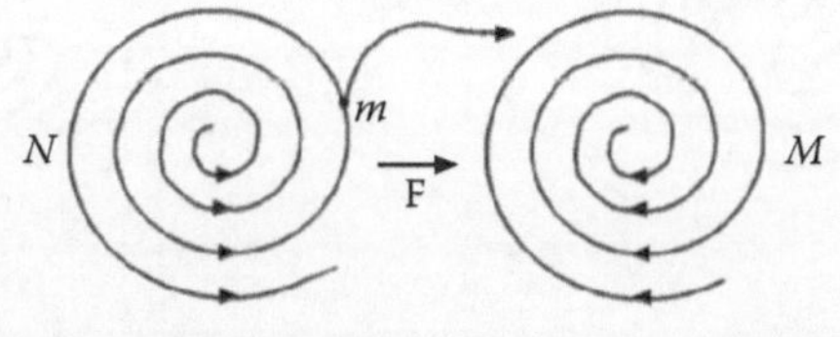

FIGURE 11.18 Opposite spins of *N* and *M*.

11.4 VALUES OF COLLEGE STUDENTS IN CHINA

Values represent the principles, beliefs, and evaluation criteria that motivate and guide a person to make rational decisions and to take relevant actions. They play an important role in the understanding of many social phenomena, such as human actions, feelings, worries, etc. The values of college students, as a special social group, to a degree determine the future of a nation and its people. This case study employs the yoyo model of systems science to construct an abstract star role model that is suitable for the study of the values of modern college students in China. On the basis of this model, an evaluation system is developed to critically look at the values of modern college students. By drawing on the methods of probability and statistics, the relevant parameters of the model are determined so that the current states of the value-related opinions of college students and key factors that influence the states are better understood. By employing the yoyo model, at the height of systems science, the relationships between these influencing factors are analyzed, and suggestions and methods are provided on how to funnel the appropriate changes in the values of college students. By comparing with the discoveries of the questionnaire, the feasibility of these suggestions is empirically confirmed so that they are not only theoretically sound but also practically usable. The presentation of this case study is based on the work of Baishun and Huang (to appear).

11.4.1 Basic Systemic Model

In terms of applications of the systemic yoyo model, assume that the whole of a society is a stable system. Within it, the totality of all things released from the big bang end is roughly the same as that absorbed into the black hole. Therefore, the internal interactions of the system will be our focus while ignoring the effects of the external world (Lin 1998).

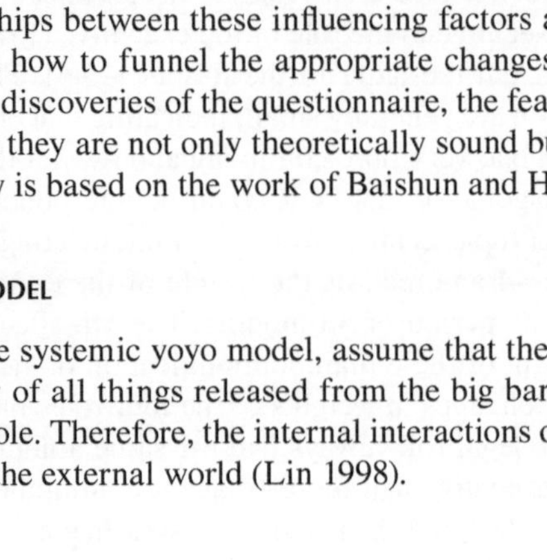

FIGURE 11.19 Opposite spins of H and K.

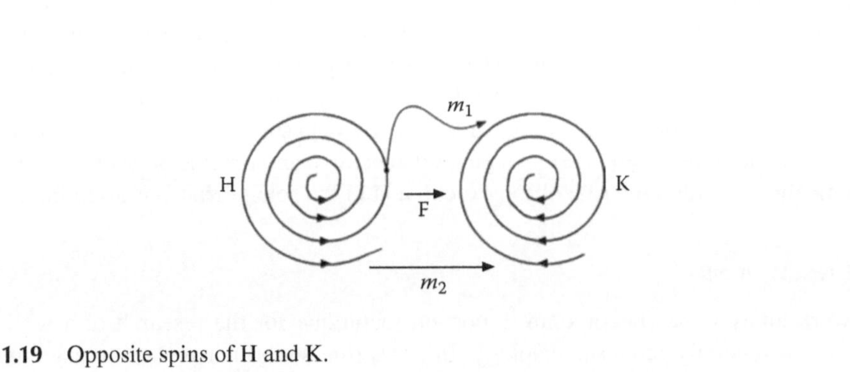

FIGURE 11.20 Family and children.

Scenario 1

As shown in Figure 11.18, the rotational directions of N and M are opposite; while one field converges, the other diverges. The symbol m stands for a loss of something from N. Its direction of motion is the same as the rotational direction of M so that M actively absorbs this potential gain. N pays the price of losing "mass" m, while M grows a little by acquiring this "mass" m, although some of its contents are released from the big bang side. This evolutionary process can be employed to analyze a family situation. The parents invest time, efforts, and money in the education of their children so that the children can potentially experience fewer worries and concerns. Here, the invested time, efforts, and money are the abstract m. The children can feel the parents' love and acquire, in the process of receiving m, additional knowledge and ability.

Scenario 2

As shown in Figure 11.19, H and K spin in opposite directions. Both m_1 and m_2 are losses of H, where the direction of movement of m_2 is against the rotational direction of K. Thus, it negatively affects and resists the rotation of K. That is like the situation in real life when the parents force their children to attend various extracurricular classes and the schools demand their students to participate in afterschool gatherings. The parents pay monetary and psychological prices, while the teachers lose their breaks, for the purse of improving the children's grades and performance on the standard examinations. However, these forces and demands are in conflict with the children's interests and wishes and might very well become obstacles in the growth and intellectual development of the children.

It can be seen that similar to the parental and school demands, the parents, teachers, and even society pay the tab "m," which spins in the same rotational direction of some children, making these children willing to accept so that they grow happily and formulate their relatively sound value systems. However, the "tab" also goes against the rotational direction of some children. It makes these children annoyed, creating obstacles in their physical and intellectual growth, and possibly distorted values. Because of the uncertainty in the direction of loss and gain of systemic yoyos, through using questionnaires, Baishun and Huang (to appear) derived the direction of loss and analyzed the process of formulating values among college students.

As a nonlinear system, the entirety of the society can be simplified into a two-body nonlinear entity by using mathematical systems theory (Lin 1999). Based on the discussion above, let us first establish the systemic model of the family and children as in Figure 11.20 (Lin 2008), assuming that the parents not only continuously transfer "mass" to the children but also influence the children through the rotation of the family, where the word "mass" stands for the parental total investment on the children in terms of money, energy, materials, values, spirits, and other physical, intellectual, and abstract items, and the phrase "acting force" stands for the effect of the parental influence on the children.

Just like the physical law that forces cannot exist independently outside objects (Lin 2008), interactions between parents and children must appear between two systems. Thus, by using analogy, these interactions can be abstracted as "forces" between two objects with internal structures of the physical world. Let us identify the abstract rotational yoyo body of a person as a star of certain mass and the effect of interference between stars as the acting forces of different persons. Analogous to the relation of mutual influences of celestial systems, the total effect a star experiences is the superposition of all force vectors of many different stars acting on the particular star. Thus, in the system of a family, the individual yoyo fields of the parents are seen as a whole and "star," and the children are seen collectively as the "planet." By doing so, we can analyze the mutual influences of two independent celestial bodies. Here, for the nonlinear problem of how the children formulate their values, only the internal interactions within the family are considered, while ignoring the effects of the society, schools, and other media. Considering the realities of the Chinese education system, this imposed constraint is quite acceptable.

By using the mathematical idea of approximating continuous functions, let us approximately treat this nonlinear problem as one that can be expressed using an explicit function so that an equation of the interacting forces can be established.

For a typical Chinese family, the parental level of education, comprehensive capability, and social status greatly influence the growth and movement of the children's yoyo field. At the same time, the children's own level of knowledge and cognitive capability also determine how well the parental forces work. The more the children know about the society, the more wide ranging their knowledge becomes, the more they understand why their parents invest so much on them, and the more they can be influenced by the parents. If children had not the slightest knowledge on their parents' capability, it would be difficult for them to accept the parental guidance. Therefore, this greatly simplified system can be defined by using the interference between two celestial bodies. Comparing to Newton's universal gravitation, we define the force F of mutual influence as follows:

$$F = \delta \frac{MN}{r^2} \tag{11.58}$$

where δ is a constant, M is the parental "mass," N is the children's "mass," and r is the degree or distance of contacts between the parents and children, which is different from spatial distance.

11.4.2 Meanings of Variables and Relevant Explanations

To understand the four variables introduced above, let us look at their determining factors. First, δ, as a prefix constant, is similar to Newton's gravitational constant or the constant of interacting electric changes. Because F describes the nonlinear interaction within a family, δ will definitely change with some of the determining factors, such as the degree of closeness between the parents and children, parents' social status, their social influence, etc. Because this case study analyzes the process of how college students formulate their values, the family system is assumed to be stable. That is, during the entire period of time the parents exert influence on the children, the family does not experience any interruptive change. Thus, this coefficient can be seen as a constant. Even so, it still changes from one family to another and is denoted $\delta(f_n)$, where f_n stands for a particular family.

Second, the "mass" M represents the totality of parents' professional titles, social status, the amount of knowledge they mastered, their level of formal education, etc. This equivalent mass M can be seen as fixed after the children are born. According to the yoyo model, when the parents exchange "materials" with the children, the amount of "mass" the children actually receive is much less than what is "given off" from the parents. In such exchanges, the parents play the role of pure givers. Therefore, in the isolated system of parents and children, $\frac{\partial M}{\partial t} < 0$, where $M = M(t)$ is a function of time t. The amount $\Delta M = \Delta M(t, \sigma_M)$ of parental "loss" also changes with time, where σ_M stands for the family's rotation constant. The magnitude of this constant σ_M determines the rotational intensity of M; the more strongly M rotates, the faster the family yoyo structure releases and absorbs everything.

At the time when the children are born, their "mass" N can be seen as an infinitesimal. As the children grow older, their yoyo structure constantly absorbs the released "mass" from the parents so that their own "mass" gradually increases. Because this family is assumed to be a stable system, for a certain period of time, the speed of "material" transfer stays constant. That is, we have the differential equation: $\frac{\partial m}{\partial t} = c$, for $t \in (\alpha, \beta)$, where c is a fixed constant, and α and β are some time moments. Thus, the children's "mass" $N = N(t)$ is a function of time t. Similar to ΔM, changes in the children's "mass" is $\Delta N = \Delta N(t, \sigma_M)$, where σ_M is the same as before, the family's rotation constant.

For the distance r of contact between the parents and children, it varies with time. The amount of time and format of contact go through a series of changes. When the children attend boarding high schools, they do not have much time to contact the parents face to face. When they communicate with parents through modern devices, interactions can still be created, even though their spatial distance might be much greater than when they talk face to face. Distant communication exerts less parental influence on the children, because it becomes difficult for the parents to observe the reactions and what the children are doing. Other than influencing by language, the parents can no longer exert direct control. It is the often-seen situation that the children might report that they are working on school work, while in fact they are playing on the Internet. Thus, $r = r(d, t)$ is a function of time t and spatial distance d.

Because most Chinese children go through 9 years of mandatory compulsory education, followed by either vocational or high school education, and then university education, let us divide the first 22 years of children's lives into four periods: 0–6 years of age (family education), 6–14 (nine years of compulsory education), 14–18 (vocational or high school), and 18–22 (work or attend college).

After analyzing the interactions between parents and children, let us now introduce the school's effect into the family system. In this case, the family, consisting of the parents and children, is seen as a yoyo structure, and the school another systemic yoyo. By doing so, through analogy with the combinations of forces in physics, the difficult problem of analyzing school, parents, and children, an impossible three-body problem, once again is simplified into the problem of a closed system with two interacting yoyo fields. Similarly, when the children interact with the society, a system consisting of two interacting yoyo fields can be obtained.

11.4.3 Questionnaire and Resolution of Model

To determine the parameters of the systemic model in Equation 11.58 and to test the validity of the model, Baishun and Huang (to appear) conducted two rounds of questionnaire surveys and analyzed the results of the surveys by applying the well-tested methods of investigation and logic of thinking (Zhou and Liu 2008; Lian and He 2008).

11.4.3.1 First Survey Sample and Model Parameters

1. Research Subjects
   A number of college students from freshman to senior years, attending schools around Changsha City, such as Hunan University, Hunan Normal University, Central South University, etc., were surveyed. A total of 120 copies of the questionnaire were distributed. Ninety of those participants completed and returned their questionnaires, which include 44 males, 46 females, 59 science majors, and 31 social science and humanity majors. There are 25 freshmen, 21 sophomores, 20 juniors, and 24 seniors. In the design of the experiment, balances in gender, age, and natural and social science majors are considered for the purpose of providing relevant data to estimate the model parameters. All the returned questionnaires are referred to as the first survey sample.
2. Dealing with the Data
   In the initial simplification of the model in Equation 11.58, r is the spatial distance, which stands for a measure of the range M could impose its influence. Generally, the distance from those being influenced changes over time. However, within a fixed time framework, this distance can be considered as a constant. In other words, for a fixed time period, r is constant. Because time is partitioned into discrete periods, r consequently takes discrete values. M stands for a relative "mass"; its influence decreases with time. The smaller the value is, the less influence M imposes.

   By using the web platform of questionnaire analysis (http://www.sojump.com/jq/610783.aspx, accessed on March 1, 2011) and the statistics software SPSS, the sample data are analyzed statistically to produce the desired causal relationships. Because of the hierarchical attribute of values (Allport and Vernon 1931), the 7 × 8 value matrix in Table 11.2 is constructed to describe personal values, where the mth row stands for the directions of n main values.

TABLE 11.2
Value Matrix

Question Type	Column 1	Column 2	Column 3	Column 4	Column 5	Column 6	Column 7	Mean Score
Basic info	Age	Gender	Political affiliation	Household registration type	Natural/social science major?			1.38
Family influence	Economic status	Attitude toward learning	Family environment	Family impact on you				2.15
Impact during compulsory education	Grade in middle school	Attendance of talent classes	Extracurricular activities	Boarding middle school?	Type of elementary/ middle school	Contact with dark sides (middle school)		2.88
Impact of high school	Participation in competitions and awards	Emotional connection with teachers	Boarding high school?	Class type in high school	Contacts with dark sides	Relationship with others	Attention to matters beyond school	2.54
Mental state	Relation with parents during compulsory education	Current degree of optimism						2.36
Values and reason (college)	Information type interested	Activities in spare time						1.89
Impact of ideological/ political education	Type of daily ideological/political education	Most influential type of ideological/ political education	Least influential type of ideological/ political education	Reason for choosing your natural/social science major				2.18
Current affairs	Plan on going abroad? Why?	You and nation	Opinion about national development	Opinion about the media	Upon graduation, plan to return to China? Why?			2.47

The questions in the questionnaire are classified into eight groups. Other than the first group on the basic information of the person who completed the questionnaire, all other groups represent different aspects of values a person takes and possible reasons he or she formulated such values. In particular, the second group looks at the family influence; the third group looks at the impacts of the time period of compulsory education; the fourth group centers on impacts experienced during the high school period; the fifth group looks at the degree of "sound" mental states; the sixth group focuses on the value directions and possible causes during the college years; the seventh group looks at the influence of ideological and political educations; and the eighth group centers on current regional, national, and international affairs.

The principle of point assignments is the following: Factors that contribute to the formation and alteration of values are divided into four tiers: "very influential," "influential," "no effect," and "opposite effect." The first tier is worth 4 points, the second 3 points, third 2 points, and the fourth tier 1 point. No weight is used for any of the factors.

After recording the scores of each question, the mean E_i, $i = 1, 2, 3, \ldots, 38$ (there are 38 questions in the questionnaire), is calculated. Then the group score S_j, $j = 1, 2, 3, \ldots, 8$, is computed as the mean of all the scores E_i of the answers to the questions in the same group. This group score S_j represents the average contribution of the jth group in the overall picture.

Now, let us see how to determine the particular forms of M, m, r, and δ. Because for different time periods, M is different, four specific timeframes T_ℓ, $\ell = 1, 2, 3, 4$, are looked at, where T_ℓ, $\ell = 1, 2, 3, 4$, respectively, stands for the period from 0 to 6 years of age, the 9 years of compulsory education until the end of the ninth grade, high school years, and college years. The symbol t represents the independent variable time. For each chosen T_ℓ, let us look at the corresponding forms of M_ℓ, m_ℓ, and δ_ℓ. To do this, define $x = [x_{ij}]_{4\times3}$, where $x = \delta, F, M, m, r, a, b, c,$ and d.

As an example, let us look at $x = \delta = (\delta_{ij})_{4\times3}$, where i (= 1, 2, 3, 4) represents one of the four time periods, and j (= 1, 2, 3) represents one of the three factors: family, school, and society. The entry δ_{ij} represents the δ value of the formula in the ith period under the influence of factor j. Similarly, define $M = (M_{ij})_{4\times3}$, $m = (m_{ij})_{4\times3}$, and the related a, b, c, and d.

a. Period from 0 to 6 Years of Age

   i. Series Approximation of the Unknown Functions

   M(t) and other functions are considered continuously differentiable with respect to time t. Then using Taylor expansion about $t_0 = 0$ produces

$$M(t) = M(0) + \frac{M'(0)}{1!}t + \frac{M^{(2)}(0)}{2!}t^2 + \frac{M^{(3)}(0)}{3!}t^3 + \ldots.$$

   Considering the reality, among all factors that influence a family, the economic status is the most important factor dominating the state of the family system (http://baike.baidu.com/view/135672.htm, accessed on March 11, 2011). For the series of $M(t)$ to converge, one must have $\lim_{n\to\infty} a_n t^n = 0$. For the choice of each coefficient of the series, the following convention is applied:

   A. For $\{a_i\}(i > 0)$, each term can be determined by one and only one factor that influences $M(t)$, except the constant term.

   B. The term a_i that is determined by the most influential factor is the coefficient of the highest-power term.

   As for the order of the polynomial approximation of $M(t)$, it is determined by the prefixed precision requirement. That is, given a certain assumption of precision, evaluate the precision after computing each order n. As soon as the required precision is reached after computing order n, the polynomial approximation of $M(t)$ can be finalized at this order.

Proposition 11.1

The first-order polynomial approximation of the function $M(t)$, established through the previous analysis, is adequate. ■

Proof

Let us first cite a well-known theorem of data modeling and analysis (Wang and Yi 1997, Theorem 2.2.8, p. 41). Let H_n be the set of all nth-order real-coefficient algebraic polynomials, $\Delta(p) = \max_{a \le t \le b} |f(t) - p(t)|$ the deviation between $p(t)$ and $f(t)$, and $E_n = E_n(f) = \inf_{p \in H_n} \{\Delta p\}$ the minimum deviation of H_n from $f(t)$. If $f^{(k)} \in C[a, b]$, $n \ge k$, then the following holds true:

$$E_n(f) \le \left(\frac{\pi}{2}\right)^k \frac{\|f^k\|_\infty}{(n+1)n\cdots(n+2-k)} \left(\frac{b-a}{2}\right)^k. \tag{11.59}$$

Because $f(t)$ is assumed to be nth-order continuously differentiable, and a_n is the coefficient of the nth power term of the nth-order polynomial approximation, the previous expression can be rewritten as follows:

$$E_n(f) = \left(\frac{\pi}{2}\right)^n \frac{\|a_n\|}{n+1} \left(\frac{b-a}{2}\right)^n. \tag{11.60}$$

From how a_n is solved for, it follows that for the series form of different orders, the coefficient of the highest-order term is the same. Thus, we assume it is c. Because only the interval (0,6) is considered, the previous equation can be simplified as follows:

$$E_n(f) = \frac{(3\pi)^n}{2^n} c \frac{1}{n+1}.$$

Considering the function

$$g(x) = \frac{(3\pi)^x}{2^x} c \frac{1}{x+1} \cdot (x > 0),$$

its derivative is

$$g'(x) = c \frac{d^x[\ln d(x+1) - 1]}{(x+1)^2}, \left(d = \frac{3\pi}{2}, x > 0\right).$$

Therefore, in its domain, $g(x)$ is strictly monotonically increasing. The smaller the n is, the smaller the $E_n(f)$ is. Therefore, when we take the first order, $E_n(f)$ is the minimum with the best approximation effect. ■

ii. Solving for Coefficient a_n of the Highest-Order Term
Because it has been objectively recognized that the family's economic state bears the most influence on children (http://baike.baidu.com/view/135672.htm, accessed on March 11, 2011), let us apply the mean score of the question related to family economic state as the coefficient of the highest-order term a_n. Because a represents the loss of the parental "mass," it should be negative. Therefore, $a_n = -1.899$ is obtained by computing the mean.

According to the interpretation of the model, it can be claimed that in this timeframe, children's values are influenced only by factors within the family (Wang 2007) with all other factors ignored. In particular,

$$M_{11} = c_{11}t + d_{11}, \text{ and } M_{12} = M_{13} = 0.$$

Also, at the time when children are born, they do not have any value, so one can set

$$d_{11} = 0,\ b_{11} = S_2 - E_5, \text{ and } a_{11} = -E_5.$$

In general, values are constrained by the outlook of life and the world view. The values of a person are formed gradually since birth under the combined influence of the family, school, and the society. With the social production mode fixed, it is the economic status that substantially makes the three environmental factors different. Thus, the family economic status plays the dominating role in the formation of values of the children. And because economic status is the key factor that influences the formation of values, and E_5 is the mean score of family economic status, one has $c_{11} = -a_{11}$, meaning that the parental "loss" is completely absorbed by the children. Also, $r_{11} = 1$; here 1 stands for a discrete value; the children basically stay along the side of the parents all the time so that their distance is the minimum. That is the sum of the radii of the two yoyo structures.

As illustrated earlier about δ, the three factors—family, school, and society—do not change and take constant values within each of the specified timeframes. For this timeframe, define δ_{i1} as the mean of E_3 and S_2, δ_{i2} the mean of S_j, $j = 3, 4, 6$, and $\delta_{i3} = S_8$, $i = 1, 2, 3, 4$. Then from $F_{ij} = \delta_{ij} \times M_{ij} \times m_{ij} / r_{ij}^2$, it follows that $F_{11} = -6.419t^2 + 0.848$.

b. Period of Compulsory Education (from First to Ninth Grade)
During this time period, Chinese students barely touch the society in general (Wang 2007). Other than spending time in classes, students live either on the school premises or at home with their parents. Their activities are very much limited to their family, school, and in between. Thus, the influence of the society can be ignored. That is, $M_{23} = 0$.

i. About the family, one has

$$M_{21} = a_{21}t + b_{21} \text{ and } m_{21} = c_{21}t + d_{21}.$$

During this timeframe, the children's "mass" is still quite small, and they are still greatly influenced by the parents. That is,

$$b_{21} = b_{11} \text{ and } a_{21} = a_{11},$$

while d_{21} is the "mass" absorbed from the parents during the first time period. Thus, $d_{21} = c_{11}t_1$, where t_i, $i = 1, 2, 3, 4$, stands for the time span of the ith period. For c_{21}, it is equal to $-k_{21}a_{21}$, where $0 < k_{21} < 1$ is a constant representing the children's rate of absorption. Because during this timeframe children start to be rebellious, parental "mass" losses generally are not entirely absorbed by the children. Thus, k_{21} is equal to the ratio of the score of the question on the relation with parents during compulsory education period over the score under the ideal circumstance, while r_{21} is the mean score of the question on whether the middle school is a boarding school.

ii. About school, one has

$$M_{22} = a_{22}t + b_{22} \text{ and } m_{22} = c_{22}t + d_{22},$$

and define $d_{22} = c_{11}t_1$ as the sum of all absorbed "masses" of all factors in the earlier time periods (for each of the following time periods, d_{ii} is defined the same way), $b_{22} = S_3$, and $a_{22} = -3.1$, where the school type is the key factor that influences the values of the children, while 3.1 is the mean score of the question on school type; $c_{22} = -k_{22}a_{22}$, where the rate k_{22} of absorption is the ratio of the mean score of the question on middle school grade over the score of the ideal circumstance (this ratio stands for effectiveness of learning, so it is reasonable to use it as the absorption rate); $r_{22} = 3 - r_{21}$, where 3 is the mean score of the question on whether the middle school is a boarding school, which represents the furthest distance from home. Thus, when subtracting the distance r_{21} from home, the result provides the distance from school. From $F_{ij} = \delta_{ij} \times M_{ij} \times m_{ij} / r_{ij}^2$, F_{21} and F_{22} are obtained.

c. High School Period (Last 3 Years of Pre-College Education)
Similar to the period of compulsory education, because the students rarely contact the society (Wang 2007) (to face the annual college entrance examinations, the absolute majority of students spend their time in school without participating in many, or any, social activities), the influence of the society can be ignored. That is, $M_{33} = 0$.

i. About the family, one has

$$M_{31} = a_{31}t + b_{31} \text{ and } m_{31} = c_{31}t + d_{31}.$$

The situation of entering into high school is similar to the timeframe of middle school, so the following can be set:

$$b_{31} = b_{21},\ a_{31} = a_{21},\ d_{31} = c_{11}t_1 + c_{21}t_2 + c_{22}t_2, \text{ and } c_{31} = -k_{31}a_{31}.$$

Because during this time period, the subjects of this study have already had a certain amount of "mass," a certain level of knowledge, and some capability of judgment, the question on the reason for choosing natural/social science major is used to illustrate this end. The four answer choices to this question are (1) recommended by parents, (2) recommended by teacher, (3) influenced by social media, and (4) personal interest. The rate k_{31} of absorption is defined as the ratio of the number of subjects who choose parents over that of subjects who choose parents, school, or society, while r_{31} is the mean score of the question of whether or not a student attended a boarding middle school.

ii. About school, one has

$$M_{32} = a_{32}t + b_{32},\ m_{32} = c_{32}t + d_{32},\ d_{32} = c_{11}t_1 + c_{21}t_2 + c_{22}t_2,$$

$$b_{32} = S_4, \text{ and } a_{32} = -2.54.$$

During this time period, the class type in school is the key school factor that influences the values of students, and 2.54 is the mean of the sum of the scores of the questions on school and class types. Let $c_{32} = -k_{32}a_{32}$ (the absorption rate k_{32} is the ratio of the number of subjects who choose school for the question on natural/social science major choices over that of those who choose parents, school, and society), and as in the period of compulsory education, $r_{32} = 3 - r_{31}$. Then from $F_{ij} = \delta_{ij} \times M_{ij} \times m_{ij}/r_{ij}^2$, F_{31} and F_{32} are obtained.

d. College Period
During this timeframe, the distance from the parents is widened; the subjects generally live on university campus and are closely in contact with society (Wang 2007).
   i. About the family, one has

$$M_{41} = a_{41}t + b_{41},\ m_{41} = c_{41}t + d_{41},\ d_{41} = c_{11}t_1 + c_{21}t_2 + c_{22}t_2 + c_{31}t_3 + c_{32}t_3.$$

During this timeframe, the subjects have acquired a good amount of "mass." They gradually become independent. However, even so, considering the fact that the majority of them are still not completely independent in terms of economics, the family influence still cannot be totally ignored. Therefore, the following can be set:

$$b_{41} = b_{31},\ a_{41} = a_{31}, \text{ and } c_{41} = -k_{41}a_{41}, \text{ and } r_{41} = 5,$$

where 3 is the defined total distance of the subjects from various factors during junior and senior high schools; considering the land mass and size of China, entering college means that the distance between the subjects and their families has changed substantially; that is how it is set to be 5. At the same time, because of the rapid increase in r_{41}, minor changes in k_{31} are ignored and set $k_{41} = k_{31}$.
   ii. About school, one has

$$M_{42} = a_{42}t + b_{42},\ m_{42} = c_{42}t + d_{42},\ d_{42} = c_{11}t_1 + c_{21}t_2 + c_{22}t_2 + c_{31}t_3 + c_{32}t_3.$$

Considering where the returned questionnaires are from, university types do not need to be investigated. Thus, let

$$b_{42} = 3 \text{ and } a_{42} = -3.$$

Referencing the high school period, a school is classified as "good," "median," and "poor," three classes with 3–1 points assigned, respectively; because our questionnaires were mainly distributed to Hunan University, Central South University, and Hunan Normal University, three of the most recognized universities in Hunan Province, they are treated as "good" and assigned the score of 3. Additionally, let us set

$$c_{42} = -k_{42}a_{42}$$

where k_{42} stands for the ratio of the number of subjects who choose professional knowledge type over the total number of subjects in the study to the question on information type interested in. Although ordinarily, students spend their time on university campus, their contacts with other students are no longer as wide ranging and frequent as in high school. The most common scenario is small-range communications. The results of the question on activities in spare time have confirmed this conjecture. Thus, the distance from school is set to be $r_{42} = 1$.

3. About the society, one has

$$M_{43} = a_{43}t + b_{43},\ m_{43} = c_{43}t + d_{43},\ d_{43} = c_{11}t_1 + c_{21}t_2 + c_{22}t_2 + c_{31}t_3 + c_{32}t_3,$$

$$b_{43} = (S_6 + S_7 + S_8)/3,$$

$a_{43} = -4$ (because the data of questions on information type interested in and activities in spare time suggest the corresponding value for school to be a little over 3, we set it to 4),

$$c_{43} = -k_{43}a_{43},\ k_{43} = 1 - k_{42},\ r_{43} = 8.$$

Next, from $F_{ij} = \delta_{ij} \times M_{ij} \times m_{ij} / r_{ij}^2$, F_{41}, F_{42}, and F_{43} are computed.
The following summarizes all the numerical results obtained above:

$$\delta = \begin{pmatrix} 1.78 & 0 & 0 \\ 1.78 & 2.44 & 0 \\ 1.78 & 2.44 & 0 \\ 1.78 & 2.44 & 2.47 \end{pmatrix}, R = \begin{pmatrix} 1 & 0 & 0 \\ 2.222 & 0.778 & 0 \\ 2.567 & 0.433 & 0 \\ 5.000 & 1.000 & 5.000 \end{pmatrix}$$

$$T = \begin{pmatrix} t_1 \\ t_2 \\ t_3 \\ t_4 \end{pmatrix} = \begin{pmatrix} 6 \\ 9 \\ 3 \\ 4 \end{pmatrix},$$

$$a = \begin{pmatrix} -1.899 & 0 & 0 \\ -1.899 & -3.1 & 0 \\ -1.899 & -2.54 & 0 \\ -1.899 & -3.00 & -4.00 \end{pmatrix}, b = \begin{pmatrix} 0.251 & 0 & 0 \\ 0.251 & 2.88 & 0 \\ 0.251 & 2.54 & 0 \\ 0.251 & 3.00 & 2.18 \end{pmatrix},$$

$$c = \begin{pmatrix} 1.899 & 0 & 0 \\ 1.619 & 2.617 & 0 \\ 0.562 & 0.747 & 0 \\ 0.562 & 1.226 & 2.312 \end{pmatrix}, d = \begin{pmatrix} 0 & 0 & 0 \\ 11.894 & 11.894 & 0 \\ 49.518 & 49.518 & 0 \\ 53.445 & 53.445 & 53.445 \end{pmatrix},$$

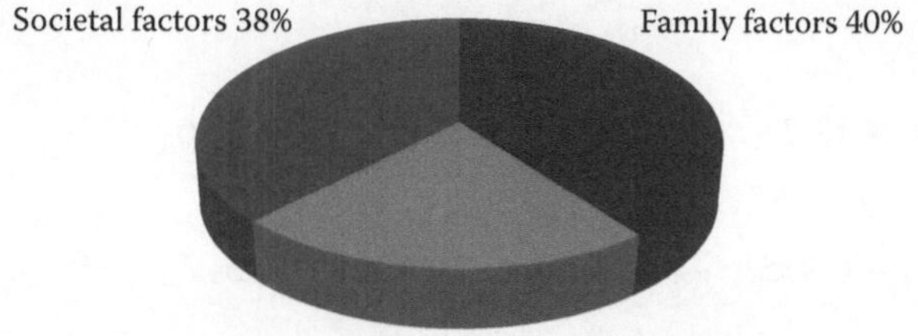

FIGURE 11.21 Major factors that influence the formulation of values.

$$M=\begin{pmatrix} -1.899t+0.251 & 0 & 0 \\ -1.899t+0.251 & -3.1t+2.88 & 0 \\ -1.899t+0.251 & -2.54t+2.54 & 0 \\ -1.899t+0.251 & -3.00t+3.00 & -4.00t+2.18 \end{pmatrix},$$

$$m=\begin{pmatrix} 1.899t & 0 & 0 \\ 1.619t+11.894 & 2.617t+11.894 & 0 \\ 0.562t+49.518 & 0.747t+49.518 & 0 \\ 0.562t+53.445 & 1.226t+53.445 & 2.312t+53.445 \end{pmatrix}, \text{ and}$$

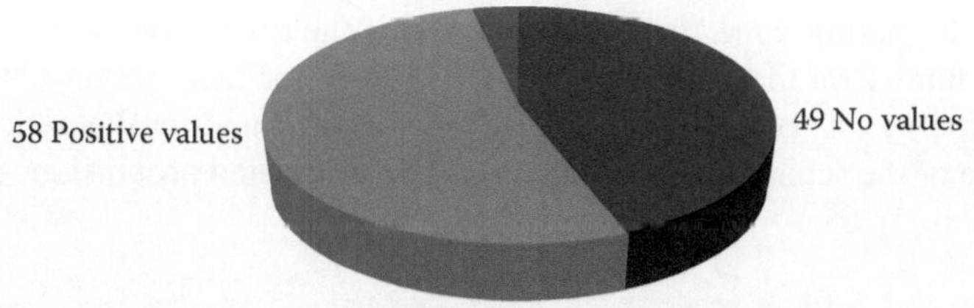

FIGURE 11.22 Proportions of the sample with different classification of values.

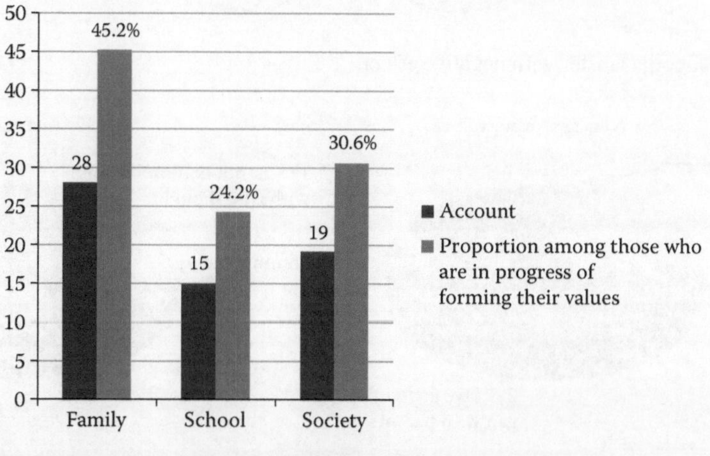

FIGURE 11.23 Proportions of various influencing factors.

$$F = \begin{pmatrix} -6.419t^2 + 0.848 & 0 & 0 \\ -1.108t^2 - 7.997t + 1.076 & -32.704t^2 - 118.252t + 138.086 & 0 \\ -0.288t^2 - 25.363t + 3.357 & -24.693t^2 - 1612.2t + 1636.9 & 0 \\ -0.076t^2 - 7.216t + 0.955 & -8.974t^2 - 382.243t + 391.217 & -0.914t^2 - 20.624t + 11.511 \end{pmatrix}.$$

11.4.3.2 Second Survey Sample and Confirmation of Model

Based on the discoveries of the first survey sample, the equations and relevant parameters of the systemic model were obtained. To test the validity of the model, double check the survey results, and analyze the statistical properties, a second round of survey was conducted.

As before, first- to fourth-year college students from Hunan University, Hunan Normal University, Central South University, and other nearby universities were applied as the pool from which to collect the sample. For this round, 150 copies of the questionnaire were distributed, and 111 completed forms were returned, where there were 53 males, 58 females, 61 natural science majors, and 51 social science majors with 30, 27, 34, and 20 freshmen, sophomores, juniors, and seniors, respectively. Again the design of the experiment considered a balance between genders, ages, and natural and social science majors. The purpose is to test the equations obtained earlier and analyze the system of concern by providing empirical data support.

Among the 111 available questionnaires, the main factors that influence the formulation of values are, respectively, family, school, and society. Their proportions are depicted in Figure 11.21.

From Figure 11.21, it follows that the effects of the family and the society are roughly the same. The formula of $F(x)$ established earlier implies that the effect of the family is in effect throughout all the time periods, while the effect of the society appears significantly only during the college time. Integrating the family "acting force" produces that during the time period of 0–22 years of age, the acting force of the family on the children is $-3.8312e+003$; for the period of 19–22 years of age, the acting force of the society on the children is $-2.3882e+003$; and for the period of 7–22 years of age, the acting force of the school is $-1.41852e+003$. The calculated proportions of these forces are,

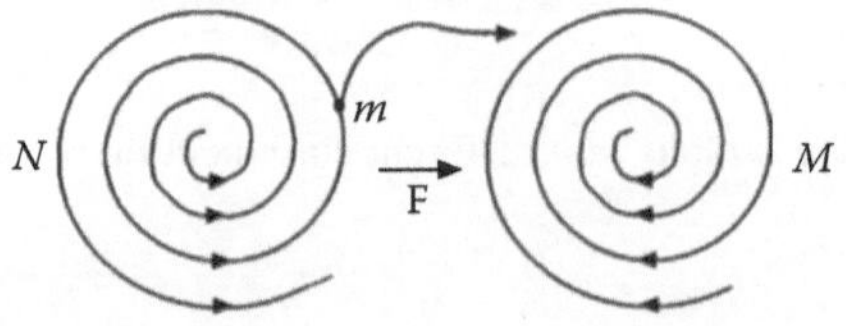

FIGURE 11.24 Smooth transfer with negative effect.

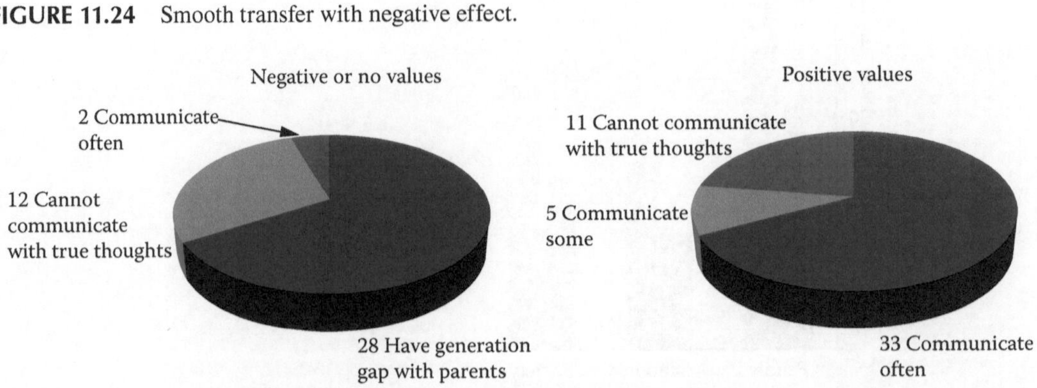

FIGURE 11.25 Comparison of causes.

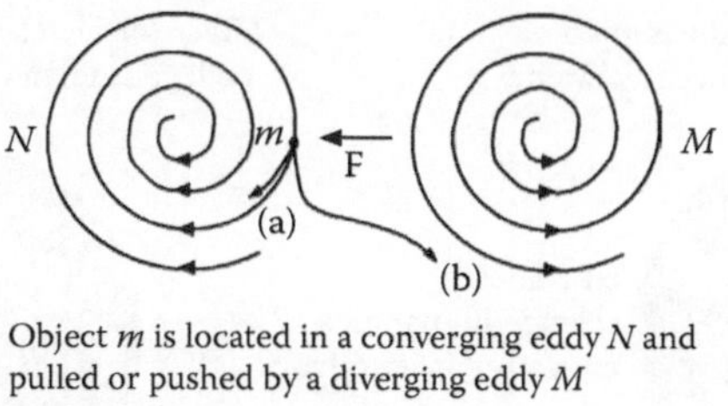

FIGURE 11.26 Responsive feedback from children.

respectively, 50% for family, 32% for society, and 18% for school. From comparing with the second survey sample, it can be seen that there are some changes in the proportions of the family and society, which are caused by the fact that when the equations were established, the effect of the society on children of the age interval 0–18 was not considered. Thus, from analyzing tables, it is found that the approximation equations from the first survey sample are quite reliable.

Among the 111 effective copies of the questionnaire, 49 subjects have positive (optimistic) values, 58 have not formed any stable value, and 4 have negative values (see Figure 11.22).

Among the 62 subjects who do not have a positive value yet, the effects of family, school, and society are, respectively, 28 (45.2%), 15 (24.2%), and 19 (30.6%); see Figure 11.23.

From the results of the second survey sample, it follows that in the process through which college students formulate their values, the main forms of influence of family, school, and society take four possibilities. In particular, for the factor of family, among the 28 subjects who contributed the cause that no values have developed in their families, 15 families focused only on the children's school works, careers, and living conditions; 9 were negatively influenced by the parents because of the parents' own reasons; 1 did not form any value due to a lack of family education; and 3 received quality family education.

Among the 49 subjects with positive (optimistic) values, 38 families paid particular attention to moral accomplishments; 10 families looked at the cultivation of real abilities; and only 1 family's education created negative effect.

From Figure 11.24 and the equation $F_{11} = \delta_{11} \times M_{11} \times m_{11}/r_{11}^2$, it follows that for the situation where the parents feed their children with too much utilitarianism, although the parental payoff might agree with the rotational direction of the children, due to the disagreement of the rotational direction of the parents and the overall, commonly accepted values of the society, the parents generate negative influence on the children, leading potentially to distorted values in the children. As for those parents with questionable self-cultivation, because of their lack of "mass," their forces acting on the children do not have much strength. Thus, in the process of formulating their values, with relatively minor parents' effect, the children might not be able to form their adequate values. For the 49 subjects with positive values, because their families' rotational direction agrees with that of the overall society, and the parents release "masses" frequently, their children formulate their positive values (Figure 11.25).

Among the 28 subjects who contributed the cause that no values have developed in their families, 14 felt the existence of generation gaps with their parents, 12 could not express their true thoughts, and only 2 had been often communicating with their parents about their true feelings.

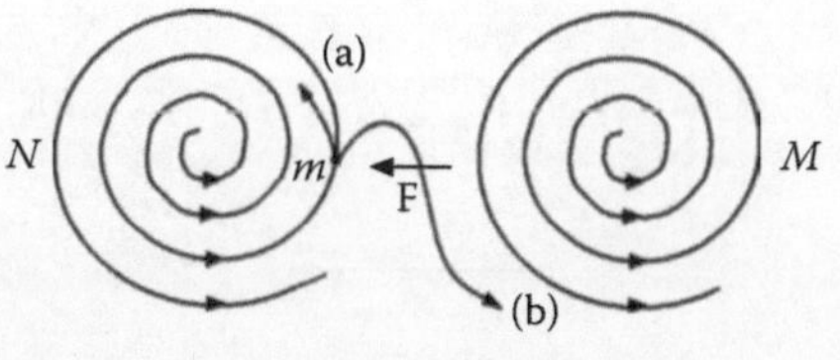

FIGURE 11.27 Interactions between students and the school.

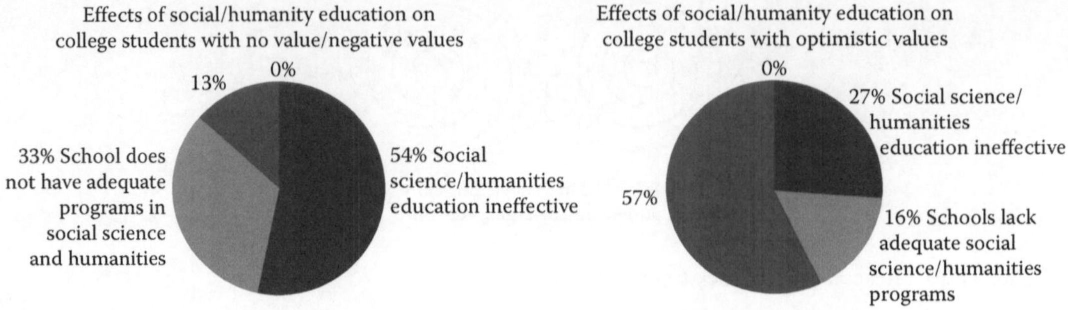

FIGURE 11.28 Impacts of social science and humanities programs.

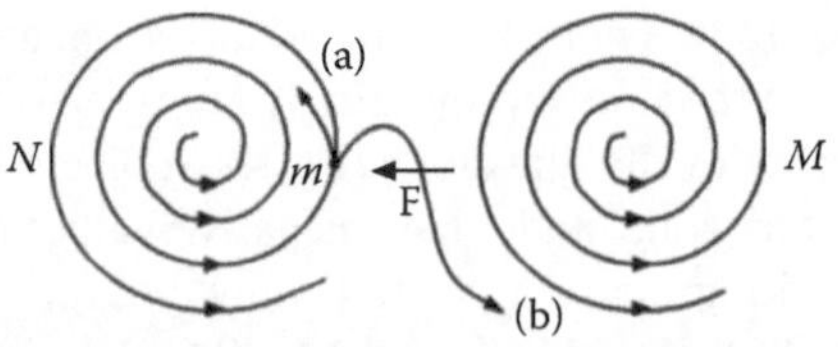

FIGURE 11.29 Form of social science and humanities program in agreement with students' direction of rotation.

Among the 49 subjects with positive (optimistic) values, 33 communicated with their parents often, 5 believed the existence of generation gap with the parents and rarely communicated, and 11 had some communication with the parents but could not express their true thoughts.

From Figure 11.26 and the equation $F_{11} = \delta_{11} \times M_{11} \times m_{11}/r_{11}^2$, it follows that when parents influence their children, they also exert acting "forces" on the "masses" the children received from other sources to help maintain the positive values while removing negative components. Because these "masses" from other sources must be visible to the parents in order for them to exert their appropriate "forces," the importance of communication becomes clear. Among the 28 subjects with negative values, 26 did not communicate with parents or did not express their true thoughts in their rare communications. Such a lack of communication makes the "masses" received from other sources invisible to the parents. Thus, even if the rotational direction of the parents agrees with that of the large society, the parents still cannot provide timely modifications for their children to alter the children's rotational directions. That leads to negative (pessimistic) values in the children. At the same time, among the 49 subjects with optimistic values, 33 often communicated with parents so that the parents could provide timely advice, helping to shape the optimistic values of the children.

In terms of the school factor, for those subjects who lack values, it is clear that they disagreed with the organized school education. The majority (13/15) were convinced that the current school education was ineffective and a big failure.

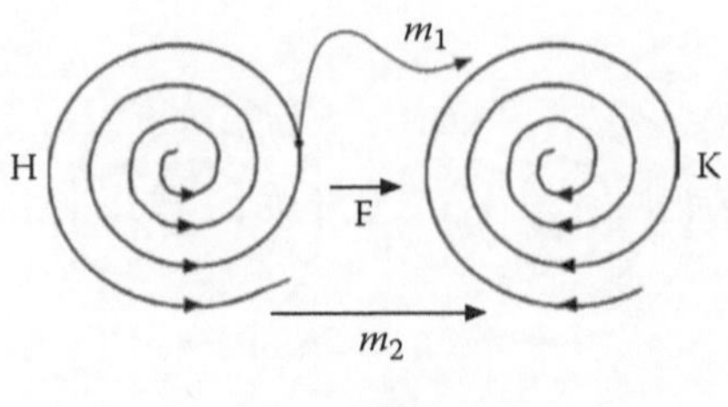

FIGURE 11.30 Interaction of college students and society.

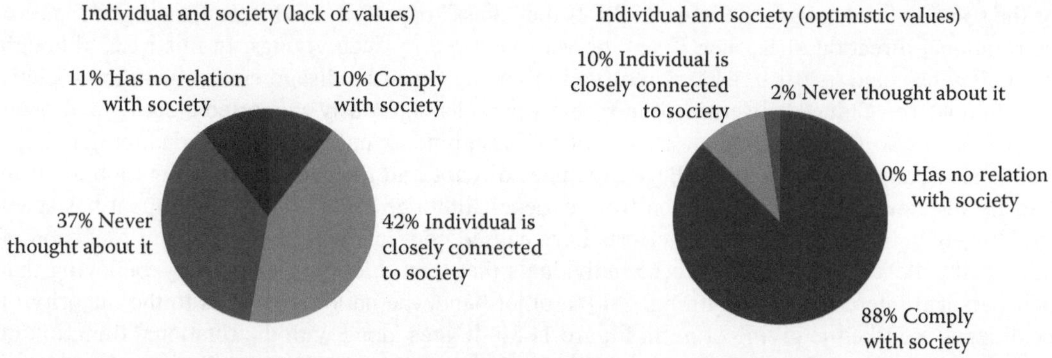

FIGURE 11.31 Relationship between an individual person and society.

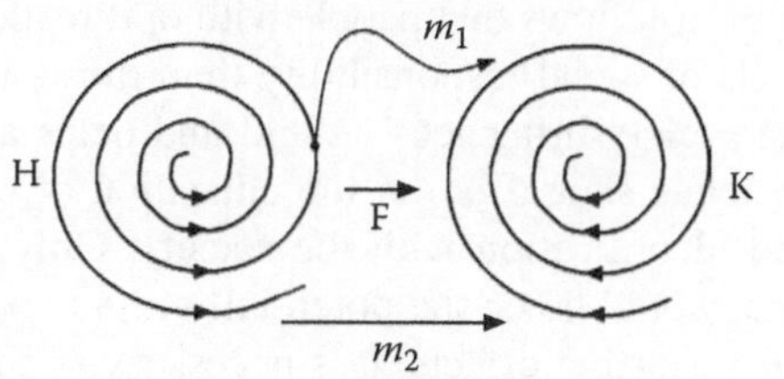

FIGURE 11.32 "Mass" transfer from society to a college student.

From Figure 11.27, it follows that if *N* stands for the yoyo structure of school, and *M* is the yoyo structure of student(s), where *m* represents a payoff (loss of "mass") of *N* toward *M*, then those students who lack any value reject the pay of the school; they instinctively dislike what school teaches and exclude all that belong to the school system. In terms of the yoyo model, this phenomenon is embodied in the mutual rejection of *M* and *N* so that *N* gradually departs from *M* and at the same time absorbs many of the things the school rejects. The psychology of weariness toward learning grows within these students. They play truant and enjoy deformity culture. However, if the payoff *m* of *M* complies with the rotational direction of *N*, then *N* will be more than happy to accept and absorb *m*, and for a following period of time, *N* will continue to accept willingly what *M* has to give. As a consequence, the rotational speed of *N* increases and becomes faster and faster, and soon, one can easily stand out from among his or her peers with recognizable achievements. It is like the scenario that when a child learns what is taught in school and becomes able to apply the knowledge to solve practical problems, his or her ability increases with adequate recognition so that he or she is more willing to learn knowledge and skills, growing toward what is expected of him or her.

As shown in Figure 11.28, the acceptance rate of social science and humanities education is low in the group of college students with pessimistic (negative) values, while the rate is high for the group of college students with optimistic values.

From Figure 11.29, it follows that schools need to develop more and stronger programs in areas of social science and humanities. However, the specific forms of these programs have to conform with the direction of the students' yoyo structures' rotation in order to prevent the possibility that "mass" transfers of the school are not accepted by the students and that relevant parts of the students' own knowledge are also being disregarded and rejected, causing difficulties in the formation of optimistic and positive values in the students.

In terms of the society factor, it can be seen that the effects of the society on college students are reflected in the students' identification with the nation and society. In particular, those college students with either pessimistic (negative) values or no values at all can be explained as follows by using the systemic yoyo model. The yoyo structure K stands for the college students, while H is for the society; for details, see Figure 11.30. If the "mass" payoff m_2 of H to K always goes against the rotational direction of K, then K will be very annoyed by such givings. In this case, although the society has paid its price, what is returned might very well be disagreements with the society. For example, the Chinese government in recent years has vigorously supported creating and owning businesses so that the original system of job assignments upon college graduation has been abandoned. Consequently, some college graduates did not find their ideal jobs or any job at all in a timely fashion upon their graduation from college. Although the Chinese government has spent "m_2" hugely on college students, its efforts have caused conflicts with the rotation of a portion of individual college graduates. It is these individuals that start to oppose the society, believing that their personal interests are everything. On the other hand, the national policies to the majority of the citizens are like the giving of m_1 in Figure 11.30. It goes along with the rotational direction of the yoyo field of K so that the intensity of K's spin is strengthened. That explains why the number of high school and college graduates who create and own businesses has been steadily increasing in recent years and why the Chinese economy has been becoming steadily more prosperous.

From Figures 11.30 and 11.31, it follows that people with optimistic values have a stronger sense of collectivism and higher levels of social responsibility than those without definite values. Figure 11.32 illustrates that when the society influences a child and helps a college student to formulate his or her worthy values, the focus should be on the education of collectivism, responsibility of individuals to the society, and identification with the society. Only by doing so, when a college student takes his or her actions, would he or she potentially think about the society in general and others in particular. Also, when exerting effects, it is necessary to select appropriate directions to offer "mass" transfers in order to prevent much of the expected "negative" effects. If the direction of transfer goes against the yoyo field of a student, even with good intention, the transfer will become m_2, which will be rejected by the yoyo pool of the intended receiving student. If the "mass" transfer might create negative effects, then even if it is m_1, it still could make the yoyo field of the student react adversely, creating distorted values.

Based on what has been discussed, in the process of formulating values, there are always individual cases that do not fit the theoretical analysis. For example, for the factor of family, out of the 28 subjects with pessimistic values, one family's role is positive; among the 49 subjects with optimistic values, 5 of them experience a generation gap in their communication with parents. Such phenomena of "inconsistence" are exactly what the nonlinearity of the entire society causes. Because the formation of value of any individual is determined collectively by various interacting systems, when the current environmental conditions become more complicated and intertwined, the appearance of such "inconsistent" phenomena will become more prevalent.

Through empirical data and analysis of equations, it is found that the family factor plays the main role in the formation or lack of values in a college student. Within the family factor, both the state of communication with the parents and the type of family education are most dominant. For the subjects who do not have well-formulated values, most of them rarely communicate with parents; even if they do, they do not or are unable to express their true mind. Additionally, these families focus more on the training of practical abilities while paying little or no attention to the nurturing of morality, creating the sense of unimportance of morality training in life. On the other hand, for subjects with optimistic values, the situation is exactly the opposite. Therefore, in order to guide college students to acquire their needed directions of value, family education and influence have to be specifically emphasized. In creating the educational opportunity of the family, the parents first need to invest in improving their self-cultivation, creating a quantity environment of learning, and focusing mainly on the training and education of morality and basic logic of thinking. At the same time, on a daily basis, the parents should initiate communications and try to understand what their children are really thinking about. When necessary, provide the children with timely guidance.

11.4.4 Statistical Analysis

There is randomness in human reactions to external stimuli. That is, human behavior can be seen as a random variable, known as probabilistic behavior. The probability of such a random variable is named as behavioral probability; its probability distribution is a behavioral probability distribution function. Based on the empirical distribution, the following is defined:

$$F_n^*(x) = \begin{cases} 0, & 0 \le x \le x_1^* \\ \dfrac{k}{n}, & x_k^* < x \le x_{k+1}^*, k = 1, 2, \cdots, n-1 \\ 1, & x > x_n^* \end{cases} \tag{11.61}$$

where the values of the probabilistic behaviors are $x_1 = 1, x_2 = 2, x_3 = 3$. According to the first sampling survey, the following is obtained:

$$F_{111}^* = \begin{cases} \dfrac{4}{111}, & x \le 1 \\ \dfrac{49}{111}, & 1 < x \le 2 \\ \dfrac{58}{111}, & 2 < x \le 3 \\ 1, & 3 < x. \end{cases} \tag{11.62}$$

To tabulate the natural science and social science majors, let the distribution function of the overall ξ be $F(x)$ and the empirical distribution $F_n^*(x)$. For any real number x, denote

$$D_n = \sup_{-\infty < x < +\infty} \left| F_n^*(x) - F(x) \right|.$$

Then the following holds:

$$P\left\{ \lim_{n \to \infty} D_n = 0 \right\} = 1.$$

Because the samples collected are medium sized, assume that $F(x) = F_n^*(x)$.

Pearson chi-square theorem (Wang and Yi 1997). Let $H_0 : F(x) = F_0(x)$. No matter what distribution $F_0(x)$ is, if H_0 holds true, then the limit distribution of the statistic

$$\eta = \sum_{i=1}^{m} \frac{(v_i - np_i)^2}{np_i} = \sum_{i=1}^{m} \frac{v_i^2}{np_i} - n$$

is a χ^2– distribution with m–1 degrees of freedom, where $F_0(x)$ does not have any unknown parameter.

By using the answers to the following three questions on the first survey questionnaire:

27. What do you currently do in your spare time? A) Study to improve myself; B) Entertain on Internet; C) Participate in social activities; D) Work on a hobby; E) Make some spending money. (collected 48, 4, 19, 19, 0)
29. If you choose to go abroad and successfully complete your education, will you consider returning home to develop your career? Why? A) Yes, China evolves fast and provides more opportunities; B) Yes, it is difficult for me to live in a foreign land; C) No, there are

more opportunities abroad; it will be easier for me to accomplish professionally; D) No, Chinese environment does not fit me well. (collected 43, 31, 8, 8)

33. If using one of the following choices to describe your relation to China, you will use: A) We are a whole, sharing weal and woe, honor and disgrace; we are closely related; B) We are in a relationship of whole and part; C) Don't know, never thought about it; D) We are in a relationship of individual. (collected 48, 38, 4, 0) the relevant frequency table is obtained:

Number	Interval	Frequency	Group Median
1	$(-\infty, 1]$	$4+3+2=9$	1
2	$(1,2]$	$38+14+66=118$	2
3	$(2,3]$	$48+73+22=143$	3

The computed theoretical frequencies are as follows:

Number	Interval	Frequency	Theoretical Frequency
1	$(-\infty, 1]$	9	9.9
2	$(1,2]$	118	119.2
3	$(2,3]$	143	142.6

Thus, the test statistic is

$$\eta = \sum_{i=1}^{3} \frac{(v_i - np_i)^2}{np_i} = \sum_{i=1}^{3} \frac{v_i^2}{np_i} - n \approx 0.092\,.$$

The significance test for H_0 is

$$P\left\{\eta > \chi^2_{(3-1)}(\alpha)\right\} = \alpha.$$

By letting $\alpha = 0.05$, it follows that $\chi_2^2(\alpha) = 5.99 > 0.092$. That is, there is not sufficient evidence to reject H_0. That is, $F(x) = F_{111}^*(x)$, where $F_{111}^*(x)$ is given in Equation 11.62. To confirm the accuracy of $F(x)$, in the following, let us test inversely. That is, from the first survey sample, the empirical function is constructed, and then Pearson χ^2-test is employed to verify the function based on the second survey sample.

Based on the second survey sample, the following is obtained:

$$F_{270}^* = \begin{cases} \frac{9}{270}, & x \le 1 \\ \frac{118}{270}, & 1 < x \le 2 \\ \frac{143}{270}, & 2 < x \le 3 \\ 1, & 3 < x. \end{cases} \tag{11.63}$$

The frequency table of the second survey sample is as follows:

Number	Interval	Frequency	Group Median
1	$(-\infty, 1]$	4	1
2	(1,2]	49	2
3	(2,3]	58	3

The computed theoretical frequencies are given in the following table:

Number	Interval	Frequency	Group Median
1	$(-\infty, 1]$	4	3.7
2	(1,2]	49	52.1
3	(2,3]	58	58.1

Constructing the test statistic leads to

$$\eta = \sum_{i=1}^{3} \frac{(v_i - np_i)^2}{np_i} = \sum_{i=1}^{3} \frac{v_i^2}{np_i} - n \approx 0.025.$$

The significance test of H_0 is

$$P\{\eta > \chi^2_{(3-1)}(\alpha)\} = \alpha.$$

By letting $\alpha = 0.05$, it follows that $\chi^2_2(\alpha) = 5.99 > 0.025$. There is no sufficient evidence to reject H_0. That is, both survey samples can be thought to satisfy the same distribution. Also, a more accurate distribution $F(x)$ can be obtained as follows:

$$F(x) = \begin{cases} \frac{13}{381}, & x \le 1 \\ \frac{167}{381}, & 1 < x \le 2 \\ \frac{201}{381}, & 2 < x \le 3 \\ 1, & 3 < x. \end{cases}$$

Therefore, it can be seen that the values in college students tend to be either optimistic or not yet finalized with a minority of pessimistic outlooks.

11.4.5 Conclusions and Suggestions

The analysis in this case study indicates that the overall trend in the values of Chinese college students is positive and optimistic; only a small proportion, less than 4%, suffers from such negative values as hating the society, pessimistic outlooks on life, etc. At the same time, a good proportion of those surveyed live in a state of value void. They can be potentially guided into a healthy, optimistic value system or be fooled by some of the society haters into committing acts that are harmful to the wellbeing of the society.

Out of the eight main forms of interaction between two yoyo fields, the surveyed samples confirm four of them, as conveyed in Figures 11.24, 11.29, 11.30, and 11.32, and it is argued that the process of value formulation of college students satisfies the following approximate equation:

$$F = \delta \frac{MN}{r^2}. \tag{11.64}$$

At the same, what is discussed includes how the variables M, N, r, and δ change with time t and how the rotation coefficient σ changes so that the effects of family, school, and society on children are analyzed for different time periods. After deriving the factors that influence the formulation of values of children, the interactions of these factors are analyzed. By borrowing the relevant knowledge in the existing literature (Huang and Zheng 2005), it is expected that the results obtained in this case study can potentially offer beneficial guidelines for the relevant practice.

In order for college students to adequately, effectively, and scientifically formulate their values, one should start with several different aspects, including family, society, school, and the students themselves. All suggestions here are provided on the basis of the rotational directions of the "mass" transferor and the receiver of the systemic yoyo model. Combined with the condition of how the "mass" transfer could produce optimistic and pessimistic effects, by analyzing how different aspects of Equation 11.64 influence the values of college students on the four layers of family, school, society, and the students themselves, the following scientifically sound suggestions are offered:

1. Parents should strengthen their own training in order to form a harmonic family environment. They should prioritize what to teach their children by placing the education of morality, thoughts, and sentiment, etc., on the top, and actively guide their children in their formulation of values.
2. Parents should frequently communicate with their children, understand their true states of mind, and provide guidance in a timely fashion.
3. In social lives, strengthen the education on team work, responsibility, and identification. For example, community agencies, health service stations, welfare institutions, etc., should recruit additional college student volunteers so that these young people would have opportunities to experience social and collective lives and understand the state of the modern society, while providing their services. Through living a real life, providing labor and intellectual efforts, these young people would be able to automatically sense togetherness with others and strengthen their responsibility and identification with the society.
4. The news media should create programs of rich real-life contents in various forms so that young college-age people could willingly accept what is shown to them. For example, more movies and TV programs should be made to advocate ordinary citizen idols of optimistic characteristics along with localized cultures and backgrounds.
5. Schools should emphasize more on the education on social science and humanities by enriching the existing educational models. The methods of knowledge transmission should include movie documentation, video materials, computer simulations, etc., in order to provide students with additional visual impacts. At the same time, schools should offer courses on interactive practice, such as opportunities where students could lecture on, perform, talk about, and play out what they truly think of, in order to form a proactive learning environment. Only in such environments could students potentially become interested in participating and actively learning the knowledge of social science and humanities.
6. Schools should create a relaxed, harmonic living/working environment and guide students with the correct purpose of learning. By placing more emphasis on the current affairs and organizing related discussion panels, students would have additional opportunities to look at the world through various angles so that they would have a better chance to formulate their values and responsibility and their identification with the culture and society they belong to.

11.4.6 Appendix: The Questionnaire

Instruction for the reader of this paper: The numbers after each question stand for the point assignments to the corresponding answer choices. The numbers after all the answer choices stand for the numbers of selections of the corresponding choices.

1. You are (1,1,1,1)
   A. Male born after 1980
   B. Female born after 1980
   C. Male born after 1990
   D. Female born after 1990 (13, 20, 21, 36)
2. Your political status is (2,3,1)
   A. Member of Communist Youth League
   B. Member of Communist Party
   C. Other (76, 12, 2)
3. You are from (1,2)
   A. The countryside
   B. A city or town (54, 36)
4. During the time period of compulsory education, how were your grades in social science and humanities? (3, 2, 1)
   A. Good
   B. General
   C. Not good (50, 38, 2)
5. Your family annual income is (not including expenses) (1, 2, 3, 4)
   A. Below 10,000
   B. Between 10,000 and 100,000
   C. Between 100,000 and 300,000
   D. Above 300,000 (14, 72, 3, 1)
6. During the time period of compulsory education, did you board? (3, 2, 1)
   A. Boarded a long time
   B. Boarded, but only a short while
   C. Boarded the entire time (43, 24, 23)
7. During your middle school, what did you do during your spare time? (3, 3, 1)
   A. Study and read
   B. Pursue interests and hobby
   C. Nothing important (41, 37, 12)
8. During the time period of compulsory education, what types of elementary and middle school did you attend? (2, 3, 4, 3)
   A. A countryside elementary school, a countryside middle school
   B. A countryside elementary school, a city/town middle school
   C. A city/town elementary school, a city/town middle school
   D. A city/town elementary school, a countryside middle school (24, 32, 33, 1)
9. During your elementary and middle school years, how much did your family (parents) emphasize on your learning? (3, 2, 1)
   A. Very much
   B. General
   C. Not at all (57, 2, 12)
10. During your elementary and middle school, did you attend any training classes for any special interest, hobby, or ability or preparing you for some kind of competition? (3, 2, 1)

A. Quite a few
B. Some
C. Never (11, 46, 33)

11. When you were in elementary and middle school, how was your family environment? (3, 2, 1)
A. I lived with my parents
B. I lived with grandparents because my parents went away to work
C. I lived in a single-parent family (72, 15, 3)

12. When you were in elementary and middle school, did you experience any unexpected event, such as assault, blackmail, robbery, etc.? (1, 2, 3)
A. Quite a few times
B. A few times
C. Never (1, 18, 71)

13. Speaking from your heart, was your childhood happy? (3, 2, 1)
A. Very happy
B. OK
C. Not happy (53, 34, 3)

14. What is the type of your high school? (2, 1)
A. Provincial focus, best
B. Ordinary (56, 34)

15. In your high school, you were in (3, 2, 1)
A. Focus (best) class
B. Ordinary class (41, 49)

16. During high school, did you participate in any competition? The outcome? (3, 2, 1)
A. Some with prizes
B. Some, but never won any prize
C. No, not at all (22, 32, 36)

17. When you were in high school, you were classified as (1, 1)
A. A student of social science and humanities
B. A student of natural science (39, 51)

18. What was your reason for choosing a social science (or natural science) major? Please select only the most important reason (3, 2, 1, 4)
A. Parents' recommendation
B. Teacher's recommendation
C. Public opinion
D. Own interest (5, 5, 7, 73)

19. Looking back, how do you feel about your high school? (3, 2, 1)
A. Grateful
B. No feeling
C. Disgusted (67, 23, 0)

20. Looking back, your feeling toward the high school teachers can be roughly described as (3, 2, 1)
A. Grateful
B. No feeling
C. Disgusted (73, 14, 8)

21. Did you board in high school? (3, 1, 2)
A. Yes, for a long time
B. No, I lived at home
C. I rented a place outside the campus (66, 15, 9)

22. How was your relationship with others in high school? (3, 2, 1)
    A. Quite good
    B. OK
    C. Not that good (56, 31, 3)
23. During high school, to what degree did you pay attention to matters such as news, current affairs, etc., that were beyond school? (3, 2, 1)
    A. Good amount
    B. General
    C. Not at all (21, 57, 12)
24. During high school, did you encounter some matters that you do not like to mention, such as assault, blackmail, robbery, etc.? (1, 2, 3)
    A. Yes, quite a few of them
    B. Some, but not many
    C. Never (3, 9, 78)
25. Did you ever participate in any of the previous matters other people are not willing to talk about or that caused harm to others? (1, 2, 3)
    A. Yes, quite a few
    B. Yes, but only some
    C. Never (1, 11, 78)
26. Currently, what kinds of information are you most interested in? (1, 1, 1, 1)
    A. Major related (scientific frontier, futures courses, etc.)
    B. Entertainment (games, music, sports)
    C. Current affairs and politics (domestic and international)
    D. Hot social issues (8, 16, 17, 19)
27. What do you currently do in your spare time? (3, 2, 3, 3, 3)
    A. Study to improve myself
    B. Entertain on Internet
    C. Participate in social activities
    D. Work on a hobby
    E. Make some spending money (48, 4, 19, 19, 0)
28. Do you consider going abroad? Why? (1, 3, 1, 2, 2, 2)
    A. No, I don't have the ability
    B. No, I want to stay in China to develop my career
    C. No, I do not have the economic means
    D. Yes, there are more opportunities for me abroad
    E. Yes, I do not fit the domestic environment well
    F. Yes, I will go abroad to be gilded; no matter what, I will be returning to China (10, 20, 37, 22, 1, 0)
29. If you choose to go abroad and successfully complete your education, will you consider returning home to develop your career? Why? (4, 2, 3, 1)
    A. Yes, China evolves fast and provides more opportunities
    B. Yes, it is difficult for me to live in a foreign land
    C. No, there are more opportunities abroad; it will be easier for me to accomplish professionally
    D. No, Chinese environment does not fit me well (43, 31, 8, 8)
30. What is your opinion about the current news media? (1, 2, 2, 3)
    A. Never watch news
    B. Too many positive reports; I need to know the real society
    C. The range of reports is too small; opinions are too narrow-minded
    D. Reports are accurate, timely, and delivered right (8, 59, 13, 10)

31. Do you feel that you are happy now? (3, 2, 1)
    A. Very happy
    B. OK
    C. Not happy (19, 66, 5)
32. In your opinion, how has China developed in the past 60 years? (4, 3, 2, 1)
    A. Economy developed fast; national defense gradually strengthened; international status gradually rising; I am full of confidence
    B. China is maintaining its development. However, many problems appeared and need to be resolved. We should understand the situation rationally
    C. Not clear, never thought about it
    D. Other (18, 67, 4, 1)
33. If using one of the following choices to describe your relation to China, you will use: (3, 2, 1, 1)
    A. We are a whole, sharing weal and woe, honor and disgrace; we are closely related
    B. We are in a relationship of whole and part
    C. Don't know, never thought about it
    D. We are in a relationship of individual (48, 38, 4, 0)
34. On which aspects does your family affect you more? Are they pessimistic or optimistic? (2, 2, 1, 1)
    A. Personal sentiment and thought training, optimistic
    B. Personal cognition and real-life abilities, optimistic
    C. Personal sentiment and thought training, pessimistic
    D. Personal cognition and real-life abilities, pessimistic (63, 20, 6, 1)
35. On which aspects does your family affect you less? Are they pessimistic or optimistic? (2, 2, 1, 1)
    A. Personal sentiment and thought training, optimistic
    B. Personal cognition and real-life abilities, optimistic
    C. Personal sentiment and thought training, pessimistic
    D. Personal cognition and real-life abilities, pessimistic (11, 38, 8, 33)
36. Ordinarily, what kind of ideological and political education do you receive? (1, 2, 2, 3, 2)
    A. No idea
    B. Book content and what is offered in school
    C. Family education
    D. Self-study, proactively learn
    E. Societal propaganda and education (6, 60, 6, 16, 0)
37. Among the previously listed methods of ideological and political education, which method has been most beneficial to you? (2, 2, 2, 2)
    A. Book contents and school offerings
    B. Family education
    C. Self-study and proactively learn
    D. Societal propaganda and education (31, 21, 33, 5)
38. Similarly, among the previously listed methods of ideological and political education, which method has been least beneficial to you? (1, 1, 1, 1)
    A. Book contents and school offerings
    B. Family education
    C. Self-study and proactively learn
    D. Societal propaganda and education (24, 14, 11, 41)

Bibliography

Adamatzky, A. (2010). *Game of Life Cellular Automata*. Berlin: Springer.

Adamic, A. L., and Adar, E. (2003). Friends and neighbors on the web. *Social Networks*, vol. 25, pp. 21–230.

Albert, R., Jeong, H., and Barabási, A. L. (2000). Attack and error tolerance of complex networks. *Nature*, vol. 406, pp. 378–382.

Albert, R., and Barabási, A. L. (2002). Statistical mechanics of complex networks. *Reviews of Modern Physics*, vol. 74, pp. 47–97.

Allen, C. W. (1976). *Astrophysical Quantities* (3rd edition). New York: The Athlone Press.

Allport, G. W., and Vernon, P. E. (1931). *A Study of Values*. Boston: Houghton Mifflin.

Angrisani, L., D'Apuzzo, M., and Moriello, R. S. L. (2006a). Unscented transform: A powerful tool for measurement uncertainty evaluation. *IEEE Transactions on Instrumentation and Measurement*, vol. 55, pp. 737–743.

Angrisani, L., Moriello, R. S. L., and D'Apuzzo, M. (2006b). New proposal for uncertainty evaluation in indirect measurements. *IEEE Transactions on Instrumentation and Measurement*, vol. 55, no. 4, pp. 1059–1064.

Arnheim, R. (2004). *Visual Thinking: Thirty-Fifth Anniversary Printing*. Berkeley, CA: The University of California Press.

Ashby, W. R. (1947). Principles of the self-organizing dynamic system. *Journal of General Psychology*, vol. 37, pp. 125–128.

Auyang, S. Y. (1999). *Foundations of Complex-System Theories: In Economics, Evolutionary Biology, and Statistical Physics*. Cambridge, England: Cambridge University Press.

Azcel, A. D., and Aczel, A. D. (2007). *Fermat's Last Theorem: Unlocking the Secret of an Ancient Mathematical Problem*. New York: Basic Books.

Bader, B. W., Harshman, R. A., and Kolda, T. G. (2007). Temporal analysis of semantic graphs using ASALSAN. In: *Proceedings of the 2007 Seventh IEEE International Conference on Data Mining*, pp. 33–42.

Baishun, C. Q., and Huang, L. (to appear). A nonlinear systemic research on the value system of college students in China. Submitted for publication.

Barabási, A. L., and Albert, R. (1999). Emergence of scaling in random networks. *Science*, vol. 286, no. 5439, pp. 509–512.

Barabási, A. L., Albert, R., and Jeong, H. (1999). Mean-field theory for scale-free random networks. *Physica A*, vol. 272, pp. 173–187.

Barana, G., and Tsuda, I. (1993). A new method for computing Lyapunov exponents. *Physics Letters A*, vol. 175, pp. 421–427.

Barnsley, M. (1988). *Fractals Everywhere*. New York: Academic Press.

Barnsley, M., and Hurd, L. P. (1989). *Fractal Image Compression*. Approximation, vol. 5, pp. 3–31. Wellesley, MA: A. K. Peters.

Barrow-Green, J. (1997). *Poincaré and the Three Body Problem*. Providence, RI: American Mathematical Society; London: London Mathematical Society.

Basdevant, J. L., and J. Dalibrad. (2000). Exact results for the three-body problem. In: *The Quantum Mechanics Solver: How to Apply Quantum Theory to Modern Physics*, Chapter 5, pp. 61–68. Berlin: Springer-Verlag.

Belliver, A. (1956). *Henri Poincaré ou la vocation souveraine*. Paris: Gallimard.

Bertalanffy, L. von (1924). *Einführung in Spengler's Werk*. Literaturblatt Kolnische Zeitung, May.

Bertalanffy, L. von (1934). *Modern Theories of Development* (Translation by J. H. Woodge). Oxford: Oxford University Press.

Bertalanffy, L. von (1937). *Das Gefüge des Lebens*. Leipzig: Teubner.

Bertalanffy, L. von (1968). *General System Theory: Foundations, Development, Applications*. New York: George Braziller (revised edition 1976).

Branden, N. (1969). *The Psychology of Self-Esteem*, p. 41. Nash Publishing Corp.

Braudel, F. (1980). *On History*. Chicago: University of Chicago Press.

Brogliato, B., Lozano, R., Maschke, B., and Egeland, O. (2007). *Dissipative Systems Analysis and Control: Theory and Applications* (2nd edition). London: Springer-Verlag.

Bu, C. J., and Luo, Y. S. (2003). *Matrix Theory*. Hei Longjiang: Press Harbin Engineer University.

Buss, D. (2004). *Evolutionary Psychology: The New Science of the Mind*. Boston: Pearson Education, Inc.

Cantor, G. (1882). Grundlagen einer allgemeinen mannichfaltigkeitslehre. *Mathematische Annalen*, vol. 21, pp. 545–591.

Carreira-Perpinan, M. A. (2001). *Continuous Latent Variable Models for Dimensionality Reduction and Sequential Data Reconstruction*. UK: Department of Computer Science University of Sheffield.

Catastrophe Theory, http://en_wikipedia.org/wiki/Catastrophe_theory.

Chen, G. (2000). *Controlling Chaos and Bifurcations in Engineering Systems*. Boca Raton, FL: CRC Press.

Chen, G., and Dong, X. (1998). *From Chaos to Order: Methodologies, Perspectives and Applications*. Singapore: World Scientific.

Cheng, S. W., and Feng, Z. Y. (1999). *Exploration on Complexity Science*. Peking: Democracy and Construction Press.

Clausius, R. (1865). *The Mechanical Theory of Heat—With Its Applications to the Steam Engine and to Physical Properties of Bodies*. London: John van Voorst, 1 Paternoster Row. MDCCCLXVII.

Clemend, E. (1988). *New Scientist*, Jan., vol. 21, p. 68.

Coffman, J. A. (unpublished). On causality in nonlinear complex systems: The developmentalist perspective, Mount Desert Island Biological Laboratory.

Coffman, J. A. (2006). Developmental ascendency: From bottom-up to top-down control. *Biological Theory*, vol. 1, no. 2, pp. 165–178.

Cohen, A. P. (1994). *Self-Consciousness*. Routledge.

Continental Drift, Dinosaur and Paleontology Dictionary, ZoomDinosaurs.com.

Conway, J. H., and Sloane, N. J. A. (1988). *Sphere Packing, Latties and Groups*. Springer-Verlag.

Cooke, E. F. (1974). *A Detailed Analysis of the Constitution*. Littlefield Adams & Co.

Cooley, J. W., and Tukey, J. W. (1965). An algorithm for the machine calculation of complex Fourier series. *Mathematics of Computation*, vol. 19, pp. 297–301.

Cooper, H., and Hedges, L. V. (1994). *The Handbook of Research Synthesis*. New York: Russell Sage.

Covey, S. R. (1989). *The 7 Habits of Highly Effective People: Powerful Lessons in Personal Change*. New York: Free Press.

Dai, W. S. (1979). *The Evolution of the Solar System* (vol. 1). Shanghai: Shanghai Science and Technology Press.

Demko, S., Hodges, L., and Naylor, B. (1985). Construction of fractal objects with iterated function systems. *Computer Graphics*, vol. 19, no. 3, pp. 271–278.

Devaney, R. L. (1989). *An Introduction to Chaotic Dynamical Systems* (2nd edition). Redwood City, CA: Addison-Wesley Co.

Dirac, P. A. M. (1937). The cosmological constants. *Nature,* vol. 139, p. 323.

Duan, X. J., and Lin, Y. (2011). Ways of fusing different types of information and how systemic yoyo model is applied in complex systems evaluation and estimation. *Kybernetes*, vol. 40, no. 1, pp. 262–274.

Duan, X. J., and Lin, Y. (to appear). Systems defined on dynamic sets and analysis of their characteristics. To appear.

Duan, X. J., Zhou, H. Y., and Yao, J. (2005). The decomposition and integration technique of fire dispersion index and conversion of impact deviation. *Journal of Ballistics*, vol. 17, no. 2, pp. 42–48.

Durkheim, E., and Mauss, M. (1971). Note on the notion of civilization. *Social Research*, vol. 38, pp. 808–813.

Dynamical system, http://en_wikipedia.org/wiki/dynamical_system, accessed on November 14, 2010.

Egan, K. (1992). *Imagination in Teaching and Learning*. Chicago: University of Chicago Press.

Eichhorst, P., and Savitch, W. J. (1980). Growth functions of stochastic Lindenmayer systems. *Information and Control*, vol. 45, pp. 217–228.

Eigen, M. (1971). Self organization of matter and the evolution of biological macromolecules. *Naturwissenschaften*, vol. 58, no. 10, pp. 465–523.

Eigen, M., and Schuster, P. (1978). Emergence of the hypercycle (part A). *Naturwissenschaften*, vol. 65, pp. 7–41.

Einstein, A. (1987). *The Collected Papers of Albert Einstein*. Princeton, NJ: Princeton University Press.

Engels, F. (1878). *Herrn Eugen Dührings Umwälzung der Wissenschaft* (commonly known as *Anti-Dühring*). Originally published in German in 1878. Also, see: Herr Eugen Dühring's revolution in science (anti-Dühring), Co-operative Pub. Society of Foreign Workers in the U.S.S.R (1934).

English, J., and Feng, G. F. (1972). *Tao De Ching*. New York: Vintage Books.

Erdos, P., and Renyi, A. (1960). On the evolution of random graphs. *Publi. Math. Inst. Hung. Acad. Sci.*, vol. 5, pp. 17–60.

Erlang, A. K. (1909). The theory of probabilities and telephone conversations. *Nyt. Tidsskrift for Matematik B*, vol. 20, pp. 33–40.

Fang, F. K., and Sanglier, M. (eds.) (1997). *Complexity and Self-Organization in Social and Economic Systems*. Berlin: Springer-Verlag.

Feynman, R. P., and Hibbs, A. R. (1965). *Quantum Mechanics and Path Integrals*. New York: McGraw-Hill.
Flood, R. L., and Carson, E. R. (1993). *Dealing with Complexity: An Introduction to the Theory and Application of Systems Science* (2nd edition). London: Springer.
Forrest, S., and Jones, T. (1994). Modeling complex adaptive systems with Echo. In: *Complex Systems: Mechanism of Adaptation*, edited by R. J. Stonier and X. H. Yu, pp. 3–21. Amsterdam: IOS Press.
Gell-Mann, M. (1995). *The Quark and the Jaguar: Adventures in the Simple and the Complex*. New York: St. Martin's Griffin.
Gerassi, J. (1989). *Jean-Paul Sartre: Hated Conscience of His Century. Volume 1: Protestant or Protester?* Chicago: University of Chicago Press
Getling, A. V. (1998). *Rayleigh–Bénard Convection: Structures and Dynamics*. Singapore: World Scientific.
Gibson, D., Kleinberg, J., and Raghavan, P. (1998). Inferring Web communities from link topology. *Proceedings of the 9th ACM Conference on Hypertext and Hypermedia*, pp. 225–234.
Glansdorff, P., and Prigogine, I. (1971). *Thermodynamic Theory of Structure, Stability and Fluctuations*. London: Wiley-Interscience.
Goode, H. H., and Machol, R. (1957). *System Engineering: An Introduction to the Design of Large-Scale Systems*. McGraw-Hill, New York.
Goodman, M. R. (1974). *Study Notes in System Dynamics*, Waltham, MA: Pegasus Communications.
Granovetter, M. (1973). The strength of weak ties. *American Journal of Sociology*, vol. 78, no. 6, pp. 1360–1380.
Grassberger, P., and Procaccia, I. (1983). Estimation of the Kolmogorov entropy from a chaotic signal. *Physical Review A*, vol. 28, pp. 2591–2593.
Gratzer, G. (1978). *Universal Algebra*. New York: Springer-Verlag.
Gregoire, N., and Prigogine, I. (1989). *Exploring Complexity: An Introduction*. New York: W. H. Freeman.
Guthrie, W. K. (1979). *A History of Greek Philosophy—The Presocratic Tradition from Parmenides to Democritus*. Cambridge, England: Cambridge University Press.
Haken, H. (1977). *Synergetics: An Introduction*. Berlin: Springer-Verlag.
Haken, H. (1982). *Synergetik*. Berlin: Springer-Verlag.
Haken, H. (1983). *Synergetics, an Introduction: Nonequilibrium Phase Transitions and Self-Organization in Physics, Chemistry, and Biology*. 3rd rev. enl. ed. New York: Springer-Verlag.
Haken, H. (1991). *Synergetic Computing and Cognition—A Top-Down Approach to Neural Nets*. Berlin: Springer-Verlag.
Haken, H. (2004). *Synergetics: Introduction and Advanced Topics*. Berlin: Springer-Verlag.
Han, W. X., and Yu, J. L. (2002). A study on the chaotic characteristics for the capital markets. *Systems Engineering: Theory and Applications*, vol. 10, pp. 43–48.
Hare, R. D. (1970). *Psychopathy: Theory and Research*. New York: John Wiley.
Harris, P. (2000). *The Work of the Imagination*. New York: Wiley-Blackwell.
Harshman, R. A. (1978). Models for analysis of asymmetrical relationships among *n* objects or stimuli. Presented at the *First Joint Meeting of the Psychometric Society and the Society for Mathematical Psychology*, McMaster University, Hamilton, Ontario, August 1978. http://publish.uwo.ca/~harshman/asym1978.pdf, accessed on June 27, 2011.
Hastings, C. S. (1909). Josiah Willard Gibbs. *Biographical Memoirs of the National Academy of Sciences*, vol. 6, pp. 372–393.
Heilbron, J. L. (2000). *The Dilemmas of an Upright Man: Max Planck and the Fortunes of German Science*. Cambridge, MA: Harvard University Press.
Hendrix, H. (2001). *Getting the Love You Want: A Guide for Couples*. New York: Owl Books.
Herman, G. T., and Rozenberg, G. (1975). *Developmental Systems and Languages*. Amsterdam: North-Holland Publishing Company.
Hess, S. L. (1959). *Introduction to Theoretical Meteorology*. New York: Holt, Rinehart and Winston.
Hewitt, C. (2008). Large-scale organizational computing requires unstratified reflection and strong paraconsistency. In: *Coordination, Organizations, Institutions, and Norms in Agent Systems III*, edited by J. Sichman, P. Noriega, J. Padget and S. Ossowski, Berlin: Springer-Verlag.
Highsmith, J. A. (2000). *Adaptive Software Development: A Collaborative Approach to Managing Complex Systems*. New York: Dorset House.
Holland, J. H. (1992). *Adaptation in Natural and Artificial Systems* (2nd edition). Cambridge, MA: MIT Press.
Holland, J. H. (1994). Echoing emergence: Objectives, rough definitions, and speculations for echo-class models. In: *Complexity: Metaphors, Models and Reality*, volume XIX of *Santa Fe Institute Studies in the Sciences of Complexity*, edited by G. A. Cowan, D. Pines, and D. Meltzer, pp. 309–342. Reading, MA: Addison-Wesley.
Holland, J. H. (1995). *Hidden Order: How Adaptation Build Complexity*. Addison-Wesley Publishing Company.

Holland, J. (1996). *Hidden Order: How Adaptation Builds Complexity*. New York: Basic Books.

Holme, P., Huss, M., and Jeong, H. (2003). Subnetwork hierarchies of biochemical pathways. *Bioinformatics*, vol. 19, pp. 532–538.

Hommes, C. H. (1991). *Chaotic Dynamics in Economic Models: Some Simple Case Studies*. Groningen: Wolter-Noordhoff.

Hsieh, D. (1989). Testing for nonlinear dependence in daily foreign exchange rates. *Journal Business*, vol. 62, no. 3, pp. 339–368.

Huang, X. T., and Zheng, Y. (2005). *Studies on the Values of Modern Chinese Youth*. Beijing: Press of People's Education.

Huang, H. Y., Duan, X. Y., and Wang, Z. M. (2008). A novel posterior-weighted Bayesian estimation method considering the credibility of the prior information. *ACAT Aeronautica ET Astronautica Sinica*, vol. 29, no. 5, pp. 1245–1251.

Hurd, L. P., Kari, J., and Culik, K. (1992). The topological entropy of cellular automata is uncomputable. *Ergodic Theory Dynamical Systems*, vol. 12, pp. 255–265.

Jackson, E. A. (1990). *Perspective of Nonlinear Dynamics*. Cambridge: Cambridge University Press.

Jones, T., and Forrest, S. (1993). An introduction to SFI Echo. Technical Report 93-12074, Santa Fe Institute, Santa Fe, NM. Available via anonymous ftp from ftp.santafe.edu:pub/Vsers/terry/echo/how-to.ps.Z.

Karp, P. D. (2001). Pathway databases: A case study in computational symbolic theories. *Science*, vol. 293, pp. 2040–2044.

Kissinger, H. A. (1994). *Diplomacy*. New York: Simon and Schuster.

Kleinberg, J. (2000a). Navigation in a small world. *Nature*, vol. 406, p. 845.

Kleinberg, J. (2000b). The small world phenomenon: An algorithmic perspective. *Proceedings of the 32nd Annual ACM Symposium on Theory of Computing*, New York, 2000, pp. 163–170.

Klir, G. (1985). *Architecture of Systems Problem Solving*. New York: Plenum Press.

Klir, G. (1989). Complexity: Some general observations. *Systems Research*, vol. 2, pp. 131–140.

Langton, C. G. (1990). Computation at the edge of chaos: Phase transitions and emergent computation. *Physica D*, vol. 42, pp. 12–37.

Lao Tzu (unknown). *Tao Te Ching: The Classic Book of Integrity and the Way*. An Entirely New Translation Based on the Recently Discovered Ma-Wang-Tui Manuscript. Translated, annotated, and with an Afterword by V. H. Mair, Bantam Books, New York, 1990.

Lao Tzu (time unknown). *Tao Te Ching, 25th-Anniversary Edition*. Translated by Jane English and Gia-Fu Feng, Vintage, New York, 1997.

Lars Onsager, http://en_wikipedia.org/wiki/Lars Onsager.

Lee, C., and Landgrebe, D. A. (1993). Analyzing high-dimensional multispectral data. *IEEE Transactions on Geoscience and Remote Sensing*, vol. 31, no. 4, pp. 792–800.

Li, C., Wang, J. N., and Yang, H. T. (2004). Accelerated mission test technology and reliability evaluation for highly reliable components. *Mechanical Science and Technology*, vol. 23, no. 7, pp. 876–882.

Li, J. W., Li, Z. H., Zhou, W., and Si, S. J. (2011). Study on the minimum of GDOP in satellite navigation and its applications. Presented at China Satellite Navigation Conference.

Li, L. Z., and Mao, J. G. (2006). Complex navigation algorithm of radar and inertial navigation. *Aerospace Control*, vol. 24, no. 1, pp. 43–48.

Li, S. H. (1997). Three begets all things of the world—Is "three" the third boundary? *Journal of Systemic Dialectics*, vol. 5, no. 4, pp. 34–37.

Li S. Y. (2006). *Nonlinear Science and Complexity Science*. Harbin: Harbin Industrial University Press.

Li, T. Y., and Yorke, J. A. (1975). Period three implies chaos. *The American Mathematical Monthly*, vol. 82, no. 10, pp. 985–992.

Lian, F. X., and He, Y. Z. (2008). Changes in college students' values. *Research of Modern Youth*, vol. 6, pp. 65–70.

Lin, Y. (1987). A model of general systems. *Mathematical Modelling*, vol. 9, pp. 95–104.

Lin, Y. (1990). Connectedness of general systems. *Systems Science*, vol. 16, pp. 5–17.

Lin, Y. (1990a). A few systems-colored views of the world. In: *Mathematics and Science*, edited by R. E. Mickens. River Edge, NJ: World Scientific, pp. 94–114.

Lin, Y. (1995). Developing a theoretical foundation for the laws of conservation. *Kybernetes: The International Journal of Systems and Cybernetics*, vol. 24, pp. 52–60.

Lin, Y. (1998). Mystery of nonlinearity and Lorenz' Chaos. *Kybernetes: The International Journal of Systems and Cybernetics*, vol. 27, nos. 6 and 7, pp. 605–854.

Lin, Y. (1999). *General Systems Theory: A Mathematical Approach*. New York: Kluwer Academic and Plenum Publishers.

Lin, Y. (2007). Systemic yoyo model and applications in Newton's, Kepler's laws, etc. *Kybernetes: The International Journal of Cybernetics, Systems and Management Science*, vol. 36, nos. 3–4, pp. 484–516.

Lin, Y. (2008). *Systemic Yoyos: Some Impacts of the Second Dimension*. New York: CRC Press, an imprint of Taylor and Francis.

Lin, Y., and Forrest, B. (2010a). The state of a civilization. *Kybernetes: The International Journal of Cybernetics, Systems and Management Science*, vol. 39, no. 2, pp. 343–356.

Lin, Y., and Forrest, B. (2010b). The life form of civilizations. *Kybernetes: The International Journal of Cybernetics, Systems and Management Science*, vol. 39, no. 2, pp. 357 – 366.

Lin, Y., and Forrest, B. (2010c). Interaction between civilizations. *Kybernetes: The International Journal of Cybernetics, Systems and Management Science*, vol. 39, no. 2, pp. 367 – 378.

Lin, Y., and Forrest, B. (2011a). Nature, human, and the phenomenon of self-awareness. *Kybernetes: The International Journal of Cybernetics, Systems and Management Science*, in press.

Lin, Y., and Forrest, B. (2011b). The mechanism behind imagination, conscience, and free will. *Kybernetes: The International Journal of Cybernetics, Systems and Management Science*, in press.

Lin, Y., and Ma, Y. H. (1993). System—A unified concept. *Cybernetics and Systems: An International Journal*, vol. 24, pp. 375–406.

Lin, Y., and OuYang, S. C. (2010). *Irregularities and Prediction of Major Disasters*. New York: CRC Press, an imprint of Taylor and Francis.

Lin, Y., and Yi, D. Y. (2010). General systems' yoyo structure and its applications in scientific mind. *Journal of Air Force Engineering University* (Natural Science Edition), vol. 11, no. 4, pp. ???.

Lindenmayer, A. (1968). Mathematical models for cellular interaction in development. *Journal of Theoretical Biology*, vol. 18, pp. 280–289.

Liu, D. Z. (eds) (1994). *Four Classes of Chinese Medicine in Modern Language*. Tianjing: Tianjing Translation Press of Science and Technology.

Lorenz, E. N. (1963). Deterministic nonperiodic flow. *Journal of the Meteorological Sciences*, vol. 20, pp. 130–141.

Lorenz, E. (1995). *The Essence of Chaos*. New York: CRC Press.

Lv, J. H., Lu, J. A., and Chen, S. H. (2001). *Analysis and Applications of Chaos Time Series*. Wuhan University Press, Wuhan.

Ma, A. N. (2006). Research on geographical system engineering. In: *Collections of Papers Presented at the Seminar of Qian Xuesen's Systems Scientific Thoughts*. Press of Shanghai Jiaotong University, pp. 126–133.

Mandelbrot, B. (1967). How long is the coast of Britain? Statistical self-similarity and fractional dimension. *Science*, vol. 156, pp. 636–638.

Mandelbrot, B. (1982). *The Fractal Geometry of Nature*. New York: W. H. Freeman and Co.

Manson, S. M. (2001). Simplifying complexity: A review on complexity theory. *Geoforum*, vol. 32, pp. 405–414.

Mao, S. S. (1999). *Bayesian Statistics*. Beijing: China Statistics Publishing House.

Martin, B. (1986). Computer recreations. *Scientific American*, September, pp. 87. Liu, Y. S. (editor) (2010). *Complete Collection of Huangdi Neijing*. Beijing: Press of Chinese Language.

Marx, C. (1957). *Economics—Philosophy Manuscripts*. Beijing: People's Publishing House.

Melko, M. (1969). *The Nature of Civilizations*. Boston: Porter Sargent.

Mickens, R. E. (1990). *Mathematics and Science*. Singapore: World Scientific.

Milgram, S. (1967). The small world problem. *Psychology Today*, May, pp. 60–67.

Miller, J. G. (1978). *Living Systems*. New York: McGraw-Hill.

Mote, F. W. (1999). *Imperial China: 900-1800*. Cambridge, MA: Harvard University Press.

Nemytskii, V. V., and Stepanov, V. V. (1989). *Qualitative Theory of Differential Equations*. Mineola, NY: Dover Publications.

Neuberger, A., Peles, S. et al. (2007). Scaling the response of circular plates subjected to large and close-range spherical explosions. Part I: Air-blast loading. *International Journal of Impact Engineering*, vol. 34, pp. 859–873.

Newman, M. E. J. (2003). The structure and function of complex networks. *SIAM Review*, vol. 45, pp. 167–256.

Newman, M. E. J., Strogatz, S. H., and Watts, D. J. (2001). Random graphs with arbitrary degree distributions and their applications. *Phys. Rev. E.*, vol. 64, pp. 026118 (???)

Newton, I. (1729). *The Mathematical Principles of Natural Philosophy*. Dawsons of Pall Mall, 1968.

Nicolis, G., and Prigogine, I. (1989). *Exploring Complexity: An Introduction* (1st edition). New York: W. H. Freeman and Co.

Norman, R. (2000). *Cultivating Imagination in Adult Education*. Proceedings of the 41st Annual Adult Education Research.

Obermark, J. (2004). Verification of simulation results using scale model flight test trajectories. *Technical Report AMR-AE-04-01*, May 2004.

Onsager, L. (1931). Reciprocal relations in irreversible processes I. *Physical Review*, vol. 37, pp. 405–426.

Packard, N. H., Crutchfield, J. P., Farmer, J. D., and Shaw, R. S. (1980). Geometry from a time series. *Physical Review Letters*, vol. 45, pp. 712–716.

Padgett, J. F., Lee, D., and Collier, N. (2003). Economic production as chemistry. *Industrial and Corporate Change*, vol. 12, p. 843.

Pan, N. B. (1980). Hubble constant. In *Large Encyclopedia of China: Astronomy*, 108. Beijing: Chinese Large Encyclopedia Press.

Parkinson, B. W., Spilker, J. J., Axelrad, P., and Enge, P. (1994). *Global Positioning System—Theory and Applications* (volume I). American Institute of Aeronautics and Astronautics, Inc.

Pastor-Satorras, R., and Vespignani, A. (2001). Epidemic spreading in scale-free networks. *Physical Review Letters*, vol. 86, no. 4, pp. 3200–3203.

Pearson, K. (1904). Mathematical contributions to the theory of evolution. *Philosophical Transactions of the Royal Society London Series A: Mathematical, Physical and Engineering Sciences*, vol. 203, pp. 53–86.

Peitgen, H.-O., Jürgens, H., and Saupe, D. (2004). *Chaos and Fractals: New Frontiers of Science*. Berlin: Springer-Verlag.

Peters, E. E. (1994). *Fractal Market Analysis: Applying Chaos Theory to Investment and Economics*. New York: Wiley.

Peters, E. E. (1996). *Chaos and Order in the Capital Markets: A New View of Cycles, Prices, and Market Volatility* (second edition). New York.

Pfaff, D. W. (2007). *The Neuroscience of Fair Play: Why We (Usually) Follow the Golden Rule*. New York: Dana Press, The Dana Foundation.

Philipatos, G. C., Pilarmu, E., and Mailliaris, A. G. (1994). Chaotic behavior in prices of European equity markets: A comparative analysis of major economic regions. *Journal of Multinational Finance Management*, vol. 3, nos. 3 and 4, pp. 5–24.

Poincaré, http://en_wikipedia.org/wiki/Poincaré.

Preston, F. W. (1948). The commonness and rarity of species. *Ecology*, vol. 29, no. 3, pp. 254–283.

Prigogine, I. http://en_wikipedia.org/wiki/Ilya Prigogine.

Prigogine, I. (1961). *Introduction to Thermodynamics of Irreversible Processes* (2nd edition). New York: Interscience.

Prusinkiewicz, P. (1986a). Graphical applications of L-systems. *Proceedings of Graphics Interface 1986—Vision Interface*, pp. 247–253.

Prusinkiewicz, P. (1986b). Application of L-systems to computer imagery. In: Lecture Notes in Computer Science vol. 291, *Proceedings of the 3rd International Workshop on Graph-Grammars and Their Application to Computer Science*. London: Springer-Verlag, pp. 534–548.

Prusinkiewicz, P., and Lindenmayer, A. (1990). *The Algorithmic Beauty of Plants*. New York: Springer-Verlag, pp. 40–50.

Prusinkiewicz, P., and Lindenmayer, A. (1996). *The Algorithmic Beauty of Plants*. London: Springer.

Prusinkiewicz, P., Hammel, M., and Mjolsness, E. (1993). Animation of plant development. *ACM SIGGRAPH*, pp. 351–360.

Qi, L. F., Feng, X. X., and Hui, X. P. (2007). Estimation of complex electronic warfare based on three-in-one combination testing method. *Modern Radar*, vol. 29, no. 6, pp. 6–8.

Qian, X. S. (1979). Science of sciences, system of science and technology, and Marxism philosophy. *Philosophical Research*, no. 1, pp. 20–27.

Qian, X. S. (1981). Systems science, cognitive science, and human body science. *Journal of Nature*, vol. 1, p. 9.

Qian, X. S. (1983a). Structure of modern science: A follow-up on the system of science and technology. *Philosophical Research*, no. 3, pp. 311–315.

Qian, X. S. (1983b). *About Systems Engineering*. Changsha: Hunan Science and Technology Press.

Qian, X. S. (1989). Basic scientific research should accept the guidance of Marxism. *Research in Philosophy*, vol. 10, pp. 3–8.

Qian, X. (1991). Automatic control and systems engineering. *Encyclopedia of China*. Press of Encyclopedia of China, Beijing.

Qian, X. S. (1996). *Human Science and Development of Modern Science and Technology*. Beijing: People's Publishing House.

Qian, X. S. (2007). *Engineering Control Theory* (New Century Edition). Shanghai: Shanghai Jiaotong University Press.

Qian, X. S., Yu, J. Y., and Dai, R. W. (1990). A new discipline of science—The study of open complex giant system and its methodology. *Nature Magazine*, vol. 13, pp. 3–10.

Quastler, H. (1965). General principles of systems analysis. In: *Theoretical and Mathematical Biology*, edited by T. H. Waterman and H. J. Morrowits. New York: Blaisdell Publishing.

Raa, T. T. (1999). *Input–Output Economics: Theory and Applications: Featuring Asian Economies*. Singapore: World Scientific.

Raa, T. T.h. (2005). *The Economics of Input-Output Analysis*. Cambridge University Press.

Ren, Z. Q. (1996). The resolution of the difficulties in the research of earth science. In: *Earth Science and Development*, edited by F. H. Wu and X. J. He. Beijing: Earthquake Press, pp. 298–304.

Ren, Z. Q., and Hu, Z. W. (1989). The product relation of mass-angular momentum in celestial body systems. In: *Advances of Comprehensive Study of Mutual Relations in Cosmos: Earth and Life*, Beijing: Chinese Science and Technology Press, pp. 269–274.

Ren, Z. Q., and Nio, T. (1994). Discussions on several problems of atmospheric vertical motion equations. *Plateau Meteorology*, vol. 13, pp. 102–105.

Ren, Z. Q., Lin, Y., and OuYang, S. C. (1998). Conjecture on law of conservation of informational infrastructures. *Kybernetes: The International Journal of Systems and Cybernetics*, vol. 27, pp. 543–552.

Rosenstein, M. T., Collins, J. J., and De Luca, C. J. (1994). Reconstruction expansion as a geometry-based framework for choosing proper delay times. *Physics Review D*, vol. 73, pp. 82–98.

Rosser, J. B., Jr. (2000). *From Catastrophe to Chaos: A General Theory of Economic Discontinuities* (volume I). *Mathematics, Microeconomics, Macroeconomics, and Finance* (2nd edition). Boston: Kluwer.

Salgado, J. F. (1997). The five factor model of personality and job performance in the European community. *Journal of Applied Psychology*, vol. 82, no. 1, pp. 30–43.

Sato, S., Sano, M., and Sawada, Y. (1985). Measurement of Lyapunov spectrum from a chaotic time series. *Physical Review Letters*, vol. 55, pp. 1082–1085.

Sato, S., Sano, M., and Sawada, Y. (1987). Practical methods of measuring the generalized dimension and largest Lyapunov exponent in high dimensional chaotic systems. *Progress of Theoretical Physics*, vol. 77, pp. 1–5.

Schiff, J. L. (2008). *Cellular Automata: A Discrete View of the World*. New York: Wiley-Interscience.

Self-organization, http://en_wikipedia.org/wiki/Self-organization.

Serletis, A., and Gogas, P. (2000). Purchasing power parity nonlinearity and chaos. *Applied Financial Economics*, vol. 10, pp. 615–622.

Shang, C. Y., and Gao, Z. H. (2006). The equivalent moment derivatives model and its usage in flight simulation. *Flight Dynamics*, vol. 24, no. 4, pp. 14–17.

Shannon, C. E. (1948). A mathematical theory of communication. *Bell System Technical Journal*, vol. 27, pp. 379–423, 623–656.

Sheng, H., Yang J. S., and Zeng, F. L. (2009). The minimum value of GDOP in pseudo-range positioning. *Fire and Command Control*, vol. 34, no. 5, pp. 22–24.

Sivanandam, S. N., and Deepa, S. N. (2007). *Introduction to Generic Algorithms*. Berlin: Springer-Verlag.

Smith, A. R. (1984). Plants, fractals, and formal languages. *Computer Graphics*, vol. 18, no. 3, pp. 1–10.

Takens, F. (1981). Detecting strange attractors in fluid turbulence. In: *Dynamical Systems and Turbulence*. Berlin: Springer, pp. 65–90.

Tang, X. M., Zhang, J. H., Shao, F. C. et al. (2001). *Test Analysis and Evaluation of Weapon Systems in Small-Sample Circumstances*. Beijing: National Defense Industry Press.

The first law of thermodynamics, http://en_wikipedia.org/wiki/the first law of thermodynamics.

The Nobel Prize in Chemistry (1968). Presentation Speech.

The second law of thermodynamics, http://en_wikipedia.org/wiki/the second law of thermodynamics.

Thom, R. (1989). *Structural Stability and Morphogenesis: An Outline of a General Theory of Models*. Reading, MA: Addison-Wesley.

Thompson, J. M., and Stenmit, H. B. (1986). *Nonlinear Dynamics and Chaos: Geometrical Methods for Engineers and Scientists*. New York: Wiley.

Tinbergen, N. (1951). *The Study of Instinct*. New York: Oxford University Press.

Toynbee, A. (1934–1961). *A Study of History*. 12 volumes. Oxford: Oxford University Press.

Umbach, E. (time unknown). The fundamental tasks of systems science. An unpublished manuscript.

U.S. Department of Defense (2008). *Global Positioning System Standard Positioning Service Performance Standard* (4th edition). Washington, DC.

Vicsek, T., Czirok, A., Ben-Jacob, E., Cohen, I., and Shochet, O. (1995). Novel type of phase transition in a system of self-driven particles. *Physical Review Letters*, vol. 75, no. 6, pp. 1226–1229.

Waley, A. (translator) (1989). *The Analects of Confucius*. New York: Vintage.

Wallerstein, I. M. (1991). *Geopolitics and Geoculture: Essays on the Changing World-System*. Cambridge: Cambridge University Press.

Walsh, D., and Downe, S. (2005). Meta-synthesis method for qualitative research: A literature review. *Journal of Advanced Nursing*, vol. 50, no. 2, pp. 204–211.

Wang, C. Z., and Yan, F. Y. (1988). *The Universe Holographic Unitics*. Shandong: Shandong People's Publishing House.

Wang, D. C., Ding, J. H., Cheng, W. D. et al. (2006). *Precise Tracking RADAR Techniques*. Beijing: Publishing House of Electronics Industry.

Wang, G. Y., Wang, L. D., and Ruan, X. X. (2002). *The Substitute Equivalent Reckoning Principle and Method for Radar ECM Test*. Beijing: National Defense Industry Press.

Wang, G., Duan, X. J., and Wang, Z. M. (2009). Conversion method of impact dispersion in substitute equivalent tests based on error propagation. *Defense Science Journal*, vol. 59, no. 1, pp. 15–21.

Wang, S. M. (2007). Thoughts on the characteristics of value of Chinese college students. *Educational Exploration*, vol. 7, pp. 109–110.

Wang, W., and Cui, M. M. (2006). Comprehensive analytic hierarchy process of complicated system evaluation. *Journal of Naval Engineering University*, vol. 18, no. 2, pp. 42–46.

Wang, Z. M., and Yi, D. Y. (1997). *Modeling Measurement Data and Estimation of Parameters*. Changsha, Hunan: Press of National University of Defense Technology.

Watts, D. J., and Strogatz, S. H. (1998). Collective dynamics of 'small-world' networks. *Nature*, vol. 393, no. 6684, pp. 440–442.

Watts, D. J., Dodds, P. S., and Newman, M. E. J. (2002). Identity and search in social networks. *Science*, vol. 296, pp. 1302–1305.

Weaver, W. (1948). Science and complexity. *American Scientist*, vol. 36, no. 4, pp. 536–544.

Wellman, B., and Berkowitz, S. D. (1988). *Social Structures: A Network Approach*. Cambridge: Cambridge University Press.

Wendt, R. P. (1974). Simplified transport theory for electrolyte solutions. *Journal of Chemical Education*, vol. 51, p. 646.

Wiener, N. (1948). *Cybernetics or Control and Communication in the Animal and the Machine*. Cambridge, MA: MIT Press.

Wigner, E. P. (1969). *Physical Science and Human Values*. Santa Barbara, CA: Greenwood Press.

Wilhalm, R., and Baynes, C. (1967). *The I Ching or Book of Changes* (3rd edition). Princeton, NJ: Princeton University Press.

Wolfram, S. (2002). *A New Kind of Science*. U.S.A.: Wolfram Media Inc.

Wu, T., and Huang, X. R. (2005). Complexity: Talking about it from "three". *Journal of Systemic Dialectics*, vol. 13, no. 1, pp. 6–11.

Wu, Y., and Lin, Y. (2002). *Beyond Nonstructural Quantitative Analysis: Blown-Ups, Spinning Currents and the Modern Science*. River Edge, New Jersey: World Scientific.

Wu, S. T., Liu, X., and Zhang, M. (2005). A multiscale approach to multisensor data fusion for terminal guidance of an anti-ship missile. *Aerospace Control*, vol. 23, no. 2, pp. 36–40.

Xian, D. C., and Wang, Z. X. (1987). Particle physics. In: *Large Encyclopedia of China: Physics*, Beijing: Chinese Large Encyclopedia Press, p. 737.

Xu, G. Z. (2000). *Systems Science*. Shanghai: Shanghai Science and Technology Press.

Xu, X. S. (2003). *Evaluation of Local Market Items Based on the Science of Complexity*. Beijing: Press of Science and Technology.

Yang, Y. F. (2004). The origin and increase of complexity. In: *1st and 2nd Conference Papers on Complexity in China*, pp. 164–175, Beijing: Science Press.

Yang, J., Shen, L. J., Huang, J., and Zhao, Y. (2008). Bayesian comprehensive assessment of reliability for electronic products by using test information of similar products. *ACTA Aeronautica et Astronatica Sinica*, vol. 29, no. 6, pp. 1550–1553.

Yun, X., Bachmann, E. R., McGhee, R. B. et al. (1999). Testing and evaluation of an integrated GPS/INS system for small AUV navigation. *IEEE Journal of Oceanic Engineering*, vol. 24, no. 3, pp. 396–404.

Zadeh, L. (1962). From circuit theory to systems theory. *Proceedings of the IRE*, vol. 50, pp. 856–865.

Zeng, Y. H., Wang, G. Y., and Wang, L. D. (2007). Substitute equivalent reckoning method for ECM test of radar netting. *Systems Engineering and Electronics*, vol. 29, no. 4, pp. 548–550.

Zhabotinsky, A. M. (1964). Periodical process of oxidation of malonic acid solution (in Russian). *Biophysics*, vol. 9, pp. 306–311.

Zhang, K., and Hu, C. S. (2006). *World Heritage in China*. Guangzhou: The Press of South China University of Technology.

Zhang, Z., Yuan, J. P., and Chen S. L. (1996). Applications of GPS in space. *Journal of Astronautics*, vol. 17, no. 3, pp. 68–71.

Zhao, J. W., Zhou, Z., Zou, D. Y., and Gong, X. Y. (2006). A method for estimating the flying-quality of airplane with complicated systems. *Computer Simulation*, vol. 23, no. 12, pp. 70–73.

Zhou, W. H., and Liu, X. X. (2008). Research on the values of Chinese college students of recent years. *Journal of Political Science Institute of China*, vol. 27, no. 2, pp. 29–32.

Zhu, X.-D., and Wu, X.-M. (1987). An approach on fixed pansystems theorems: Panchaos and strange panattractor. *Applied Mathematics and Mechanics*, vol. 8, no. 4, 339–344.

Zhu, X. F., and Gu, S. Q. (2003). The degree of chaos in stock prices of Shanghai and Shenzhen Stock Exchanges and methods for its control. *Journal of Management Science*, vol. 3, no. 1, pp. 53–56.

[illegible]

Index

Page numbers followed by *f* and *t* indicate figures and tables, respectively.

D

E

F

G

H

T